Topics in Current Genetics

Series Editor: *Stefan Hohmann*

Springer

Berlin
Heidelberg
New York
Hong Kong
London
Milan
Paris
Tokyo

Eckhard Boles · Reinhard Krämer (Eds.)

Molecular Mechanisms Controlling Transmembrane Transport

With 61 Figures, 5 of Them in Color; and 9 Tables

 Springer

Professor Dr. ECKHARD BOLES
Institute of Microbiology
J. W. Goethe-University
Marie-Curie-Straße 9
60439 Frankfurt/Main
Germany

Professor Dr. REINHARD KRÄMER
Institute of Biochemistry
University of Köln
Zülpicher Straße 47
50674 Köln
Germany

The cover illustration depicts pseudohyphal filaments of the ascomycete *Saccharomyces cerevisiae* that enable this organism to forage for nutrients. Pseudohyphal filaments were induced here in a wild-type haploid MATa Σ1278b strain by an unknown readily diffusible factor provided by growth in confrontation with an isogenic petite yeast strain in a sealed petri dish for two weeks and photographed at 100X magnification (provided by Xuewen Pan and Joseph Heitman).

ISSN 1610-2096
ISBN 3-540-21837-8 Springer-Verlag Berlin Heidelberg New York

Library of Congress Control Number: 2004105719

Springer-Verlag is a part of Springer Science + Business Media
springeronline.com

Springer-Verlag Berlin Heidelberg 2004
Printed in Germany

The use of general descriptive names, registered names, trademarks, etc. in this publication does not imply, even in the absence of a specific statement, that such names are exempt from the relevant protective laws and regulations and therefore free for general use.

Typesetting: Camera ready by editors
Data-conversion: PTP-Berlin, Stefan Sossna e.K.
Cover Design: Design & Production, Heidelberg
39/3150-WI - 5 4 3 2 1 0 - Printed on acid-free paper

Table of contents

Transcriptional regulation of intestinal nutrient transporters............1
  Soraya P. Shirazi-Beechey...1
    Abstract 1
    1 Introduction 1
    2 Gene transcription 5
    3 Organisation of the intestinal epithelium and gene expression 5
    4 Transcriptional regulation of nutrient transporters in the small intestine ...6
      4.1 Intestinal iron absorption 7
      4.2 Intestinal peptide transport..7
      4.3 Intestinal bile salt transport...8
      4.4 Intestinal glucose transport 9
    5 Regulation of nutrient transport in the large intestine 12
      5.1 Adaptive response of the colonic epithelium in short-bowel
      syndrome...15
    Acknowledgements 17
    References...18

Generation of transporter isoforms by alternative splicing............23
  Gerardo Gamba...23
    Abstract 23
    1 Introduction 23
    2 Alternative splicing with physiological consequences in membrane
    transporters 24
      2.1 Electroneutral Cl$^-$-coupled cotransporters (SLC12 family)...............24
      2.2 The urea transporters (SLC14 family) 28
      2.3 Na$^+$/bile acid transporter (SLC10A family) 30
      2.4 Na$^+$/*myo*-Inositol cotransporters (members of the SLC5 family)......32
      2.5 Na$^+$/bicarbonate cotransporters and exchangers (members of the
      SLC4 family) 34
      2.6 The divalent metal transporter DMT1 (member of the SLC11
      family) 37
    3 Conclusion...38
    References...39

Transport-dependent gene regulation by sequestration of transcriptional regulators 47
  Alex Böhm and Winfried Boos...47
    Abstract 47
    1 Introduction 47
    2 Signaling pathways: two examples 48
      2.1 The transcription activator MalT 48

2.2 The global regulator Mlc 56
3 The PutA, GlnK, and NifL signal transduction proteins 58
4 Conclusion 60
Acknowledgements 61
References 61
Abbreviations 66

Regulation of carrier-mediated sugar transport by transporter quaternary
structure **67**
Kara B. Levine and Anthony Carruthers 67
Abstract 67
1 Introduction 67
2 Human red blood cell glucose transport 69
2.1 GluT1 quaternary structure 69
2.2 Determinants of GluT1 quaternary structure 76
2.3 Is quaternary structure the sole determinant of GluT1 function?...... 78
3 Glucose transporter quaternary structure and regulation of GluT1 80
3.1 A model for nucleotide regulation of sugar transport 82
Acknowledgements 85
References 85

Regulation and function of ammonium carriers in bacteria, fungi,
and plants **95**
Nicolaus von Wirén and Mike Merrick 95
Abstract 95
1 Introduction 95
2 The ammonium transport protein family 97
2.1 Distribution and phylogeny 97
2.2 Membrane topology 98
2.3 Structure-function relationships 100
2.4 Mode of action 102
3 Bacterial ammonium transporters 104
3.1 Regulation of Amt activity in bacteria 104
3.2 The role of Amt in ammonium sensing 106
4 Fungal ammonium transporters 107
4.1 Regulation of Amt/Mep protein activity in fungi 107
4.2 Amt proteins as ammonium sensors in fungi 108
5 Plant ammonium transporters 109
5.1 The function of Amt proteins in plants 109
5.2 Biochemical transport properties of AMTs 110
5.3 Transcriptional regulation of AMT transporters in plants 110
6 Ammonium/ammonia transport by other (than AMT-type)
transporters in plants 113
References 114

Role of transporter-like sensors in glucose and amino acid signalling in yeast..121

Eckhard Boles and Bruno André 121
Abstract 121
1 Introduction 121
2 The Snf3, Rgt2, and Ssy1 sensors are members of large transporter families...122
2.1 The yeast hexose transporter family 122
2.2 The yeast amino acid transporters.................................123
3 The yeast glucose and amino acid sensors: initial description and functions...125
3.1 The Snf3 and Rgt2 glucose sensors 125
3.2 The Ssy1 amino acid sensor..127
3.3 Mechanisms of sensing: models 129
4 Signal transduction pathways...134
4.1 The glucose-sensing pathway 134
4.2 The amino-acid-sensing pathway..................................140
5 Are there transporter-like sensors in higher eukaryotes?.........144
6 Prospects and future directions...145
Acknowledgement..146
References 146

Osmoregulation and osmosensing by uptake carriers for compatible solutes in bacteria 155

Susanne Morbach and Reinhard Krämer 155
Abstract 155
1 Introduction 155
2 Osmodependent transport systems in bacteria..........................156
3 Osmoregulated solute carriers of C. glutamicum: osmosensing and osmoregulation 158
3.1 BetP of C. glutamicum: structural properties...............159
3.2 BetP of C. glutamicum: catalytic properties 160
3.3 BetP of C. glutamicum: regulatory properties...............160
3.4 BetP of C. glutamicum: sensory properties...................164
4 Other osmoregulated carrier systems 168
4.1 ProP of E. coli...169
4.2 OpuA / BusA of L. lactis 170
4.3 Kdp of E. coli..171
4.4 Osmoregulation and osmosensing by mechanosensitive channels .172
5 General concepts and conclusions...173
References 174

The bacterial phosphotransferase system: a perfect link of sugar transport and signal transduction..179

Jörg Stülke and Matthias H. Schmalisch 179
Abstract 179

1 Introduction 179
2 The phosphorylation state of PTS proteins is tightly controlled 182
   2.1 Control of EIIA^{Crr} phosphorylation in *E. coli* 182
   2.2 Control of HPr phosphorylation in Gram-positive bacteria 184
   2.3 Regulation of HPr kinase/phosphorylase activity 185
   3 Control of transporter and enzyme activities by PTS components 187
   3.1 Control of transporter activities by the PTS in Gram-negative
   bacteria 187
   3.2 Control of transporter activities by the PTS in Gram-positive
   bacteria 190
   3.3 Control of enzymatic activities by the PTS 191
4 Regulation of transcription by PTS components 193
   4.1 PTS-dependent generation of cofactors for transcription regulators 193
   4.2 Control of transcription regulators by direct phosphorylation 194
5 Conclusion 197
Acknowledgements 197
References 197

Ancillary proteins in membrane targeting of transporters 207
   Tomas Nyman, Jhansi Kota, and Per O. Ljungdahl 207
   Abstract 207
   1 Introduction 207
   2 The secretory pathway 209
   3 Vesicle-mediated transport 211
   4 Formation of COPII-coated vesicles 213
   5 Cargo selection – sorting motifs bind Sec24p 215
   6 Sorting signals 217
   7 Sec24p homologs – cargo selection by combinatorial mechanisms 218
   8 Ancillary proteins – packaging chaperones 219
   9 Shr3p - the original packaging chaperone 220
   10 Additional packaging chaperones – Pho86p, Gsf2p, and Chs7p 222
   11 Packaging chaperones - presentation of sorting motifs 223
   12 Coatomer asymmetry may influence the differential packaging
   of cargo 225
   13 Concluding remarks 226
   Acknowledgements 228
   References 228

Regulation of transporter trafficking by the lipid environment 235
   Miroslava Opekarová 235
   Abstract 235
   1 Introduction 235
   2 Itineration of transporters to the plasma membrane 236
   3 Topology of lipid synthesis along the secretory pathway 237
   4 Lipid rafts and associated transporters 239
   5 Raft-dependent sorting of transporters in the secretory pathway 240

5.1 Location of transporter-raft association 240
5.2 Lipid requirements for transporter-raft association......................242
5.3 Asymmetric distribution of transporters in polarized cells 242
5.4 Molecular models for raft-based protein sorting............................243
5.5 Determination of raft association..244
6 Clustering and oligomerization of transporter molecules in the
plasma membrane...245
7 Lipid function as molecular chaperons..246
8 Conclusions 247
Acknowledgement...247
References 248

Trafficking of vesicular transporters to secretory vesicles 255
Vania F. Prado, Marc G. Caron, and Marco A.M. Prado............................255
Abstract 255
1 Classes of vesicular transporters 255
2 Classes of secretory vesicles 256
3 Cellular routes taken by vesicular proteins.......................................257
3.1 SV proteins 257
3.2 LDCV proteins...259
4 Localization of vesicular transporters in neurons...............................260
5 Sorting of vesicular transporters in PC12 cells...................................261
6 Sorting motifs identified in vesicular transporters..............................262
7 Concluding remarks 265
Acknowledgements 266
References...266

**Membrane trafficking of yeast transporters: mechanisms and
physiological control of downregulation..273**
Rosine Haguenauer-Tsapis and Bruno André..273
Abstract 273
1 Introduction 273
2 Trafficking of plasma membrane transporters along the secretory
pathway 274
2.1 ER-associated events 275
2.2 Posttranslational modifications...277
2.3 ER-to-Golgi trafficking, and beyond 280
3 Downregulation of plasma membrane transporters and channels 281
3.1 Discovery of ubiquitin-dependent internalization of yeast
plasma membrane transporters 281
3.2 Ubiquitin-dependent internalization of mammalian channels.........284
3.3 Mechanisms involved in the ubiquitylation of yeast plasma
membrane transporters...285
3.4 Plasma membrane-to-vacuole targeting of yeast transporters.........290
4 Regulation of transporters and channels at membrane trafficking
levels: signals and mechanisms 296

4.1 Physiological control of the rate of internalization of
transporters 296
4.2 Physiological control of the sorting and recycling of newly
synthesized transporters...301
4.3 Mechanisms and signalling pathways governing the regulated
trafficking of transporters 308
5 Conclusion 311
Acknowledgements 311
References..312

Regulated transport of the glucose transporter GLUT4.................................325
Hadi Al-Hasani 325
Abstract..325
1 Mammalian glucose transport proteins 325
2 Insulin regulated glucose uptake 326
3 The GLUT4 translocation cycle 326
3.1 The insulin-sensitive GLUT4 storage compartment.......................328
3.2 Retention of GLUT4 in the insulin-sensitive compartment...........329
3.3 Recycling of GLUT4 in heterologous cells 331
3.4 Components of GLUT4 vesicles...331
4 Signal transduction for the insulin-stimulated GLUT4-translocation 332
4.1 The IRS\PI3K\AKT\aPKC pathway...333
4.2 The Cbl-CAP\C3G\TC10 pathway...335
5 Insulin-stimulated exocytosis of GLUT4 337
6 Mechanism of GLUT4 endocytosis 338
6.1 Role of dynamin GTPase in the endocytosis of GLUT4 339
6.2 Clathrin but not caveolin mediate GLUT4 internalization 339
7 Targeting signals in GLUT4 340
8 Outlook...343
Acknowledgements 344
References..344
Abbreviations 352

Aquaporin-2 trafficking...353
Sebastian Frische, Tae-Hwan Kwon, Jørgen Frøkiær, and Søren Nielsen 353
Abstract..353
1 Vasopressin induced AQP2-trafficking...353
1.1 Historical overview...353
1.2 Role of AQP2-phosphorylation in vasopressin stimulated AQP2
recruitment to the plasma membrane..355
1.3 Constitutive AQP2-recycling and turnover rate of AQP2 in the
plasma membrane 358
1.4 Concluding remarks on regulation mechanisms in vasopressin
controlled AQP2-trafficking...359
2 AQP2 trafficking and the cytoskeleton 360
2.1 Modulation of the actin cytoskeleton...360

2.2 Microtubules and AQP2 vesicle transport 363
3 Vasopressin induced intracellular Ca^{2+}-signalling 364
   3.1 Studies in isolated perfused IMCD's indicate a role of intracellular
   Ca^{2+} in AQP2 trafficking 364
   3.2 Ca^{2+} is not important for AQP2 traffic in primary cultures of
   IMCD cells...366
   3.3 Comparison of the results from IMCD's and primary cultured
   IMCD-cells 366
   3.4 Which vasopressin sensitive receptors are involved in calcium
   signalling in AQP2 containing collecting duct cells? 367
   3.5 Myosins as targets for calcium signalling in IMCD?.....................368
4 Evidence for role of vesicle targeting receptors in AQP2 trafficking 368
5 Concluding remarks 370
References 370
Abbreviations: 376

**Molecules in motion: multiple mechanisms that regulate the GABA
transporter GAT1**...379
   Michael W. Quick..379
   Abstract 379
   1 Introduction 379
   2 The rat brain GABA transporter GAT1..381
      2.1 Molecular properties 381
      2.2 Physiological properties..382
      2.3 Introduction to regulation 383
   3 GAT1 trafficking...384
      3.1 Basal recycling of the transporter 384
      3.2 Molecular mechanisms regulating GAT1 expression 388
      3.3 Signaling cascades that regulate trafficking..................................390
   4 Regulating transport 391
      4.1 Intermolecular interactions 391
      4.2 Transporter ligands 394
   5 Summary 394
   References 394

Index **401**

List of contributors

Al-Hasani, Hadi
German Institute for Human Nutrition, Arthur-Scheunert-Allee 114-116, D-14558 Potsdam-Rehbrücke, Germany
al-hasani@mail.dife.de

André, Bruno
Université Libre de Bruxelles CP300, Institut de Biologie et de Médecine Moléculaires (IBMM), Laboratoire de Physiologie Moléculaire de la Cellule, Rue des Pr. Jeener et Brachet, 12 6041 Gosselies Belgium
bran@ulb.ac.be

Böhm, Alex
Department of Biology, University of Konstanz, Universitätsstrasse 10, 78457 Konstanz, Germany

Boles, Eckhard
Institut fuer Mikrobiologie, Goethe-Universitaet Frankfurt, Marie-Curie-Str. 9, D-60439 Frankfurt am Main, Germany
E.Boles@em.uni-frankfurt.de

Boos, Winfried
Department of Biology, University of Konstanz, Universitätsstrasse 10, 78457 Konstanz, Germany
winfried.boos@uni-konstanz.de

Caron, Marc G.
Department of Cell Biology and Howard Hughes Medical Institute, Duke University Medical Center, Durham NC

Carruthers, Anthony
Department of Biochemistry & Molecular Pharmacology, University of Massachusetts Medical School, Lazare Research Building, 364 Plantation Street, Worcester, MA 01605, USA
anthony.carruthers@umassmed.edu

Frische, Sebastian
The Water and Salt Research Center, Building 233/244, Institute of Anatomy, University of Aarhus, DK-8000 Aarhus C, Denmark

Frøkiær, Jørgen
The Water and Salt Research Center, Building 233/244, Institute of Anatomy, University of Aarhus, DK-8000 Aarhus C, Denmark

Gamba, Gerardo
Molecular Physiology Unit, Instituto Nacional de Ciencias Médicas y Nutrición Salvador Zubirán and Instituto de Investigaciones Biomédicas, Universidad Nacional Autónoma de México, Mexico City CP 14000, Mexico
gamba@sni.conacyt.mx

Haguenauer-Tsapis, Rosine
Institut Jacques Monod-CNRS, Universités Paris VI and VII, 2 place Jussieu 75251 Paris Cedex 05, France
haguenauer@ijm.jussieu.fr

Kota, Jhansi
Ludwig Institute for Cancer Research, Box 240, S-171 77 Stockholm

Krämer, Reinhard
Institute of Biochemistry, Universität Köln, Zülpicher Str. 47, 50674 Köln, Germany
r.kraemer@uni-koeln.de

Kwon, Tae-Hwan
The Water and Salt Research Center, Building 233/244, Institute of Anatomy, University of Aarhus, DK-8000 Aarhus C, Denmark, and
Department of Biochemistry, School of Medicine, Kyungpook National University, Taegu, Korea

Levine, Kara B.
Department of Biochemistry & Molecular Pharmacology, University of Massachusetts Medical School, Lazare Research Building, 364 Plantation Street, Worcester, MA 01605, USA

Ljungdahl, Per O.
Ludwig Institute for Cancer Research, Box 240, S-171 77 Stockholm
plju@licr.ki.se

Merrick, Mike
Department of Molecular Microbiology, John Innes Centre, Norwich NR4 7UH, UK
mike.merrick@bbsrc.ac.uk

Morbach, Susanne
Institute of Biochemistry, Universität Köln, Zülpicher Str. 47, 50674 Köln, Germany
s.morbach@uni-koeln.de

Nielsen, Søren
The Water and Salt Research Center, Building 233/244, Institute of Anatomy,

University of Aarhus, DK-8000 Aarhus C, Denmark
sn@ana.au.dk

Nyman, Tomas
Ludwig Institute for Cancer Research, Box 240, S-171 77 Stockholm

Opekarová, Miroslava
Institute of Microbiology, Czech Academy of Sciences, 142 20 Prague, Czech
Republic
opekaro@biomed.cas.cz

Prado, Marco A.M.
Departamento de Farmacologia, ICB, Universidade Federal de Minas Gerais,
Av. Antonio Carlos 6627, Belo Horizonte, 31270-901 Brazil
mprado@icb.ufmg.br

Prado, Vania F.
Departamento de Bioquímica-Imunologia, ICB, Universidade Federal de Minas Gerais, Av. Antonio Carlos 6627, Belo Horizonte, 31270-901 Brazil

Quick, Michael W.
Department of Biological Sciences, University of Southern California, HNB
228, 3641 Watt Way, Los Angeles CA 90089-2520, USA
mquick@usc.edu

Schmalisch, Matthias H.
Department of General Microbiology, Institute for Microbiology and Genetics,
Georg-August University Göttingen, Grisebachstr. 8, D-37077 Göttingen,
Germany

Shirazi-Beechey, Soraya P.
Epithelial Function and Development Group, Department of Veterinary Preclinical Sciences, The University of Liverpool, Brownlow Hill, Liverpool L69
7ZJ, UK
spsb@liv.ac.uk

Stülke, Jörg
Department of General Microbiology, Institute for Microbiology and Genetics,
Georg-August University Göttingen, Grisebachstr. 8, D-37077 Göttingen,
Germany
jstuelk@gwdg.de

von Wirén, Nicolaus
Institut für Pflanzenernährung, Fachbereich "Mineralstoffwechsel und -
transport", Universität Hohenheim, D-70593 Stuttgart, Germany

Transcriptional regulation of intestinal nutrient transporters

Soraya P. Shirazi-Beechey

Abstract

Nutrient transport across the apical plasma membrane of enterocytes is mediated by highly specialised membrane proteins, the majority of which are adaptively regulated by dietary substrates. The nutrient signals are either transmitted by the transporter itself, regulating the external nutrient access to the intracellular environment, or via distinct luminal membrane nutrient sensors. This chapter addresses the importance of the transcriptional regulation of intestinal nutrient transporter genes, and highlights its relevance to nutrition, health and disease. Examples are given of nutrient transporters from both the small and large intestine for which the underlying transcriptional mechanisms have been identified. These include the Na^+/glucose cotransporter (SGLT1) and the bile salt transporter (ABST) expressed in the small intestine, and the monocarboxylate transporter (MCT1), residing in the large intestine. A better understanding of transcriptional regulation of intestinal nutrient transporters will undoubtedly have important clinical and nutritional implications.

1 Introduction

Nutrient transport across the plasma membrane of cells is accomplished by highly specialised integral membrane proteins; their function and regulation are essential for maintenance of metabolism and homeostasis.

Nutrient transporters expressed on the luminal membrane of intestinal absorptive cells are directly exposed to an environment with massive fluctuations in the types and the levels of nutrients entering the lumen of the intestine. It is not surprising therefore that the majority of intestinal nutrient transporters are adaptively regulated by dietary substrates. Positive and negative regulatory patterns, assessed by the rates of nutrient transport, have been observed as mechanisms of adaptation. The positive response to the level of specific dietary substrates, 'upregulation' in the activity of some nutrient transporters, was proposed to ensure economy of biosynthetic costs, while at the same time maximising dietary energy (or other) gain. Examples of nutrient transporters upregulated in response to dietary substrates are the small intestinal luminal membrane, Na^+/glucose cotransporter and peptide transporter, and the colonic monocarboxylate transporter. The

Topics in Current Genetics, Vol. 9
E. Boles, R. Krämer (Eds.): Molecular Mechanisms Controlling Transmembrane Transport
DOI 10.1007/b96814 / Published online: 9 March 2004
© Springer-Verlag Berlin Heidelberg 2004

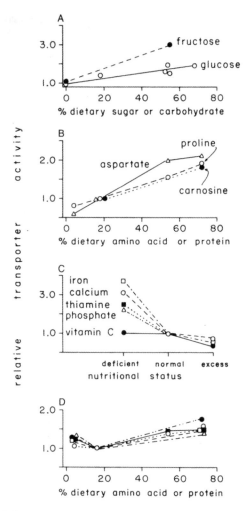

Fig. 1. The effect of dietary substrate concentration on transporter activity in the small intestine. Reprinted with permission from Diamond and Karasov (1987).

negative response, to high dietary substrate levels, 'downregulation' of other intestinal nutrient transporters for substrates, such as vitamins and trace minerals, and bile salts is observed. This was suggested to ensure adequate absorption of essential nutrients, and to minimise the effects of nutrients with potential toxicity when in excess (Diamond and Karasov 1987) (see Fig. 1).

Adaptation to the nutritional environment through the modulation of gene expression is an essential requirement for all cell types. In mammalian species, the potential effect of nutrients has been concealed by the importance devoted to the nutritionally induced hormones. It is only recently that a more direct role of nutrients has been appreciated.

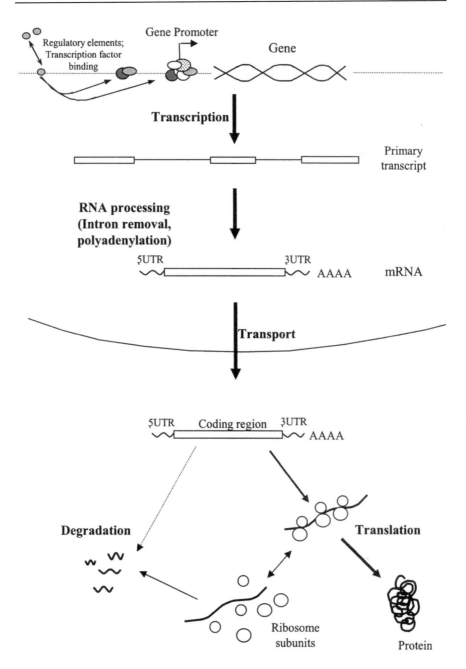

Fig. 2. The gene expression pathway: control points. The schematic representation of the pathway from gene to protein, indicating potential points of control (bold).

With the advent of molecular biology, allowing the isolation of cDNAs encoding mammalian intestinal nutrient transporters, and the use of appropriate *in vivo* and *in vitro* models, the molecular mechanisms by which nutrients regulate the expression of some intestinal nutrient transporters have been elucidated. In most instances, the modulation in mRNA content has been shown to be associated not only with transcription, but also with mRNA stability. The nutrient signals are either transmitted by the nutrient transporter itself (regulating the external nutrient access to the intracellular environment) or via some luminal membrane nutrient sensors.

It has been shown that in addition to regulating the expression of their own transporters, certain luminal nutrients have the capability of controlling the expression of genes regulating proliferation, apoptosis and differentiation of intestinal epithelial cells.

The expression of intestinal nutrient transporters, by dietary substrates, may be regulated at multiple control points (see Fig. 2). This includes synthesis and degradation of mRNA, translation of mRNA into protein, and the processing or modification of synthesised proteins. Control of RNA synthesis (gene transcription) is one critical point. It is the first step in any gene expression pathway, and its modulation represents an economical, and hence, attractive method of regulating gene expression. Therefore, it is not surprising that the control of transcription is a principal method of regulating most genes (Tjian 1996).

This chapter will examine the transcriptional control of intestinal nutrient transporter genes by dietary nutrients, and highlight its nutritional and clinical relevance. I shall first outline the principal features of transcriptional regulation. I shall then describe molecular mechanisms involved in the transcriptional regulation of intestinal (both small and large intestine) nutrient transporters by luminal nutrients. I shall also highlight how the alterations in composition and levels of luminal nutrients, brought about by either dietary manipulation or the ectopic presence of nutrients after surgical resection of the intestine, affects intestinal nutrient transporter expression.

The approach is selective rather than comprehensive. My overall task is to highlight the pathways by which dietary nutrients regulate the expression of intestinal nutrient transporters through transcriptional mechanisms, without diminishing the roles of other regulatory events. At this stage, my task is not too laborious, as presently there are few examples of intestinal nutrient transporters for which transcriptional regulation by dietary substrates has been studied in detail. However, this is an intense area of investigation and it is predicted that further examples will soon become available.

One crucial point to be borne in mind is that, in most cases, dietary manipulation and assessment of its effects on intestinal adaptation in humans is not ethically possible. Therefore, the majority of studies aimed to identify the regulation of intestinal nutrient transporters by dietary nutrients have relied on the use of appropriate animal (and cell line) models. While in some cases there may be species differences in the adaptive response, very often these differences provide us with important clues.

2 Gene transcription

The transcriptional control regions of eukaryotic genes can be separated into two basic categories: (i) the "core promoter region", which typically encompasses DNA sequences approximately −40 and +50 respective to the transcription initiation site; and (ii) upstream (or downstream) regulatory elements (Tjian 1996). The core promoter functions to bind RNA polymerase and its accessory factors, which together constitute the general transcriptional machinery, and in doing so ensures transcription is initiated from the correct site (Carey and Smale 1999). As such, the core promoter is obligatory for the basic process of transcription. Although the core promoter is necessary for transcription to occur, it is not always sufficient, and as such, transcription may require activation. Immediately upstream of the core promoter are proximal regulatory elements, which function to bind proteins (transcription factors) that may activate, enhance or repress transcription (Zawel and Reinberg 1995). This modulation is thought to occur by influencing recruitment of the general machinery to the core promoter. Still further away, either upstream or downstream, are other groups of control sequences, termed enhancers (or distal sequences). These too function by binding transcription factors that may influence transcription either positively or negatively.

Some transcription factors are ubiquitously expressed, whereas others are restricted to specific cell types. Furthermore, each gene contains a unique array of proximal and distal regulatory sequences, and the activity or expression of their transcription factors may be responsive to specific signals (Latchman 1995). This arrangement allows appropriate regulation of the genes required for particular cellular functions or in response to specific stimuli (e.g. nutrient signal). Thus, transcription is controlled by a complicated interplay between distinct genomic DNA control elements and transcription factors, which interact with each other and the general transcription machinery at the gene promoter to influence transcription. This in turn influences the abundance of specific mRNAs available for translation in the cytoplasm.

3 Organisation of the intestinal epithelium and gene expression

An understanding of intestinal gene transcription must be placed in the context of the intestinal epithelium, a structure with an exquisitely regulated spatial organisation.

Intestinal mucosal absorptive cells (enterocytes) originate from committed stem cells positioned above the base of the crypt (Potten and Loeffler 1990). Crypt cells undergo several rounds of cell division while they reside in the crypt compartment. Proliferation ceases in the upper third of the crypt, and cell division does not occur once the cells migrate onto the villus. Enterocytes move out of the crypt in a cohesive band up the side of the villus to the point at the villus tip where they are lost, by exfoliation and apoptosis, into the lumen. This process takes 3-4 days. As such, the complete process of differentiation occurs within the confined architecture of the crypt-villus unit.

Gradients in gene expression within the adult mammalian intestinal epithelium are maintained in several dimensions, resulting in regional differences in function and morphology. Along the longitudinal axis of the gut, variations in specific absorptive functions are found between duodenum, jejunum, ileum and colon. Gradients in gene expression also exist in another spatial dimension, namely between the crypt and the villus tip. Many mRNAs, proteins and enzymatic activities, that define important enterocytic functions, have been shown to be expressed first near the crypt-villus junction (Traber 1990; Rings et al. 1992; Freeman et al. 1993). This geographic regulation of specific gene expression is established and maintained in an epithelium that is regenerated perpetually. Diet has been proposed as one of the factors involved in maintaining this regulation.

The importance of spatial assignment for gene regulation along the radial (crypt-villus) and longitudinal (cranial-caudal) axes of the gut has been addressed, using transgenic mice (Hermiston and Gordon 1993). These studies suggest that transcriptional induction of genes involved in differentiation is intimately associated with the cessation of proliferation of cells in the upper third of the crypt. The onset of expression of enterocytic genes at the crypt-villus junction appears to be regulated by a combination of positive and negative DNA regulatory elements located outside the promoters (Sweetser et al. 1988). The importance of repressors for expression along the crypt-villus axis has been shown for the intestinal fatty acid binding protein (*fabpi*) gene. Nucleotides –277 to +28 of the *fabpi* gene are capable of recapitulating the normal pattern of *fabpi* expression, with transcription initiated in the upper third of the crypt near the crypt-villus junction (Cohn et al. 1992). However, the use of nucleotides –130 to +28 in transgenic mice resulted in precocious expression of the transgene in mid-crypt cells, although enterocyte-specific expression was maintained. There is a similar situation for the sucrase-isomaltase gene, although the elements have been less precisely defined (Traber and Silberg 1996).

An interesting concept emerging from transgenic studies is that multiple elements are required for regulation along the longitudinal axis of the intestine. Genes that have been studied, such as *fabpi*, ileal lipid binding protein (*ilbp*) and sucrase isomaltase (*SI*), have been found to have different DNA regulatory elements that are required for modulating transcription in different bowel segments (Traber and Silberg 1996). The specific molecular mechanisms underlying this regulation are however unknown. Dietary signals have been proposed as important factors.

4 Transcriptional regulation of nutrient transporters in the small intestine

Small intestine is the major site of digestion and absorption of dietary nutrients. Much evidence indicates that although the expression of intestinal nutrient transporters, by dietary substrates, may be regulated at multiple control points, transcription is a critical step in the regulation.

4.1 Intestinal iron absorption

The main site of intestinal iron absorption is in the duodenum where the slightly acidic environment not only promotes solubilisation of iron transformed to Fe^{2+} by ferrireductase and ascorbate, but also provides the inwardly-directed proton electrochemical gradient to drive H^+/Fe^{2+} co-transport across the enterocyte brushborder membrane by the Divalent Metal Transporter 1, DMT1 (SLC11A1) (Gunshin et al. 1997). Iron is then transported across the basolateral membrane into the systemic system by ferroportin 1 (FPN1) (McKie et al. 2000). Mammals have no regulated pathway for iron excretion, and therefore, iron absorption is carefully regulated to maintain body stores.

For the maintenance of iron homeostasis, the rate of intestinal iron absorption is adapted to the body's demand for iron (Philpott 2002). Iron uptake is, therefore, increased in iron deficiency and decreased in iron overload. This is a good example of nutrient regulation of a nutrient, which is potentially toxic at high concentrations (see above). Accordingly, in patients with hereditary haemochromatosis and progressive iron deposition in parenchymal organs, intestinal iron acquisition is inappropriately high (Britton et al. 2002).

The effect of extracellular iron perturbations on DMT1 and FPN1 expression has been investigated in intestinal epithelial cells and in primary tissue cultures of human duodenal biopsies. It has been shown that iron deprivation results in increased protein and mRNA expression of DMT1 and FPN1. Nuclear run off analysis demonstrated that the change in mRNA abundance is due to the modulation of transcription of these genes. To date, promoter analysis of the putative DMT1 promoter and results from promoter truncation assays are inconclusive in terms of iron dependent regulation of DMT1 transcription. It is proposed that this may be due to the fact that the known promoter length of -1849 bp is not the complete promoter (Lee et al. 1998). In addition, recently published data suggest a complex splicing of DMT1 mRNA, with different regulatory properties of different splicing variants (Hubert and Hentze 2002). Further work is needed to characterise the complete regulatory network for the transcription.

4.2 Intestinal peptide transport

Absorption of small peptides makes a significant contribution to total dietary protein assimilation (Matthews 1991). In mammals, PEPT1 (SLC15A1), an integral membrane protein expressed on the luminal membrane of proximal intestinal epithelial cells, is responsible for the uptake of bulk quantities of di- and tripeptides as end products of the luminal and membrane-associated digestion of dietary proteins. The ability of PEPT1 to transport a variety of peptidomimetics, such as antibiotics of the aminocephalosporin and aminopenicillin classes, or selected peptide inhibitors, such as bestatin or captopril, makes PEPT1 an excellent route for the oral absorption of drugs (Adibi 1997).

In vivo feeding studies in rats and mice have suggested that PEPT1 is regulated by dietary peptides (Ferraris et al. 1988; Erickson et al. 1995). Using the human intestinal cell line Caco-2, Walker et al. (1998) have shown that the Vmax for the luminal uptake of the dipeptide glycyl-sarcosine was increased several fold after incubation of Caco-2 cells in a peptide rich medium, indicating that this is a direct effect of nutrient supply on epithelial function, without a requirement for neural and/or hormonal mediators. It was proposed that the underlying molecular mechanism for this upregulation of functional activity is increased levels of PEPT1 mRNA and concomitant increases in PEPT1 protein expression. The close agreement in the magnitude of the increases in PEPT1 activity, mRNA and protein levels demonstrated control of mRNA accumulation, rather than translation or post-translational modification as the primary mechanism of regulation. Although an increase in PEPT1 mRNA stability accounts for some degree of observed upregulation (half-life increasing from 8.9- to 12.5 h), a discrepancy between the calculated level of PEPT1 mRNA accumulation and the measured increase in PEPT1 mRNA indicates that there is also a contribution from increased PEPT1 gene transcription (Walker et al. 1998). Direct demonstration of a transcriptional component to the regulatory response to the peptide, however awaits investigation.

4.3 Intestinal bile salt transport

The intestinal apical sodium dependent bile salt transporter (ASBT, SLC10A2) plays a critical role in the intestinal reclamation of bile salts that are secreted by the liver. The expression of ASBT is restricted to the terminal ileum (Coppola et al. 1998; Schneider 2001). Complete disruption of bile salt transport leads to congenital primary bile acid malabsorption (Oelkers et al. 1997). Partial inhibition, by either ileal exclusion or pharmacological blockade leads to bile salt wasting and provides an important approach to the treatment of hypercholesterolaemia and cholestasis (Oelkers et al. 1997; Hollands et al. 1998).

It has been shown that bile acid feeding in mice results in decreased expression of ASBT protein and mRNA, suggesting that in mouse intestine ASBT is under negative feedback regulation (Torchia et al. 1996; Ma et al. 2000). However, in rat, ileal ASBT does not respond to alterations in lumenal bile acid concentration (Stravitz et al. 1997). ASBT regulation appears to be both transcriptional and post-transcriptional (Chen et al. 2001).

Analysis of the mechanisms underlying the transcriptional regulation of *ASBT* expression have been performed using a cloned 3.1- kb DNA fragment containing 2.7 kb of *ASBT* 5' genomic sequence (Chen et al. 2001). A 208-bp region of the proximal promoter was sufficient to impart basic promoter activity and cell line specificity. These effects appeared to be mediated in large part through AP-1 *cis* elements. Mutations of the AP-1 *cis* elements led to suppression of *ASBT* directed gene expression. The *trans*-acting factors involved in *ASBT* expression were identified and shown to include at least four nuclear proteins ABP1-4, amongst which

ABP2 has been identified as an AP-1 element binding both c-Jun and c-Fos. Recent knockout studies indicate that hepatocyte nuclear factor 1α (HNF-1α) also plays a key role in the transcriptional regulation of ASBT (Shih et al. 2001). The HNF-1α knockout mice have increased faecal and urinary excretion of bile salts, which is coupled with reduced ileal expression of ASBT protein and mRNA (Shih et al. 2001). Analysis of the human *ASBT* promoter revealed a functional HNF-1 binding site. This site is highly conserved in rat, rabbit and human *ASBT* 5' sequences, and is immediately proximal to a conserved AP-1 binding site (Shih et al.2001). These findings suggested that the expression of *ASBT* may be controlled by combinatorial regulation utilising both HNF-1α and AP-1 (Flores et al. 2000; Ghazi and VijayRaghhavan 2000). It has been shown recently that the liver receptor homologue-1 (LRH-1) protein is also a transcriptional activator of *ASBT*. LRH-1 does not appear to be essential for basal promoter activity, but plays a crucial role in mediating bile acid responsiveness. This is supported by the enhanced activity of the mouse *ASBT* promoter versus that of the rat promoter; the mouse but not rat *ASBT* promoter has a functional LRH-1 *cis*-acting element (Chen et al. 2003). In addition, mouse but not rat ileum expresses the LRH-1 protein. The divergent and coordinated evolution of both the LRH-1 *cis*- and *trans*-acting elements between these two closely related rodents has been speculated to have evolved in conjunction with the presence (mouse) or absence (rat) of the gallbladder (Chen et al. 2003). Bile acid responsiveness of the human intestinal *ASBT* gene, if proved to be under negative feed back regulation, will have important clinical implications (Chen et al. 2003).

4.4 Intestinal glucose transport

The Na$^+$/glucose cotransporter, SGLT1 (SLC5A1), transports dietary sugars, D-glucose and D-galactose, from the lumen of the intestine into the enterocytes by coupling the movement of sugar to that of Na$^+$ and the associated electrochemical gradients (Hediger and Rhoads 1994; Wright et al. 1994; Shirazi-Beechey 1995). The major site of glucose absorption is in jejunum>duodenum>ileum. cDNAs encoding SGLT1 in several species (rabbit, human, rat, mouse, pig, sheep, cow and horse) have been cloned and sequenced and have been shown to exhibit a high degree of sequence homology (Hediger et al. 1987, 1989; Shirazi-Beechey 1995; Tarpey et al. 1995; Wood et al. 2000; Dyer et al. 2002).

The expression of SGLT1 is adaptively regulated by dietary sugars in the majority of mammalian species studied (Ferraris et al. 1990; Shirazi-Beechey 1996). Feeding mice and rats with diets containing high levels of either D-glucose or non-metabolisable analogues of D-glucose leads to a two to threefold enhancement of the activity of intestinal SGLT1 (Solberg and Diamond 1987; Ferraris and Diamond 1989). The expression of SGLT1 in human intestine has also been shown to be positively modulated by dietary carbohydrates (Dyer et al. 1997).

Work in the laboratory of the author has shown that in response to luminal monosaccharides the expression of SGLT1 is regulated at both transcriptional and post-transcriptional levels. These studies have been facilitated by the recognition that ovine intestine is an excellent, naturally occurring, model for studies of the regulation of intestinal SGLT1 by dietary sugars (Shirazi-Beechey et al. 1991; Vayro et al. 2001). Rumen development in sheep is a natural and efficient way of blocking the delivery of monosaccharides to the small intestine. In pre-ruminant lambs (birth to three weeks) milk sugar, lactose, is hydrolysed by intestinal lactase into D-glucose and D-galactose, and these sugars are transported by SGLT1 (being highly expressed in lamb small intestine). Lambs are weaned between 3-10 weeks of age and, as the diet changes from milk to grass, the rumen develops. Dietary carbohydrates are fermented by rumen microflora to short chain fatty acids, selectively blocking the delivery of monosaccharides to the small intestine. Associated with the decline in luminal sugars there is >50-fold decrease in the activity and the expression of SGLT1 (Wood et al. 2000; Vayro et al. 2001). Infusion of either D-glucose or non-metabolisable analogues of D-glucose, via duodenal cannulae, into the intestinal luminal contents of ruminant sheep enhances the levels of both SGLT1 mRNA and functional protein, over 50-fold, reaching the levels detected in the intestines of pre-ruminant lambs (Wood et al. 2000; Vayro et al. 2001).

The similarity in the magnitude of the increase in SGLT1 mRNA and protein suggested that the exposure of the intestine to luminal glucose leads to an enhancement in abundance of the SGLT1 transcript, and that this, in turn, gives rise to a corresponding increase in the level of SGLT1 protein. Nuclear run on assays indicated that the induction of SGLT1 mRNA by glucose is in part due to an increase in transcription. In nuclei isolated from the intestine of D-glucose infused animals, the level of newly synthesised RNA encoding SGLT1 was two to threefold higher than that for control ruminant sheep, and was similar to that seen in pre-ruminant lambs (Vayro et al. 2001) (see Fig. 3). This two to threefold increase cannot account entirely for the observed, >50-fold, enhancement in steady-state levels of SGLT1 mRNA detected by northern analysis, and it is apparent that SGLT1 expression is also controlled at the level of post-transcription.

Analysis of the mechanisms underlying the transcriptional regulation of *SGLT1* expression have been performed using a cloned 3.2 kb DNA fragment of the ovine *SGLT1* 5'-flanking region (Wood et al. 1999; Vayro et al.2001). This DNA fragment contains a TATA box motif, 28bp upstream of the transcriptional start site and proximal binding sites for HNF-1 and SP-1.

A –66/+21 region of the ovine proximal *SGLT1* promoter was sufficient to confer basic promoter activity and cell line specificity (Wood et al. 1999). These effects appeared to be mediated in large part through the HNF-1 *cis* element. Mutations of the HNF-1 consensus motif within this region led to suppression of *SGLT1* promoter directed gene expression (Vayro et al. 2001). In human intestine the *SGLT1* core promoter region has been recognised, and it has been shown that HNF-1, acting synergistically with SP-1 family members, is important in the basal promoter function (Martin et al. 2000).

As previously mentioned, it is well documented that in the majority of species studied *SGLT1* expression is positively regulated by dietary sugars. We have shown that HNF-1, in addition to being essential for basal promoter activity, is an important player in mediating the glucose -responsiveness of ovine *SGLT1* (Vayro

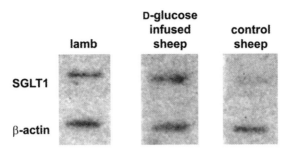

Fig. 3. Effect of lumenal D-glucose on *in vitro* transcription of *SGLT1* in ovine small intestine. D-glucose in the lumen of the intestine, specifically, leads to an increase in the rate of *SGLT1* transcription. Using nuclear run-on assays, nuclei isolated from the jejunal mucosa of ruminant sheep whose intestines were infused with D-glucose showed levels of newly synthesised SGLT1 mRNA two to threefold higher than for control age-matched ruminant sheep. Similar levels were detected in the jejunum of pre-ruminant lambs. The level of β-actin mRNA synthesis was identical in all three samples. Reprinted from Vayro et al. (2001) with permission.

et al. 2001). This is supported by transfection studies showing enhanced activity of the ovine *SGLT1* promoter in response to increases in media glucose concentration, and electrophoretic mobility shift assays demonstrating an increase in the specific binding of HNF-1 to an *SGLT1* promoter fragment in response to glucose (Fig. 4). The latter showed that the binding of HNF-1 to the promoter consensus sequence is increased twofold in nuclear extracts isolated from the intestine of ruminant sheep whose intestines had been infused with D-glucose, compared to age-matched control ruminant sheep (Fig. 4). Furthermore, the expression of HNF-1α was higher in the intestinal tissue of glucose infused sheep, and pre-ruminant lambs (with high SGLT1 expression) compared to controls.

We questioned the mechanism by which the luminal sugar signal interacts with the intestinal mucosal cells. To this end, we synthesised a membrane impermeable D-glucose analogue, di(glucos-6-yl)poly(ethylene glycol) 600 (di(glucos-6-yl)PEG$_{600}$) (Dyer et al. 2003a). Infusion of the intestines of ruminant sheep with di(glucos-6-yl)PEG$_{600}$ led specifically to induction of functional SGLT1. However, di(glucos-6-yl)PEG$_{600}$ did not inhibit Na$^+$-dependent glucose transport into intestinal brush-border membrane vesicles (Dyer et al. 2003a). Studies using intestinal cells showed that increased medium glucose upregulated SGLT1 protein abundance and *SGLT1* promoter activity, and increased intracellular cyclic AMP (cAMP) levels. Furthermore, the glucose-induced activation of the *SGLT1* promoter was mimicked by the protein kinase A (PKA) agonist, 8Bromo-cAMP, and was inhibited by H-89, a PKA inhibitor. Pertussis toxin, a G-protein (G$_i$)-specific inhibitor, enhanced SGLT1 protein abundance to the levels observed in response to glucose or 8-Br-cAMP. Accordingly, we have concluded that luminal glucose is sensed by a glucose sensor, distinct from SGLT1, residing on the external face of the luminal membrane. The glucose sensor initiates a signalling pathway, involving a G-protein coupled-receptor linked to a cAMP-PKA pathway, resulting in an enhancement in SGLT1 transcription (Dyer et al. 2003a, 2003b).

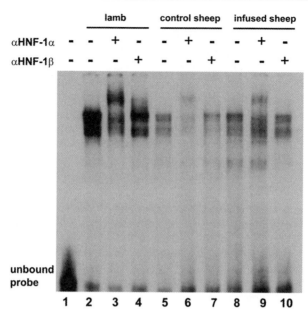

Fig. 4. Involvement of HNF-1 in glucose-responsive *SGLT1* expression. Nuclear extracts were isolated from the intestinal mucosa of pre-ruminant lambs, three-year-old sheep and age-matched animals whose intestines had been infused with D-glucose through duodenal cannulae. EMSA's were carried with a probe corresponding to the HNF-1 binding motif. Antibodies to HNF-1α (lanes 3, 6 and 9) and HNF-1β (lanes 4, 7 and 10) were included in the binding reactions. Lane 1, no nuclear extracts. Reprinted from Vayro et al. (2001) with permission.

A better understanding of the regulatory/signalling pathways, which control the transcription of SGLT1, the gatekeeper of glucose absorption, will have clinical implications in diseases such as diabetes, obesity, and diarrhoeal conditions.

5 Regulation of nutrient transport in the large intestine

Study of nutrient transport in the large intestine has traditionally been neglected in favour of that in the small intestine. Consequently, less is known of nutrient transport, and the mechanisms by which dietary components modulate transporter expression in the large intestine. Nevertheless, interest is expanding in this area, as it becomes increasingly evident that the large intestine serves as an important site for the salvage of nutrients escaping digestion in the small intestine (Macfarlane and Cummings 1991; Danielsen and Jackson 1992, Cuff et al. 2002). Considerable attention has (and remains) focused on the transport of short chain fatty acids (SCFA; acetate, propionate, and butyrate) produced naturally in the lumen of the colon by microbial fermentation of dietary fibre and resistant starch. Much evidence suggests that these SCFA, particularly butyrate, are of fundamental importance to the health of the normal colonic mucosa. Indeed, butyrate serves as the primary respiratory fuel of colonocytes (Cummings 1984) and regulates expression of genes involved in their proliferation, differentiation, and apoptosis. These butyrate responsive genes include the cell cycle progression inhibitor p21$^{wafl/cip1}$, a number of cyclins and members of the Bcl-2 family (Archer et al. 1998; Bai and Merchant 2000, Siavoshian et al. 2000; Hague et al. 1997).

At the colonic luminal pH, SCFA exist almost entirely in anionic form; the cellular entry of which is dependent upon a specific carrier protein(s). Accordingly, we have shown that butyrate is transported across the human colonic luminal membrane by the monocarboxylate transporter isoform 1, MCT1 (SLC16A1) (Ritzhaupt et al. 1998). Subsequently, other laboratories using the human colon cancer cell line, Caco-2, have also demonstrated MCT1-mediated butyrate transport (Stein et al. 2000). Further work has shown that MCT1 has the ability to transport, in addition to butyrate, acetate and propionate (Tamai et al. 1999). Acetate and propionate, once transported across the colonocytes into the circulation are used by the peripheral tissues as respiratory fuel (acetate) or for gluconeogenesis (propionate) (Macfarlane and Cummings 1991).

The central role of butyrate in cellular metabolism and the maintenance of colonic tissue homeostasis make an understanding of the regulation of its transport of particular importance. Accordingly, we have examined the regulation of MCT1 expression in response to its substrate, butyrate. Treatment of human colonic epithelial cells (AA/C1) with butyrate resulted in concentration- and time-dependent upregulation of both MCT1 mRNA and protein. The magnitude of induction of mRNA (5.7-fold) entirely accounted for the 5.2-fold increase in protein abundance and was paralleled by a corresponding increase in the rate (V_{max}) of butyrate transport, suggesting (i) that the primary mechanism of control is at the level of mRNA abundance, and (ii) that the increased transport reflects an elevation in the number of MCT1 molecules rather than their affinity for substrate (Cuff et al. 2002).

Nuclear run on assays, employing nuclei isolated from human colonic AA/C1 cells, demonstrate that the induction of MCT1 mRNA is at least in part due to an increase in transcription (see Fig. 5). However, the magnitude of the increase (threefold) is insufficient to fully account for the induction of overall transcript abundance (5.7-fold) and, therefore, indicates additional post-transcriptional control. This is confirmed by the finding that butyrate also increases the half-life of the MCT1 transcript by a factor of about 2. Taken together, these increases in transcript stability (twofold) and transcription (threefold) correlate well with the overall increase in MCT1 mRNA levels detected by northern analysis. Thus, as for SGLT1 in the small intestine, upregulation of MCT1 expression in the colon is achieved by the dual control of mRNA synthesis and degradation.

To gain an insight into the transcriptional control of MCT1 we sought to isolate nucleotide regions with the potential to be involved in this regulation. Transcriptional control elements may be located a considerable distance from the core promoter region that drives basal transcription. Indeed, *cis*-acting elements may be present not only upstream of the transcriptional start site but also in the 5'-UTR (Mittanck et al. 1997), downstream introns (Polakowska et al. 1999), and within

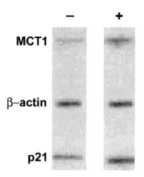

Fig. 5. Effect of butyrate on *in vitro* transcription of MCT1. Nuclear run-on assays indicated that the transcription rates of MCT1 and p21 (used as a positive control) in colonic epithelial cells, HT-29, are enhanced in response to increased levels of media butyrate (+). The transcription rate of β-actin is unaffected. Reprinted from Cuff et al. (2002) with permission.

the coding region itself (Carey and Smale 1999). Accordingly, to gain an insight into the mechanisms underlying the transcriptional regulation of MCT1, we have isolated more than 100 kb of genomic DNA corresponding to the *MCT1* gene locus on chromosome 1. This region encompasses the entire MCT1 transcription unit and extends more than 50 kb upstream into the 5'-flanking region. We have determined the intron/exon organisation, the site of transcription initiation, and have isolated and characterised the core promoter region that drives basal transcription (Cuff and Shirazi-Beechey 2002). The MCT1 transcription unit (including untranslated regions, UTR, introns and coding sequence) encompasses almost 44 kb, and consists of 5 exons intervened by 4 introns. The first of these introns is located in the 5'-UTR encoding DNA, spans >26 kb and, thus, accounts for approximately 60% of the entire transcription unit.

Isolation of the *MCT1* gene promoter was demonstrated by transient transfection reporter assays employing fragments of the *MCT1* 5'-flanking region to drive luciferase expression in AA/C1 cells that endogenously express MCT1 (Cuff and Shirazi-Beechey 2002). The mean reporter activity resulting from a region extending approximately 1.5 kb upstream of the transcription initiation site (-1525/+213) was more than 25-fold greater than controls (promoterless vector; see Fig. 6). Deletion analyses indicate that the *cis*-acting elements necessary for basal transcription reside within the −70/+213 proximal region of the *MCT1* gene promoter (Cuff and Shirazi-Beechey 2002). Although the 5'-flanking region lacks a classical TATA box motif, it contains potential binding sites for a variety of transcription factors with known association with butyrate's action in the colon. As such, the significance of these and other potential regulatory regions at the *MCT1* gene locus to act as specific butyrate response elements is currently under investigation.

Although the mechanisms underlying the specific regulation of nutrient transporter expression are beginning to be identified, a major challenge that remains is to gain an insight into the significance of this control. This is perhaps especially important for transporters of nutrients that perform functions beyond their

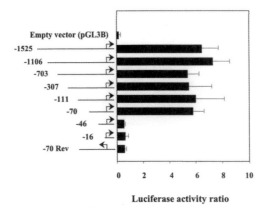

Fig. 6. Functional analysis of the human *MCT1* promoter. Deletion analysis indicated that the -70/+213 region contains the essential *cis*-acting elements necessary for basal transcription of *MCT1*. *Rev* indicates the reverse orientation. Reprinted from Cuff and Shirazi-Beechey (2002) with permission.

recognised role in metabolism. The transport of butyrate by MCT1 falls into such a category. Indeed, we have recently reported that MCT1 expression is dramatically reduced during the transition from normality to malignancy (Lambert et al. 2002), and have proposed that this decline in transporter expression may result in a reduction in the intracellular availability of butyrate required to regulate expression of genes associated with the processes maintaining tissue homeostasis within the colonic mucosa. To test this hypothesis, and so examine the significance of our previous findings, we have employed the technique of RNA interference (RNAi) (Yu et al. 2002) to specifically inhibit MCT1 expression, and examined the consequences of this inhibition on the ability of butyrate to exert its effects on target gene expression and cellular function *in vitro*. Using this approach, we have found that inhibition of MCT1 expression, and hence the rate of butyrate uptake, has profound inhibitory effects on the ability of butyrate to regulate expression of key target genes associated with the processes of cellular proliferation and differentiation (see Fig. 7). It is clear, therefore, that the identification of detailed regulatory and signalling networks controlling the regulation of colonic MCT1 may have both nutritional and clinical significance.

5.1 Adaptive response of the colonic epithelium in short-bowel syndrome

Short bowel syndrome (SBS) is a condition where severe malabsorption occurs after extensive resection of the small bowel. Clinical features of SBS include chronic diarrhoea, dehydration, electrolyte abnormalities, and malnutrition (Seidner 2002). All symptoms of SBS occur because of the failure of the intestine to absorb nutrients. Morphological and physiological adaptation in residual small

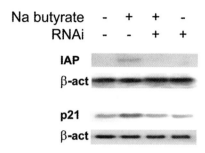

Fig. 7. Effect of inhibition of MCT1 expression by RNAi on the butyrate-induced expression of alkaline phosphatase (IAP) and p21. Colonic epithelial HT29 cells were transfected with siRNA targeted to the MCT1 transcript. Cells were then treated with butyrate or maintained under standard culture conditions. siRNA specifically inhibited the butyrate-induced expression of IAP and p21. This treatment had no effect on β-actin expression.

intestine occurs after massive enterectomy. The mechanism of adaptation occurs through hypertrophy and hyperplasia as well as in nutrient absorption capacity. The adaptation process evolves over time, and for any of these adaptive changes to occur, food must be presented to the gut (Wang et al. 1997). It appears that ileal and colonic epithelial cells can adapt, with time, and express functional characteristics of the proximal small intestine, while duodenum and jejunum are not capable of expressing the functions of the lower bowel. There are, however, exceptions, for example, lactase deficiency will occur with the resection of jejunum, in spite of the presence of an intact ileum (Seidner 2002). The molecular basis of the differential adaptive response of various regions of the gut to express digestive and absorptive functions is not known.

We have shown in our laboratory that the ectopic presence of nutrients in the colon has resulted in the expression of a unique nutrient transporter. Amino acids are absorbed normally in the small intestine by several different Na^+-dependent and independent mechanisms; with the major site of absorption occurring in the ileum (Palacin et al. 1998). To date, no amino acid transporter has been detected in normal healthy human colon. However, we have shown that ATB^{0+} (SLC6A14), a broad range neutral and cationic amino acid transporter, capable of transporting both D- and L-amino acids was expressed in colonic biopsies of an adult individual with SBS (see Fig. 8) whose ileum and part of jejunum had been removed after birth due to a congenital defect (Zibrik et al. 2003). It has also been demonstrated that the peptide transporter, PEPT1, is expressed in the colon of patients with SBS (Ziegler et al. 2002). These are very good examples of the adaptive ability of the colon to salvage unabsorbed nutrients. A better understanding of the molecular basis of the adaptive response of the residual intestine to recover intestinal function, will allow the development of strategies to alleviate the patient's symptoms and to enhance the quality of life markedly.

It is clear that the transport of nutrients across the plasma membrane of the intestinal epithelial cells, accomplished by highly specialised membrane proteins, is an important process. It is the primary step in the provision of nutrients to the body, and can be exploited for drug delivery. Furthermore, the regulation of the

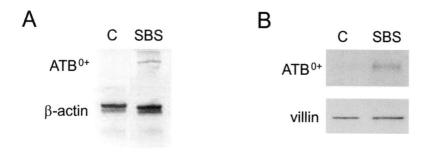

Fig. 8. Amino acid transporter ATB^{0+} is expressed in the colon of a short-bowel patient. A: Ribonuclease protection assays indicating the presence of an ATB^{0+} protected fragment in RNA isolated from the colon of an SBS patient (SBS), but not in healthy controls (C). B: Western blot analysis showing the presence of ATB^{0+} protein in colonic luminal membrane vesicles of the SBS patient. C: Results of the combined densitometric analyses of RPAs and western blots. Data are presented as mean ± S.E. n.d.: not detected.

expression of nutrient transport proteins provides a level of control, which is essential for the maintenance of metabolism and homeostasis.

The complete understanding of the integrated mechanisms controlling the transcription of intestinal nutrient transporters is a challenging task. The identification of all *cis* elements, and *trans* factors, including their regulation and post-translational modifications, and the cloning of the transcription factors and cofactors will be necessary to understand the linkage of cell signalling to the transcriptional machinery. This knowledge will assist the design of novel approaches to the therapeutic intervention in various abnormalities that are associated with deregulated gene transcription.

It is evident that this area of research will entertain membrane transport and intestinal biologists for many years to come.

Acknowledgements

It is a pleasure to acknowledge the contributions of my colleagues and associates Jane Dyer, Mark Cuff, Tony Ellis, Steven Vayro, Stuart Wood, Daniel Lambert, Armin Ritzhaupt, and Lea Zibrik. In particular I wish to thank Jane and Mark for their help and support. The studies carried out in the laboratory of the author were supported by grants from the Biotechnology and Biological Sciences Research Council, The Wellcome Trust, and Tenovus Cancer Charity.

References

Adibi SA (1997) The oligopeptide transporter (Pept-1) in human intestine: biology and function. Gastroenterology 113:332-340

Archer SY, Meng S, Shei A, Hodin RA (1998) p21(WAF1) is required for butyrate-mediated growth inhibition of human colon cancer cells. Proc Natl Acad Sci USA 95:6791-6796

Bai L, Merchant JL (2000) Transcription factor ZBP-89 cooperates with histone acetyl-transferase p300 during butyrate activation of p21waf1 transcription in human cells. J Biol Chem 275:30725-30733

Britton RS, Fleming RE, Parkkila S, Waheed A, Sly WS, Bacon BR (2002) Pathogenesis of hereditary hemochromatosis: genetics and beyond. Semin Gastrointest Dis 13:68-79

Carey M, Smale ST (1999) In Transcriptional regulation in eukaryotes. Cold Spring Harbor Laboratory Press, New York, pp. 1-50

Chen F, Ma L, Al-Ansari N, Schneider B (2001) The role of AP-1 in the transcriptional regulation of the rat apical sodium-dependent bile acid transporter. J Biol Chem 276:38703-38714

Chen F, Ma L, Dawson PA, Sinal CJ, Sehayek E, Gonzalez FJ, Breslow J, Ananthana-ratanan M, Schneider BL (2003) Liver receptor homologue-1 mediates species- and cell line-specific bile acid-dependent negative feedback regulation of the apical sodium dependent bile acid transporter. J Biol Chem 278:19909-19916

Cohn SM, Simon TC, Roth KA, Birkenmeier EH, Gordon JI (1992) Use of transgenic mice to map cis-acting elements in the intestinal fatty acid binding protein gene (Fabpi) that control its cell lineage-specific and regional patterns of expression along the duodenal-colonic and crypt-villus axes of the gut epithelium. J Cell Biol 119:27-44

Coppola CP, Gosche JR, Arrese M, Ancowitz B, Madsen J, Vanderhoof J, Schneider BL (1998) Molecular analysis of the adaptive response to intestinal bile acid transport after ileal resection. Gastroenterology 115:1172-1178

Cuff MA, Shirazi-Beechey SP (2002) The human monocarboxylate transporter, MCT1: genomic organisation and promoter analysis. Biochem Biophys Res Commun 262:1048-1056

Cuff MA, Lambert DW, Shirazi-Beechey SP (2002) Substrate-induced regulation of the human colonic monocarboxylate transporter, MCT1. J Physiol (Lond) 539.2:361-371

Cummings JH (1984) The importance of SCFA in man. Scand J Gastroenterology 19:89-99

Danielsen M, Jackson AA (1992) Limits of adaptation to a diet low in protein in normal man: urea kinetics. Clin Sci 83:103-108

Diamond JM, Karasov WH (1987) Adaptive regulation of intestinal nutrient transporters. Proc Natl Acad Sci USA 84:2242-2245

Dyer J, Hosie KB, Shirazi-Beechey SP (1997) Nutrient regulation of human intestinal sugar transporter (SGLT1) expression. Gut 41:56-59

Dyer J, Fernandez-Castaño Merediz E, Salmon KSH, Proudman CJ, Edwards GB, Shirazi-Beechey SP (2002) Molecular characterisation of carbohydrate digestion and absorption in equine small intestine. Equine Vet J 34:349-358

Dyer J, Vayro S, King TP, Shirazi-Beechey SP (2003a) Glucose sensing in the intestinal epithelium. Eur J Biochem 270:3377-3388

Dyer J, Vayro S, Shirazi-Beechey SP (2003b) Mechanism of glucose sensing in the small intestine. Biochem Soc Trans 31:1140-1142

Erickson RH, Gum JRG, Lindstrom MM, McKean D, Kim YS (1995) Regional expression and dietary regulation of rat small intestinal peptide and amino acid transporter mRNAs. Biochem Biophys Res Commun 216:249-257

Ferraris RP, Diamond JM, Kwan WW (1988) Dietary regulation of the intestinal transport of the dipeptide carnosine. Am J Physiol 255:G143-G149

Ferraris RP, Diamond JM (1989) Specific regulation of intestinal nutrient transporters by their dietary substrates. Annu Rev Physiol 51:125-141

Ferraris RP, Villenas SA, Hirayama BA, Diamond J (1990) Effect of diet on glucose transporter site density along the intestinal crypt-villus axis. Am J Physiol 262:G1060-G1068

Flores G, Duan H, Yan H, Nagaraj R, Fu W, Zou Y, Noll M, Banerjee U (2000) Combinatorial signaling in the specification of unique cell fates. Cell 103:75-85

Freeman TC, Wood IS, Sirinathsinghji DJS, Beechey RB, Dyer J, Shirazi-Beechey SP (1993) The expression of Na$^+$-glucose cotransporter (SGLT1) gene in lamb intestine during postnatal development. Biochim Biophys Acta 1146:203-212

Ghazi A, VijayRaghhavan K (2000) Developmental biology. Control by combinatorial codes. Nature 408:419-420

Gunshin H, MacKenzie B, Berger UV, Gunshin Y, Romero MF, Boron WF, Nussberger S, Gollan JL, Hediger MA (1997) Cloning and characterization of a mammalian proton-coupled metal-ion transporter. Nature 388:482-488

Hague A, Diaz GD, Hicks D, Krajewski S, Reed JC, Paraskeva C (1997) bcl-2 and bak may play a pivotal role in sodium butyrate-induced apoptosis in colonic epithelial cells; however, overexpression of bcl-2 does not protect against bak-mediated apoptosis. Int J Cancer 72:898-905

Hediger MA, Coady MJ, Ikeda TS, Wright EM (1987) Expression cloning and cDNA sequencing of the Na$^+$-glucose co-transporter. Nature 330:379-381

Hediger MA, Turk E, Wright EM (1989) Homology of the human intestinal Na$^+$/glucose and *Escherichia coli* Na$^+$/proline cotransporters. Proc Natl Acad Sci USA 86:5748-5742

Hediger MA, Rhoads DB (1994) Molecular physiology of sodium-glucose cotransporters. Physiol Rev 74:993-1026

Hermiston ML, Gordon JI (1993) Use of transgenic mice to characterize the multipotent intestinal stem cell and to analyze regulation of gene expression in various epithelial cell lineages as a function of their position along the cephalo-caudal and crypt-to-villus (or crypt-to-surface epithelial cuff) axes of the gut. Semin Dev Biol 4:275-291

Hollands CM, Rivera-Pedrogo J, Gonzalez-Vallina R, Loret-de-Mola O, Nahmad M, Brunweit CA (1998) Ileal exclusion for Byler's disease: an alternative surgical approach with promising early results for pruritus. J Pediatr Surg 33:220-224

Hubert N, Hentze MW (2002) Previously uncharacterized isoforms of divalent metal transporter (DMT)-1: implications for regulation and cellular function. Proc Natl Acad Sci USA 99:12345-12350

Lambert DW, Wood IS, Ellis A, Shirazi-Beechey SP (2002) Molecular changes in the expression of human colonic nutrient transporters during the transition from normality to malignancy. Br J Cancer 86:1262-1269

Latchman DS (1995) Eukaryotic transcription factors, 2nd Edition. Academic Press Limited

Lee PL, Gelbart T, West C, Halloran C, Beutler E (1998) The human Nramp2 gene: characterization of the gene structure, alternative splicing, promoter region and polymorphisms. Blood Cells Mol Dis 24:199-215

Ma L, Sehayek E, Breslow J, Schneider B (2000) Discoordinate regulation of the ileal bile acid transporter (ASBT) and bile acid binding protein (ILBP) in mouse ileum. Gastroenterology 118:165 A 934

Martin M, Wang J, Solorzano-Vargas S, Lam JT, Turk E, Wright EM (2000) Regulation of the human Na^+-glucose cotransporter gene, SGLT1, by HNF-1 and Sp1. Am J Physiol 278:G591-G603

Matthews DM (1991) Protein absorption. Wiley-Liss, New York

Macfarlane GT, Cummings JH (1991) The colonic flora, fermentation, and large bowel digestive function. In: Phillips SF, Pemberton JH, Shorter RG, eds. The large intestine; physiology, pathophysiology and disease. Raven Press Ltd, New York, pp. 51-92

McKie AT, Marciani P, Rolfs A, Brennan K, Wehr K, Barrow D, Miret S, Bomford A, Peters TJ, Farzaneh F, Hediger MA, Hentze MW, Simpson RJ (2000) A novel duodenal iron-regulated transporter, IREG1, implicated in the basolateral transfer of iron to the circulation. Mol Cell 5:299-309

Mittanck DW, Kim SW, Rotwein P (1997) Essential promoter elements are located within the 5' untranslated region of human insulin-like growth factor-I exon I. Mol Cell Endocrinol 126:153-163

Oelkers P, Kirby LC, Heubi JE, Dawson PA (1997) Primary bile acid malabsorption caused by mutations in the ileal sodium-dependent bile acid transporter gene (SLC10A2). J Clin Invest 99:1880-1887

Palacin M, Estevez R, Bertran J, Zorzano A (1998) Molecular biology of mammalian plasma membrane amino acid transporters. Physiol Rev 78:969-1054

Philpott CC (2002) Molecular aspects of iron absorption: insights into the role of HFE in hemochromatosis. Hepatology 35:993-1001

Polakowska RR, Graf BA, Falciano V, LaCelle P (1999) Transcription regulatory elements of the first intron control human transglutaminase type I expression in epidermal keratinocytes. J Cell Biochem 73:355-369

Potten CS Loeffler M (1990) Stem cells: attributes, cycles, spirals, pitfalls and uncertainties. Lessons from the crypt. Development 110:1001-1020

Rings EHHM, DeBoer PAJ, Moorman AFM, VanBeers EH, Dekker J, Montgomery RK, Grand RJ, Buller HA (1992) Lactase gene expression during early development of rat small intestine. Gastroenterology 103:1154-1161

Ritzhaupt A, Ellis A, Hosie KB, Shirazi-Beechey SP (1998) The characterization of butyrate transport across pig and human colonic luminal membrane. J Physiol (Lond) 507:819-830

Schneider B (2001) Intestinal bile acid transport: biology, physiology, and pathophysiology. J Pediatr Gastroenterol Nutr 32:407-417

Seidner DL (2002) Short bowel syndrome: etiology, pathophysiology and management. Practical Gastroenterology 25:63-72

Shih D, Bussen M, Sehayek E, Ananthanarayanan M, Schneider B, Suchy F, Shefer S, Bollileni J, Gonzalez F, Breslow J, Stoffel M (2001) Hepatocyte nuclear-factor 1 alpha is an essential regulator of bile acid and plasma cholesterol metabolism. Nat Genet 27:375-382

Shirazi-Beechey SP, Hirayama BA, Wang Y, Scott D, Smith MW, Wright EM (1991) Ontogenic development of lamb intestinal sodium-glucose co-transporter is regulated by diet. J Physiol (Lond) 437:699-708

Shirazi-Beechey SP (1995) Molecular biology of intestinal glucose transport. Nutr Res Rev 8:27-41

Shirazi-Beechey SP (1996) Intestinal sodium-dependent D-glucose cotransporter: dietary regulation. Proc Nutr Soc 55:167-178

Siavoshian S, Segain JP, Kornprobst M, Bonnet C, Cherbut C, Galmiche JP, Blottiere HM (2000) Butyrate and trichostatin A effects on the proliferation/differentiation of human intestinal epithelial cells: induction of cyclin D3 and p21 expression. Gut 46:507-514

Solberg DH, Diamond JM (1987) Comparison of different dietary sugars as inducers of intestinal sugar transporters. Am J Physiol 252:G574-G584

Stein J, Zores M, Schroder O (2000) Short–chain fatty acid (SCFA) uptake into Caco-2 cells by a pH dependent and carrier mediated transport mechanism. European J Nutr 39:121-125

Stravitz RT, Sanyal AJ, Pandak WM, Vlahoevia ZR, Beeta JW, Dawson PA (1997) Induction of sodium-dependent bile acid transporter messenger RNA, protein, and activity in rat ileum by cholic acid. Gastroenterology 113:1599-1608

Sweetser DA, Birkenmeier EH, Hoppe PC, McKeel DW, Gordon JI (1988) Mechanisms underlying generation of gradients in gene expression within the intestine: an analysis using transgenic mice containing fatty acid binding protein-human growth hormone fusion genes. Genes Dev 2:1318-1332

Tamai I, Sai Y, Ono A, Kido Y, Yabuuchi H, Takanaga H, Satoh E, Ogihara T, Amano O, Izeki S, Tsuji A (1999) Immunohistochemical and functional characterisation of pH-dependent intestinal absorption of weak organic acids by the monocarboxylic acid transporter MCT1. J Pharm Pharmacol 51:1113-1121

Tarpey PS, Wood IS, Shirazi-Beechey SP, Beechey RB (1995) Amino acid sequence and the cellular location of the Na(+)-dependent D-glucose symporters (SGLT1) in the ovine enterocyte and the parotid acinar cell. Biochem J 312:293-300

Tjian R (1996) The biochemistry of transcription in eukaryotes: a paradigm for multisubunit regulatory complexes. Philos Trans R Soc Lond B 351:491-499

Torchia E, Cheema S, Agellon L (1996) Coordinate regulation of bile acid biosynthetic and recovery pathways. Biochem Biophys Res Commun 225:128-133

Traber PG (1990) Regulation of sucrase-isomaltase gene expression along the crypt-villus axis of rat small intestine. Biochem Biophys Res Commun 173:765-773

Traber PG, Silberg DG (1996) Intestinal-specific gene transcription. Ann Rev Physiol 58:275-297

Vayro S, Wood IS, Dyer J, Shirazi-Beechey SP (2001) Transcriptional regulation of the ovine intestinal Na^+/glucose cotransporter SGLT1 gene: role of HNF-1 in glucose activation of promoter function. Eur J Biochem 268:5460-5470

Walker D, Thwaites DT, Simmons NL, Gilbert HJ, Hirst BH (1998) Substrate upregulation of the human small intestinal peptide transporter, hPepT1. J Physiol 507:697-706

Wang HT, Miller JH, Iannoli P, Sax HC (1997) Intestinal adaptation and amino acid transport following massive enterectomy. Frontiers in Biosciences 2:116-122

Wood IS, Allison GG, Shirazi-Beechey SP (1999) Isolation and characterization of a genomic region upstream from the ovine Na^+/D-glucose cotransporter (SGLT1) cDNA. Biochem Biophys Res Comm 257:533-537

Wood IS, Dyer J, Hofmann RR, Shirazi-Beechey SP (2000) Expression of the Na^+/glucose co-transporter (SGLT1) in the intestine of domestic and wild ruminants. Pflügers Arch 441:155-162

Wright EM, Loo DDF, Panayotova-Heiermann M, Lostao MP, Hirayama BA, Mackenzie B, Boorer K, Zampighi G (1994) "Active" sugar transport in eukaryotes. J Exp Biol 196:197-212

Yu J-Y, DeRuiter S, Turner DL (2002) RNA interference by expression of short interfering RNAs and hairpin RNAs in mammalian cells, Proc Natl Acad Sci USA 99:6047-6052

Zawel L, Reinberg D (1995) Common themes in assembly and function of eukaryotic transcription complexes. Ann Rev Biochem 64:533-561

Zibrik L, Dyer J, Ellis T, Shirazi-Beechey SP (2003) Amino acid transport in human colon in short bowel syndrome. Gastroenterology 124:204, A31

Ziegler TR, Fernandez-Estivariz C, Gu LH, Bazargan N, Umeakunne K, Wallace TM, Diaz EE, Rosado KE, Pascal RR, Galloway JR, Wilcox JN, Leader LM (2002) Distribution of the H^+/peptide transporter PepT1 in human intestine: upregulated expression in the colonic mucosa of patients with short-bowel syndrome. Am J Clin Nutr 75:922-930

Shirazi-Beechey, Soraya P.

Epithelial Function and Development Group, Department of Veterinary Preclinical Sciences, The University of Liverpool, Brownlow Hill, Liverpool L69 7ZJ, UK

spsb@liv.ac.uk

Generation of transporter isoforms by alternative splicing

Gerardo Gamba

Abstract

Post-translational gene processing is an important strategy to increase the complexity of the vertebrate proteome. The membrane transporters are one clear example in which alternative splicing has been shown to be an important tool to increase transporter diversity. This chapter is an overview of alternative splicing of membrane transporter genes in which a variety of physiological consequences are the results of the splicing isoforms. In most transporter genes, splicing variants are functional, resulting in changes of the pharmacological or kinetic properties of the transporters, in the polarization of isoforms to apical or basolateral membranes, in the tissue distribution or subcellular localization, and in the regulation by specific signaling pathways or elements. In some cases, although the splicing isoform is functional, the physiological consequences are still unknown. Finally, in other examples, splicing variants are not functional but possess regulatory properties. Thus, membrane transporter diversity is clearly enhanced by alternative splicing.

1 Introduction

A surprising conclusion that came together with the complete draft of the human genome project (Lander et al. 2001) was that humans have just twice the number of genes of the worm and fly, suggesting that the obviously larger complexity of the vertebrate proteome is accomplished with a limited number of genes. This observation pointed out to the mechanisms for post-transcriptional gene processing, such as alternative pre-mRNA splicing, the use of multiple transcription start sites, polyadenylation, pre-mRNA editing, and protein post-translational modifications, as a fundamental strategy used by higher organisms to achieve their complexity.

Alternative pre-mRNA splicing is the mechanism by which at least two or more closely related polypeptide chains are obtained from a single gene. In this process, the pre-mature mRNA containing all introns and exons encoded by a given gene is exposed to the spliceosome that removes the introns and in doing so the exons can be alternatively splicing in order to produce different, but related proteins. Some exons are constitutive, while others are optional. Still other exons are mutually exclusive. That is, one or another exon is included, but not both at the same time. Alternative splicing can also modulate the availability of certain gene products by generating transcripts with divergent 5' or 3' untranslated regions (UTRs), affect-

Topics in Current Genetics, Vol. 9
E. Boles, R. Krämer (Eds.): Molecular Mechanisms Controlling Transmembrane Transport
DOI 10.1007/b95780 / Published online: 9 March 2004
© Springer-Verlag Berlin Heidelberg 2004

ing the stability of the mature mRNA. To learn about the mechanisms of alternative pre-mRNA splicing, the reader is referred to several excellent up-to-date reviews on this subject by Black (2003), Maniatis and Tasic (2002), Roberts and Smith (2002).

The bioinformatics analysis between the high-throughput sequencing of the human genome and the expressed sequence tag sequences support the hypothesis that vertebrate proteome can be several times larger than the genome due to alternative splicing. Between 35 to 59 % of the human genes undergo alternative splicing, to produce at least one splicing isoform. In addition, it is also known that 70-88 % of the splicing products change the primary sequence of the protein (Modrek and Lee 2002; Lander 2001; Kan et al. 2001). Interestingly enough, alternative splicing itself is a process that can be modulated. Splicing requires exon recognition, followed by intron cleavage and exon rejoining. In this process, the splicing machinery is provided with several exonic and intronic enhancers or silencers that working together defines the exons that will be part of the mature mRNA (Graveley 2001; Nissim-Rafinia and Kerem 2002; Lopez 1998). Some splicing events are tissue specific, occur at a certain time in development or are clearly regulated in response to several physiological or biochemical stimuli. Finally, alternative splicing is also a source of human disease. The molecular pathophysiology of 10 - 15 % of mutations in the human genome associated with disease includes disruption of the pre-mRNA splicing (Nissim-Rafinia and Kerem 2002; Caceres and Kornblihtt 2002).

Bioinformatics analysis is producing a catalog of alternative splicing possibilities in several genes. Thus, one of the major challenges for research is to define the real splicing possibilities for each gene and then to find out what functional consequences the splicing variants may have. This line of research has been very active in the last years in the membrane transport field, emerging as an interesting strategy to increase the transport proteins repertoire and regulatory mechanisms. This chapter presents an overview of alternative splicing in membrane transporter genes. Because most transport families exhibit at least one splicing isoform, rather than presenting a catalog of membrane transporter splicing variants, this chapter concentrates on some transporter gene families in which the extent of study of their alternative splicing isoforms has shown that splicing is physiologically relevant (Table 1).

2 Alternative splicing with physiological consequences in membrane transporters

2.1 Electroneutral Cl⁻-coupled cotransporters (SLC12 family)

Physiological studies have demonstrated the existence of four types of electroneutral Cl⁻ coupled cotransporters. The thiazide-sensitive Na^+:Cl^- cotransporter (Ellison et al. 1987), the loop-diuretic sensitive Na^+:K^+:$2Cl^-$ and Na^+:Cl^-

Table 1. Physiological consequences of spliced isoforms in membrane transporters

Membrane transporter/gene family	Spliced variants	Splicing mechanism	Physiological consequence
Electroneutral cotransporters/SLC12			
Apical Na^+:K^+:$2Cl^-$ cotransporter	A, B, and F	Mutually exclusive exons	Different kinetic properties
Apical Na^+:K^+:$2Cl^-$ cotransporter	Shorter or large C-terminal domain	Poly-adenylation site and internal donor site	Change in stoichiometry of the transport process and dominant negative effect
Basolateral Na^+:K^+:$2Cl^-$ cotransporter	Shorter C-terminal domain	Skipping of exon 21	Tissue distribution
K^+:Cl^- cotransporter KCC3	Different N-terminal domain	Use of alternative exons 1a and 1b	Not known. Possible regulation due to loss of PKC sites
Urea transporters/SLC14			
Urea transporter UT-A	Five isoforms	Diverse exons combination	Intra nephron localization. Signalling regulation
Na^+/bile salt cotransporters/SLC10			
NTCP transporter	Different C-terminal end	Skipping or inclusion of intron 4	Different kinetic properties
ASBT transporter	Short and large isoforms	Skipping of exon 2 and frame shift	Polarization to apical or basolateral membrane
Na^+/myo-inositol cotransporter/SLC5			
SMIT1	Three C-terminal ends	Different usage of exons 2, 3, 4, and 5	Signalling regulation
Na^+/bicarbonate cotransporter/SLC4			
NBC-1	Two N-terminal domains	Alternative promoters before exons 1 and 3	Tissue distribution
NBC-1	Two C-terminal ends	97 pb deletion, possible exon skipping	Different distribution within the central nervous system cells
NBC-4*	Four different C-terminal domains	Combination of exon skipping and frame shift	Not known. Possible regulation due to loss of PKA sites
Iron/proton antiporter/SLC11	Two N-terminal domains and two C-terminal ends	Alternative exons 1a and 1b and different usage of 3' exons 16 and 17	Regulation by iron. Tissue and sub-cellular distribution

*This gene encodes two different proteins: the membrane transporter NBC-4 and the cytoplasmic P150^{Glued}

cotransporters (Geck et al. 1980; Eveloff and Calamia 1986), and the $K^+:Cl^-$ cotransporter (Lauf 1983). Eight genes encoding members of this family have been identified: two for the apical and basolateral bumetanide-sensitive $Na^+:K^+:2Cl^-$ cotransporters BSC1 (SLC12A1) and BSC2 (SLC12A2), respectively; one for the thiazide-sensitive $Na^+:Cl^-$ cotransporter (TSC; SLC12A3), and four that encode the $K^+:Cl^-$ cotransporter isoforms KCC1 to KCC4 (SLC12A4 – SLC12A7).

The gene encoding BSC1 is localized on human chromosome 15 and encodes a $Na^+:K^+:2Cl^-$ cotransporter that is exclusively expressed in the apical membrane of the thick ascending limb of Henle's loop (TALH) in the mammalian kidney (Kaplan et al. 1996). Inactivating mutations of this gene results in Bartter disease (inherited hypokalemic metabolic alkalosis syndrome). BSC1 gives rise to six alternatively splicing isoforms due to combination of two splicing mechanisms. Both splicing mechanisms produce transporters with distinct functional properties. As Figure 1A shows, one splicing is due to the presence of three 96 bp (31 amino acid residues) mutually exclusive cassette exons (A, B, and F) that encode the putative transmembrane domain 2 and the connecting segment between the transmembrane segments 2 and 3 (Payne and Forbush 1994; Igarashi et al. 1995; Mount et al. 1999). The other splicing mechanism arises from the utilization of a poly-adenylation site and an internal donor site, which results in two different carboxyl-terminal domains (Mount et al. 1999): a long one with 457 amino acid residues and a short one with 129. Note in Figure 1A that this splicing also confers differences in putative PKC and PKA phosphorylation sites at the carboxyl-terminal domain.

As shown in Figure 1B, the three long isoforms A, B, and F exhibit an axial distribution along the TALH and distinct functional properties. The F isoform is predominantly expressed at the inner stripe of the outer medulla, exhibits the lower affinity for Na^+, K^+, Cl^-, and the loop diuretic bumetanide, and is the isoform with the highest response to change in extracellular osmolarity. The A isoform is expressed along all the TALH and is the isoform with the higher transport capacity, whereas the B isoform, expressed only in the cortical segment of the TALH, is the isoform with the highest affinity for Na^+, K^+, Cl^-, and bumetanide (Payne and Forbush 1994; Igarashi et al. 1995; Yang et al. 1996; Plata et al. 2002). Thus, the well known heterogeneity of the transport properties along the TALH (Rocha and Kokko 1973; Reeves et al. 1988) is explained by the axial distribution of alternative splicing isoforms of the $Na^+:K^+:2Cl^-$ cotransporter with distinct functional properties.

The shorter isoform is also exclusively expressed at the apical membrane of the TALH, with little to no expression in the cortical segment (Mount et al. 1999) (Fig. 1B). This splicing isoform exerts a dominant negative effect upon the function of the $Na^+:K^+:2Cl^-$ cotransporter, that is abrogated by cAMP (Plata et al. 1999), by a mechanism that involve trafficking modulation of the $Na^+:K^+:2Cl^-$ cotransporter containing vesicles by the shorter splice variant (Meade et al. 2003). Thus, this splicing isoform could be important in the regulation of the $Na^+:K^+:2Cl^-$ cotransporter function by hormones, such as vasopressin, which generates cAMP via their respective Gs-coupled receptors (Hebert et al. 1981). In addition, the

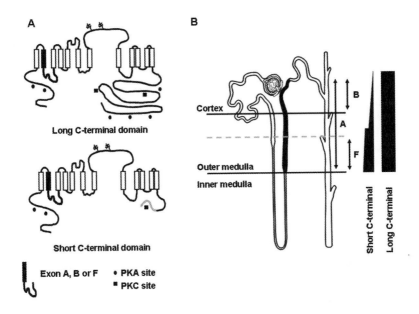

Fig. 1. Proposed topology and distribution of the SLC12A1 gene splicing variants within the mammalian nephron. 1A) The Na^+:K^+:$2Cl^-$ cotransporter features a central hydrophobic domain with twelve putative transmembrane segments and a glycosylated hydrophilic loop between segments 7 and 8. The central domain is flanked by hydrophilic amino and carboxyl-terminal domains. Six splicing variants are possible due to combination of two splicing mechanisms: the existence of three mutually exclusive cassette exons A, B, and F (in black) and two distinct carboxyl-terminal domains, one large and one short (in gray). 1B) Distribution of the SLC12A1 gene isoforms along the thick ascending limb of Henle's loop (shown in black). The black bars in the right represent relative amounts of the short and long isoform. While the long isoform is heavily expressed all along the thick ascending limb, the shorter isoform is less abundant and almost not present in the cortical fraction of the thick limb.

shorter isoform also perform as cotransporter, with dramatic differences with the functional properties of the longer isoform. While this last one encodes a Na^+:K^+:$2Cl^-$ cotransporter that is activated by hypertonicity and cAMP (Gamba et al. 1994; Plata et al. 1999), the shorter isoform behaves as a hypotonically activated, loop-diuretic sensitive, K^+-independent Na^+:Cl^- cotransporter that can be inhibited with cAMP or activated by protein kinase A inhibitors (Plata et al. 2001). In this regard, it has been shown in mouse and rabbit TALH that vasopressin and extracellular osmolarity modulate the NaCl transport mode (Eveloff and Calamia 1986; Sun et al. 1991). In the absence of vasopressin and the presence of low tonicity, NaCl is transported by a K^+-independent, furosemide-sensitive Na^+:Cl^- pathway, whereas the presence of vasopressin or hypertonicity switches the $Na^+:Cl^-$ transport mode to the $Na^+:K^+:2Cl^-$ cotransporter (Eveloff and Calamia 1986). Thus, the shorter splicing isoform could provide the explanation for the switching between $Na^+:Cl^-$ and $Na^+:K^+:2Cl^-$ cotransporters to the TALH. Thus, SLC12A1 is an example of a gene that uses splicing mechanisms to produce several isoforms that explain the heterogeneity of the salt reabsorption properties of the TALH and provides an intriguing mechanism for cAMP-regulation of the cotransporter function (Burg 1982; Rocha and Kokko 1973; Eveloff and Calamia 1986; Sun et al. 1991).

Other members of the SLC12 gene family express alternative splicing transcripts. In the brain, an isoform lacking exon 21 has been reported for the basolateral isoform of the $Na^+:K^+:2Cl^-$ cotransporter BSC2 (Randall et al. 1997) and it was shown that this splicing variant functions as $Na^+:K^+:2Cl^-$ cotransporter and is expressed in several tissues, with an up to 68-fold variation in the isoforms ratio among 14 tested tissues (Vibat et al. 2001). The functional consequences of this splicing form are not known. Finally, in the $K^+:Cl^-$ cotransporter subfamily, it has been shown that KCC3 gives rise to two splicing isoforms generated by transcriptional initiation 5' of two separate first coding exons (Mount and Gamba 2001). The longer isoform, KCC3a (Mount et al. 1999), utilizes exon 1a, whereas KCC3b (Hiki et al. 1999) uses exon 1b, situated ~23 kb 3' within the human KCC3 gene on chromosome 15 (Pearson et al. 2001). The predicted KCC3a and KCC3b proteins of 1150 and 1099 amino acids respectively, differ dramatically in content and distribution of putative phosphorylation sites for protein kinases. Both isoforms are functional, but the physiological consequences of this splicing are needed to be defined. Interestingly, inactivating mutations of KCC3 produce hereditary motor and sensory neuropathy with agenesis of the corpus callosum syndrome (Howard et al. 2002).

2.2 The urea transporters (SLC14 family)

Urea is one of the most abundant waste products that the body must eliminate every day. As with many other molecules, the permeability of the cellular membrane to urea is very low, but it is dramatically enhanced by the presence of urea transporters (Sands 2003b). Because the amount of urea to be excreted is very high, to avoid the loss of water that would accompany the excretion of nitrogenous waste, the urea transporters helps to concentrate this molecule in the renal medullary interstitum, contributing to the characteristic hypertonicity of this region of the kidney. Two different genes encoding urea transporters have been identified: SLC14A1 and SLC14A2 that encode the urea transporters UT-B and UT-A, respectively (Sands 2003a, 2003b). UT-A, also known as the vasopressin-regulated urea transporter is predominantly expressed in the kidney (Shayakul et al. 1996), whereas UT-B is present in erythrocytes and also in several tissues including the kidney in which its expression is limited to the endothelial cells of the vasa recta in the renal medulla (Xu et al. 1997; Olives et al. 1994). The identity degree between UT-A and UT-B is 63 % and interestingly, both genes are located at the same locus on human chromosome 18. Two splicing mRNA transcripts of UT-B have been documented due to alternative polyadenylation sites. The splicing, however, has no effect on the protein sequence (Lucien et al. 1998; Sands 2003a).

In contrast to the SLC14A1 gene, five alternatively splicing isoforms of the mammalian SLC14A2 gene have been identified and named UT-A1 (Shayakul et al. 1996), UT-A2 (You et al. 1993), UT-A3 (Karakashian et al. 1999), UT-A4 (Karakashian et al. 1999), and UT-A5 (Fenton et al. 2000). As shown in Figure 2, UT-A1 is the complete isoform of the urea transporter, whereas the other four are truncated isoforms. The complete transporter UT-A1 consists of two halves, each one containing 6 hydrophobic membrane-spanning domains and a putative extracellular hydrophilic loop, with at least one glycosylation site, between transmembrane segments 3 and 4 in the first half and between segments 9 and 10 in the second half. The halves are connected by a putative intracellular hydrophilic loop with several PKA, PKC and a tyrosine phosphorylation sites. UT-A1 is expressed exclusively at the apical membrane of the inner medullary collecting duct cells (Nielsen et al. 1996; Shayakul et al. 1996). UT-A2 transporter corresponds to the carboxyl-terminal half of UT-A1, is expressed in the descending thin limb of Henle's loop and is the only isoform that is not regulated by cAMP or forskolin (You et al. 1993). UT-A3 is basically the amino-terminal half of UT-A1 and has also been immunolocalized only to the apical membrane of the inner medullary collecting duct cells (Karakashian et al. 1999; Terris et al. 2001), whereas rat UT-A4 corresponds to the first quarter of UT-A1, spliced into the last quarter of UT-A1 (Karakashian et al. 1999). The precise localization of UT-A4 expression is not known. In addition to all these protein isoforms, UT-A1, UT-A2, and UT-A3 also exhibit splicing transcripts with a distinct 3' UTRs known as UT-A1b, UT-A2b, and UT-A3b (Bagnasco et al. 2000; Sands 2003a). Finally, a fifth isoform known as UT-A5 that was isolated from mouse testis is the shortest one and is the only SLC14A2 splicing isoform that is not present in kidney (Fenton et al. 2000). It is identical to UT-A3, but lacking the first 139 amino acid residues and with a distinct 5' UTR. In the testis, this isoform has been localized in the peritubular myoid cells surrounding the seminiferous tubules.

The SLC14A2 gene that encodes for UT-A transcripts contains 24 exons and the origin of the splicing isoforms is the following (Bagnasco et al. 2001; Nakayama et al. 2001; Bagnasco et al. 2003): the longer isoform UT-A1 is encoded by exons 1-12, spliced into exons 14-23. UT-A3 is encoded only by exons 1-12, while UT-A4 is made from exons 1-7 that are spliced into exons 18-23. Thus, UT-A1, UT-A3, and UT-A4 exhibit the same transcription start site that is located in exon 4 (Bagnasco et al. 2000). UT-A2 is unique since it is the only isoform that starts in exon 13, and is made up from exons 13 to 23 (Bagnasco et al. 2001). Despite the extensive splicing of UT-A transporters (Fig. 2), when expressed in *Xenopus* oocytes or transiently transfected into HEK-293 cells, all five splicing isoforms induced the expression of a phloretin-sensitive urea transport mechanism, with similar functional properties. The major difference between these isoforms, in addition to their localization, is in the response to regulatory stimuli since they exhibit divergent responses to acute or chronic exposure to vasopressin, angiotensin II, and hyperosmolarity (for excellent up-to-date reviews see Sands 2003a, 2003b). Thus, the SLC12A2 gene represents an example of a gene in

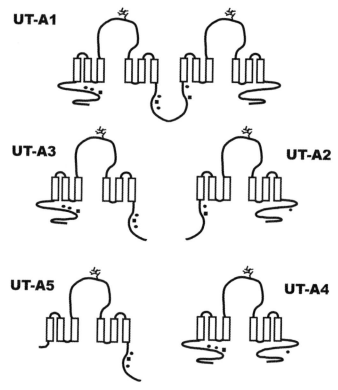

Fig. 2. Proposed topology of the five alternatively spliced isoforms of the SLC14A2 gene encoding the UT-A urea transporter. UT-A1 is the complete transporter featuring two similar halves, with six putative transmembrane segments in each one that are connected by an intracellular hydrophilic loop. UT-A2, UT-A3, UT-A4, and UT-A5 are shorter splicing variants.

which alternative splicing produces five isoforms that thanks to their different distribution and regulation, work together in order to achieve the urea recycling that is critical for the concentration of urine in mammalian kidney (Berliner and Bennett 1967).

2.3 Na⁺/bile acid transporter (SLC10A family)

Bile acids are essential components of bile that is required for lipid digestion and absorption in the gut, cholesterol homeostasis, and for the hepatic excretion of lipid-soluble drugs. Bile is produced and secreted by the liver and once in the intestinal lumen is reabsorbed to be reused, constituting the entero-hepatic circulation (St Pierre et al. 2001). The transport of bile acid through plasma membrane is possible due to the existence of the Na^+-dependent bile acid transporters, for which two genes have been identified. SLC10A1 and SLC10A2 encode the liver

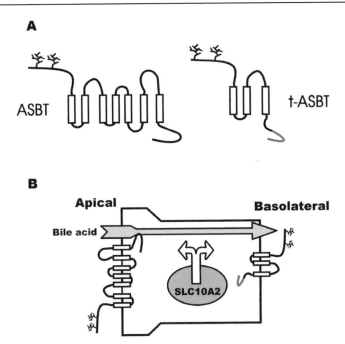

Fig. 3. Proposed topology and effect of splicing in the SLC10A2 gene encoding the Na⁺/bile acid cotransporter ASBT. 1A) The complete ASBT isoform features seven transmembrane segments, while the splicing isoform t-ASBT, exhibits only the first three segments and a unique short carboxyl-terminal domain. 1B) Effect of splicing in cell surface polarization. The longer ASBT isoform is expressed in the apical membrane, while the shorter isoform t-ASBT is directed to the basolateral membrane.

basolateral Na⁺/taurocholate cotransporter (NTCP) (Hagenbuch et al. 1991) and the ileum apical Na⁺/bile transporter (ASBT) (Wong et al. 1994), respectively. The identity between NTCP and ASBT is about 35%. Transcripts of the ASBT cotransporter are heavily expressed in the ileum epithelia and renal proximal tubule (Christie et al. 1996) in which this cotransporter plays a critical role for the efficient conservations of bile salts.

Two splice variants of the NTCP transporter due to retention or splicing of intron 4 have been identified from mouse liver (Cattori et al. 1999). NTCP1, in which intron 4 is spliced, is made up by 362 amino acids, while the less-abundant NTCP2, in which intron 4 is retained, is composed by 317 amino acids and exhibits a shorter C-terminal domain. When injected into *X. laevis* oocytes, both isoforms induced saturable Na⁺-dependent taurocholate influx, but with distinct kinetic transport properties, suggesting that NTCP1 encodes the low affinity, high capacity isoform, while NTCP2 represents the high affinity, low capacity transporter. Thus, SLC10A1 is another example of a gene that using alternative splicing produces two similar cotransporters, with distinct kinetics properties.

As shown in Figure 3A, two alternatively splicing isoforms of the SLC10A2 gene have also been identified. ASBT, the complete isoform encoding 348 amino acid residues (50 kDa) protein (Shneider et al. 1995), featuring seven membrane spanning domains, and a shorter isoform t-ASBT due to skipping of exon 2 (Lazaridis et al. 2000), producing a protein with 154 residues (19 kDa), that features only the first three transmembrane segments. Despite the truncation of about half of the cotransporter, the two isoforms are functional. As shown in Figure 3B, the physiological relevance of this splicing seems to be the polarization of each isoform. The complete cotransporter is expressed in the apical membrane of bile-acid transporting epithelia and functions as a Na^+/bile acid cotransporter (Christie et al. 1996), whereas the shorter isoform is expressed on the basolateral membrane and exhibits activity as a bile acid efflux carrier (Lazaridis et al. 2000). Thus, these splicing variants are complementary. One does the uptake of Na^+/bile acid at the apical membrane and the other one performs the efflux in the basolateral membrane. SLC10A2 represents an example of a solute carrier gene that uses alternative splicing for encoding the solute uptake and efflux mechanisms.

2.4 Na$^+$/*myo*-Inositol cotransporters (members of the SLC5 family)

Myo-inositol is one of the most important osmolytes that plays a critical role in cell volume regulation, particularly in the cells of the renal medulla that are exposed to a wide fluctuation of extracellular solute concentrations. These cells are able to maintain normal cell volume, despite the extremely high NaCl and urea concentration of the environment (up to 1200 mOsm/kg H_2O in human kidney), thanks to their ability to accumulate osmolytes, such as *myo*-inositol, sorbitol, and others (Nakanishi et al. 1988). Several extra renal cells also accumulate *myo*-inositol (Ashizawa et al. 2000). In fact, mammalian serum levels of *myo*-inositol are about 70 µM, whereas in some cells concentration can achieve up to 30 mM (Coady et al. 2002). The accumulation of *myo*-inositol in renal and other cells (i.e. retinal pigment epithelial cells) is due to the function of the Na^+- *myo*-inositol cotransporters. The importance of inositol transport in brain metabolism has been clearly defined by the early death, after birth of the Na^+-myo-inositol cotransporter (SMIT1) null mice due to central apnea (Berry et al. 2003).

There are two genes that encode for the Na^+-*myo*-inositol cotransporters: SMIT1 (SLC5A3) (Kwon et al. 1992) and SMIT2 (Coady et al. 2002; Roll et al. 2002). SMIT1 was cloned from the Madin-Darby canine kidney cells following a functional expression strategy in *X. laevis* oocytes (Kwon et al. 1992) and was observed to be one of the many members of the SLC5 carrier family that include the Na^+-dependent glucose cotransporter. SMIT1 gene is located on chromosome 21, within the Down's syndrome region. SMIT2, the second gene encoding a Na^+-*myo*-inositol cotransporter, was recently identified when Coady et al (Coady et al. 2002) observed that when injected in *X. laevis* oocytes, rkST1 cRNA, an orphan member of the SLC5 family with 43 % identity with SMIT1, induced the expression of a phlorizin-inhibitable Na^+-dependent *myo*-inositol pathway. SMIT2 gene is localized on human chromosome 16 (Roll et al. 2002).

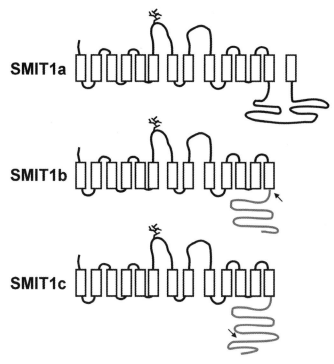

Fig. 4. Splicing isoforms of the Na⁺-*myo*-inositol cotransporter, SMIT1. SMIT1a is a cotransporter encoded entirely by exon 2 and features 14 transmembrane segments. SMIT1b and SMIT1c are identical to SMIT1a, until the very end in which the 14th membrane segment is absent and they show unique carboxyl-terminal sequences that are encodes by exons 3, 4, and 5. The arrows highlight the differences between them.

The use of SMIT1 cDNA as a probe revealed the existence of several transcripts within the range of 1.0 to 13.5 kb in RNA from rat kidney (Kwon et al. 1992), lens (Zhou et al. 1994), endothelial cells (Wiese et al. 1996), and central nervous system (Paredes et al. 1992; Wiese et al. 1996) some of which are up-regulated by hypertonicity (Kwon et al. 1992). Interestingly, the entire 718 amino acid residues that make SMIT1 protein are encoded by a single intronless exon (Berry et al. 1995), while the human SMIT1 gene is composed of at least five exons, which give rise to several splicing transcripts, from which most of them exhibit only differences in the 5' and 3'UTRs, without apparent functional significance (Porcellati et al. 1999, 1998). There are two splicing variants, however, that do result in changes in the coding region. Thus, as shown in Figure 4, the complete protein of 718 residues is known as SMIT1a (Porcellati et al. 1998) and the alternative isoforms as SMIT1b and SMIT1c (Coady et al. 2002). SMIT1b isoform uses an alternate splice donor site that is upstream of the exon 2 stop codon, with the same acceptor site on the 5' of exon 3, thus, generating a new stop codon in exon 4, while SMIT1c uses the same alternate donor site as SMIT2, but with different acceptor site on exon 4, also generating a different stop codon. Therefore, SMIT1b and SMIT1c are predicted to lack the last transmembrane domain and contain new sequences of the C-terminal domain that are encoded by exons 3, 4, and 5. Functional expression studies have shown that *X. laevis* oocytes injected with SMIT1a, SMIT1b, or SMIT1c cRNA, exhibit the appearance of a robust Na^+-dependent *myo*-inositol transporter activity, with interesting differences in the response to PKA activation. While SMIT1a is activated by IBMX-forskolin combination, SMIT1b and SMIT1c are partially inhibited (Porcellati et al. 1999). Therefore, alternative splicing of SMIT1 is an example of a gene in humans that give rise to functional isoforms that are differentially regulated by second messengers and by hypertonicity.

2.5 Na^+/bicarbonate cotransporters and exchangers (members of the SLC4 family)

One of the major homeostatic functions of membrane transporters is to maintain the intracellular pH. Therefore, several families of membrane transporters are made up of proteins that are involved in this function. One of these families deals with bicarbonate transport through the cellular membranes. In addition, at the systemic level, bicarbonate transport in the kidney plays a critical role in the whole body acid base metabolism.

At least four genes encoding Na^+-dependent bicarbonate cotransporters, known as the NBC family, have been described as part of the SLC4 superfamily. The first one to be identified was the NBC-1 gene (SLC4A4) (Romero et al. 1997; Burnham et al. 1997) that is localized on human chromosome 4. Inactivating mutations of SLC4A4 gene produce an inherited syndrome characterized by proximal renal tubular acidosis and ocular abnormalities (Igarashi et al. 1999). As shown in Figure 5, this gene gives rise to three alternatively splicing isoforms. Two variants exhibit different amino-terminal domains: NBC-1A, exclusively expressed in the proximal tubular cells from human kidney (Burnham et al. 1997) and NBC-1B cloned from human pancreas and expressed in several tissues (Abuladze et al. 1998) are identical in 994 amino acid residues, but differ at the amino-terminal domain. NBC-1A is transcribed from an alternative promoter in intron 3 (Abuladze et al. 2000) and contains 41 unique amino acids at the amino-terminal domain. In contrast, NBC-1B is transcribed from exon 1 (Abuladze et al. 2000) and contains 85 distinct amino acid residues at the amino-terminal end. Because the functional properties are similar in both isoforms, it is possible that the major consequence of this splicing is tissue distribution. The third variant of NBC-1 cotransporter is due to the existence of a carboxyl-terminal splicing isoform. This is known as rb2NBC and was described from rat brain (Bevensee et al. 2000). This shorter cotransporter is the result of a 97 bp deletion near the end of the open reading frame, causing a frame shift that change and extends the end of the reading frame by 61 amino acid residues, which differ from the last 46 residues in the NBC-1A or NBC-1B isoforms. The carboxyl-terminal splicing isoform rb2NBC is fully functional and is predominantly present in neurons, whereas NBC-1B is

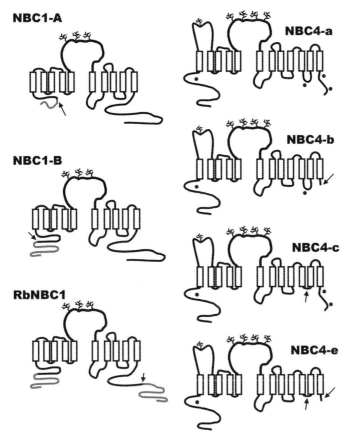

Fig. 5. Alternatively spliced isoforms of the SLC4A4 (left panel) and SLC4A5 (right panel) genes that encode the Na^+/bicarbonate cotransporters NBC-1 and NBC-4, respectively. For the NBC-1 cotransporter the hydrophobicity analysis predicts the existence of a central hydrophobic domain with 10 putative transmembrane segments, flanked by hydrophilic amino and carboxyl-terminal domains that are presumably located within the cell. Arrows show the differences in amino and carboxyl-terminal domains. For NBC-4, the hydropathy analysis predicts twelve transmembrane segments. There is a glycosylated hydrophilic loop between the fifth and six membrane domains. The differences between the NBC-4 isoforms are between transmembrane segments 10 and 11, and at the carboxyl-terminal domains, as marked and highlighted by the arrows. The black circles in NBC4 isoforms depict the localization of the putative protein kinase A sites.

predominantly expressed in astrocytes, suggesting that SLC4A4 is an example of a gene that uses alternative splicing mainly to regulate tissue distribution.

The human genes SLC4A7 and SLC4A8 encode for the NBC-3 and NBC-2 cotransporters, respectively (Ishibashi et al. 1998; Pushkin et al. 1999ᵃ; Choi et al. 2000; Amlal et al. 1999). The identity between NBC-2 and NBC-3 with NBC-1 is about 50%. NBC-3 was mapped to human chromosome 3 (Pushkin et al. 1999b) and functions as a DIDS-insensitive, EIPA inhibitable, electroneutral sodium bicarbonate cotransporter (Pushkin et al. 1999a; Choi et al. 2000) and in contrast to NBC-1, in the kidney NBC-3 is expressed at the basolateral membrane of the outer medulla thick ascending limb cells and intercalated cells of the medullary collecting duct (Vorum et al. 2000). While in most tissues NBC-3 is expressed as a 7.5 kb mRNA transcript, in testis the transcript is ~ 4.0 kb. NBC-2 in contrast was localized to human chromosome 12 and encodes an electrogenic Na^+-driven Cl^-/HCO_3^- exchanger, which is robustly expressed in the central nervous system, as three different size transcripts (9.5, 4.4 and 3 kb), and with lesser intensity in other tissues, as a unique 4.4 kb transcript (Grichtchenko et al. 2001; Amlal et al. 1999). It is not known, however, if the different size transcripts in SLC4A7 and SLC4A8 represent alternatively splicing isoforms.

The human SLC4A5 gene encodes a fourth Na^+-bicarbonate cotransporter, identified by Pushkin et al (2000a) from a heart cDNA library and was localized on human chromosome 2. NBC-4 exhibits 58, 43 and 67 % identity with NBC-1, NBC-2 and NBC-3, respectively. Three or four alternatively splicing isoforms from this gene have been identified (Pushkin et al. 2000b; Sassani et al. 2002) (Fig. 5). NBC-4a and NBC-4b contain 1137 and 1047 amino acid residues, respectively. The difference is at the end of the C-terminal domain, due to a 16 bp insertion that shifts the open reading frame, increasing the protein by 90 residues. The consequence of this splicing mechanism on NBC-4 is not known, however, it is likely to play a role in subcellular targeting or functional regulation of the protein because the extra 90 residues in NBC-4a contain two protein interacting domains (Pushkin et al. 2000b) and two putative PKA sites that are missing in NBC-4b (Fig. 5). In addition, while NBC-4a and NBC-4b are expressed in the heart, only NBC-4a is present in the testis. The two other splicing variants are known as NBC-4c and NBC-4e. As shown in Figure 5, NBC-4c isoform is identical to NBC-4a, but is missing 16 amino acid residues that are part of the intracellular interconnecting segment between transmembrane domains 10 and 11. Interestingly, there is a putative PKA site in this exon that is, thus, not present in NBC-4c. NBC-4e is similar to NBC-4c (is missing the same 16 residues), but in addition, this isoform exhibits a truncated C-terminal domain few amino acids after the end of transmembrane segment 12. Thus, NBC-4e lacks three of the four putative PKA sites (Fig. 5). One can speculate that these small variations could be responsible for an interesting difference in the functional properties between these two isoforms, since NBC-4c performs as an electrogenic Na^+-bicarbonate cotransporter, when expressed in HEK-293 cells (Sassani et al. 2002), whereas NBC-4e performed as an electroneutral Na^+-bicarbonate cotransporter when expressed in *X. laevis* oocytes (Xu et al. 2003).

Finally, the SLC4A5 gene represents one of the few examples in eukaryotes in which two completely distinct proteins are encoded by one gene. In addition to NBC-4, the protein known as p150[Glued] is encoded by this gene (Pushkin et al. 2001). p150[Glued] is a 150 kDa cytoplasmic protein that is part of the dynactin heteromultimeric complex of proteins that are necessary for dynein-mediated transport of vesicles and organelles along microtubules (Waterman-Storer and Holzbaur 1996; Holzbaur et al. 1991). Because SLC4A5 and p150[Glued] were localized to the same locus on human chromosome 2, by independent groups (Holzbaur and Tokito 1996; Pushkin et al. 2000b), Pushkin et al. (2001) carefully studied this region of the genome, concluding that there is a single large gene that spans ~230 kb and contains 66 exons. The translational start site of p150^{Glued} was localized in exon 1 and the stop codon in exon 32, whereas the NBC-4 mRNA is encoded from exons 18 to 66, from which the 5'UTR correspond to exons 18 to 38, with the translational start site in exon 38.

2.6 The divalent metal transporter DMT1 (member of the SLC11 family)

Iron is an essential nutrient that is required for oxygen transport and for the proper function of several heme and non-heme enzymes. A minimum level of iron in the body is required to produce the necessary number of healthy erythrocytes to assure tissue oxygenation. People with iron deficiency develop microcytic anemia. On the other hand, iron overload is harmful because iron induces the production of free radicals. Thus, iron in the body must be tightly regulated by a precise balance between absorption, utilization and losses. The three major organs involved in iron metabolism are the intestine, particularly the very first portion, the lymphoreticular system and the kidney.

Iron is transported through cellular membranes by an antiporter that interchanges iron with H$^+$. The DMT1 gene encoding this transporter was simultaneously cloned by two independent groups. Using an expression cloning strategy in *X. laevis* oocytes, Gunshin et al (1997) identified a cDNA from rat duodenum that encodes an electrogenic Fe^{2+}/H$^+$ antiporter that is capable to translocate a variety of divalent metals such as Zn^{2+}, Mn^{2+}, Cu^{2+}, Co^{2+}, Ni^{2+}, and Pb^{2+}. On the other hand, by positional cloning, Fleming et al. (1998) demonstrated that the gene responsible for the microcytic anemia in the *mk* mouse model (Fleming et al. 1997) and the rat Belgrade model (Lee et al. 1998) was the Nramp2 gene that was previously identified with no function from humans, due to its homology with Nramp1 (Natural resistance-associated macrophage protein), a phagocyte specific protein that helps to control infections by organisms like *Salmonela* or *Mycobacterium* (Vidal et al. 1995). DMT1 is involved in iron transport at both, the plasma membrane and at subcellular locations. Supporting this conclusion it has been shown that this transporter is expressed in the apical membrane of duodenal enterocytes (Canonne-Hergaux et al. 1999) and renal proximal tubule (Ferguson et al. 2001), as well as in intracellular vesicle compartments in several tissues (Su et al. 1998; Gruenheid et al. 1997; Tabuchi et al. 2000).

As shown in Figure 6, DMT1 features a central hydrophobic domain containing 12 transmembrane segments that are flanked by short hydrophilic amino- and carboxyl-terminal domains. Four splicing variants have been identified in DMT1 (Lee et al. 1998; Hubert and Hentze 2002). The four isoforms result from combination of two independent splicing mechanisms that confer DMT1 with an interesting mechanism for regulation. One splicing is due to differential usage of two 3'exons (exons 16 and 17) that confer different ends to the carboxyl-terminal domain and a totally different 3'UTR, which differ by the presence or absence of an iron regulatory element (IRE). The last 18 amino acid residues of the IRE variant are unique, while the last 25 residues in the non-IRE isoform are unique. Initial studies showed that the isoform containing IRE is more abundant in the duodenum, a place in which it is regulated by dietary iron (Canonne-Hergaux et al. 1999), while the non-IRE variant is more abundant in erythroid cell precursors and is regulated by erythropoietin (Canonne-Hergaux et al. 2001). In another study, it was observed that IRE containing transcripts are more frequently observed in epithelial cell lines, with subcellular localization to late endosomes and lysosome, while the non-IRE containing transcripts are typical of blood cell lines and within the cell are expressed in early endosomes (Tabuchi et al. 2002). Finally, a recent report revealed a second splicing mechanism that is due to the existence of two alternative promoters with two exons 1, denominated 1a and 1b. Because the ATG in exon 1a is in frame with exon 2, the difference between DMT-1a and DMT-1b is only the length of the amino-terminal domain, which is 29 residues longer in DMT-1a than in DMT1-b variant (Hubert and Hentze 2002). It was shown that this splicing mechanism can be combined with the splicing at the 3'end, thus, producing four distinct proteins: DMT1a-non-IRE, DMT1a-IRE, DMT1b-non IRE, and DMT1b-IRE (Fig. 6). This study also showed that exon 1a is regulated by iron and that DMT-1a isoform is expressed in duodenum and kidney, while exon 1b is not regulated by iron and its expression is ubiquitous. Thus, in general, DMT1 variants containing exon 1a and/or the IRE in the 3'UTR are regulated by iron, while isoforms with exon 1b and/or without IRE are not sensitive to iron. Therefore, DMT1 constitute a clear example in which the role of alternative splicing is mainly the modulation of the transporter expression by a nutrient.

3 Conclusion

A growing number of genes encoding membrane transporters give rise to alternatively spliced isoforms with important physiological consequences. Sometimes, the splicing isoforms are not functional, but possess a dominant negative effect on the transporter function that could be important in the transporter regulation. Most of the time, however, the splice variants are functional resulting in a variety of physiological effects such as changes in the stoichiometry or kinetics of a transport process, different tissue, cell or subcellular distribution, and changes in the polarization of the splicing variant to the apical or basolateral membrane, to mention some examples. There are still several examples in which splicing isoforms of some transporters exhibit tissue distribution, polarization, and functional properties that are similar, in which the physiological consequence of splicing is still unknown. One good possibility in these examples is distinct regulation of the splicing isoforms. As the scientific community defines the purpose of each splicing, it will be important to begin to analyze the differential regulation of the splicing mechanisms in order to understand how the cells decide which isoform to

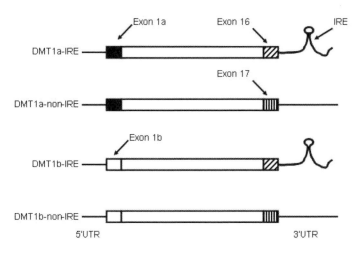

Fig. 6. Linear representation of the mature mRNAs encoding the four isoforms of the divalent metal transporter 1, DMT-1. The boxes represent the open reading frame. Alternative exons 1a and 1b are shown in black or white boxes, respectively.

produce, how much of each isoform is to be produced at a particular moment, and how this is regulated by intracellular or extracellular physiological stimuli.

References

Abuladze N, Song1 M, Pushkin A, Newman D, Lee I, Nicholas S and Kurtz I (2000) Structural organization of the human NBC1 gene: kNBC1 is transcribed from an alternative promoter in intron 3. Gene 251:109-122

Abuladze N, Lee I, Newman D, Hwang J, Boorer K, Pushkin A and Kurtz I (1998) Molecular cloning, chromosomal localization, tissue distribution, and functional expression of the human pancreatic sodium bicarbonate cotransporter. J Biol Chem 273:17689-17695

Amlal H, Burnham CE, Soleimani M (1999) Characterization of Na+/HCO-3 cotransporter isoform NBC-3. Am J Physiol Renal 276:F903-F913

Ashizawa N, Yoshida M, Aotsuka T (2000) An enzymatic assay for myo-inositol in tissue samples. J Biochem Biophys Methods 44:89-94

Bagnasco SM (2003) Gene structure of urea transporters. Am J Physiol Renal 284:F3-F10

Bagnasco SM, Peng T, Janech MG, Karakashian A, Sands JM (2001) Cloning and characterization of the human urea transporter UT-A1 and mapping of the human Slc14a2 gene. Am J Physiol Renal 281:F400-F406

Bagnasco SM, Peng T, Nakayama Y, Sands JM (2000) Differential expression of individual UT-A urea transporter isoforms in rat kidney. J Am Soc Nephrol 11:1980-1986

Berliner RW, Bennett CM (1967) Concentration of urine in the mammalian kidney. Am J Med 42:777-789

Berry GT, Mallee JJ, Kwon HM, Rim JS, Mulla WR, Muenke M and Spinner NB (1995) The human osmoregulatory Na+/myo-inositol cotransporter gene (SLC5A3): molecular cloning and localization to chromosome 21. Genomics 25:507-513

Berry GT, Wu S, Buccafusca R, Ren J, Gonzales LW, Ballard PL, Golden JA, Stevens MJ and Greer JJ (2003) Loss of murine Na+/myo-inositol cotransporter leads to brain myo-inositol depletion and central apnea. J Biol Chem 278:18297-18302

Bevensee MO, Schmitt BM, Choi I, Romero MF, Boron WF (2000) An electrogenic Na(+)-HCO(-)(3) cotransporter (NBC) with a novel COOH- terminus, cloned from rat brain. Am J Physiol Cell 278:C1200-C1211

Black DL (2003) Mechanisms of alternative pre-messenger RNA splicing. Annu Rev Biochem 72:291-336

Burg MB (1982) Thick ascending limb of Henle's loop. Kidney Int 22:454-464

Burnham CE, Amlal H, Wang Z, Shull GE, Soleimani M (1997) Cloning and functional expression of a human kidney Na+:HCO3- cotransporter. J Biol Chem 272:19111-19114

Caceres JF, Kornblihtt AR (2002) Alternative splicing: multiple control mechanisms and involvement in human disease. Trends Genet 18:186-193

Canonne-Hergaux F, Gruenheid S, Ponka P, Gros P (1999) Cellular and subcellular localization of the Nramp2 iron transporter in the intestinal brush border and regulation by dietary iron. Blood 93:4406-4417

Canonne-Hergaux F, Zhang AS, Ponka P, Gros P (2001) Characterization of the iron transporter DMT1 (NRAMP2/DCT1) in red blood cells of normal and anemic mk/mk mice. Blood 98:3823-3830

Cattori V, Eckhardt U, Hagenbuch B (1999) Molecular cloning and functional characterization of two alternatively spliced Ntcp isoforms from mouse liver. Biochim Biophys Acta 1445:154-159

Choi I, Aalkjaer C, Boulpaep EL, Boron WF (2000) An electroneutral sodium/bicarbonate cotransporter NBCn1 and associated sodium channel. Nature 405:571-575

Christie DM, Dawson PA, Thevananther S, Shneider BL (1996) Comparative analysis of the ontogeny of a sodium-dependent bile acid transporter in rat kidney and ileum. Am J Physiol (Gastroint) 271:G377-G385

Coady MJ, Wallendorff B, Gagnon DG, Lapointe JY (2002) Identification of a novel Na+/myo-inositol cotransporter. J Biol Chem 277:35219-35224

Ellison DH, Velazquez H, Wright FS (1987) Thiazide-sensitive sodium chloride cotransport in early distal tubule. Am J Physiol (Renal Fluid Electrolyte) 253:F546-F554

Eveloff J, Calamia J (1986) Effect of osmolarity on cation fluxes in medullary thick ascending limb cells. Am J Physiol (Renal Fluid Electrolyte) 250:F176-F180

Fenton RA, Howorth A, Cooper GJ, Meccariello R, Morris ID, Smith CP (2000) Molecular characterization of a novel UT-A urea transporter isoform (UT- A5) in testis. Am J Physiol Cell 279:C1425-C1431

Ferguson CJ, Wareing M, Ward DT, Green R, Smith CP, Riccardi D (2001) Cellular localization of divalent metal transporter DMT-1 in rat kidney. Am J Physiol Renal 280:F803-F814

Fleming MD, Romano MA, Su MA, Garrick LM, Garrick MD, Andrews NC (1998) Nramp2 is mutated in the anemic Belgrade (b) rat: evidence of a role for Nramp2 in endosomal iron transport. Proc Natl Acad Sci USA 95:1148-1153

Fleming MD, Trenor CC, III, Su MA, Foernzler D, Beier DR, Dietrich WF and Andrews NC (1997) Microcytic anaemia mice have a mutation in Nramp2, a candidate iron transporter gene. Nat Genet 16:383-386

Gamba G, Miyanoshita A, Lombardi M, Lytton J, Lee WS, Hediger MA and Hebert SC (1994) Molecular cloning, primary structure and characterization of two members of the mammalian electroneutral sodium-(potassium)-chloride cotransporter family expressed in kidney. J Biol Chem 269:17713-17722

Geck P, Pietrzyk C, Burckhardt B C, Pfeiffer B, Heinz E (1980) Electrically silent cotransport of Na+, K+ and Cl- in Ehrlich cells. Biochim Biophys Acta 600:432-447

Graveley BR (2001) Alternative splicing: increasing diversity in the proteomic world. Trends Genet 17:100-107

Grichtchenko II, Choi I, Zhong X, Bray-Ward P, Russell JM, Boron WF (2001) Cloning, characterization and chromosomal mapping of a human electroneutral Na+- driven Cl-HCO3 exchanger. J Biol Chem 276:8358-8363

Gruenheid S, Pinner E, Desjardins M, Gros P (1997) Natural resistance to infection with intracellular pathogens: the Nramp1 protein is recruited to the membrane of the phagosome. J Exp Med 185:717-730

Gunshin H, Mackenzie B, Berger UV, Gunshin Y, Romero MF, Boron WF, Nussberger S, Gollan JL and Hediger MA (1997) Cloning and characterization of a mammalian proton-coupled metal-ion transporter. Nature 388:482-488

Hagenbuch B, Stieger B, Fouget M, Lubbert H, Meier PJ (1991) Functional expression cloning and characterization of the hepatocyte Na$^+$/bile acid cotransport system. Proc Natl Acad Sci USA 88:10629-10633

Hebert SC, Culpepper RM, Andreoli TE (1981) NaCl transport in mouse medullary thick ascending limbs. I. Functional nephron heterogeneity and ADH-stimulated NaCl cotransport. Am J Physiol (Renal Fluid Electrolyte) 241:F412-F431

Hiki K, D'Andrea R J, Furze J, Crawford J, Woollatt E, Sutherland G R, Vadas M A and Gamble J R (1999) Cloning, characterization, and chromosomal location of a novel human K$^+$-Cl$^-$ cotransporter. J Biol Chem 274:10661-10667

Holzbaur EL, Hammarback JA, Paschal BM, Kravit NG, Pfister KK, Vallee RB (1991) Homology of a 150K cytoplasmic dynein-associated polypeptide with the *Drosophila* gene Glued. Nature 351:579-583

Holzbaur EL, Tokito MK (1996) Localization of the DCTN1 gene encoding p150Glued to human chromosome 2p13 by fluorescence in situ hybridization. Genomics 31:398-399

Howard HC, Mount DB, Rochefort D, Byun N, Dupre N, Lu J, Fan X, Song L, Riviere JB, Prevost C, Horst J, Simonati A, Lemcke B, Welch R, England R, Zhan FQ, Mercado A, Siesser WB, George AL, Jr , McDonald MP, Bouchard JP, Mathieu J, Delpire E and Rouleau GA (2002) The K-Cl cotransporter KCC3 is mutant in a severe peripheral neuropathy associated with agenesis of the corpus callosum. Nat Genet 32:384-392

Hubert N, Hentze MW (2002) Previously uncharacterized isoforms of divalent metal transporter (DMT)-1: implications for regulation and cellular function. Proc Natl Acad Sci USA 99:12345-12350

Igarashi P, Vanden Heuver GB, Payne JA, Forbush III B (1995) Cloning, embryonic expression, and alternative splicing of a murine kidney-specific Na-K-Cl cotransporter. Am J Physiol (Renal Fluid Electrolyte) 269:F406-F418

Igarashi T, Inatomi J, Sekine T, Cha SH, Kanai Y, Kunimi M, Tsukamoto K, Satoh H, Shimadzu M, Tozawa F, Mori T, Shiobara M, Seki G and Endou H (1999) Mutations in SLC4A4 cause permanent isolated proximal renal tubular acidosis with ocular abnormalities. Nat Genet 23:264-266

Ishibashi K, Sasaki S, Marumo F (1998) Molecular cloning of a new sodium bicarbonate cotransporter cDNA from human retina. Biochem Biophys Res Commun 246:535-538

Kan Z, Rouchka EC, Gish WR, States DJ (2001) Gene structure prediction and alternative splicing analysis using genomically aligned ESTs. Genome Res 11:889-900

Kaplan MR, Plotkin MD, Lee W-S, Xu Z-C, Lytton J, Hebert SC (1996) Apical localization of the Na-K-Cl cotransporter, rBSC1, on rat thick ascending limbs. Kidney Int 49:40-47

Karakashian A, Timmer RT, Klein JD, Gunn RB, Sands JM, Bagnasco SM (1999) Cloning and characterization of two new isoforms of the rat kidney urea transporter: UT-A3 and UT-A4. J Am Soc Nephrol 10:230-237

Kwon H M, Yamauchi A, Uchida S, Preston A S, Garcia-Perez A, Burg MB and Handler JS (1992) Cloning of the cDNA for a Na^+/myoinositol cotransporter, a hypertonicity stress protein. J Biol Chem 267:6297-6301

Lander ES et al. (2001) Initial sequencing and analysis of the human genome. Nature 409:860-921

Lauf PK (1983) Thiol-dependent passive K/Cl transport in sheep red cells: I. Dependence on chloride and external ions. J Membr Biol 73:237-246

Lazaridis KN, Tietz P, Wu T, Kip S, Dawson PA, LaRusso NF (2000) Alternative splicing of the rat sodium/bile acid transporter changes its cellular localization and transport properties. Proc Natl Acad Sci USA 97:11092-11097

Lee PL, Gelbart T, West C, Halloran C, Beutler E (1998) The human Nramp2 gene: characterization of the gene structure, alternative splicing, promoter region and polymorphisms. Blood Cells Mol Dis 24:199-215

Lopez AJ (1998) Alternative splicing of pre-mRNA: developmental consequences and mechanisms of regulation. Annu Rev Genet 32:279-305

Lucien N, Sidoux-Walter F, Olives B, Moulds J, Le Pennec PY, Cartron JP and Bailly P (1998) Characterization of the gene encoding the human Kidd blood group/urea transporter protein. Evidence for splice site mutations in Jknull individuals. J Biol Chem 273:12973-12980

Maniatis T, Tasic B (2002) Alternative pre-mRNA splicing and proteome expansion in metazoans. Nature 418:236-243

Meade P, Hoover RS, Plata C, Vazquez N, Bobadilla NA, Gamba G and Hebert SC (2003) cAMP-dependent activation of the renal-specific Na+-K+-2Cl- cotransporter is mediated by regulation of cotransporter trafficking. Am J Physiol Renal 284:F1145-F1154

Modrek B, Lee C (2002) A genomic view of alternative splicing. Nat Genet 30:13-19

Mount D B, Baekgard A, Hall A E, Plata C, Xu J, Beier DR, Gamba G and Hebert SC. (1999) Isoforms of the Na-K-2Cl transporter in murine TAL I. Molecular characterization and intrarenal localization. Am J Physiol Renal 276:F347-F358

Mount D B, Mercado A, Song L, Xu J, Geroge Jr AL, Delpire E and Gamba G (1999) Cloning and characterization of KCC3 and KCC4, new members of the cation-chloride cotransporter gene family. J Biol Chem 274:16355-16362

Mount DB, Gamba G (2001) Renal potassium-chloride cotransporters. Curr Opin Nephrol Hypertens 10:685-691

Nakanishi T, Balaban R S, Burg MB (1988) Survey of osmolytes in renal cell lines. Am J Physiol Cell 255:C181-C191

Nakayama Y, Naruse M, Karakashian A, Peng T, Sands JM, Bagnasco SM (2001) Cloning of the rat Slc14a2 gene and genomic organization of the UT-A urea transporter. Biochim Biophys Acta 1518:19-26

Nielsen S, Terris J, Smith CP, Hediger MA, Ecelbarger CA, Knepper MA (1996) Cellular and subcellular localization of the vasopressin-regulated urea transporter in rat kidney. Proc Natl Acad Sci USA 93:5495-5500

Nissim-Rafinia M, Kerem B (2002) Splicing regulation as a potential genetic modifier. Trends Genet 18:123-127

Olives B, Neau P, Bailly P, Hediger MA, Rousselet G, Cartron JP and Ripoche P (1994) Cloning and functional expression of a urea transporter from human bone marrow cells. J Biol Chem 269:31649-31652

Paredes A, McManus M, Kwon HM, Strange K (1992) Osmoregulation of Na(+)-inositol cotransporter activity and mRNA levels in brain glial cells. Am J Physiol Cell 263:C1282-C1288

Payne JA, Forbush III B (1994) Alternatively spliced isoforms of the putative renal Na-K-Cl cotransporter are differentially distributed within the rabbit kidney. Proc Natl Acad Sci USA 91:4544-4548

Pearson MM, Lu J, Mount DB, Delpire E (2001) Localization of the K(+)-Cl(-) cotransporter, KCC3, in the central and peripheral nervous systems: expression in the choroid plexus, large neurons and white matter tracts. Neuroscience 103:481-491

Plata C, Meade P, Hall A E, Welch RC, Vazquez N, Hebert SC and Gamba G (2001) Alternatively spliced isoform of the apical Na-K-Cl cotransporter gene encodes a furosemide sensitive Na-Cl cotransporter. Am J Physiol Renal 280:F574-F582

Plata C, Mount DB, Rubio V, Hebert SC, Gamba G (1999) Isoforms of the Na-K-2Cl cotransporter in murine TAL. II. Functional characterization and activation by cAMP. Am J Physiol Renal 276:F359-F366

Plata C, Meade P, Vazquez N, Hebert SC, Gamba G (2002) Functional properties of the apical Na+-K+-2Cl- cotransporter isoforms. J Biol Chem 277:11004-11012

Porcellati F, Hlaing T, Togawa M, Stevens MJ, Larkin DD, Hosaka Y, Glover TW, Henry DN, Greene DA and Killen PD (1998) Human Na(+)-myo-inositol cotransporter gene: alternate splicing generates diverse transcripts. Am J Physiol Cell 274:C1215-C1225

Porcellati F, Hosaka Y, Hlaing T, Togawa M, Larkin DD, Karihaloo A, Stevens MJ, Killen PD and Greene DA (1999) Alternate splicing in human Na+-MI cotransporter gene yields differentially regulated transport isoforms. Am J Physiol Cell 276:C1325-C1337

Pushkin A, Abuladze N, Lee I, Newman D, Hwang J, Kurtz I (1999a) Cloning, tissue distribution, genomic organization, and functional characterization of NBC3, a new member of the sodium bicarbonate cotransporter family. J Biol Chem 274:16569-16575

Pushkin A, Abuladze N, Lee I, Newman D, Hwang J, Kurtz I (1999b) Mapping of the human NBC3 (SLC4A7) gene to chromosome 3p22. Genomics 58:321-322

Pushkin A, Abuladze N, Newman D, Lee I, Xu G, Kurtz I (2000a) Cloning, characterization and chromosomal assignment of NBC4, a new member of the sodium bicarbonate cotransporter family. Biochim Biophys Acta 1493:215-218

Pushkin A, Abuladze N, Newman D, Lee I, Xu G, Kurtz I (2000b) Two C-terminal variants of NBC4, a new member of the sodium bicarbonate cotransporter family: cloning, characterization, and localization. IUBMB Life 50:13-19

Pushkin A, Abuladze N, Newman D, Tatishchev S, Kurtz I (2001) Genomic organization of the DCTN1-SLC4A5 locus encoding both NBC4 and p150(Glued). Cytogenet Cell Genet 95:163-168

Randall J, Thorne T, Delpire E (1997) Partial cloning and characterization of Slc12a2: the gene encoding the secretory $Na^+-K^+-2Cl^-$ cotransporter. Am J Physiol Cell 273:C1267-C1277

Reeves WB, Molony DA, Andreoli TE (1988) Diluting power of thick limbs of Henle. III. Modulation of in vitro diluting power. Am J Physiol Renal 255:F1145-F1154

Roberts GC, Smith CW (2002) Alternative splicing: combinatorial output from the genome. Curr Opin Chem Biol 6:375-383

Rocha AS, Kokko JP (1973) Sodium chloride and water transport in the medullary thick ascending limb of Henle. Evidence for active chloride transport. J Clin Invest 52:612-623

Roll P, Massacrier A, Pereira S, Robaglia-Schlupp A, Cau P, Szepetowski P (2002) New human sodium/glucose cotransporter gene (KST1): identification, characterization, and mutation analysis in ICCA (infantile convulsions and choreoathetosis) and BFIC (benign familial infantile convulsions) families. Gene 285:141-148

Romero MF, Hediger MA, Boulpaep EL, Boron WF (1997) Expression cloning and characterization of a renal electrogenic Na^+/HCO_3^- cotransporter. Nature 387:409-413

Sands JM (2003a) Mammalian urea transporters. Annu Rev Physiol 65:543-566

Sands JM (2003b) Molecular mechanisms of urea transport. J Membr Biol 191:149-163

Sassani P, Pushkin A, Gross E, Gomer A, Abuladze N, Dukkipati R, Carpenito G and Kurtz I (2002) Functional characterization of NBC4: a new electrogenic sodium-bicarbonate cotransporter. Am J Physiol Cell 282:C408-C416

Shayakul C, Steel A, Hediger MA (1996) Molecular cloning and characterization of the vasopressin-regulated urea transporter of rat kidney collecting ducts. J Clin Invest 98:2580-2587

Shneider BL, Dawson PA, Christie DM, Hardikar W, Wong MH, Suchy FJ (1995) Cloning and molecular characterization of the ontogeny of a rat ileal sodium-dependent bile acid transporter. J Clin Invest 95:745-754

St Pierre MV, Kullak-Ublick GA, Hagenbuch B, Meier PJ (2001) Transport of bile acids in hepatic and non-hepatic tissues. J Exp Biol 204:1673-1686

Su MA, Trenor CC, Fleming JC, Fleming MD, Andrews NC (1998) The G185R mutation disrupts function of the iron transporter Nramp2. Blood 92:2157-2163

Sun A, Grossman EB, Lombardi M, Hebert SC (1991) Vasopressin alters the mechanism of apical Cl^- entry from $Na^+:Cl^-$ to $Na^+:K^+:2Cl^-$ cotransport in mouse medullary thick ascending limb. J Membr Biol 120:83-94

Tabuchi M, Tanaka N, Nishida-Kitayama J, Ohno H, Kishi F (2002) Alternative splicing regulates the subcellular localization of divalent metal transporter 1 isoforms. Mol Biol Cell 13:4371-4387

Tabuchi M, Yoshimori T, Yamaguchi K, Yoshida T, Kishi F (2000) Human NRAMP2/DMT1, which mediates iron transport across endosomal membranes, is localized to late endosomes and lysosomes in HEp-2 cells. J Biol Chem 275:22220-22228

Terris JM, Knepper MA, Wade JB (2001) UT-A3: localization and characterization of an additional urea transporter isoform in the IMCD. Am J Physiol Renal 280:F325-F332

Vibat CR, Holland MJ, Kang JJ, Putney LK, O'Donnell ME (2001) Quantitation of Na+-K+-2Cl- cotransport splice variants in human tissues using kinetic polymerase chain reaction. Anal Biochem 298:218-230

Vidal S, Belouchi AM, Cellier M, Beatty B, Gros P (1995) Cloning and characterization of a second human NRAMP gene on chromosome 12q13. Mamm Genome 6:224-230

Vorum H, Kwon TH, Fulton C, Simonsen B, Choi I, Boron W, Maunsbach AB, Nielsen S and Aalkjaer C (2000) Immunolocalization of electroneutral Na-HCO(3)(-) cotransporter in rat kidney. Am J Physiol Renal 279:F901-F909

Waterman-Storer CM, Holzbaur EL (1996) The product of the *Drosophila* gene, Glued, is the functional homologue of the p150Glued component of the vertebrate dynactin complex. J Biol Chem 271:1153-1159

Wiese TJ, Dunlap JA, Conner CE, Grzybowski JA, Lowe WL, Yorek MA (1996) Osmotic regulation of Na-myo-inositol cotransporter mRNA level and activity in endothelial and neural cells. Am J Physiol Cell 270:C990-C997

Wong MH, Oelkers P, Craddock AL, Dawson PA (1994) Expression cloning and characterization of the hamster ileal sodium- dependent bile acid transporter. J Biol Chem 269:1340-1347

Xu J, Wang Z, Barone S, Petrovic M, Amlal H, Conforti L, Petrovic S and Soleimani M (2003) Expression of the Na+-HCO-3 cotransporter NBC4 in rat kidney and characterization of a novel NBC4 variant. Am J Physiol Renal 284:F41-F50

Xu Y, Olives B, Bailly P, Fischer E, Ripoche P, Ronco P, Cartron JP and Rondeau E (1997) Endothelial cells of the kidney vasa recta express the urea transporter HUT11. Kidney Int 51:138-146

Yang T, Huang YG, Singh I, Schnermann J, Briggs JP (1996) Localization of bumetanide- and thiazide-sensitive Na-K-Cl cotransporters along the rat nephron. Am J Physiol (Renal Fluid Electrolyte) 271:F931-F939

You G, Smith CP, Kanai Y, Lee WS, Stelzner M, Hediger MA (1993) Cloning and characterization of the vasopressin-regulated urea transporter. Nature 365:844-847

Zhou C, Agarwal N, Cammarata PR (1994) Osmoregulatory alterations in myo-inositol uptake by bovine lens epithelial cells. Part 2: Cloning of a 626 bp cDNA portion of a Na+/myo- inositol cotransporter, an osmotic shock protein. Invest Ophthalmol Vis Sci 35:1236-1242

Gamba, Gerardo

Molecular Physiology Unit, Instituto Nacional de Ciencias Médicas y Nutrición Salvador Zubirán and Instituto de Investigaciones Biomédicas, Universidad Nacional Autónoma de México, Mexico City CP 14000, Mexico

gamba@sni.conacyt.mx

Transport-dependent gene regulation by sequestration of transcriptional regulators

Alex Böhm and Winfried Boos

Abstract

We describe a new principle of signal transduction in prokaryotes: certain transport systems have an additional function in signaling. Depending on the activity of the transporter, transcription factors are sequestered, thus shuttling between the inner face of the cytoplasmic membrane and their target promoters. The change in their subcellular address corresponds to the on/off state of the respective signal transduction cascade. For example, *Escherichia coli* MalT is kept in an inactive form by the idling maltodextrin ABC transporter. When the transporter is activated by maltodextrin transport, MalT is released and binds to its cognate operators. *E. coli* Mlc is bound to its operators when the glucose-specific enzyme II (EIICB^{Glc}) of the phosphoenolpyruvate:carbohydrate phosphotransferase system is not transporting; when glucose transport starts, Mlc is captured by the dephosphorylated form of EIICB^{Glc}, thus allowing transcription of its cognate operons. *Salmonella typhimurium* PutA, *E. coli* GlnK and *Klebsiella pneumoniae* NifL are discussed as additional examples.

1 Introduction

In prokaryotes, gene regulation in response to the external availability of nutrients has traditionally been viewed as being simple. The first model of gene regulation by Jacob and Monod proposed a repressor that would block the promoter by binding to an operator (Jacob and Monod 1961). If an inducer – be it the transported nutrient itself or a metabolite of the substrate – would bind to the repressor, the operator is freed and transcription of the structural genes that encode transport proteins and breakdown enzymes can be initiated. This paradigm, as it has been exemplified with the *E. coli* lactose system, still holds true. Later, it became evident that this was just one, among many, mechanisms of gene regulation. Since then, we have learned about positive regulation (Englesberg and Wilcox 1974; Schleif 1996; Schwartz 1987) where an intracellular inducer activates transcription activators rather than to inactivate repressors, and we learned about two-component systems where the inducer is often bound extracytoplasmically and the signal is transduced via a phosphorelay to affect transcription (Parkinson and Kofoid 1992; West and Stock 2001). Global gene regulation was discovered that involves, for example, the use of novel sigma factors (Lloyd et al. 2001) or small

Topics in Current Genetics, Vol. 9
E. Boles, R. Krämer (Eds.): Molecular Mechanisms Controlling Transmembrane Transport
DOI 10.1007/b95774 / Published online: 9 March 2004
© Springer-Verlag Berlin Heidelberg 2004

second messengers. For the latter, catabolite repression in *E. coli* was the first example: based on controlling the concentration of cAMP a global regulatory system affecting more than one regulon is established by the phosphotransferase system (PTS) (Postma et al. 1996).

Here we are presenting another principle in bacterial signal transduction that has only recently been recognized. In this scheme, the signaling proteins are shuttling between the cytoplasm and the inner face of the cytoplasmic membrane. This change in their subcellular localization corresponds to the on/off state of their signal transduction cascades. The underlying molecular principle of switching on or off these signaling pathways is thus to sequester away or adding back critical signaling components rather than to activate or inactivate transcriptional regulators by binding of small compounds or covalent modification. While this type of regulation is common in eukaryotes – where, for example, some transcription factors are traveling between the cytoplasm and the nucleus (Vandromme et al. 1996) – it has not attained much attention in the prokaryotic world until recently. With the development of cell biology tools for the small bacterial cells (like the GFP fusion technique) it has become clear that also prokaryotes possess very dynamic proteins which can change their cellular addresses in a highly coordinated fashion (Margolin 2000; Shapiro and Losick 2000), revealing the bacterial cytoplasm, in contrast to the traditional views, as highly dynamic.

We will discuss this signal transduction scheme with the maltose system and the Mlc regulatory protein of *Escherichia coli* as prime examples. Yet, this regulatory principle has been recognized in other systems as well. For instance, the regulatory function of the GlnK protein in its association with the ammonium transporter (see chapter by Merrick, this book) or of the PutA protein as membrane-bound degradative enzyme in one mode and cytoplasmic regulator in the other will be mentioned briefly as additional examples (Coutts et al. 2002; Muro-Pastor et al. 1997).

2 Signaling pathways: two examples

2.1 The transcription activator MalT

2.1.1 The E. coli maltose system

The *E. coli* maltose system is composed of 10 genes encoding proteins that are engaged in the effective utilization of maltose and maltodextrins (Boos and Shuman 1998). The two cytoplasmic enzymes amylomaltase and maltodextrin phosphorylase convert maltodextrins into glucose and glucose-1-phosphate. After conversion by glucokinase and phosphoglucomutase (which are not encoded by *mal* genes) into glucose-6-P, both compounds are subsequently funneled into glycolysis.

While the structural genes for amylomaltase and maltodextrin phosphorylase are in one operon at 74 min on the *E. coli* chromosome, all genes encoding transport proteins are in two divergently transcribed operons at 91 minutes. The *malEFG* gene products, together with the *malK* gene product, make up the specific maltose/maltodextrin binding protein-dependent ABC transporter (ATP-Binding Cassette transporter) (Boos and Lucht 1996; Dassa et al. 1999; Higgins 2001). Gene *lamB* encodes maltoporin (or glycoporin), which facilitates the specific diffusion of maltodextrins through the outer membrane and, in addition, is the receptor for phage lambda (Schirmer et al. 1995). Once in the periplasm, maltodextrins are specifically recognized by the high affinity maltose-binding protein (MBP or MalE) and are delivered to the membrane-bound MalF/MalG heterodimer that establishes the translocation pore. Unidirectional transport through the pore is energized by the MalK dimer, the ABC subunit, which is attached at the so-called EAA loops of MalF and MalG on the cytoplasmic side of the membrane (Mourez et al. 1997). The MalK dimer hydrolyzes ATP in response to the docking of substrate-loaded MBP at the periplasmic loops of the transport complex (Covitz et al. 1994). It is this multicomponent, binding protein-dependent ABC transporter that is at the center of the proposed novel regulatory scheme. Its structure and function will be discussed in more detail below.

All *mal* genes are under positive control of the central transcriptional activator MalT, a 103 kDa protein, which has to bind ATP and the inducer maltotriose to activate transcription of the *mal* genes (Débarbouillé et al. 1978; Richet and Raibaud 1989). Maltotriose is a metabolite that is always formed by amylomaltase when maltodextrins are degraded. Therefore, the classical textbook regulation of the *E. coli mal* regulon is simple: when maltodextrins are used as carbon source, the internal concentration of maltotriose rises, MalT is activated and in turn activates transcription of the *mal* genes.

Like most degradative systems in *E. coli*, the maltose system is subjected to the global gene regulation of catabolite repression (see also chapter by Stülke, this book). This additional level of control is exerted by the cAMP/CAP (catabolite activator protein) system. cAMP-loaded CAP has to bind in the upstream region of the *malEFG, malK-lamB,* and *malT* operons in order to allow full *mal* gene transcription. The concentration of cAMP is in turn controlled in response to the activity of the phosphotransferase system (Postma et al. 1996). Here, in the absence of glucose in the medium, all PTS proteins, in particular the $EIIA^{Gluc}$ protein, will be in a phosphorylated form ($EIIA^{Gluc}$-P). According to the dogma of catabolite repression in *E. coli*, $EIIA^{Gluc}$-P stimulates adenylate cyclase, thus yielding high concentrations of cAMP and the cAMP/CAP complex, respectively (Postma et al. 1996). With glucose present in the medium, $EIIA^{Gluc}$-P is effectively dephosphorylated and cAMP levels are low. Neither *malT* nor any MalT-dependent *mal* gene is significantly expressed. In addition to this transcriptional control, PTS-mediated transport of glucose directly controls the transport of maltose: $EIIA^{Gluc}$ in its dephosphorylated form reduces the activity of the maltodextrin ABC transporter, presumably by inhibiting the ATPase activity of MalK (Dean et al. 1990; Kühnau et al. 1991). This effect is well known as inducer exclusion (Postma et al. 1996).

2.1.2 The role of MalK in mal gene regulation

About two decades ago, it had been recognized that MalK is involved in *mal* gene expression. It was found that loss-of-function mutations in *malK* strongly increased the expression of the remaining *mal* genes despite of the fact that these strains are unable to grow on maltodextrins (Bukau et al. 1986). Inversely, the separate overexpression of *malK* (for instance from a plasmid) represses *mal* gene expression (Reyes and Shuman 1988). When transformed into a wild type strain, plasmid-encoded MalK renders the strain unable to utilize maltose. This allowed the isolation of mutants that became resistant to the repressive effect of overproduced MalK (Böhm et al. 2002; Dean et al. 1990; Kühnau et al. 1991). The mutations were mostly in *malK* but also extragenic suppressors could be isolated. Among the latter was a mutation in *mlc* which ultimately led to the discovery of another sequestration type regulator which will be discussed below.

The regulatory function of MalK in *mal* gene expression remained mysterious for a long time. Even though phenotypically behaving as a repressor, its sequence did not reveal any indications of a DNA-binding protein. Constitutivity of *mal* gene expression in *malK* null mutants (which are unable to transport maltodextrins) still required the presence of MalT and, therefore, the presence of internally synthesized maltotriose, the so-called endogenous inducer whose metabolic origin is still unclear (Decker et al. 1993)[1]. Mutations in *malK* that result in the loss of repression (but not of the transport function of MalK) were preferentially affecting residues near the C-terminus of the protein. This extended C-terminal domain is typical for a large subset of ATP-hydrolyzing subunits of prokaryotic ABC transporters and contains several conserved sequence motifs, but it is not essential for ATP hydrolysis since most bacterial ABC transporters lack it. An attractive but unprovable hypothesis at the time was that the C-terminal domain of MalK would act as an enzyme degrading the internal inducer (Bukau et al. 1986; Kühnau et al. 1991).

The first indication for a molecular mechanism by which MalK regulates the *mal* genes was the observation that matrix-attached purified MalT protein could specifically bind MalK *in vitro* (Panagiotidis et al. 1998). If such an interaction would also occur *in vivo*, then an interaction of MalK with MalT could control the activity of MalT as transcriptional activator and the interaction itself could be controlled by the transport activity. Meanwhile it was found that MalT surprisingly interacts also with other proteins which inhibit its function in *mal* gene expression. For example, overexpression of MalY, a cytoplasmic βC-S lyase (cystathionase) (Zdych et al. 1995) strongly reduces MalT-dependent *mal* gene expression *in vivo* (Clausen et al. 2000; Reidl and Boos 1991). Indeed, MalY reduces the ability of MalT to act as a transcription activator *in vitro* (Schreiber et al. 2000). Similarly, Aes, an acyl esterase of unknown function (Peist et al. 1997), interacts with MalT curbing its regulatory activity (Joly et al. 2002). In these studies, it could be shown that the interaction of the effector (MalY or Aes) competes with the inducer mal-

---

[1] This is a completely different situation to maltotriose formation in strains that take up maltodextrins and have amylomaltase, cf. 3.1.1, second paragraph.

totriose. In addition, mutants in *malT* have been analyzed that were isolated in a selection for inducer-independence. Some of these mutations became partially resistant to the repressing effect of MalY and Aes as well as MalK (Schlegel et al. 2002). In conclusion, it became clear that the activity of MalT is not only controlled by maltotriose and ATP, but in addition by a number of proteins, leading to a modulation of its transcription-enhancing activity. However, while the regulatory effect of MalK makes obvious sense (see below), the physiological roles of the effects exerted on MalT by MalY and Aes are not clear at present (Boos and Böhm 2000).

2.1.3 MalT shuttles between the chromosome and the maltodextrin ABC transporter

The biochemically observed interaction of MalT with MalK posed an important question: is *in vivo* binding of MalT to MalK taking place at the transport complex (sequestration of MalT) or is MalK dissociating from the transporter to form a soluble, transcriptionally inactive cytoplasmic complex with MalT? In the wild type, MalK has never been observed in significant amounts in the membrane-free cellular extract (our unpublished observation); in addition, the transport complex appears to be very stable (Chen et al. 2001; Davidson and Nikaido 1991). This indicates that MalK in the wild type is synthesized in stoichiometric amounts and never dissociates from the transporter. Also, in the absence of transportable substrate MalT is largely inactive despite of the presence of endogenous inducer (Decker et al. 1993). We conclude that this inactivity is caused by the sequestretion of MalT by the idling maltodextrin transporter (Boos and Böhm 2000). Nevertheless, MalT must also be able to recognize soluble MalK since overexpression of MalK (beyond the stoichiometry of the transport complex) is very effective in MalT inactivation, even in the presence of maltodextrins in the medium. Yet, chromosomally encoded MalK in the absence of MalF and MalG is less effective in repressing a reporter *mal* gene fusion than in the presence of MalF and MalG where transport complex formation takes place (our unpublished observation). This demonstrates that the physiologically relevant location of the MalK-MalT interaction is at the transport complex.

The other important question is whether or not transport activity of the ABC transporter indeed controls the interaction of MalK with MalT. There are mutants in *malF* or *malG* that render the transporter independent of the maltose binding protein (Covitz et al. 1994). Unlike the wild type, they are characterized by a binding protein independent – constitutive – ATPase activity of MalK in the *in vitro* system of membrane vesicles or reconstituted liposomes (Covitz et al. 1994; Dean et al. 1989). Despite the fact that they contain wild type MalK, these mutants are partially constitutive for *mal* gene expression (Panagiotidis et al. 1998) indicating that the uncoupled ATPase activity (and thus transport activity) interferes with the ability of MalK to bind MalT. Since an isolated peptide from the C-terminal part of the protein encompassing the entire so-called regulatory domain of MalK is fully active in repressing *mal* gene expression (Böhm et al. 2002), we conclude

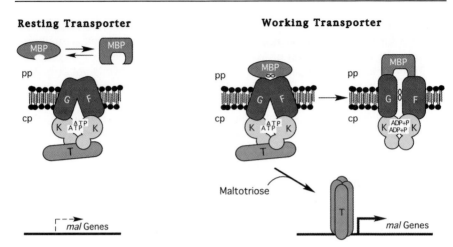

Fig. 1. The MalT cycle of *mal* gene regulation. In the resting state, the maltose binding protein is free of substrate and oscillates between the open and closed form. The transport complex has ATP bound but does not hydrolyze it. The ATPase domains of MalK are in the open conformation, the MalF/G channel is closed and the regulatory subunit of MalK sequesters the transcriptional activator MalT, rendering it inactive and thus allowing only basal *mal* gene expression. In the transporting state, substrate loaded binding protein binds to the closed complex, triggering the opening of the channel and subsequent ATP hydrolysis. The transition state is characterized by the empty binding protein tightly bound to the opened channel. The conformational change of the MalF/G channel also changes the conformation of the MalK dimer and thus alters the affinity of the regulatory domains of MalK (possibly by changing the relative positioning of the two regulatory domains) towards MalT. MalT is released and together with binding the inducer maltotriose activates gene expression of all 10 *mal* genes. pp, periplasm; cp, cytoplasm. One letter abbreviations correspond to the respective Mal proteins, e.g. K, MalK. MBP, maltose binding protein. Two small ovals indicate maltose.

that the default conformational state of the regulatory domain of MalK is the one which is able to bind MalT, thereby inactivating its transcriptional activity.Transport activity, with its concomitant ATPase activity, would then alter the conformation of the C-terminal domain, or its position relative to the ATPase domain, releasing MalT from MalK and thus allowing *mal* gene expression (Fig. 1).

2.1.4 The structure of MalK

To get a more precise picture of this scheme at the atomic level, the structure of MalK from the hyperthermophilic archaeon *Thermococcus litoralis* has been solved in 2000 (Diederichs et al. 2000). In contrast to the *E. coli* system, the *T. litoralis* ABC transporter is responsible for the uptake of trehalose and maltose but not of longer maltodextrins. Nevertheless, both systems are very similar on the level of the amino acid sequence (47% identical residues). The crystal structure obtained for *T. litoralis* MalK showed that the C-terminal portion that has been implicated in the regulatory functions of the *E. coli* MalK formed a largely independent domain composed of a β-barrel (Diederichs et al. 2000). This C-terminal domain is also present in the recently crystallized GlcV from the glucose ABC transporter from *Sulfolobus solfataricus*. Despite of a rather low degree of sequence conservation in the C-terminal part it exhibits a fold that is strikingly similar to the MalK structures from both *E. coli* and *T. litoralis* (Verdon et al. 2003). The domain also shows high structural homology to a subdomain (domain II) in the ModE protein, the transcriptional regulator of the molybdate-dependent regulon in *E. coli*. This domain supposedly binds molybdate (Hall et al. 1999) and is discussed as a general substrate recognition element. The structure of the ATPase domain in the monomeric chain of MalK is almost identical to the previously published structure of HisP, the ATPase of the histidine transporter from *E. coli* (Hung et al. 1998), a member of the ABC family that does not possess the C-terminal domain. Even though *T. litoralis* MalK behaved in solution as a monomeric protein (Greller et al. 1999), the crystal packing suggested dimer formation that we interpreted as relevant to the MalK dimer structure in the native transport complex (Diederichs et al. 2000). Meanwhile, several conclusions based on biochemical (Fetsch and Davidson 2002; Hopfner et al. 2000; Moody et al. 2002; Samanta et al. 2003) and, more stringently, on structural data (Locher et al. 2002) became available that made it clear that the dimer interface proposed for the *T. litoralis* MalK structure was most likely due to crystallographic constraints and is unlikely to represent the dimer interface occurring in the complete transport complex *in vivo*. The most likely dimer interface necessitates the close juxtapositioning of the signature motif of one subunit with the Walker A and B motifs of the other subunit, as clearly seen in the complete BtuCD structure (Locher et al. 2002). The remodeling of the more likely *T. litoralis* dimer interface based on the one seen in the LolA structure (Karpowich et al. 2001) moved the C-terminal regulatory domain underneath the ATP-hydrolyzing domain and in close proximity to it (Samanta et al. 2003). A similar model for the *T. litoralis* MalK dimer using the dimer interface of BtuCD as guidance gave the same picture (Böhm 2002). Therefore, it became clear that the C-terminal regulatory domains in the dimer are in close contact and the C-terminal part of one subunit is in close proximity to the ATPase domain of the other.

Very recently, the structure of *E. coli* MalK has also been solved, revealing an almost identical monomeric structure to *T. litoralis* MalK (Chen et al. 2003). In this study, three different dimeric structures have been found. A "closed" dimer form – with two molecules ATP bound – is similar to the BtuD-like, modeled MalK dimer. The structure of MalK and the analysis of the different mutations that had been isolated in *E. coli malK* permitted the following conclusions or predictions:

- The crystal structures in combination with a large body of biochemical data led to a model of the working ABC transporter (Chen et al. 2003, 2001): The resting transporter is in an open conformation and has two ATP molecules bound. The transmembrane channel is open towards the cytoplasm and closed towards the periplasm. When maltodextrin-bound MBP docks to the periplasmic lobes of MalF/G, major conformational changes take place and the dimer interface is closed, accompanied by a rotational movement of the helical part of the AT-Pase domain (Karpowich et al. 2001). Since the helical domain makes direct contact with the EAA loops of MalF/G, this rotation is transformed into opening of the channel towards the periplasm. This, in turn, results in a deformation of the binding protein, thus strongly decreasing its affinity for maltodextrins, which can now be bound by a low affinity binding site in the channel, allowing for their subsequent diffusion into the cytoplasm. In this transition state (open channel, tightly bound MBP, closed dimer interface), ATP is hydrolyzed, which leads to channel closing, release of the periplasmic binding protein and, after exchange of ADP for ATP, opening of the dimer interface. This transition state intermediate was defined by vanadate trapping experiments that freeze the transporter after a single ATP hydrolysis event (Chen et al. 2001).

- The regulatory domain of one monomer contacts the ATPase domain of the other during the transport cycle. Indeed, mutations in the regulatory domain have been isolated that affect the transport and ATPase activity (Böhm et al. 2002; Hunke et al. 2000) (Fig. 2). Since all mutations leading to resistance against inhibition by EIIA^{Glc} are positioned at a contiguous surface between the C-terminal domain of one monomer and the ATPase of the other, this surface must be the EIIA^{Glc}-binding site. The concave structure of this surface suggests that EIIA^{Glc} blocks the movement of the ATPase and regulatory domains against each other, thus, preventing ATP hydrolysis and transport.

Fig. 2. The structure of dimeric *E. coli* MalK with bound ATP. Top: The MalK dimer is shown in a ribbon representation. The N-terminal ATPase domains of the two monomers are given in yellow and green. The helical domain of the yellow monomer is in the upper foreground, the corresponding helical domain of the green monomer in the upper background. These helical domains interact with the EAA motif of the MalF/G channel. The energy-driven movement of these domains is likely to cause the opening of the MalF/G channel (not shown). The C-terminal regulatory domain belonging to the green ATPase monomer is shown in the foreground in light green (lower portion of the picture), the C-terminal domain of the yellow ATPase monomer is in the lower background in a brownish color. Two ATP molecules are clamped into the dimer interface (stick representation).
Bottom: View of the dimer turned by 90° to the right along the vertical axis showing the helical domain of the yellow ATPase at the right upper corner in the foreground. Cloring of the ribbon structure is as above. Highlighted as blue spheres are residues implicated in the interaction with MalT, highlighted as brown spheres are residues implicated in the interaction with EIIA^{Glc}. It is very obvious that the interaction surface with EIIA^{Glc} constitutes a continuous surface formed by the C-terminal regulatory domain of one monomer and the ATPase domain of the other monomer. Therefore, the two subdomains must move relative to each other during the transport cycle and open up to the interfering access of EIIA^{Glc}. The structure does not offer an obvious explanation for the release of MalT during transport. The structures were constructed with PyMol (DeLano 2002) using the atomic coordinates of the closed form of dimeric *E. coli* MalK as published (Chen et al. 2003). We gratefully acknowledge the authors' permission to use their coordinates prior to publication.

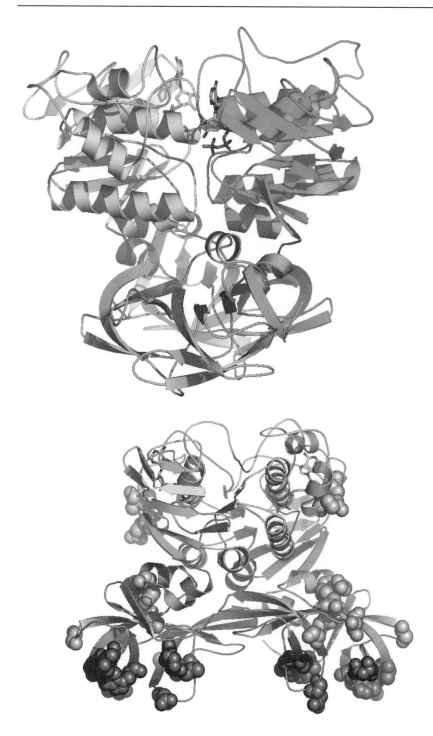

- The position for the interaction of the regulatory domain with MalT, characterized by the position of mutations causing a regulation-negative phenotype *in vivo* (Böhm et al. 2002; Kühnau et al. 1991) are at the "bottom" of the structure in Figure 2 (lower panel). It is possible that both regulatory domains participate in a common MalT-binding surface capable of binding a single MalT polypeptide. Indeed, the present understanding of MalT activation by maltotriose and ATP is based on multimerization of the MalT protein (Schreiber and Richet 1999). Thus, sequestration of the inactive, monomeric MalT by dimeric MalK (the closed form) that has ATP bound is a plausible explanation for the inhibitory mode of MalK. But at present the three crystal structures of MalK do not provide an obvious clue to MalT release. All three structures are attached at their C-terminal portions, opening only the interface between the ATPase dimers and allowing little or no movement of the C-terminal parts (Chen et al. 2003). Obviously, only the structure of the complete transport complex will bring us closer to the revelation of the mechanism of MalT release.

2.2 The global regulator Mlc

2.2.1 Discovery and characterization of the regulator Mlc

Mlc is a protein that acts as a global repressor of gene expression in *E. coli*. Its gene had originally been identified in a multicopy screen for genes that would increase the growth rate of *E. coli* in rich medium containing high concentrations of glucose and was thus named *mlc* for makes large colonies (Hosono et al. 1995). Paradoxically, *mlc* overexpression was actually found to reduce the glucose transport rate. Its growth-stimulating effect has been explained by the decrease in glucose fermentation and, in consequence, a decrease in the massive formation of acetate, which in the weakly buffered rich medium, becomes growth-inhibitory. Later, as mentioned above, the gene was rediscovered in a different context. In a screen for insertion mutants that would relieve the strong *mal* gene repression by plasmid-expressed *malK*, an *mlc* insertion mutant was isolated (Decker et al. 1998). It became clear that the *mlc* mutation caused enhanced transcription of *malT*, thus overcoming the repressing effect of MalK. At the same time it was found that the *manXYZ* operon (encoding a PTS transport system for mannose) was derepressed in *mlc* insertion mutants (Plumbridge 1998). In both systems, the overexpression of plasmid-encoded Mlc strongly repressed *mal* and *man* gene expression. Subsequently, Mlc was found to repress its own expression as well as the expression of *ptsG* encoding the glucose-specific EIIBC transporter and of the genes encoding the general PTS components EI and Hpr (Plumbridge 1999). Thus, Mlc appeared to act as a global regulator involved in sugar metabolizing systems with the preference for PTS sugars. The protein was shown by footprint analysis to bind directly to the individual promoter regions, acting as a classical transcriptional repressor by blocking access of RNA polymerase. An Mlc operator sequence could be defined that is palindromic and characterized by its close similarity to the consensus operator site of the NagC repressor (Plumbridge 2001). A characteristic feature of Mlc-dependent genes was their simultaneous dependency on the cAMP/CAP complex, which is known to be the basis for catabolite repression (Plumbridge 2002).

2.2.2 Mlc shuttles between PtsG and the chromosome

Because classical repressors such as the LacI protein bind an inducer that inactivates their repressive function, glucose or a metabolite of it was thought to act as the Mlc-specific inducer. Indeed, when glucose is present in the growth medium the Mlc-dependent repression of *ptsG* (Plumbridge 1999) as well as of *malT* (Decker et al. 1998) was relieved. Yet, all attempts to demonstrate inducer-dependent inactivation of Mlc in band shifts or footprint experiments failed to identify a low molecular weight inducer for Mlc. However, it became evident that transport per se of glucose through the *ptsG*-encoded EIIBC transporter for glucose was necessary and sufficient for Mlc-dependent regulation. Mutants in *ptsG* were isolated that had lost this regulation (Notley-McRobb and Ferenci 2000; Plumbridge 2000; Zeppenfeld et al. 2000) indicating that it was controlled by the state of the PtsG transporter.

Several scenarios could explain the mechanics of PtsG-dependent Mlc regulation. One of them, in analogy to the *bgl* antitermination scheme (Amster-Choder and Wright 1993; Chen and Amster-Choder 1998), involved a possible phosphotransfer from the EIIBC transporter to Mlc: in the absence of glucose transport Mlc would constantly receive phosphate groups from EIIBC^{Glc} and would, thus, be kept in the repressor mode. If glucose were transported, phosphoryl groups would be drained from EIIBC^{Glc}, Mlc-P would become dephosphorylated and therefore inactived as a repressor, thus allowing transcription of Mlc-controlled operons. Yet, all attempts to detect PtsG-dependent phosphorylation of Mlc failed. But another scheme emerged by a finding made in several laboratories: Mlc was effectively bound by actively transporting (desphosphorylated) EIIBC^{Glc} but was not bound when EIIBC^{Glc} was idling and hence in the fully phosphorylated state (Lee et al. 2000; Nam et al. 2001; Tanaka et al. 2000). Thus, the sequestration of Mlc by unphosphorylated PtsG is the underlying principle that leads to inactivation of Mlc and to the release of Mlc-dependent genes from transcriptional repression. Sequestration is mediated by the EIIB domain of PtsG. The extreme C-terminal portion of Mlc, which is likely to form an amphipathic helix, is required for the tetramerization of the protein as well as for its interaction with PtsG (Seitz et al. 2003). Surprisingly, only the membrane-bound EIIB domain, either connected to its natural partner EIIC or another protein that serves as a membrane anchor, is able to cause Mlc inactivation. In contrast, Mlc bound to the soluble EIIB domain is fully active as a repressor. Mutational analysis of EIIB involving an exchange of the phosphorylation site, Cys421, as well as of amino acids in the vicinity of Cys421 demonstrated that it is not the charge alteration due to phosphorylation but the change in the conformational state of neighboring residues that leads to release of Mlc when EIIB is phosphorylated (Seitz et al. 2003).

The message emerging from these findings is summarized in Figure 3. The glucose transport activity of the cell, signaled through the phosphorylation state of EIIB, is sensed by Mlc and leads to either membrane-associated Mlc (by unphosphorylated EIIB) or DNA-bound Mlc. There is yet another player in Mlc-dependent regulation. By looking for mutants with reduced PtsG-mediated glucose transport, Becker and Jahreis have isolated an insertion in *yeeI*, encoding a soluble protein of unknown function. It turned out that the *yeeI* mutation entails reduced *ptsG* expression and that this effect is dependent on functional Mlc (VAAM Annual Meeting 2003, poster KC002). Indeed, *in vitro* experiments revealed a tight interaction between Mlc and YeeI (Becker et al., unpublished observations) indicating that the affinity of Mlc for membrane-bound PtsG might be modulated by YeeI.

Other, less well understood observations concern control circuits that affect the expression of *mlc* itself. It was found that an increase in the level of σ^{32}, the classical heat shock sigma factor, be it by heat shock or by plasmid-encoded overexpression, dramatically enhances expression of *mlc* (Shin et al. 2001).

However, the "physiological purpose" of Mlc function is difficult to understand. It appears to be contradictory to impose (by glucose transport) strong catabolite repression onto e.g. the *mal* regulon and, at the same time, lift the Mlc-mediated repression of *malT*. It is possible that this seemingly paradoxical scheme is instrumental in fine tuning the bacterial cell's response to subtle changes in the environment.

3 The PutA, GlnK, and NifL signal transduction proteins

To our knowledge, the principle of membrane sequestration of a signal transduction protein has first been recognized for the *Salmonella typhimurium* proline utilization (Put) system (Muro-Pastor et al. 1997; Ostrovsky de Spicer et al. 1991). Proline can be used as carbon, nitrogen, and energy source or as an osmoprotectant by *S. typhimurium* and is, among other transporters, taken up by the Na^+-symporter PutP (Hanada et al. 1992). Its subsequent conversion into glutamate is catalyzed in a two-step oxidation reaction by PutA, a multifunctional 144 kDa proline dehydrogenase (Menzel and Roth 1981). Oxidation is coupled to reduction of the FAD cofactor and soluble NAD^+ and proceeds via a pyrroline-5-carboxylate intermediate (Surber and Maloy 1998). Subsequent recycling of FAD requires electron transfer to the electron transport chain and, consequently, the association of PutA with the cytoplasmic membrane. In addition to its enzymatic functions, PutA has also been shown to regulate transcription of the *putAP* operon (Ostrovsky de Spicer et al. 1991). The seemingly contradictory finding of a membrane-bound transcriptional regulator was explained by a mechanism in which PutA shuttles between a membrane-associated state where it is active as a proline dehydrogenase, and a cytoplasmic DNA-bound state where it is active as a transcriptional repressor (Ostrovsky de Spicer and Maloy 1993). The signaling proteins MalT and Mlc are bound by the respective transport proteins and the primary

Resting Transporter   **Working Transporter**

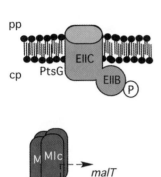
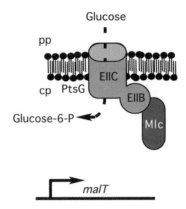

Fig. 3. Model for the sequestration of Mlc. Left, resting transporter: when no glucose is transported, PtsG remains in its phosphorylated state. In this conformation, PtsG does not recognize Mlc allowing it to repress Mlc-dependent operators (only *malT* is shown as an example). Right, working transporter: when glucose is transported, PtsG transfers phosphate groups to glucose and is therefore predominantly in its unphosphorylated state that is able to sequester Mlc. Mlc bound to PtsG cannot recognize its cognate operators and transcription is allowed to proceed. Mlc has been characterized as a tetramer in solution and may bind to the operators as such. The quaternary structure of Mlc when bound to PtsG is unknown. The putative amphipathic helix at the extreme of the C-terminal part of Mlc is necessary for both, repression of responsive operators and binding to PtsG. pp, periplasm; cp, cytoplasm; encircled "P", phosphate.

signal leading to their sequestration is the substrate transport activity per se. In contrast, membrane-sequestration of PutA does not require PutP or any other protein, and the signal that causes the change in the subcellular address is the redox state of the FAD cofactor, which in turn, is dependent on the intracellular proline concentration (Surber and Maloy 1999; Zhu and Becker 2003).

Yet another variation of this scheme is exemplified by the sophisticated regulation of the nitrogen assimilation systems of *E. coli* and *Klebsiella pneumoniae*. The GlnK signal transduction protein from *E. coli*, a member of the P_{II} protein family, switches between a cytoplasmic free state and a membrane-bound state where it is associated with the ammonium transporter AmtB (Coutts et al. 2002) (see also the chapter by Merrick, this book). There are two potential physiological roles for this switch as discussed by Merrick and coworkers. In one scheme, GlnK is responsible for fine tuning of the transport rate of AmtB in response to the intracellular nitrogen status, shutting off ammonium transport when the cell is saturated with nitrogen. The second scheme is more similar to what we describe here for MalT and Mlc: accordingly, sequestration of GlnK by AmtB in an ammonium transport rate-dependent manner is a means to affect downstream signaling events that lead to a transcriptional output (Coutts et al. 2002). A clue to what such

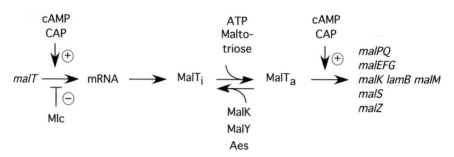

Fig. 4. Regulation of the *E. coli* maltose system. $MalT_i$ and $MalT_a$ indicate the inactive and active state of MalT, respectively, the latter being oligomeric. The cAMP/CAP complex is controlled by PTS-mediated glucose transport and symbolizes the basis of catabolite repression. PTS-mediated glucose transport also controls the activity of Mlc, which in turn, controls *malT* expression. The activity of MalT is positively influenced by maltotriose and inhibited by MalK but also by other, non-transport-related cytoplasmic proteins. Ten *mal* genes in five transcriptional units are under the control of MalT.

downstream signaling components are comes from *K. pneumoniae*. Here, the NifL protein, the antagonist of the NifA transcription factor (Dixon 1998), has been shown to travel between the cytoplasm and the inner membrane, possibly in a GlnK-dependent fashion (Klopprogge et al. 2002).

4 Conclusion

The complex regulation of the maltose regulon is summarized in Figure 4. At the center is MalT, the transcriptional activator of all *mal* genes. It occurs in two conformations, active and inactive, hence the major control is exerted on the activity of MalT. The protein is stimulated by its inducer maltotriose and inhibited by a number of proteins of which MalK is the most prominent. The second level of control is on *malT* expression. There again, a dual control is seen: stimulation of *malT* transcription occurs by the cAMP/CAP complex and inhibition by Mlc. These opposite effects are controlled by glucose transport. The most interesting aspect of *mal* gene regulation is the novel principle of sequestrating (and inactivating) the regulator by a transport protein. Transport of substrate by the binding protein-dependent ABC transporter controls MalT activity, and transport of glucose by the PTS-mediated PstG transporter controls the activity of Mlc and thus the expression of *malT*. It is significant that a large portion of binding protein-dependent prokaryotic ABC transporters do contain the C-terminal domain that is typical for MalK. So far, the structures of only a few MalK homologs have been analyzed. All have been found to contain identical regulatory domains as *E. coli* MalK. From the fact that these C-terminal portions contain conserved motifs one might be inclined to associate regulatory functions to these domains in all transporters that possess them. Whether or not they have dual functions as in the maltose system (passive regulation by EII^{Glc} and active regulation of MalT) will be seen in the future.

Acknowledgements

We thank Amy Davidson and Jue Chen for sharing coordinates of *E. coli* MalK prior to publication and Erika Oberer-Bley for her help with the preparation of the manuscript. This work was supported by grants from the Deutsche Forschungs-gemeinschaft and the Fonds der Chemischen Industrie.

References

Amster-Choder O, Wright A (1993) Transcriptional regulation of the *bgl* operon of *Escherichia coli* involves phosphotransferase system-mediated phosphorylation of a transcriptional antiterminator. J Cell Biochem 51:83-90

Böhm A (2002) Mechanismus und physiologische Bedeutung der MalK-vermittelten Repression des *Escherichia coli* Maltose Regulons. PhD thesis. Universität Konstanz

Böhm A, Diez J, Diederichs K, Welte W, Boos W (2002) Structural model of MalK, the ABC subunit of the maltose transporter of *Escherichia coli*: implications for *mal* gene regulation, inducer exclusion, and subunit assembly. J Biol Chem 277:3708-3717

Boos W, Böhm A (2000) Learning new tricks from an old dog. MalT of the *Escherichia coli* maltose system is part of a complex regulatory network. Trends Genet 16:404-409

Boos W, Lucht JM (1996) Periplasmic binding protein-dependent ABC transporters. In: Neidhardt FC et al. (eds) *Escherichia coli* and *Salmonella typhimurium*; cellular and molecular biology. vol. 1, American Society of Microbiology, Washington, DC, pp 1175-1209

Boos W, Shuman HA (1998) The maltose/maltodextrin system of *Escherichia coli*; transport, metabolism and regulation. Microbiol Mol Biol Rev 62:204-229

Bukau B, Ehrmann M, Boos W (1986) Osmoregulation of the maltose regulon in *Escherichia coli*. J. Bacteriol. 166:884-891

Chen J, Lu G, Lin J, Davidson AL, Quiocho FA (2003) A tweezers-like motion of the ATP-binding cassette dimer in an ABC transport cycle. Mol Cell 12:651-661

Chen J, Sharma S, Quiocho FA, Davidson AL (2001) Trapping the transition state of an ATP-binding cassette transporter: evidence for a concerted mechanism of maltose transport. Proc Natl Acad Sci USA 98:1525-1530

Chen Q, Amster-Choder O (1998) BglF, the sensor of the *bgl* system and the β-glucosides permease of *Escherichia coli*: Evidence for dimerization and intersubunit phosphotransfer. Biochemistry 37:8714-8723

Clausen T, Schlegel A, Peist R, Schneider E, Steegborn C, Chang Y-S, Haase A, Bourenkov GP, Bartunik HD, Boos W (2000) X-ray structure of MalY from *Escherichia coli*: a pyridoxal 5'-phosphate-dependent enzyme acting as a modulator in *mal* gene expression. EMBO J 19:831-842

Coutts G, Thomas G, Blakey D, Merrick M (2002) Membrane sequestration of the signal transduction protein GlnK by the ammonium transporter AmtB. EMBO J 21:536-545

Covitz KMY, Panagiotidis CH, Hor LI, Reyes M, Treptow NA, Shuman HA (1994) Mutations that alter the transmembrane signalling pathway in an ATP binding cassette (ABC) transporter. EMBO J 13:1752-1759

Dassa E, Hofnung M, Paulsen IT, Saier MH (1999) The *Escherichia coli* ABC transporters: an update. Mol Microbiol 32:887-889

Davidson AL, Nikaido H (1991) Purification and characterization of the membrane-associated components of the maltose transport system from *Escherichia coli*. J Biol Chem 266:8946-8951

Dean DA, Davidson AL, Nikaido H (1989) Maltose transport in membrane vesicles of *Escherichia coli* is linked to ATP hydrolysis. Proc Natl Acad Sci USA 86:9134-9138

Dean DA, Reizer J, Nikaido H, Saier M (1990) Regulation of the maltose transport system of *Escherichia coli* by the glucose-specific enzyme III of the phosphoenolpyruvate-sugar phosphotransferase system. Characterization of inducer exclusion-resistant mutants and reconstitution of inducer exclusion in proteoliposomes. J Biol Chem 265:21005-21010

Débarbouillé M, Shuman HA, Silhavy TJ, Schwartz M (1978) Dominant constitutive mutations in *malT*, the positive regulator of the maltose regulon in *Escherichia coli*. J Mol Biol 124:359-371

Decker K, Peist R, Reidl J, Kossmann M, Brand B, Boos W (1993) Maltose and maltotriose can be formed endogenously in *Escherichia coli* from glucose and glucose-1-phosphate independently of enzymes of the maltose system. J Bacteriol 175:5655-5665

Decker K, Plumbridge J, Boos W (1998) Negative transcriptional regulation of a positive regulator: the expression of *malT*, encoding the transcriptional activator of the maltose regulon of *Escherichia coli*, is negatively controlled by Mlc. Mol Microbiol 27:381-390

DeLano WL (2002) The PyMol molecular graphics system. http://www.pymol.org

Diederichs K, Diez J, Greller G, Müller C, Breed J, Schnell C, Vonrhein C, Boos W, Welte W (2000) Crystal structure of MalK, the ATPase subunit of the trehalose/maltose ABC transporter of the archaeon *Thermococcus litoralis*. EMBO J 19:5951-5961

Dixon R (1998) The oxygen-responsive NIFL-NIFA complex: a novel two-component regulatory system controlling nitrogenase synthesis in gamma-proteobacteria. Arch Microbiol 169:371-380

Englesberg E, Wilcox G (1974) Regulation: positive control. Annu Rev Genet 8:219-242

Fetsch EE, Davidson AL (2002) Vanadate-catalyzed photocleavage of the signature motif of an ATP-binding cassette (ABC) transporter. Proc Natl Acad Sci USA 99:9685-9690

Greller G, Horlacher R, DiRuggiero J, Boos W (1999) Molecular and biochemical analysis of MalK, the ATP-hydrolyzing subunit of the trehalose maltose transport system of the hyperthermophilic archaeon *Thermococcus litoralis*. J Biol Chem 274:20259-20264

Hall DR, Gourley DG, Leonard GA, Duke EMH, Anderson LA, Boxer DH, Hunter WN (1999) The high-resolution crystal structure of the molybdate-dependent transcriptional regulator (ModE) from *Escherichia coli*: a novel combination of domain folds. EMBO J 18:1435-1446

Hanada K, Yoshida T, Yamato I, Anraku Y (1992) Sodium ion and proline binding sites in the Na+/proline symport carrier of *Escherichia coli*. Biochim Biophys Acta 1105:61-66

Higgins CF (2001) ABC transporters: physiology, structure, and mechanism. An overview. Res Microbiol 152:205-210

Hopfner K-P, Karcher A, Shin DS, Craig L, Arthur LM, Carney JP, Tainer JA (2000) Structural biology of Rad50 ATPase: ATP-driven conformational control in DNA double-strand break repair and the ABC-ATPase superfamily. Cell 101:789-800

Hosono K, Kakuda H, Ichihara S (1995) Decreasing accumulation of acetate in a rich medium by *Escherichia coli* on introduction of genes on a multicopy plasmid. Biosci Biotech Biochem 59:256-261

Hung L-W, Wang IX, Nikaido K, Liu P-Q, Ames GF-L, Kim S-H (1998) Crystal structure of the ATP-binding subunit of an ABC transporter. Nature 396:703-707

Hunke S, Landmesser H, Schneider E (2000) Novel missense mutations that affect the transport function of MalK, the ATP-binding-cassette subunit of the *Salmonella enterica* serovar typhimurium maltose transport system. J Bacteriol 182:1432-1436

Jacob F, Monod J (1961) Genetic regulatory mechanisms in the synthesis of proteins. J Mol Biol 3:318-356

Joly N, Danot O, Schlegel A, Boos W, Richet E (2002) The Aes protein directly controls the activity of MalT, the central transcriptional activator of the *Escherichia coli* maltose regulon. J Biol Chem 277:16606-16613

Karpowich N, Martsinkevich O, Millen L, Yuan YR, Dai PL, MacVey K, Thomas PJ, Hunt JF (2001) Crystal structures of the MJ1267 ATP binding cassette reveal an induced-fit effect at the ATPase active site of an ABC transporter. Structure 9:571-586

Klopprogge K, Grabbe R, Hoppert M, Schmitz RA (2002) Membrane association of *Klebsiella pneumoniae* NifL is affected by molecular oxygen and combined nitrogen. Arch Microbiol 177:223-234

Kühnau S, Reyes M, Sievertsen A, Shuman HA, Boos W (1991) The activities of the *Escherichia coli* MalK protein in maltose transport, regulation and inducer exclusion can be separated by mutations. J Bacteriol 173:2180-2186

Lee S-J, Boos W, Bouché J-P, Plumbridge J (2000) Signal transduction between a membrane bound transporter, PtsG, and a soluble transcription factor, Mlc, of *Escherichia coli*. EMBO J 19:5353-5361

Lloyd G, Landini P, Busby S (2001) Activation and repression of transcription initiation in bacteria. Essays Biochem 37:17-31

Locher KP, Lee AT, Rees DC (2002) The *E. coli* BtuCD structure: a framework for ABC transporter architecture and mechanism. Science 296:1091-1098

Margolin W (2000) Green fluorescent protein as a reporter for macromolecular localization in bacterial cells. Methods 20:62-72

Menzel R, Roth J (1981) Purification of the *putA* gene product. A bifunctional membrane-bound protein from *Salmonella typhimurium* responsible for the two-step oxidation of proline to glutamate. J Biol Chem 256:9755-9761

Moody JE, Millen L, Binns D, Hunt JF, Thomas PJ (2002) Cooperative, ATP-dependent association of the nucleotide binding cassettes during the catalytic cycle of ATP-binding cassette transporters. J Biol Chem 277:21111-21114

Mourez M, Hofnung M, Dassa E (1997) Subunit interactions in ABC transporters: a conserved sequence in hydrophobic membrane proteins of periplasmic permeases defines an important site of interaction with the ATPase subunits. EMBO J 16:3066-3077

Muro-Pastor AM, Ostrovsky P, Maloy S (1997) Regulation of gene expression by repressor localization: biochemical evidence that membrane and DNA binding by the PutA protein are mutually exclusive. J Bacteriol 179:2788-2791

Nam T-W, Cho S-H, Shin D, Kim J-H, Jeong J-Y, Lee J-H, Roe J-H, Peterkofsky A, Kang S-O, Ryu S, Seok Y-J (2001) The *Escherichia coli* glucose transporter enzyme IICBGlc recruits the global repressor Mlc. EMBO J 20:491-498

Notley-McRobb L, Ferenci T (2000) Substrate specificity and signal transduction pathways in the glucose-specific enzyme II (EIIGlc) component of the *Escherichia coli* phosphotransferase system. J Bacteriol 182:4437-4442

Ostrovsky de Spicer P, Maloy S (1993) PutA protein, a membrane-associated flavin dehydrogenase, acts as a redox-dependent transcriptional regulator. Proc Natl Acad Sci USA 90:4295-4298

Ostrovsky de Spicer P, O'Brien K, Maloy S (1991) Regulation of proline utilization in *Salmonella typhimurium*: a membrane-associated dehydrogenase binds DNA *in vitro*. J Bacteriol 173:211-219

Panagiotidis CH, Boos W, Shuman HA (1998) The ATP-binding cassette subunit of the maltose transporter MalK antagonizes MalT, the activator of the *Escherichia coli mal* regulon. Mol Microbiol 30:535-546

Parkinson JS, Kofoid EC (1992) Communication modules in bacterial signaling proteins. Annu Rev Genet 26:71-112

Peist R, Koch A, Bolek P, Sewitz S, Kolbus T, Boos W (1997) Characterization of the *aes* gene of *Escherichia coli* encoding an enzyme with esterase activity. J Bacteriol 179:7679-7686

Plumbridge J (1998) Control of the expression of the *manXYZ* operon in *Escherichia coli*: Mlc is a negative regulator of the mannose PTS. Mol Microbiol 27:369-380

Plumbridge J (1999) Expression of the phosphotransferase system both mediates and is mediated by Mlc regulation in *Escherichia coli*. Mol Microbiol 33:260-273

Plumbridge J (2000) A mutation which affects both the specificity of PtsG sugar transport and the regulation of *ptsG* expression by Mlc in *Escherichia coli*. Microbiology 146:2655-2663

Plumbridge J (2001) DNA binding sites for the Mlc and NagC proteins: regulation of *nagE*, encoding the N-acetylglucosamine-specific transporter in *Escherichia coli*. Nucl Acid Res 29:506-514

Plumbridge J (2002) Regulation of gene expression in the PTS in *Escherichia coli*: the role and interactions of Mlc. Curr Opin Microbiol 5:187-193

Postma PW, Lengeler JW, Jacobson GR (1996) Phosphoenolpyruvate:carbohydrate phosphotransferase system. In: Neidhardt FC et al. (eds) *Escherichia coli* and *Salmonella typhimurium*; cellular and molecular biology, American Society of Microbiology, Washington, DC, pp 1149-1174

Reidl J, Boos W (1991) The *malX malY* operon of *Escherichia coli* encodes a novel enzyme II of the phosphotransferase system recognizing glucose and maltose and an enzyme abolishing the endogenous induction of the maltose system. J Bacteriol 173:4862-4876

Reyes M, Shuman HA (1988) Overproduction of MalK protein prevents expression of the *Escherichia coli mal* regulon. J Bacteriol 170:4598-4602

Richet E, Raibaud O (1989) MalT, the regulatory protein of the *Escherichia coli* maltose system, is an ATP-dependent transcriptional activator. EMBO J 8:981-987

Samanta S, Ayvaz T, Reyes M, Shuman HA, Chen J, Davidson AL (2003) Disulfide cross-linking reveals a site of stable interaction between C-terminal regulatory domains of the two MalK subunits in the maltose transport complex. J Biol Chem 278:35265-35271

Schirmer T, Keller TA, Wang YF, Rosenbusch JP (1995) Structural basis for sugar translocation through maltoporin channels at 3.1 Å resolution. Science 267:512-514

Schlegel A, Danot O, Richet E, Ferenci T, Boos W (2002) The N terminus of the *Escherichia coli* transcription activator MalT is the domain of interaction with MalY. J Bacteriol 184:3069-3077.

Schleif R (1996) Two positively regulated systems, *ara* and *mal*. In: Neidhardt FC et al. (eds) *Escherichia coli* and *Salmonella typhimurium*; cellular and molecular biology. vol. 1, American Society of Microbiology, Washington, DC, pp 1300-1309

Schreiber V, Richet E (1999) Self-association of the *Escherichia coli* transcription activator MalT in the presence of maltotriose and ATP. J Biol Chem 274:33220-33226

Schreiber V, Steegborn C, Clausen T, Boos W, Richet E (2000) A new mechanism for the control of a prokaryotic transcriptional regulator: antagonistic binding of positive and negative effectors. Mol Microbiol 35:765-776

Schwartz M (1987) The maltose regulon. In: Neidhardt FC et al. (eds) *Escherichia coli* and *Salmonella typhimurium*: cellular and molecular biology. vol. 2, American Society of Microbiology, Washington D. C., pp 1482-1502

Seitz S, Lee SJ, Pennetier C, Boos W, Plumbridge J (2003) Analysis of the interaction between the global regulator Mlc and EIIBGlc of the glucose-specific phosphotransferase system in *Escherichia coli*. J Biol Chem 278:10744-10751

Shapiro L, Losick R (2000) Dynamic spatial regulation in the bacterial cell. Cell 100:89-98

Shin D, Lim S, Seok YJ, Ryu S (2001) Heat shock RNA polymerase (Eσ32) is involved in the transcription of *mlc* and crucial for induction of the Mlc regulon by glucose in *Escherichia coli*. J Biol Chem 276:25871-25875

Surber MW, Maloy S (1998) The PutA protein of *Salmonella typhimurium* catalyzes the two steps of proline degradation via a leaky channel. Arch Biochem Biophys 354:281-287

Surber MW, Maloy S (1999) Regulation of flavin dehydrogenase compartmentalization: requirements for PutA-membrane association in *Salmonella typhimurium*. Biochim Biophys Acta 1421:5-18

Tanaka Y, Kimata K, Aiba H (2000) A novel regulatory role of glucose transporter of *Escherichia coli*: membrane sequestration of a global repressor Mlc. EMBO J 19:5344-5352

Vandromme M, Gauthier-Rouviere C, Lamb N, Fernandez A (1996) Regulation of transcription factor localization: fine-tuning of gene expression. Trends Biochem Sci 21:59-64

Verdon G, Albers SV, Dijkstra BW, Driessen AJ, Thunnissen AM (2003) Crystal structures of the ATPase subunit of the glucose ABC transporter from *Sulfolobus solfataricus*: nucleotide-free and nucleotide-bound conformations. J Mol Biol 330:343-358

West AH, Stock AM (2001) Histidine kinases and response regulator proteins in two-component signaling systems. Trends Biochem Sci 26:369-376

Zdych E, Peist R, Reidl J, Boos W (1995) MalY of *Escherichia coli* is an enzyme with the activity of a βC-S lyase (cystathionase). J Bacteriol 177:5035-5039

Zeppenfeld T, Larisch C, Lengeler JW, Jahreis K (2000) Glucose transporter mutants of *Escherichia coli* K-12 with changes in substrate recognition of the IICBGlc and induction behavior of the *ptsG* gene. J Bacteriol 182:4443-4452

Zhu W, Becker DF (2003) Flavin redox state triggers conformational changes in the PutA protein from *Escherichia coli*. Biochemistry 42:5469-5477

Abbreviations

ABC transporter: $\underline{A}$TP-$\underline{B}$inding $\underline{C}$assette transporter
CAP: $\underline{c}$atabolite $\underline{a}$ctivator $\underline{p}$rotein
MBP: $\underline{m}$altose-$\underline{b}$inding $\underline{p}$rotein
PTS: $\underline{p}$hospho$\underline{t}$ransferase $\underline{s}$ystem

Böhm, Alex
    Department of Biology, University of Konstanz, Universitätsstrasse 10, 78457 Konstanz, Germany

Boos, Winfried
    Department of Biology, University of Konstanz, Universitätsstrasse 10, 78457 Konstanz, Germany
    winfried.boos@uni-konstanz.de

Regulation of carrier-mediated sugar transport by transporter quaternary structure

Kara B. Levine and Anthony Carruthers

Abstract

Protein mediated solute transport across cell membranes is catalyzed by a large, diverse family of polytopic integral membrane proteins called transporters. Transporters fall into two classes. The channels are characterized by high catalytic throughput approaching the diffusional limit. The carriers undergo slow, substrate-induced conformational changes resulting in lower rates of solute transport. Recent advances in channel and carrier structure emphasize the challenges presented to this field. Several transporters have been crystallized as monomeric transport proteins. These structures provide enormous insight into individual channel and carrier function. Other transporters have been crystallized as ordered oligomers in which each subunit presents an individual transport pathway. While revealing much about transporter function, it is unclear why these structures assemble as oligomers. Low-resolution analyses show that some transporters form oligomeric complexes within the lipid bilayer but may retain oligomeric structure or dissociate into monomeric forms in detergents. Genetic and biophysical analysis of a large number of transport proteins of a common subgroup (e.g. Major Facilitator Superfamily) indicates that specific proteins may form monomeric, dimeric, trimeric or tetrameric complexes. While the minimal MFS transport pathway appears to be a monomer containing 12 membrane-spanning domains, some transporters may require dimers to catalyze transport. Why such diversity when solute specificity could evolve by re-engineering a common catalytic scaffold? The answer may be related to function and the complexity of transport regulation. We describe the oligomeric structure of an MFS transporter GluT1 – the glucose transport protein of human red blood cells. We show that the kinetics of GluT1-mediated sugar transport and its regulation by intracellular nucleotides are determined by transporter oligomeric structure.

1 Introduction

Protein-mediated solute transport across cell membranes is mediated by a family of polytopic integral membrane proteins called transporters. The transporters are comprised of channels and carriers. Channels form membrane spanning pores, which, when activated by ligands, mechano-transduction, or by altered membrane potential, catalyze extremely high throughput of cations or anions ($\leq 10^7$ transport

Topics in Current Genetics, Vol. 9
E. Boles, R. Krämer (Eds.): Molecular Mechanisms Controlling Transmembrane Transport
DOI 10.1007/b97183 / Published online: 9 March 2004
© Springer-Verlag Berlin Heidelberg 2004

events per channel per sec; (Stein 1986)). Carriers catalyze relatively low rates of substrate-induced solute transport ($\leq 10^3$ events per sec) and seem not to present transport pathways that are simultaneously accessible to extra- and intracellular substrates (Stein 1986).

Some carriers catalyze active transport in which net solute transport against an electrochemical gradient is coupled to and driven by an exergonic reaction such as ATP-hydrolysis or co-solute transport down an electrochemical gradient. Other carriers and all small substrate channels are passive transporters providing alternative transport pathways which, in the absence of solute transformation by cellular metabolism, allow solutes to rapidly attain electrochemical equilibrium between cytoplasmic and interstitial water (Stein 1986).

The past ten years have seen great strides in our understanding of channel- and carrier-mediated transport. Several channels and carriers have yielded to structural analysis by X-ray crystallography and electron diffraction methodologies (Walz et al. 1997; Fu et al. 2000; Dutzler et al. 2002; Jiang et al. 2002, 2003; Locher et al. 2002; Toyoshima and Nomura 2002; Abramson et al. 2003; Bass et al. 2003; Huang et al. 2003; MacKinnon 2003; Toyoshima et al. 2003). These studies provide new insight into channel and carrier architecture and herald a new phase in our understanding of transporter function.

A related structural problem requiring systematic evaluation is the question of transporter quaternary structure. Channels appear to be comprised of single or multiple subunits which form monomeric or multimeric structures capable of catalyzing transport (Table 1, for review see Veenhoff et al. 2002). In some instances, e.g. the MIPs, each channel is comprised of a single polypeptide but is assembled to form a tetrameric structure (Manley et al. 2000; Duchesne et al. 2002). The functional advantages conferred by MIP tetramerization are unclear but may be less related to catalysis and more related to post-translational folding issues in which intermediates may be stabilized by MIP self-association. In other instances, e.g. the acetylcholine receptor, the channel is formed from four separate polypeptides assembled as an $\alpha_2,\beta,\gamma,\delta$ complex (Unwin 2003). Some carriers appear to be monomeric (functionally and structurally) while other closely related proteins appear to require oligomerization for full function (e.g. the mammalian glucose transport proteins GluT3 and GluT1 and the prokaryotic LacT and LacS proteins; see Table 1). Some carriers appear to transition between different oligomeric states depending on the specific conformations presented to available substrates (e.g. Na,KATPase; see Table 1). Yet other carriers are reported to adopt multimeric states only in the presence of substrate gradients (SGLT; Table 1). Some carriers appear to form hetero-oligomers with closely related proteins while others do not (e.g. SUT1, 2 and 3 *versus* GluT1, 3 and 4; Table 1). In some cases, transporter quaternary structure appears to be fixed and insensitive to transporter concentration (e.g. GluT1, AE1 and DAT – the mammalian glucose, anion, and dopamine transporters; see Table 1). In some cases, transporter oligomeric structure and functional cooperativity are coincident (e.g. GluT1 and the Ca and NaKATPases; Table 1).

A recent review has critically examined methodologies used to determine membrane protein oligomeric structure in membranes and in detergent micelles

(Veenhoff et al. 2002). This present review examines the oligomeric structure of one specific transport system in some detail - the erythroid glucose transporter GluT1. We examine how GluT1 function and regulation are affected by GluT1 quaternary structure. We conclude that each GluT1 protein is capable of catalyzing sugar transport but that the properties of GluT1-mediated sugar transport and transport regulation are critically dependent upon GluT1 quaternary structure.

2 Human red blood cell glucose transport

2.1 GluT1 quaternary structure

The facilitated diffusion of sugars across cell membranes is mediated by a family of integral membrane proteins called glucose transporters. At least ten mammalian glucose transporters have been characterized (Joost and Thorens 2001). Of these transporters, the erythroid transporter GluT1 (Mueckler et al. 1985) has been studied most extensively with respect to structure (Sogin and Hinkle 1978; Baldwin et al. 1982) and function (Naftalin and Holman 1977; Baker et al. 1978; Baker and Naftalin 1979; Krupka and Devés 1981; Helgerson and Carruthers 1987; Appleman and Lienhard 1989; Carruthers 1991; Carruthers and Helgerson 1991; Cloherty et al. 1996; Hamill et al. 1999; Sultzman and Carruthers 1999; Cloherty et al. 2001b).

The mechanism of GluT1-mediated glucose translocation is unknown. In the absence of stable GluT1-substrate intermediates for analysis, investigators have turned to the analysis of transport data in order to characterize intermediates in GluT1-mediated sugar transport. The most influential quantitative proposal resulting from this approach is the simple carrier or alternating conformer hypothesis (Widdas 1952; Vidaver 1966; Lieb and Stein 1974). According to this model, the transporter sequentially presents a sugar influx site (e_2) and then a sugar efflux site (e_1). Extracellular sugar binding to the influx site promotes a conformational change, resulting in the translocation of bound sugar to the cytosol and the creation of the sugar efflux site. Sugar binding to the efflux site promotes the reverse conformational change, causing translocation of sugar to the extracellular water and the regeneration of the sugar influx site. Conversion between e_1 and e_2 states in the absence of sugar occurs by a process called relaxation, but this is much slower than conversion in the presence of sugar (*7, 8*).

Table 1: Control of membrane transporters

Substrates	Transport System	Oligomeric state	Functional unit	Allostery	Regulation via oligomeric state	References
acylCoA	PMP70 (ABC transporter)	dimer/oligomer				(Imanaka et al. 2000)
ADP/ATP	Nucleotide (mitochondria)	Dimer Each monomer has 6 TM domains	dimer			(Hackenberg and Klingenberg 1980; Zimmermann and Neupert 1980; Klingenberg 1981)
Ammonium	AmtB (*E. coli*)	trimer (detergent)	unknown			(Blakey et al. 2002)
Ammonium	Rh proteins RhAG, RhCE and RHD (detergent solubilized human red cell)	hetero-tetramer ($\alpha2$-$\beta2$)	monomer			(Hartel-Schenk and Agre 1992; Eyers et al. 1994; Marini et al. 2000)
Anions	AE1 (RBC)	dimers & tetramers			synthesized as tetramers becomes dimer at plasma membrane	(Hymel et al. 1985; Casey and Reithmeier 1991; Sekler et al. 1995; Hanspal et al. 1998)
Anions	AE2	homo-oligomer				(Zolotarev et al. 1999)
Anions	CFTR	dimer				(Ko et al. 1993; Eskandari et al. 1998)
Ca	CaATPase (Sarcoplasmic Reticulum)	monomer (detergent) & higher (membrane)	monomer (ATPase)			(Martin and Tanford 1984)
Ca	CaATPase	dimer/tetramer		at high [ATP]. e2 state forms multimers e1 state forms monomers	Cooperative CaATPase and phospholambam (PL) display reciprocal aggregation (ATPase)$_n$.PL$_i$ is inactive, (ATPase)$_i$.PL$_5$ is active	(Yeh et al. 1978; Martin and Tanford 1984; Chadwick et al. 1987; Boldyrev 1988, 1995; Boldyrev and Quinn 1994; Thomas et al. 1998)
Ca	Channels	heteroligomers				(Nargeot et al. 1997)
Ca	P2X receptor	homo & hetero oligomers				(Surprenant 1996)

Table 1: Continued

Substrates	Transport System	Oligomeric state	Functional unit	Allostery	Regulation via oligomeric state	References
Ca	DHP-sensitive Ca Channels (muscle)	hetero-oligomers				(Gutierrez et al. 1991)
Ca	Trp1 & 3	hetero-oligomers				(Imanaka et al. 2000)
cations	h5HT3A	homo-oligomer				(Van Hooft and Vijverberg 1997; Eskandari et al. 1998)
cations	VDAC channels (S. cerevisiae)		monomer			(Peng et al. 1992)
cations	AMPA receptors	hetero-oligomers of GluR1-4 subunits				(Cotton and Partin 2000)
cations/anions/nonelectrolytes	connexin 50	hexamer				(Eskandari et al. 1998; Kocabas et al. 2003)
citrate	CitS	dimer				(Heuberger et al. 2002)
Cystine Renal	NBAT	hetero-oligomers				(Wang and Tate 1995)
Dopamine Na/Cl cotransport	DAT	homo-oligomer	dimer		oligomers formed in ER and trafficked to plasma membrane	(Norgaard-Nielsen et al. 2002; Sorkina et al. 2003)
Fructose	Frc permease (E. coli)	IIB dimers				(Charbit et al. 1996)
Glc	GluT1	dimer/tetramer	monomer	tetramer		(Sogin and Hinkle 1978; Baldwin et al. 1982; Lundahl et al. 1991; Pessino et al. 1991; Hebert and Carruthers 1992)
Glc	Glut3		monomer			(Burant and Bell 1992)
Glc	IIGlc (bacteria)	dimer				(Werner and Reithmeier 1985)
glycerol	aquaglyceroporin	monomer (detergent) & higher (membrane)	monomer		TM5 of α interacts with TM1 of β	(Manley et al. 2000; Borgnia and Agre 2001; Duchesne et al. 2002)

Table 1: Continued

Substrates	Transport System	Oligomeric state	Functional unit	Allostery	Regulation via oligomeric state	References
glycerol phosphate	GlpT (E. coli)	dimer				(Larson et al. 1982)
glycerol phosphate	GlpT (E. coli)	monomer				(Huang et al. 2003)
H	yeast H ATPase	monomer in bilayers lacking ceramide multimers in bilayers containing ceramide				(Lee et al. 2002)
H	HATPase (Neurospora)	hexamer				(Erni 1986)
H_2O	aquaporin	tetramer				(Eskandari et al. 1998)
K	K channels	multisubunit complexes				(Deutsch 2002)
K	KefC (E. coli)	multimer				(Douglas et al. 1994)
lactose/H	LacS (S. thermophilus)	dimer			subunit contacts at extracellular regions of TM 5 & 8 and intracellular regions of TM 6 & 7	(Spooner et al. 2000; Heuberger et al. 2002)
lactose/H	lac permease (E. coli)	monomer	monomer			(Sahin-Toth et al. 1994)
Maltose	maltose transport complex	malF:malG:malK 1:1:2				(Kennedy and Traxler 1999)
mannitol	mannitol permease (E. coli)	heterodimer				(Jacobson 1992; Lintschinger et al. 2000)
Mannose	E. coli mannose permease	IIC:IID 1:2				(Rhiel et al. 1994)
monoamines		oligomers	monomer			(Horschitz et al. 2003)

Table 1: Continued

Substrates	Transport System	Oligomeric state	Functional unit	Allostery	Regulation via oligomeric state	References
Multiple substrates (cationic, neutral & anionic)	AcrB mulitdrug exporter	trimer	monomer			(Murakami et al. 2002, 2004; Elkins and Nikaido 2003; Yu et al. 2003)
Na & Glc	SGLT	monomer; In presence of Na gradient = tetramer. In absence of Na gradient but presence of Glc = monomer	dimer (phlorizin binding)			(Gerardi-Laffin et al. 1993; Jette et al. 1996, 1997; Eskandari et al. 1998)
Na & H	NHE1-6	dimer, oligomers				(Kinsella et al. 1998; Williams 2000)
Na & K	NaKATPase	hetero-oligomer. (α,β,γ) oligomer in membrane. $(\alpha,\beta)_4$	α Monomer (ATPase) heterodimer (α,β) required for ouabain binding	Cooperative at high [ATP]. e2 forms multimers. e1 forms monomers	oligomer needed for folding. α subunit monomer has high affinity for ATP; multimer has low affinity for ATP. one half binds ATP, the other is intermediate	(Craig 1982; Askari and Huang 1984; Chadwick et al. 1987; Boldyrev 1988, 1995; Norby 1988; Horowitz et al. 1990; Beguin et al. 1998a, 1998b; Lopina 2000; Taniguchi et al. 2001)
Na/PO4	rat kidney	monomers, dimers & tetramers				(Jette et al. 1996; Xiao et al. 1997)
Norepinephrine	hNET	homo-oligomer				(Delisle et al. 1994)
Peptide/H	peptide symporter	tetramer				(Kocabas et al. 2003)
Phosphate	PIC (mitochondrial)	homodimer Each monomer has 6 TM domains	dimer			(Schroers et al. 1998)
serotonin	h5HT3	homo-oligomer				(Boll and Daniel 1995)
Sucrose	Sucrose permease (soybean cotyledon)	homo-dimer & trimer				(Falk 2000)
sucrose/H	SUT1, 2 & 3	homo & hetero oligomers				(Overvoorde et al. 1997)
Tetracycline	TetA	trimeric	uncertain			(Yin et al. 2000)
water	aquaporin	tetramer	monomer			(Borgnia and Agre 2001; Reinders et al. 2002)
Xylose	XylP	dimer			TM5 of α interacts with TM1 of β	(Duchesne et al. 2002; Heuberger et al. 2002)

The simple carrier hypothesis was later modified to account for unequal affinities for glucose at the influx and efflux sites (Hankin et al. 1972; Karlish et al. 1972). Several groups have demonstrated that $K_{m(app)}$ and V_{max} for sugar efflux into sugar-free medium (zero-trans efflux) are five-tenfold greater than $K_{m(app)}$ and V_{max} for sugar uptake into sugar-depleted cells (zero-trans uptake) (Hankin et al. 1972; Karlish et al. 1972; Baker and Naftalin 1979; Carruthers and Melchior 1983; Lowe and Walmsley 1986; Wheeler 1986; Carruthers 1989, 1991; Cloherty et al. 1996). An asymmetric simple carrier model can, in principle, account for this behavior (Stein 1986; Carruthers 1991).

This general mechanism for carrier-mediated sugar transport has enjoyed broad success in describing the catalytic behavior of many other carrier transfer systems (for reviews see Stein 1986). Most carrier-mediated transport processes are consistent with a mechanism in which the carrier isomerizes between e2 and e1 states. In many instances, the quantitative agreement between predicted and experimental catalytic behavior is so close as to preclude the need for alternative explanations. This success - when coupled with biochemical characterizations of carrier e1 and e2 states through ligand binding or differential chemical modification studies - strongly supports the hypothesis that membrane carriers alternate or rock between import and export states. The recent high resolution structures of the lacY lactose/H^+ symporter (Abramson et al. 2003) and the GlpT glycerol phosphate antiporter (Huang et al. 2003) suggest a single translocation pathway for substrate that is alternately accessible to extracellular and intracellular water.

Why then, would we consider other possible carrier transfer mechanisms? Erythrocyte glucose transport and anion exchange transport have been studied so extensively that multiple, significant deviations from simple-carrier-mediated transport behavior have been catalogued and cannot be ignored (Harris 1964; Hankin et al. 1972; Lieb and Stein 1977; Baker and Naftalin 1979; Carruthers and Melchior 1985; Carruthers 1991; Cloherty et al. 1996)

Baker and Widdas (1973) proposed a fundamentally different model for erythrocyte glucose transport that was later termed the fixed-site carrier (Naftalin and Holman 1977). According to this model, the transporter *simultaneously* presents a sugar efflux site *and* a sugar influx site. The fixed-site carrier is, thus, capable of forming an "extracellular sugar-carrier-intracellular sugar" ternary complex. Several studies from this laboratory demonstrate that the human erythrocyte sugar transporter presents e1 and e2 sites simultaneously (Helgerson and Carruthers 1987; Carruthers 1991; Carruthers and Helgerson 1991; Hebert and Carruthers 1992; Coderre et al. 1995; Cloherty et al. 1996, 2001b; Hamill et al. 1999). This appears to fly in the face of reason. How could a carrier mechanism permit relatively large substrates to simultaneously pass in opposite directions through a single pathway and yet retain specificity?

Clues to the answer to this question were obtained from analyses of GluT1 quaternary structure (Hebert and Carruthers 1991, 1992; Zottola et al. 1995). These studies demonstrated that the transporter is a complex of four GluT1 proteins. More precisely, the glucose carrier is a dimer of GluT1 dimers. In any given dimer, when one subunit presents an e2 (import) state, its neighbor must present an e1 (export) state and *vice versa*. Recent studies demonstrate negative cooperativity between sugar import and sugar export sites within each dimer and positive cooperativity between dimers (Hamill et al. 1999; Cloherty et al. 2001b). Thus, the glucose carrier of red cells is a fixed-site carrier formed from individual alternating carrier subunits whose functional states are constrained by subunit-subunit interactions.

GluT1 reduction by extracellular reductants causes the transporter to dissociate into GluT1 dimers in which each GluT1 protein functions independently of its neighbor as a simple carrier and with lower catalytic turnover (Appleman and Lienhard 1985; Hebert and Carruthers 1992; Zottola et al. 1995). GluT1 reduction may result in the loss of a single intramolecular disulfide bridge between GluT1 cysteine residues 347 and 421, which stabilizes GluT1 oligomeric structure (Zottola et al. 1995).

These observations are consistent with target size analyses suggesting GluT1 forms a tetramer in red cell membranes (Jung et al. 1980) and with freeze-fracture electron microscopy suggesting that GluT1 forms a dimer upon purification under reducing conditions (Sogin and Hinkle 1978; Cloherty et al. 2001a) and a tetramer in the absence of reductant (Cloherty et al. 2001a). At this time, monomeric GluT1 has not been demonstrated physically in reconstituted or biological membranes. Some studies have suggested that octylglucoside-solubilized, reduced, purified human GluT1 is monomeric (Lundahl et al. 1991). The significance of this is unclear, however, since detergents with short (C8) alkyl chains are known to denature the native structure of the erythrocyte anion transporter while detergents with longer (C12) alkyl chains preserve transporter native structure (Casey and Reithmeier 1991). Our dynamic light scattering studies suggest the opposite for GluT1. Octylglucoside-solubilized GluT1 is a stable tetramer but dodecylmaltoside-solubilized GluT1 rapidly dissociates at room temperature to form GluT1 dimers (Hamill et al. 1999).

Studies of GluT1 expressed in *Xenopus* oocytes indicate that Cys-less GluT1 is functional (Wellner et al. 1995) and that wild type GluT1 forms a reductant-insensitive, nontetrameric (possibly dimeric) state (Zottola et al. 1995). It is not possible to state definitively that the catalytic turnover of wild type and Cys-less GluT1 expressed in *Xenopus* oocytes approaches that observed for GluT1 in human red cells. Neither are data available to assess whether GluT1 expressed in *Xenopus* oocytes presents e1 and e2 sites simultaneously as in the red cell or, like reduced GluT1, functions as a simple carrier. However, these observations do establish that fundamental GluT1 function (sugar translocation) does not require GluT1 assembly into a homotetramer.

These and other studies have led to the suggestion that each GluT1 protein functions in isolation as a simple carrier but that cooperative interactions between GluT1 proteins within the GluT1 tetramer cause the transport complex to present two e1 sites and two e2 sites at any instant (Hebert and Carruthers 1992; Zottola et al. 1995). Thus, while each GluT1 protein is a fundamental simple carrier unit, GluT1 quaternary structure may be the major determinant of overall carrier kinetic behavior.

2.2 Determinants of GluT1 quaternary structure

The primary sequence and putative topology of the human GluT1 protein are illustrated in Figure 1. GluT1 quaternary structure has been investigated by freeze-fracture electron microscopy of purified, reconstituted GluT1 (Sogin and Hinkle 1978; Cloherty et al. 2001a), by chemical cross-linking (Hebert and Carruthers 1992) and by hydrodynamic size analysis of detergent-solubilized GluT1 (Hebert and Carruthers 1992; Zottola et al. 1995; Hamill et al. 1999). The available data indicates that reduced GluT1 forms smaller membrane particles (GluT1 dimers) than does nonreduced GluT1 (tetramers) and that cholic acid- or octylglucoside-solubilized GluT1 form larger protein/lipid/detergent micelles (hydrodynamic radius, R_h = 8 nm) in the absence of reductant than in its presence (R_h = 6 nm). Exhaustive hydrodynamic analysis of cholic acid-solubilized GluT1 indicates that the larger (nonreduced) 8 nm R_h particle contains four copies of GluT1 while the smaller (reduced) 6 nm R_h particle contains two copies of GluT1 (Hebert and Carruthers 1992; Zottola et al. 1995; Hamill et al. 1999). Some caution should be exercised when interpreting results obtained with detergents, however, since our own studies demonstrate that the stability of GluT1/lipid/detergent micelles is determined by the nature of the detergent. Some detergents (e.g. dodecyl-maltoside) are almost as effective as reductant in causing dissociation of 8 nm GluT1 particles into 6 nm particles (Hamill et al. 1999; Cloherty et al. 2001a)

The interactions that determine GluT1 quaternary structure have been investigated by chemical and molecular approaches. GluT1 quaternary structure is reductant-sensitive yet when nonreduced GluT1 is resolved by denaturing electrophoresis, its mobility is identical to that of denatured reduced GluT1 (monomeric) and is insensitive to reductant (Hebert and Carruthers 1992). These findings suggest the presence of an intramolecular GluT1 disulfide bridge not an intermolecular covalent linkage. Differential alkylation studies demonstrate that reduced, denatured GluT1 exposes six –SH groups (GluT1 contains six cysteine residues) while nonreduced GluT1 exposes only two free thiols. GluT1 is comprised of two domains (amino and carboxyl terminal domains each containing six membrane spanning alpha-helices) linked by a large cytoplasmic loop. Both N- and C-terminal domains of nonreduced GluT1 contain one free and two occluded thiols (Zottola et al. 1995). Two-dimensional (nonreducing followed by reducing) electrophoresis of chymotrypic digests of GluT1 indicates that the GluT1 carboxyl-terminal domain contains a putative intramolecular disulfide bridge between GluT1 cysteines 347 and 421 (Fig. 2 – Graybill and Carruthers, unpublished). Mutagenesis of GluT1 cysteines (cysteine to serine) demonstrates that GluT1 cysteine residues 347 and 421 are essential for exposure of non-reduced GluT1 specific epitopes when mutants are stably expressed in CHO cells (Zottola et al. 1995).

Although GluT1 displays close primary structural homology to the insulin-sensitive glucose transporter GluT4 (Birnbaum 1992) and to the neuron-specific transporter GluT3 (Kayano et al. 1990), it does not form heterocomplexes with these transporters when expressed in 3T3-L1 adipocytes or *Xenopus* oocytes (Burant and Bell 1992). Construction of a GluT1-GluT4 chimeric protein (GluT1-4c) in which the cytoplasmic carboxy-terminal 50 amino acids (see Fig. 1) of

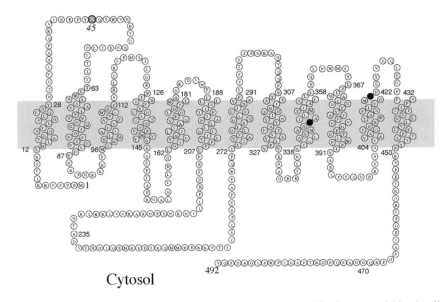

Fig. 1. Primary sequence and putative membrane topology of the human red blood cell membrane protein GluT1. GluT1 contains 12 putative transmembrane domains based on hydropathy and hydrophobic moment analysis (Mueckler et al. 1985). Individual amino acids are indicated using the one letter code. Amino acids at membrane boundaries are numbered. The membrane bilayer is indicated by the grey rectangle. The site of N-linked glycosylation (Asn 45) is found in loop 1. Cysteine residues 347 and 421 are indicated by the filled circles. These residues are thought to form an intramolecular disulfide bridge.

GluT1 are substituted with the equivalent domain of GluT4 permits subsequent immunoprecipitation of the chimera with a GluT4 carboxyl-terminal specific antibody. Overexpression of the chimera in GluT1 expressing CHO cells allows co-immunoprecipitation of endogenous GluT1 and the GluT1/GluT4 chimera using the GluT4-specific antibody (Pessino et al. 1991). Endogenous GluT1 is not pulled down by anti-GluT4 antibody in cells transfected with wild type GluT4 (Pessino et al. 1991). These results indicate that GluT1 residues N-terminal to the carboxyl-terminal 50 residues mediate GluT1-GluT1 self-association. When the entire carboxyl-terminal half of GluT1 plus the large, middle, cytoplasmic loop (GluT1 residues 207 - 492) are substituted by equivalent GluT4 sequence, expression of this GluT1n-4c chimera still permits co-immunoprecipitation of endogenous GluT1 using anti-GluT4 IgGs (Pessino et al. 1991). These findings demonstrate that primary structural determinants of GluT1 homo-oligomerization reside in the GluT1 N-terminal domain.

GluT1-4c/detergent/lipid micelles obtained from cholic acid solubilized cells expressing GluT1-4c and GluT1 are resolved as particles of Stokes radius (R_h) = 8 nm. GluT1n-4c/GluT1/detergent/lipid micelles are resolved as particles of R_h = 6 nm. Recalling that detergent solubilized, purified tetrameric and dimeric GluT1 are resolved as particles of R_h = 8 and 6 nm respectively, these findings suggest

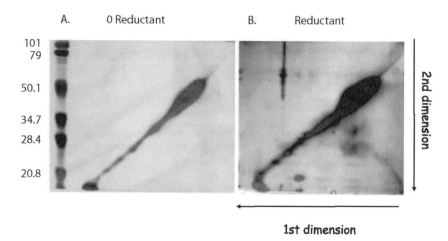

Fig. 2. Two-dimensional (nonreducing followed by A: nonreducing or B: reducing) SDS PAGE of chymotryptic digests of purified GluT1. GluT1 chymotryptic digests (100 μg GluT1, 2 μg chymotrypsin) were resolved on 15% acrylamide gels in the absence of reductant then each lane was excised and soaked in tank buffer lacking (A) or containing (B) 50 mM DTT. The below diagonal spots in B were excised and sequenced by Mass Spectrometry at the UMMS proteomics facility. The two vertically separated spots running at ≈ 20.8 kDa correspond to Glut1 fragments 363 – 492 and 251 – 350. Because GluT1 cys429 is accessible to exofacial alkylating agents (Zottola et al. 1995), this result suggests that GluT1 cysteines 347 and 421 are linked by an internal disulfide bridge (Graybill and Carruthers, unpublished results).

that the N-terminal half of GluT1 contains GluT1 dimerization determinants while the carboxyl-terminal half of GluT1 contains a GluT1 tetramerization motif. This hypothesis is further supported by the observation that the intramolecular disulfide bridge whose formation promotes GluT1 tetramerization lies in the carboxyl-terminal half of GluT1 (Zottola et al. 1995).

2.3 Is quaternary structure the sole determinant of GluT1 function?

Neither the asymmetric simple carrier nor the fixed-site carrier can account for the steady-state sugar transport properties of human erythrocytes (Carruthers 1991; Cloherty et al. 1996). Unidirectional sugar efflux from red cells into medium containing saturating sugar levels is characterized by a $K_{m(app)}$ that is almost one order of magnitude lower than that predicted by either carrier mechanism (Hankin et al. 1972; Baker and Naftalin 1979; Carruthers 1991; Cloherty et al. 1996). $K_{m(app)}$ for unidirectional 3-O-methylglucose (3OMG) uptake by red cells containing saturating 3OMG levels is significantly greater than $K_{i(app)}$ for extracellular 3OMG inhibition of net 3OMG exit (Baker and Widdas 1988). These and related observations have prompted Naftalin and co-workers (Naftalin and Holman 1977; Baker and Naftalin 1979; Naftalin 1997, 1998) to suggest that human erythrocyte sugar transport may be intrinsically symmetric but that intracellular sugar complexation or the presence of a cytosolic unstirred sugar layer just below the plasma membrane could give rise to inaccuracies in initial transport rate determinations, thereby leading to apparent transport asymmetry and complexity.

This hypothesis is supported by several observations. 1) Human erythrocyte sugar transport asymmetry can be reduced or even lost in red blood cell ghosts (Jung et al. 1971; Carruthers and Melchior 1983; Carruthers 1986b, 1986a; Carruthers and Melchior 1986; Carruthers and Helgerson 1991; Cloherty et al. 1996). 2) GluT1-mediated sugar transport in rabbit erythrocytes, in metabolically poisoned pigeon erythrocytes, and in rat adipocytes is symmetric (Regen and Morgan 1964; Taylor and Holman 1981; Simons 1983). Rat erythrocyte sugar transport has been reported to be either symmetric (Helgerson and Carruthers 1989; Whitesell et al. 1989) or asymmetric (Whitesell et al. 1989). 3) Transport measurements in human erythrocytes indicate that the transporter is asymmetric while CCB binding measurements indicate transporter symmetry (Cloherty et al. 1996). 4) Not all intracellular D-glucose is freely diffusible in erythrocytes (Naftalin et al. 1985; Helgerson and Carruthers 1989; Naftalin and Rist 1991). 5) Reconstituted, purified glucose transporter displays 1.4-fold catalytic asymmetry versus five-tenfold asymmetry *in situ* (Carruthers and Melchior 1984; Carruthers 1986b). 6) If transport were rate-limited by a cytosolic diffusion barrier, net sugar uptake or exit at low sugar concentrations ($<< K_m$) should be a bi-exponential process reflecting the movement of sugar through two compartments in series. Our studies confirm this prediction showing that net uptake is consistent with the rapid filling of a small intracellular compartment followed by slower filling of a larger compartment (Cloherty et al. 1995).

Closer analysis indicates that the rapid component of sugar uptake cannot result from the simple filling of a small aqueous compartment because the absolute volume of the rapid space falls in a saturable manner with increasing intracellular sugar concentration (Heard et al. 2000). The "rapid compartment" may, thus, represent sugar binding to a high-affinity, intracellular sugar binding complex.

These findings lead to the proposal (Heard et al. 2000) that newly imported sugar interacts rapidly with an intracellular sugar binding complex from which it dissociates slowly into bulk cytosol. This complex was hypothesized to consist of GluT1 and two nucleotide-sensitive enzymes of the glycolytic pathway – hexokinase and glyceraldehyde phosphate dehydrogenase (Lachaal et al. 1990; Lachaal and Jung 1993). Subsequent studies from this laboratory (Heard et al. 1998) indicated that neither protein forms a specific complex with GluT1. Neither does their association with the cell membrane promote glucose complexation (Heard et al. 1998). Further experiments demonstrated that the sugar binding complex is not a membrane constituent that interacts with GluT1 with high affinity. The sugar binding complex persists in red cell membranes depleted of peripheral membrane proteins and co-purifies with the glucose transport protein GluT1 (Heard et al. 2000). These findings suggest that the "glucose binding complex" is none other than the glucose transporter.

Fig. 3 (overleaf). Model for ATP-regulation of sugar transport. **(A)** The glucose transporter (viewed from above the cell membrane) is a complex of four GluT1 proteins (subunits). Two subunits present sugar uptake (e2) conformations while the remaining subunits present sugar exit (e1) conformers. Upon exposure to extracellular reductant, tetrameric GluT1 dissociates into two GluT1 dimers. **(B)** The transporter is sectioned (normal to the plane of the bilayer) through the catalytic center of two of the four subunits. Each subunit contains a water-filled channel or vestibule penetrating the interior of the carrier and extending either to the cytoplasmic water (the e1 configuration as observed for LacY (Abramson et al. 2003) and GlpT (Huang et al. 2003)) or to the interstitium (the e2 configuration). ATP binds only to the nonreduced (tetrameric) transporter. In the nonreduced carrier, when one subunit presents an e1 conformation, its neighbor must present the e2 conformer. Thus, each dimer in the nonreduced carrier consists of an e2.e1 pair or of an e1.e2 pair. Reduction of tetrameric GluT1 causes the complex to dissociate into 2 dimers of GluT1. Now each subunit can isomerize between e1 and e2 states independently of its partner subunit and the dimer may adopt e1.e1, e1.e2, e2.e1, or e2.e2 states. Sugar may bind to e1 and e2 conformers to form the transport competent (catalytically active) states eS1 or eS2. For simplicity, only one subunit of each dimer in this figure is shown to undergo transport. However, both subunits can catalyze sugar transport. In the absence of ATP, the cytosolic domains of the transporter present a relaxed conformation through which any newly imported sugar molecule can gain unrestricted access to bulk cytosol. When ATP binds to tetrameric GluT1, cytosolic domains of the transporter undergo a conformational change, forming a cage around the water-filled vestibule leading from the exit site to bulk cytosol and evolving a sugar binding site at this location. Sugar binding at this noncatalytic site forms the e1Sb or e2Sb complex. A newly imported sugar now has a greater probability of binding to this non catalytic site or being recycled to the interstitium via translocation (eS1.e2 → eS2.e1 conformational change) than release into cytosol. For simplicity, only one subunit of each dimer in this figure is shown to bind sugar at this noncatalytic site. Both subunits can present the noncatalytic site in the presence of ATP. When tetrameric GluT1 dissociates in the presence of reductant, the nucleotide binding site is lost, the non-catalytic sugar binding site disappears, subunits within each dimer are functionally uncoupled and the transporter behaves as a simple carrier.

3 Glucose transporter quaternary structure and regulation of GluT1

Equilibrium binding of the nonmetabolizable sugar 3MG to red cell membranes and to purified reconstituted GluT1 is consistent with one high affinity ($K_{d(app)}$ = 200 μM) 3MG binding site per GluT1 protein (Heard et al. 2000). In the presence of ATP (2 mM), GluT1 3MG binding capacity increases to two sites per subunit (Heard et al. 2000). ATP half-maximally increases 3MG binding to GluT1 at 400 μM nucleotide. ATP hydrolysis is not required for modulation of 3MG binding to GluT1 because several nonhydrolyzable analogs of ATP can substitute for ATP in this action (Heard et al. 2000). GluT1 does not possess intrinsic ATP-ase activity

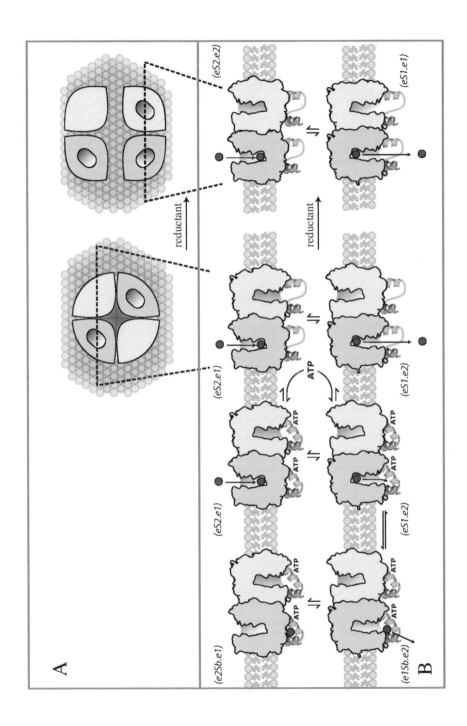

(Carruthers and Helgerson 1989). In transport experiments using 100 µM extracellular 3MG, these actions of ATP on GluT1 equilibrium 3MG binding are reflected as an ATP-dependent doubling of the size of the rapid component of sugar uptake.

When human red cells are exposed to extracellular reductant, net 3MG uptake is inhibited (Zottola et al. 1995), tetrameric GluT1 dissociates into GluT1 dimers (Zottola et al. 1995) and the size of the rapid component of sugar uptake is halved (Heard et al. 2000). The size of the rapid component of sugar uptake in ATP-free ghosts is restored by intracellular ATP (Heard et al. 2000) but not in the presence of extracellular reductant.

GluT1 is also a nucleotide binding protein binding ATP, ADP or AMP at a single site per GluT1 monomer (Carruthers and Helgerson 1989). ATP interaction with GluT1 reduces V_{max} and K_m for net sugar uptake. AMP and ADP competitively inhibit ATP binding to GluT1 but are unable to mimic the action of ATP on sugar transport (Carruthers and Helgerson 1989; Helgerson et al. 1989). When tetrameric GluT1 is exposed to reductant, ATP binding is also inhibited (Levine et al. 1998), the transporter is unable to respond to ATP and one of the two sugar binding sites is lost (Heard et al. 2000).

3.1 A model for nucleotide regulation of sugar transport

Figure 3 presents a model for regulation of GluT1 function by ATP and its dependence on GluT1 quaternary structure. Nonreduced GluT1 is a nucleotide binding protein that presents two sugar binding sites at the cytoplasmic surface. The "catalytic" sugar binding site is proposed to lie at the apex of a water-filled channel or vestibule penetrating the interior of the carrier and extending to the cytoplasmic water (Fig. 3). This site is the sugar export site mediating sugar efflux from cytosol to interstitium. The "noncatalytic" sugar binding site is ATP-dependent. This is a more peripheral vestibular site that lies between the export site and cytosol. This site does not catalyze sugar transport. However, when occupied by sugar, this site may limit diffusion of sugar from the export site to cytosol. The nucleotide binding site is exposed to the cytoplasm. Occupancy by ATP promotes a conformational change revealing the noncatalytic sugar binding site. Reduced GluT1, however, is unable to interact with ATP and loses the noncatalytic sugar binding site.

3.1.1 Catalytic (transport) sites

The tetrameric GluT1 complex is proposed to be a dimer of GluT1 dimers. Each dimer presents both a sugar import and a sugar export site at any instant. Since the GluT1 subunit can present import and export sites only alternately (Gorga and Lienhard 1982; Hebert and Carruthers 1992), this requires that when one subunit of the dimer presents an e2 conformation, its partner must present an e1 conformation (Fig. 3B). If one subunit undergoes an e2 → e1 conformational change, its neighbor must undergo the reverse, e1 → e2 conformational change. In this way, each GluT1 tetramer presents two sugar import (e2) and two sugar export (e1) sites at all times. Reduction of the GluT1 intramolecular disulfide (cys 347-421) by dithiothreitol results in dissociation of GluT1 tetramers into GluT1 dimers in which the obligate antiparallel arrangement of subunit function is now lost (Fig. 3B). Thus, reduced GluT1 dimers may present two e1 or two e2 conformations simultaneously. These changes in GluT1 subunit interactions have a profound impact on GluT1 function. The tetramer behaves as if a fixed site carrier while the dimer functions as a simple carrier (Hebert and Carruthers 1992).

3.1.2 Nucleotide binding sites

A further distinction between tetrameric and dimeric GluT1 is the presence of noncatalytic, nucleotide- and sugar-binding sites in the tetramer. Sequence analysis of peptides released upon proteolysis of azido ATP-photolabeled GluT1 indicates that GluT1 residues 301 through 364 form an integral component of the GluT1 ATP binding domain (Levine et al. 1998, 2001). GluT1 residues 307-327 and 338-358 are strongly hydrophobic, are predicted to form membrane-spanning domains (3) and are, therefore, unlikely to be accessible to cytosolic ATP. An adenylate kinase nucleotide binding homology domain (a hydrophobic pocket that flanks the adenine ribose and triphosphate chain of liganded Mg.ATP (Fry et al. 1986)) is positioned in the intervening (cytosolic) sequence between these membrane-spanning regions and is a stronger candidate for direct interaction with ATP. This GluT1 sequence (326–338) has been subjected to alanine scanning mutagenesis (Levine et al. 2001, 2002) and disruption of a putative, proton-sensitive salt-bridge (Glu329-Arg333/334) by mutagenesis (Levine et al. 2002) or by acidification (Cloherty et al. 2001b) profoundly impacts ATP binding to and regulation of GluT1.

3.1.3 Non catalytic (glucose-sensing) sugar binding sites

ATP binding to tetrameric GluT1 increases the sugar binding capacity of GluT1 from 0.8 to 1.7 mol 3MG per mol GluT1. In the absence of ATP, 3MG binding to GluT1 proceeds with relatively high affinity ($K_{d(app)}$ = 270 µM). In the presence of ATP, $K_{d(app)}$ for 3MG binding increases to 690 µM. Assuming that binding to the first site is unaffected by ATP, these data are also consistent with ATP-dependent exposure of a second site of significantly lower 3MG affinity ($K_{d(app)}$ = 1760 µM). The 3MG binding data do not permit distinction between two sites of identical affinity or two sites of widely different affinity. Cytochalasin B (an e1-reactive ligand) abolishes GluT1 3MG binding while extracellular maltose (an e2-reactive ligand) reduces GluT1 3MG binding by 66 ± 20% (Heard et al. 2000). This suggests the interesting possibility that both sugar binding sites are sensitive to intracellular inhibitor whereas only one site can be inhibited by extracellular inhibitor.

We propose that the ATP-insensitive sugar binding site represents the catalytic center of each GluT1 subunit whereas the ATP-dependent site is a non catalytic, glucose-sensing or regulatory site that is rendered inaccessible in the presence of cytochalasin B. This site is exposed at the cytoplasmic surface of GluT1 (Heard et al. 2000) and, based upon a presumed homology with the lacY structure, may be formed at the mouth of or within the relatively large, water-filled vestibule leading from cytoplasm to the exit (e1) site. Occupancy of the site is dependent upon cytosolic sugar levels and may inhibit net sugar transport (see below).

Maltose is expected to lock GluT1 in the e2 state and, thus, competitively inhibit 3MG binding to e2 but not to the noncatalytic, cytoplasmic site. Cytochalasin B locks GluT1 in the e1 state and competitively inhibits 3MG binding to e1. Because cytochalasin B also penetrates the vestibule, 3MG binding to the noncatalytic, cytoplasmic site is also inhibited.

ATP does not increase V_{max} for net sugar exit from human red blood cells (Carruthers and Melchior 1983) indicating that the ATP dependent, second intracellular sugar binding site does not contribute quantitatively to net sugar transport. Two findings suggest that sugar-occupancy of the ATP-dependent site modulates transport and, thereby, serves a regulatory "sensor" role. 1) Inhibition of net sugar uptake by intracellular sugar occurs with high affinity in the presence of ATP but with low affinity in the absence of nucleotide (Carruthers 1986a). 2) $K_{m(app)}$ for net sugar exit is increased by intracellular ATP (Carruthers and Melchior 1983; Carruthers 1986a).

3.1.4 Nucleotide dependent GluT1 conformational changes

ATP binding to tetrameric GluT1 promotes significant conformational changes in GluT1 cytoplasmic domains. In the presence of ATP, the GluT1 carboxyl-terminal domain is relatively less-accessible to antibodies (C-Ab) directed against the C-terminal 13 amino acids of GluT1. Equilibrium binding of C-Ab is unaffected by ATP but the time required to achieve equilibrium binding is significantly extended (Carruthers and Helgerson 1989). This indicates a motional flexibility in the GluT1 carboxyl terminus and that this domain may alternate between several conformational states - one of which is less accessible to cytoplasm in the presence of ATP. The large cytoplasmic loop linking GluT1 transmembrane domains 6 and 7 also appears to undergo ATP-induced conformational change. Middle loop directed antibodies interact very differently with GluT1 in the presence of cytoplasmic ATP (Afzal et al. 2002).

GluT1 exposure to trypsin results in significant carrier proteolysis (Carruthers and Helgerson 1989). This proteolysis, which releases GluT1 cytoplasmic domains (Mueckler et al. 1985), is significantly inhibited by the presence of cytoplasmic ATP (Cloherty et al. 2001a). GluT1 equilibrium quaternary structure and the rate of reductant-promoted tetrameric GluT1 dissociation into GluT1 dimers are unaffected by the presence of ATP (Cloherty et al. 2001a, 2001b).

These observations suggest that GluT1 quaternary structure is unaffected by ATP (although quaternary structure does affect GluT1 nucleotide binding capacity) but that GluT1 cytoplasmic domains undergo significant conformational change when ATP is complexed to GluT1.

We propose that GluT1 presents a vestibule to cytoplasmic water. This vestibule carries cytoplasmic substrate to the exit (e1) site located at the apex of the vestibule. In the presence of ATP (which binds only to tetrameric GluT1), two changes occur within the vestibule. 1) GluT1 cytoplasmic domains reorient to generate a non catalytic sugar binding site at the vestibule mouth. 2) When occupied by sugar, this site may physically restrict sugar exchange between vestibular and cytoplasmic water. This site persists when GluT1 undergoes the e1 to e2 conformational change. Thus, extracellular maltose will inhibit 3MG binding to e1 but not to the cytoplasmic site. Cytochalasin B will inhibit binding to both e1 and the cytoplasmic site.

When tetrameric GluT1 dissociates in the presence of reductant, the nucleotide binding site is lost, the non-catalytic sugar binding site disappears, subunits within each dimer are functionally uncoupled and the transporter behaves as a simple carrier. This hypothesis predicts that GluT1 sugar transport kinetics should be significantly simplified in the absence of ATP or in the presence of reductant. Experiments designed to test these predictions are currently underway in this laboratory.

Acknowledgements

This work was supported by NIH grants DK 36081 and DK 44888

References

Abramson J, Smirnova I, Kasho V, Verner G, Kaback HR, Iwata S (2003) Structure and mechanism of the lactose permease of *Escherichia coli*. Science 301:610-615

Afzal I, Cunningham P, Naftalin RJ (2002) Interactions of ATP, oestradiol, genistein and the anti-oestrogens, faslodex (ICI 182780) and tamoxifen, with the human erythrocyte glucose transporter, GLUT1. Biochem J 365:707-719

Appleman JR, Lienhard GE (1985) Rapid kinetics of the glucose transporter from human erythrocytes. Detection and measurement of a half-turnover of the purified transporter. J Biol Chem 260:4575-4578

Appleman JR, Lienhard GE (1989) Kinetics of the purified glucose transporter. Direct measurement of the rates of interconversion of transporter conformers. Biochemistry 28:8221-8227

Askari A, Huang WH (1984) Reaction of (Na+ + K+)-dependent adenosine triphosphatase with inorganic phosphate. Regulation by Na+, K+, and nucleotides. J Biol Chem 259:4169-4176

Baker GF, Basketter DA, Widdas WF (1978) Asymmetry of the hexose transfer system in human erythrocytes. Experiments with non-transportable inhibitors. J Physiol (Lond) 278:377-388

Baker GF, Naftalin RJ (1979) Evidence of multiple operational affinities for D-glucose inside the human erythrocyte membrane. Biochim Biophys Acta 550:474-484

Baker GF, Widdas WF (1973) The asymmetry of the facilitated transfer system for hexoses in human red cells and the simple kinetics of a two component model. Biochim Biophys Acta (Lond) 231:143–165

Baker GF, Widdas WF (1988) Parameters for 3-O-methyl glucose transport in human erythrocytes and fit of asymmetric carrier kinetics. J Physiol (Lond) 395:57-76

Baldwin SA, Baldwin JM, Lienhard GE (1982) The monosaccharide transporter of the human erythrocyte. Characterization of an improved preparation. Biochemistry 21:3836–3842

Bass RB, Locher KP, Borths E, Poon Y, Strop P, Lee A, Rees DC (2003) The structures of BtuCD and MscS and their implications for transporter and channel function. FEBS Lett 555:111-115

Beguin P, Hasler U, Beggah A, Geering K (1998a) Regulation of expression and function by subunits of oligomeric P-type ATPases. Acta Physiol Scand Suppl 643:283-287

Beguin P, Hasler U, Beggah A, Horisberger JD, Geering K (1998b) Membrane integration of Na,K-ATPase alpha-subunits and beta-subunit assembly. J Biol Chem 273:24921-24931

Birnbaum MJ (1992) The insulin-sensitive glucose transporter. Int Rev Cytol 137:239-297

Blakey D, Leech A, Thomas GH, Coutts G, Findlay K, Merrick M (2002) Purification of the *Escherichia coli* ammonium transporter AmtB reveals a trimeric stoichiometry. Biochem J 364:527-535

Boldyrev AA (1988) Physiological significance of oligomeric associations of ionic pumps in biomembranes. Acta Physiol Pharmacol Bulg 14:3-9

Boldyrev AA (1995) [Oligomeric assemblies of transport adenosine triphosphatases in a membrane bilayer]. Zh Evol Biokhim Fiziol 31:375-386

Boldyrev AA, Quinn PJ (1994) E1/E2 type cation transport ATPases: evidence for transient associations between protomers. Int J Biochem 26:1323-1331

Boll M, Daniel H (1995) Target size analysis of the peptide/H(+)-symporter in kidney brush-border membranes. Biochim Biophys Acta 1233:145-152

Borgnia MJ, Agre P (2001) Reconstitution and functional comparison of purified GlpF and AqpZ, the glycerol and water channels from *Escherichia coli*. Proc Natl Acad Sci USA 98:2888-2893

Burant CF, Bell GI (1992) Mammalian facilitative glucose transporters: evidence for similar substrate recognition sites in functionally monomeric proteins. Biochemistry 31:10414-10420

Carruthers A (1986a) Anomalous asymmetric kinetics of human red cell hexose transfer: role of cytosolic adenosine 5'-triphosphate. Biochemistry 25:3592-3602

Carruthers A (1986b) ATP regulation of the human red cell sugar transporter. J Biol Chem 261:11028-11037

Carruthers A (1989) Hexose transport across human erythrocyte membranes. In: Raess BU, G. Tunnicliff G (eds) The Red Cell Membrane. Humana Press, Clifton, N.J., pp 249-279

Carruthers A (1991) Mechanisms for the facilitated diffusion of substrates across cell membranes. Biochemistry 30:3898-3906

Carruthers A, Helgerson AL (1989) The human erythrocyte sugar transporter is also a nucleotide binding protein. Biochemistry 28:8337-8346

Carruthers A, Helgerson AL (1991) Inhibitions of sugar transport produced by ligands binding at opposite sides of the membrane. Evidence for simultaneous occupation of the carrier by maltose and cytochalasin B. Biochemistry 30:3907-3915

Carruthers A, Melchior DL (1983) Asymmetric or symmetric? Cytosolic modulation of human erythrocyte hexose transfer. Biochim Biophys Acta 728:254-266

Carruthers A, Melchior DL (1984) A rapid method of reconstituting human erythrocyte sugar transport proteins. Biochemistry 23:2712-2718

Carruthers A, Melchior DL (1985) Transport of α– and β-D-glucose by the intact human red cell. Biochemistry 24:4244-4250

Carruthers A, Melchior DL (1986) How bilayer lipids affect membrane protein activity. TIBS 11:331–335

Casey JR, Reithmeier RA (1991) Analysis of the oligomeric state of Band 3, the anion transport protein of the human erythrocyte membrane, by size exclusion high performance liquid chromatography. Oligomeric stability and origin of heterogeneity. J Biol Chem 266:15726-15737

Chadwick CC, Goormaghtigh E, Scarborough GA (1987) A hexameric form of the Neurospora crassa plasma membrane H+-ATPase. Arch Biochem Biophys 252:348-356

Charbit A, Reizer J, Saier MH Jr (1996) Function of the duplicated IIB domain and oligomeric structure of the fructose permease of Escherichia coli. J Biol Chem 271:9997-10003

Cloherty EK, Hamill S, Levine K, Carruthers A (2001a) Sugar Transporter Regulation by ATP and quaternary structure. Blood Cells, Molecules and Disease 27:102-107

Cloherty EK, Heard KS, Carruthers A (1996) Human erythrocyte sugar transport is incompatible with available carrier models. Biochemistry 35:10411-10421

Cloherty EK, Levine KB, Carruthers A (2001b) The red blood cell glucose transporter presents multiple, nucleotide sensitive sugar exit sites. Biochemistry 40

Cloherty EK, Sultzman LA, Zottola RJ, Carruthers A (1995) Net Sugar Transport is a Multi-Step Process. Evidence for cytosolic sugar binding sites in erythrocytes. Biochemistry 34:15395–15406

Coderre PE, Cloherty EK, Zottola RJ, Carruthers A (1995) Rapid substrate translocation by the multi-subunit, erythroid glucose transporter requires subunit associations but not cooperative ligand binding. Biochemistry 34:9762-9773

Cotton JL, Partin KM (2000) The contributions of GluR2 to allosteric modulation of AMPA receptors. Neuropharmacology 39:21-31

Craig WS (1982) Monomer of sodium and potassium ion activated adenosinetriphosphatase displays complete enzymatic function. Biochemistry 21:5707-5717

Delisle MC, Giroux S, Vachon V, Boyer C, Potier M, Beliveau R (1994) Molecular size of the functional complex and protein subunits of the renal phosphate symporter. Biochemistry 33:9105-9109

Deutsch C (2002) Potassium channel ontogeny. Annu Rev Physiol 64:19-46

Douglas RM, Ritchie GY, Munro AW, McLaggan D, Booth IR (1994) The K(+)-efflux system, KefC, in Escherichia coli: genetic evidence for oligomeric structure. Mol Membr Biol 11:55-61

Duchesne L, Pellerin I, Delamarche C, Deschamps S, Lagree V, Froger A, Bonnec G, Thomas D, Hubert JF (2002) Role of C-terminal domain and transmembrane helices 5 and 6 in function and quaternary structure of major intrinsic proteins: analysis of aquaporin/glycerol facilitator chimeric proteins. J Biol Chem 277:20598-20604

Dutzler R, Campbell EB, Cadene M, Chait BT, MacKinnon R (2002) X-ray structure of a ClC chloride channel at 3.0 A reveals the molecular basis of anion selectivity. Nature 415:287-294

Elkins CA, Nikaido H (2003) 3D structure of AcrB: the archetypal multidrug efflux transporter of Escherichia coli likely captures substrates from periplasm. Drug Resist Updat 6:9-13

Erni B (1986) Glucose-specific permease of the bacterial phosphotransferase system: phosphorylation and oligomeric structure of the glucose-specific IIGlc-IIIGlc complex of Salmonella typhimurium. Biochemistry 25:305-312

Eskandari S, Wright EM, Kreman M, Starace DM, Zampighi GA (1998) Structural analysis of cloned plasma membrane proteins by freeze-fracture electron microscopy. Proc Natl Acad Sci USA 95:11235-11240

Eyers SA, Ridgwell K, Mawby WJ, Tanner MJ (1994) Topology and organization of human Rh (rhesus) blood group-related polypeptides. J Biol Chem 269:6417-6423

Falk MM (2000) Biosynthesis and structural composition of gap junction intercellular membrane channels. Eur J Cell Biol 79:564-574

Fry DC, Kuby SA, Mildvan AS (1986) ATP-binding site of adenylate kinase: Mechanistic implications of its homology with ras-encoded p21, F1-ATPase, and other nucleotide-binding proteins. Proc Natl Acad Sci USA 83:907-911

Fu D, Libson A, Miercke LJ, Weitzman C, Nollert P, Krucinski J, Stroud RM (2000) Structure of a glycerol-conducting channel and the basis for its selectivity. Science 290:481-486

Gerardi-Laffin C, Delque-Bayer P, Sudaka P, Poiree JC (1993) Oligomeric structure of the sodium-dependent phlorizin binding protein from kidney brush-border membranes. Biochim Biophys Acta 1151:99-104

Gorga FR, Lienhard GE (1982) Changes in the intrinsic fluorescence of the human erythrocyte monosaccharide transporter upon ligand binding. Biochemistry 21:1905-1908

Gutierrez LM, Brawley RM, Hosey MM (1991) Dihydropyridine-sensitive calcium channels from skeletal muscle. I. Roles of subunits in channel activity. J Biol Chem 266:16387-16394

Hackenberg H, Klingenberg M (1980) Molecular weight and hydrodynamic parameters of the adenosine 5'-diphosphate-adenosine 5'-triphosphate carrier in triton X-100. Biochemistry 19:548-555

Hamill S, Cloherty EK, Carruthers A (1999) The human erythrocyte sugar transporter presents two sugar import sites. Biochemistry 38:16974-16983

Hankin BL, Lieb WR, Stein WD (1972) Rejection criteria for the asymmetric carrier and their application to glucose transport in the human red blood cell. Biochim Biophys Acta 288:114-126

Hanspal M, Golan DE, Smockova Y, Yi SJ, Cho MR, Liu SC, Palek J (1998) Temporal synthesis of band 3 oligomers during terminal maturation of mouse erythroblasts. Dimers and tetramers exist in the membrane as preformed stable species. Blood 92:329-338

Harris EJ (1964) An analytical study of the kinetics of glucose movement in human erythrocytes. J Physiol 173:344-353

Hartel-Schenk S, Agre P (1992) Mammalian red cell membrane Rh polypeptides are selectively palmitoylated subunits of a macromolecular complex. J Biol Chem 267:5569-5574

Heard KS, Diguette M, Heard AC, Carruthers A (1998) Membrane-bound glyceraldehyde-3-phosphate dehydrogenase and multiphasic erythrocyte sugar transport. Exp Physiol 83:195-201

Heard KS, Fidyk N, Carruthers A (2000) ATP-dependent substrate occlusion by the human erythrocyte sugar transporter. Biochemistry 39:3005-3014

Hebert DN, Carruthers A (1991) Cholate-solubilized erythrocyte glucose transporters exist as a mixture of homodimers and homotetramers. Biochemistry 30:4654-4658

Hebert DN, Carruthers A (1992) Glucose transporter oligomeric structure determines transporter function. Reversible redox-dependent interconversions of tetrameric and dimeric GLUT1. J Biol Chem 267:23829-23838

Helgerson AL, Carruthers A (1987) Equilibrium ligand binding to the human erythrocyte sugar transporter. Evidence for two sugar-binding sites per carrier. J Biol Chem 262:5464-5475

Helgerson AL, Carruthers A (1989) Analysis of protein-mediated 3-O-methylglucose transport in rat erythrocytes: rejection of the alternating conformation carrier model for sugar transport. Biochemistry 28:4580-4594

Helgerson AL, Hebert DN, Naderi S, Carruthers A (1989) Characterization of two independent modes of action of ATP on human erythrocyte sugar transport. Biochemistry 28:6410-6417

Heuberger EH, Veenhoff LM, Duurkens RH, Friesen RH, Poolman B (2002) Oligomeric state of membrane transport proteins analyzed with blue native electrophoresis and analytical ultracentrifugation. J Mol Biol 317:591-600

Horowitz B, Eakle KA, Scheiner-Bobis G, Randolph GR, Chen CY, Hitzeman RA, Farley RA (1990) Synthesis and assembly of functional mammalian Na,K-ATPase in yeast. J Biol Chem 265:4189-4192

Horschitz S, Hummerich R, Schloss P (2003) Functional coupling of serotonin and noradrenaline transporters. J Neurochem 86:958-965

Huang Y, Lemieux MJ, Song J, Auer M, Wang DN (2003) Structure and mechanism of the glycerol-3-phosphate transporter from *Escherichia coli*. Science 301:616-620

Hymel L, Nielsen M, Gietzen K (1985) Target sizes of human erythrocyte membrane Ca2+-ATPase and Mg2+-ATPase activities in the presence and absence of calmodulin. Biochim Biophys Acta 815:461-467

Imanaka T, Aihara K, Suzuki Y, Yokota S, Osumi T (2000) The 70-kDa peroxisomal membrane protein (PMP70), an ATP-binding cassette transporter. Cell Biochem Biophys 32 Spring:131-138

Jacobson GR (1992) Interrelationships between protein phosphorylation and oligomerization in transport and chemotaxis via the *Escherichia coli* mannitol phosphotransferase system. Res Microbiol 143:113-116

Jette M, Vachon V, Potier M, Beliveau R (1996) The renal sodium/phosphate symporters: evidence for different functional oligomeric states. Biochemistry 35:15209-15214

Jette M, Vachon V, Potier M, Beliveau R (1997) Radiation-inactivation analysis of the oligomeric structure of the renal sodium/D-glucose symporter. Biochim Biophys Acta 1327:242-248

Jiang Y, Lee A, Chen J, Cadene M, Chait BT, MacKinnon R (2002) Crystal structure and mechanism of a calcium-gated potassium channel. Nature 417:515-522

Jiang Y, Lee A, Chen J, Ruta V, Cadene M, Chait BT, MacKinnon R (2003) X-ray structure of a voltage-dependent K+ channel. Nature 423:33-41

Joost HG, Thorens B (2001) The extended GLUT-family of sugar/polyol transport facilitators: nomenclature, sequence characteristics, and potential function of its novel members (review). Mol Membr Biol 18:247-256

Jung CY, Carlson LM, Whaley DA (1971) Glucose transport carrier activities in extensively washed human red cell ghosts. Biochim Biophys Acta 241:613–627

Jung CY, Hsu TL, Hah JS, Cha C, Haas MN (1980) Glucose transport carrier of human erythrocytes. Radiation-target size of glucose-sensitive cytochalasin B binding protein. J Biol Chem 255:361-364

Karlish SJD, Lieb WR, Ram D, Stein WD (1972) Kinetic Parameters of glucose efflux from human red blood cells under zero-trans conditions. Biochim Biophys Acta 255:126-132

Kayano T, Burant CF, Fukumoto H, Gould GW, Fan YS, Eddy RL, Byers MG, Shows TB, Seino S, Bell GI (1990) Human facilitative glucose transporters. Isolation, functional characterization, and gene localization of cDNAs encoding an isoform (GLUT5) expressed in small intestine, kidney, muscle, and adipose tissue and an unusual glucose transporter pseudogene-like sequence (GLUT6). J Biol Chem 265:13276-13282

Kennedy KA, Traxler B (1999) MalK forms a dimer independent of its assembly into the MalFGK2 ATP-binding cassette transporter of Escherichia coli. J Biol Chem 274:6259-6264

Kinsella JL, Heller P, Froehlich JP (1998) Na+/H+ exchanger: proton modifier site regulation of activity. Biochem Cell Biol 76:743-749

Klingenberg M (1981) Membrane protein oligomeric structure and transport function. Nature 290:449-454

Ko YH, Thomas PJ, Delannoy MR, Pedersen PL (1993) The cystic fibrosis transmembrane conductance regulator. Overexpression, purification, and characterization of wild type and delta F508 mutant forms of the first nucleotide binding fold in fusion with the maltose-binding protein. J Biol Chem 268:24330-24338

Kocabas AM, Rudnick G, Kilic F (2003) Functional consequences of homo- but not hetero-oligomerization between transporters for the biogenic amine neurotransmitters. J Neurochem 85:1513-1520

Krupka RM, Devés R (1981) An experimental test for cyclic versus linear transport models. The mechanism of glucose and choline transport in erythrocytes. J Biol Chem 256:5410-5416

Lachaal M, Berenski CJ, Kim J, Jung CY (1990) An ATP-modulated specific association of glyceraldehyde-3-phosphate dehydrogenase with human erythrocyte glucose transporter. J Biol Chem 265:15449-15454

Lachaal M, Jung CY (1993) Interaction of facilitative glucose transporter with glucokinase and its modulation by ADP and glucose-6-phosphate. J Cell Physiol 156:326-329

Larson TJ, Schumacher G, Boos W (1982) Identification of the glpT-encoded sn-glycerol-3-phosphate permease of Escherichia coli, an oligomeric integral membrane protein. J Bacteriol 152:1008-1021

Lee MC, Hamamoto S, Schekman R (2002) Ceramide biosynthesis is required for the formation of the oligomeric H+-ATPase Pma1p in the yeast endoplasmic reticulum. J Biol Chem 277:22395-22401

Levine KB, Cloherty EK, Fidyk NJ, Carruthers A (1998) Structural and physiologic determinants of human erythrocyte sugar transport regulation by adenosine triphosphate. Biochemistry 37:12221-12232

Levine KB, Cloherty EK, Hamill S, Carruthers A (2002) Molecular determinants of sugar transport regulation by ATP. Biochemistry 41:12629-12638

Levine KB, Hamill S, Cloherty EK, Carruthers A (2001) Alanine scanning mutagenesis of the human erythrocyte glucose transporter putative ATP binding domain. Blood Cells, Molecules and Disease 27:139-142

Lieb WR, Stein WD (1974) Testing and characterizing the simple carrier. Biochim Biophys Acta 373:178–196

Lieb WR, Stein WD (1977) Is there a high affinity site for sugar transport at the inner face of the human red cell membrane? J Theor Biol 69:311-319

Lintschinger B, Balzer-Geldsetzer M, Baskaran T, Graier WF, Romanin C, Zhu MX, Groschner K (2000) Coassembly of Trp1 and Trp3 proteins generates diacylglycerol- and Ca2+-sensitive cation channels. J Biol Chem 275:27799-27805

Locher KP, Lee AT, Rees DC (2002) The E. coli BtuCD structure: a framework for ABC transporter architecture and mechanism. Science 296:1091-1098

Lopina OD (2000) Na+,K+-ATPase: structure, mechanism, and regulation. Membr Cell Biol 13:721-744

Lowe AG, Walmsley AR (1986) The kinetics of glucose transport in human red blood cells. Biochim Biophys Acta 857:146-154

Lundahl P, Mascher E, Andersson L, Englund AK, Greijer E, Kameyama K, Takagi T (1991) Active and monomeric human red cell glucose transporter after high performance molecular-sieve chromatography in the presence of octyl glucoside and phosphatidylserine or phosphatidylcholine. Biochim Biophys Acta 1067:177-186

MacKinnon R (2003) Potassium channels. FEBS Lett 555:62-65

Manley DM, McComb ME, Perreault H, Donald LJ, Duckworth HW, O'Neil JD (2000) Secondary structure and oligomerization of the E. coli glycerol facilitator. Biochemistry 39:12303-12311

Marini AM, Matassi G, Raynal V, Andre B, Cartron JP, Cherif-Zahar B (2000) The human Rhesus-associated RhAG protein and a kidney homologue promote ammonium transport in yeast. Nat Genet 26:341-344

Martin DW, Tanford C (1984) Solubilized monomeric sarcoplasmic reticulum Ca pump protein. Phosphorylation by inorganic phosphate. FEBS Lett 177:146-150

Mueckler M, Caruso C, Baldwin SA, Panico M, Blench I, Morris HR, Allard WJ, Lienhard GE, Lodish HF (1985) Sequence and structure of a human glucose transporter. Science 229:941-945

Murakami S, Nakashima R, Yamashita E, Yamaguchi A (2002) Crystal structure of bacterial multidrug efflux transporter AcrB. Nature 419:587-593

Murakami S, Tamura N, Saito A, Hirata T, Yamaguchi A (2004) Extramembrane central pore of multidrug exporter AcrB in Escherichia coli plays an important role in drug transport. J Biol Chem 279:3743-3748

Naftalin RJ (1997) Evidence from studies of temperature-dependent changes of D-glucose, D- mannose and L-sorbose permeability that different states of activation of the human erythrocyte hexose transporter exist for good and bad substrates. Biochim Biophys Acta 1328:13-29

Naftalin RJ (1998) Evidence from temperature studies that the human erythrocyte hexose transporter has a transient memory of its dissociated ligands. Exp Physiol 83:253-258

Naftalin RJ, Holman GD (1977) Transport of sugars in human red cells. In: Ellory JC, Lew VL (eds) Membrane transport in red cells. Academic Press, New York, pp 257-300

Naftalin RJ, Rist RJ (1991) 3-O-methyl-D-glucose transport in rat red cells: effects of heavy water. Biochim Biophys Acta 1064:37-48

Naftalin RJ, Smith PM, Roselaar SE (1985) Evidence for non–uniform distribution of D–glucose within human red cells during net exit and counterflow. Biochim Biophys Acta 820:235–249

Nargeot J, Lory P, Richard S (1997) Molecular basis of the diversity of calcium channels in cardiovascular tissues. Eur Heart J 18 Suppl A:A15-26

Norby JG (1988) The reaction mechanism of Na,K-ATPase: problems regarding the oligomeric/monomeric state and the phosphorylated intermediates of the system. Braz J Med Biol Res 21:1251-1259

Norgaard-Nielsen K, Norregaard L, Hastrup H, Javitch JA, Gether U (2002) Zn(2+) site engineering at the oligomeric interface of the dopamine transporter. FEBS Lett 524:87-91

Overvoorde PJ, Chao WS, Grimes HD (1997) A plasma membrane sucrose-binding protein that mediates sucrose uptake shares structural and sequence similarity with seed storage proteins but remains functionally distinct. J Biol Chem 272:15898-15904

Peng S, Blachly-Dyson E, Colombini M, Forte M (1992) Determination of the number of polypeptide subunits in a functional VDAC channel from *Saccharomyces cerevisiae*. J Bioenerg Biomembr 24:27-31

Pessino A, Hebert DN, Woon CW, Harrison SA, Clancy BM, Buxton JM, Carruthers A, Czech MP (1991) Evidence that functional erythrocyte-type glucose transporters are oligomers. J Biol Chem 266:20213-20217

Regen DM, Morgan HE (1964) Studies of the glucose-transport system in the rabbit erythrocyte. Biochim Biophys Acta 79:151-166

Reinders A, Schulze W, Kuhn C, Barker L, Schulz A, Ward JM, Frommer WB (2002) Protein-protein interactions between sucrose transporters of different affinities colocalized in the same enucleate sieve element. Plant Cell 14:1567-1577

Rhiel E, Flukiger K, Wehrli C, Erni B (1994) The mannose transporter of *Escherichia coli* K12: oligomeric structure, and function of two conserved cysteines. Biol Chem Hoppe Seyler 375:551-559

Sahin-Toth M, Lawrence MC, Kaback HR (1994) Properties of permease dimer, a fusion protein containing two lactose permease molecules from *Escherichia coli*. Proc Natl Acad Sci USA 91:5421-5425

Schroers A, Burkovski A, Wohlrab H, Kramer R (1998) The phosphate carrier from yeast mitochondria. Dimerization is a prerequisite for function. J Biol Chem 273:14269-14276

Sekler I, Kopito R, Casey JR (1995) High level expression, partial purification, and functional reconstitution of the human AE1 anion exchanger in *Saccharomyces cerevisiae*. J Biol Chem 270:21028-21034

Simons TJB (1983) Characterization of sugar transport in the pigeon red blood cell. J Physiol 338:477-500

Sogin DC, Hinkle PC (1978) Characterization of the glucose transporter from human erythrocytes. J Supramolec Struct 8:447-453

Sorkina T, Doolen S, Galperin E, Zahniser NR, Sorkin A (2003) Oligomerization of dopamine transporters visualized in living cells by fluorescence resonance energy transfer microscopy. J Biol Chem 278:28274-28283

Spooner PJ, Friesen RH, Knol J, Poolman B, Watts A (2000) Rotational mobility and orientational stability of a transport protein in lipid membranes. Biophys J 79:756-766

Stein WD (1986) Transport and diffusion across cell membranes. Academic Press, New York

Sultzman LA, Carruthers A (1999) Stop-flow analysis of cooperative interactions between GLUT1 sugar import and export sites. Biochemistry 38:6640-6650

Surprenant A (1996) Functional properties of native and cloned P2X receptors. Ciba Found Symp 198:208-219; discussion 219-222

Taniguchi K, Kaya S, Abe K, Mardh S (2001) The oligomeric nature of Na/K-transport ATPase. J Biochem (Tokyo) 129:335-342

Taylor LP, Holman GD (1981) Symmetrical kinetic parameters for 3-O-methyl-D-glucose transport in adipocytes in the presence and in the absence of insulin. Biochim Biophys Acta 642:325-335

Thomas DD, Reddy LG, Karim CB, Li M, Cornea R, Autry JM, Jones LR, Stamm J (1998) Direct spectroscopic detection of molecular dynamics and interactions of the calcium pump and phospholamban. Ann N Y Acad Sci 853:186-194

Toyoshima C, Nomura H (2002) Structural changes in the calcium pump accompanying the dissociation of calcium. Nature 418:605-611

Toyoshima C, Nomura H, Sugita Y (2003) Crystal structures of Ca2+-ATPase in various physiological states. Ann N Y Acad Sci 986:1-8

Unwin N (2003) Structure and action of the nicotinic acetylcholine receptor explored by electron microscopy. FEBS Lett 555:91-95

Van Hooft JA, Vijverberg HP (1997) RS-056812-198: partial agonist on native and antagonist on cloned 5-HT3 receptors. Eur J Pharmacol 322:229-233

Veenhoff LM, Heuberger EH, Poolman B (2002) Quaternary structure and function of transport proteins. Trends Biochem Sci 27:242-249

Vidaver GA (1966) Inhibition of parallel flux and augmentation of counterflux shown by transport models not involving a mobile carrier. J Theor Biol 10, 301–306

Walz T, Hirai T, Murata K, Heymann JB, Mitsuoka K, Fujiyoshi Y, Smith BL, Agre P, Engel A (1997) The three-dimensional structure of aquaporin-1. Nature 387:624-627

Wang Y, Tate SS (1995) Oligomeric structure of a renal cystine transporter: implications in cystinuria. FEBS Lett 368:389-392

Wellner M, Monden I, Keller K (1995) From triple cysteine mutants to the cysteine-less glucose transporter GLUT1: a functional analysis. FEBS Lett 370:19-22

Werner PK, Reithmeier RA (1985) Molecular characterization of the human erythrocyte anion transport protein in octyl glucoside. Biochemistry 24:6375-6381

Wheeler TJ (1986) Kinetics of glucose transport in human erythrocytes: zero-trans efflux and infinite-trans efflux at 0°C. Biochim Biophys Acta 862:387-398

Whitesell RR, Regen DM, Beth AH, Pelletier DK, Abumrad NA (1989) Activation energy of the slowest step in the glucose carrier cycle: Break at 23°C and correlation with membrane lipid fluidity. Biochemistry 28:5618-5625

Widdas WF (1952) Inability of diffusion to account for placental glucose transfer in the sheep and consideration of the kinetics of a possible carrier transfer. J Physiol (Lond) 118:23–39

Williams KA (2000) Three-dimensional structure of the ion-coupled transport protein NhaA. Nature 403:112-115

Xiao Y, Boyer CJ, Vincent E, Dugre A, Vachon V, Potier M, Beliveau R (1997) Involvement of disulphide bonds in the renal sodium/phosphate co-transporter NaPi-2. Biochem J 323 (Pt 2):401-408

Yeh Y, Selser JC, Baskin RJ (1978) Dynamic light scattering characterization of the detergent-free, delipidated (Ca2+ + Mg2+)-ATPase from sarcoplasmic reticulum. Biochim Biophys Acta 509:78-89

Yin CC, Aldema-Ramos ML, Borges-Walmsley MI, Taylor RW, Walmsley AR, Levy SB, Bullough PA (2000) The quaternary molecular structure of TetA, a secondary tetracycline transporter from Escherichia coli. Mol Microbiol 38:482-492

Yu EW, McDermott G, Zgurskaya HI, Nikaido H, Koshland DE Jr (2003) Structural basis of multiple drug-binding capacity of the AcrB multidrug efflux pump. Science 300:976-980

Zimmermann R, Neupert W (1980) Transport of proteins into mitochondria. Posttranslational transfer of ADP/ATP carrier into mitochondria *in vitro*. Eur J Biochem 109:217-229

Zolotarev AS, Shmukler BE, Alper SL (1999) AE2 anion exchanger polypeptide is a homooligomer in pig gastric membranes: a chemical cross-linking study. Biochemistry 38:8521-8531

Zottola RJ, Cloherty EK, Coderre PE, Hansen A, Hebert DN, Carruthers A (1995) Glucose transporter function is controlled by transporter oligomeric structure. A single, intramolecular disulfide promotes GLUT1 tetramerization. Biochemistry 34:9734-9747

Carruthers, Anthony
    Department of Biochemistry & Molecular Pharmacology, University of Massachusetts Medical School, Lazare Research Building, 364 Plantation Street, Worcester, MA 01605, USA
    anthony.carruthers@umassmed.edu

Levine, Kara B.
    Department of Biochemistry & Molecular Pharmacology, University of Massachusetts Medical School, Lazare Research Building, 364 Plantation Street, Worcester, MA 01605, USA

Regulation and function of ammonium carriers in bacteria, fungi, and plants

Nicolaus von Wirén and Mike Merrick

Abstract

The ammonium transport (Amt) family of proteins comprises a unique and ubiquitous group of integral membrane proteins found in all domains of life. They are present in bacteria, archaea, fungi, plants, and animals, including humans where they are represented by the Rhesus proteins. The Amt proteins have a variety of functions. In bacteria and fungi, they act to scavenge ammonium and to recapture ammonium lost from cells by diffusion across the cell membrane. In fungi, they have also been proposed to act as ammonium sensors to control filamentous growth. In plants, they make a major contribution to nitrogen nutrition and in higher animals; they are involved in ammonium fluxes in kidney and liver. In this paper, we review current knowledge of the biology of Amt proteins in bacteria, fungi, and plants with particular attention to the different functions of the proteins and their modes of regulation.

1 Introduction

Ammonium[1] plays a key role in the nitrogen metabolism of most cells. For many organisms, notably bacteria and eukaryotic microbes, ammonium is the preferred nitrogen source and although organisms can often acquire a variety of nitrogen sources most of these are transformed into ammonium before they are utilised in biosynthetic pathways. The importance of ammonium as a nitrogen source also means that a number of microbes show chemo-attraction towards ammonium, which raises the concept of ammonium sensors in both bacteria and fungi. Ammonium is important in plant metabolism not only as a major nitrogen source for plant growth and development, but also as a major form for nitrogen retrieval in leaves. Since ammonium is generated by photorespiration in the mitochondria and subsequently assimilated in the chloroplast, significant ammonium fluxes are likely to occur across a number of plant cell membranes. Finally, in animals, transmembrane fluxes of ammonium have been studied primarily in the mammalian kidney where excretion of ammonium constitutes the major component of acid

---

[1] The term ammonium is used to denote both NH_3 and NH_4^+ and chemical symbols are used when specificity is required.

Topics in Current Genetics, Vol. 9
E. Boles, R. Krämer (Eds.): Molecular Mechanisms Controlling Transmembrane Transport
DOI 10.1007/b95775 / Published online: 9 March 2004
© Springer-Verlag Berlin Heidelberg 2004

excretion to maintain acid/base balance. However, a flux of ammonium from neurons to glial cells is also found in most nervous tissues implicating ammonium transport in brain function.

The beneficial effect of ammonium as the key nitrogen form for acquisition, intracellular transport and retrieval of nitrogen is reversed at high ammonium concentrations when ammonium becomes harmful. Sensitivity to ammonium is a universal phenomenon and occurs in animals, plants, and microorganisms, although toxicity levels depend strongly on the type of organism and can vary largely between closely related species (Bai et al. 2001; Britto et al. 2001a). Several explanations have been provided for the toxic effect of ammonium, such as the dissipation of proton gradients across membranes, the acidification of the external medium in response to ammonium uptake, or a disequilibrium in the acid/base balance (Gerendas et al. 1997). However these explanations can at best only partially explain the observed ammonium toxicity syndromes and in plants membrane transport of ammonium, in particular efflux from the most sensitive cellular compartments, appears to be crucial for alleviating ammonium toxicity (Britto et al. 2001a).

Despite the fact that these varied biological systems all involve flux of ammonium across cell membranes the view was held for many decades that these fluxes can be satisfied by the free diffusion of NH_3 and that there is no requirement for specific ammonium transport systems. Quantitative determinations of the permeability coefficient for NH_3 have been carried out for a variety of membranes from bacterial, plant, and animal cells as well as for artificial bilayers. The accumulated data suggest that NH_3 is indeed capable of permeating through membranes at a rate similar to that for water. However as ammonium has a pK_a of 9.25, both NH_3 and NH_4^+ will normally be present in biological systems and at a physiological pH of around 7 (or below) 99% will be present as NH_4^+. Consequently, both the diffusion of NH_3 and the specific transport of NH_4^+ can potentially occur in biological systems.

In plants, attempts have been made to predict the likelihood and direction of NH_4^+ and NH_3 transport across membranes by measuring ammonium concentrations in different compartments and by calculating concentration gradients according to compartmental pH and membrane potential (Britto et al. 2001b; Howitt and Udvardi 2000). Assuming ammonium concentrations in the cytosol in the millimolar range, thermodynamic considerations argue in favour of cytosolic ammonium import in form of NH_4^+, which requires secondary active transport, as long as external concentrations are below millimolar. By contrast, loading of the plant vacuole with NH_3 might be thermodynamically more favourable though it could occur with NH_4^+ when vacuolar acidification is inhibited (Britto et al. 2001b; Plant et al. 1999). Thus, cells may have adapted to varying ammonium concentrations by evolving transporters for both substrates, NH_4^+ and NH_3 thereby allowing cellular import and compartmentalisation of ammonium whilst minimising energy demand.

The first proposal for the existence of active ammonium transport was obtained in 1970 in the fungus *Penicillium chrysogenum* and since then evidence for ammonium carriers has been reported in a wide range of organisms, mostly bacteria.

In neutral environments most bacteria maintain a higher intracellular than extracellular pH, which would promote NH_3 efflux down an NH_3 gradient. Hence, in many organisms, the measurement of ammonium gradients with significantly higher concentrations inside the cell than outside strongly argues that transport of either NH_3 or NH_4^+ should take place. Studies in a number of organisms have suggested the presence of more than one ammonium uptake system. In prokaryotes, kinetic studies in both proteobacteria e.g. *Rhodobacter sphaeroides* (Cordts and Gibson 1987) and cyanobacteria e.g. *Anacystis nidulans* (Cordts and Gibson 1987) could be interpreted in this way. However, these data could also reflect kinetic differences between ammonium uptake and assimilation. In plants, ammonium depletion studies indicated that concentration-dependent ammonium uptake into *Lemna* cells followed bi- or multiphasic kinetics (Ullrich et al. 1984), and uptake studies using ^{13}N-labelled ammonium showed that substrate affinity and maximum transport capacity of rice roots varied significantly depending on the nitrogen status (Wang et al. 1993).

Initial genetic approaches to the identification of potential ammonium transporters in bacterial systems were unsuccessful. Early reports of an ammonium transport gene (*amtA*) in *E. coli* later proved to be incorrect and the gene was correctly identified as *cysQ*, which plays a role in sulphite synthesis (Fabiny et al. 1991; Neuwald et al. 1992). Indeed it was not until 1994 that the cloning of two genes, one from *Saccharomyces cerevisiae* and one from *Arabidopsis thaliana*, led to the recognition of the first two members of the ammonium transporter (Amt) family of proteins. In this review, we will discuss the very considerable progress that has been made since 1994 in characterising this family of proteins from a wide variety of organisms and in beginning to understand the ways in which Amt proteins function.

2 The ammonium transport protein family

2.1 Distribution and phylogeny

The cloning of the first two genes that encode ammonium transport proteins was facilitated by the availability of a novel mutant of *S. cerevisiae* that grew very slowly on minimal medium containing only 1mM ammonium as sole nitrogen source (Marini et al. 1994; Ninnemann et al. 1994). Genomic and cDNA libraries from both *S. cerevisiae* and *A. thaliana*, respectively, were used to complement the growth defect of the mutant leading to the isolation of *S. cerevisiae MEP1*[2] and *A. thaliana AMT1;1*. A number of potential homologous proteins were identified from the, then relatively sparse, gene databases and these included proteins from *Bacillus subtilis*, *Rhodobacter capsulatus*, *Corynebacterium glutamicum,* and *Caenorhabditis elegans* suggesting that comparable transporters were present in

---

[2] The designation MEP derives from methylammonium permease because the ammonium analogue [14][C] – methylammonium is used as a substrate for transport assays.

bacteria and in nematode worms. With the exponential growth of the databases, Amt homologues have now been identified in all the domains of life. Representatives are found in eubacteria, archaea, fungi, nematode worms, insects, fish, and primates.

Amongst published genomes Amt proteins are notably absent in certain bacterial pathogens many of which rely on their hosts for the provision of nitrogen-containing compounds and consequently do not need systems for ammonium uptake (Thomas et al. 2000a). Nevertheless, these organisms are the exceptions and Amt proteins are encoded in nearly all genomes such that in excess of 200 members of the family can now be identified. Furthermore, in very many cases multiple copies of *amt* genes are present. Two or more *amt* genes are found in many bacteria and archaea, three copies are present in *S. cerevisiae* and ten putative *amt* genes are found in rice (Marini et al. 1997a; Suenaga et al. 2003; Thomas et al. 2000a).

A major extension of the Amt family occurred with the recognition that the human Rhesus (Rh) proteins, that are found both in erythrocytes and in the kidney and liver, show significant homology at the primary amino acid sequence level to Amt proteins (Marini et al. 1997b). Furthermore the human RhAG and RhGK genes can rescue the growth of a *mep* deficient strain of yeast on medium with low ammonium (Marini et al. 2000a). Phylogenetic analysis shows that the Rh proteins constitute a discreet subfamily of the Amt proteins (Liu et al. 2000; Ludewig et al. 2001). Furthermore, Rh proteins are not restricted to primates but are also found in fish, insects, slime moulds, and marine sponges. It is interesting to note that *C. elegans* and *Drosophila melanogaster* encode both Amt and Rh homologues in their genomes (Ludewig et al. 2001).

2.2 Membrane topology

The ammonium transport proteins were expected to be integral membrane proteins and computer analyses of the first cloned sequences revealed them to encode highly hydrophobic proteins with a molecular mass of 50-55 kDa. Hydrophathy plots have been reported for a number of homologues and have identified between 9 and 12 putative transmembrane (TM) regions (Javelle et al. 2003a; Marini et al. 1994; Monahan et al. 2002a; Montanini et al. 2002; Ninnemann et al. 1994; Siewe et al. 1996; Thomas et al. 2000b; Vermeiren et al. 2002). The variation in the predicted number of TM helices arises in part because of differences between the various algorithms used but may also reflect some real differences between Amt proteins (see below). The majority of analyses predict that the C-terminal region of Amt proteins constitutes a discreet cytoplasmic region of at least 30 amino acids.

A more definitive model of Amt topology can be derived from a combination of empirical analysis and multiple alignment of sequences derived from databases. A detailed empirical analysis of the *Escherichia coli* Amt protein (AmtB) used fusions between AmtB and the reporter proteins alkaline phosphatase (which is active in the periplasm) and β-galactosidase (which is active in the cytoplasm) to

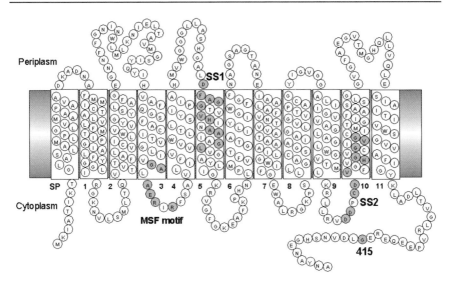

Fig. 1. Topology model for Amt proteins. Model based on empirical analysis of the topology of the E. coli AmtB protein. The signal peptide (SP) is apparently only present in Gram negative bacteria and is cleaved off in the mature protein suggesting that all Amt proteins could have a common membrane topology with an extracytoplasmic N-terminus and a cytoplasmic C-terminus. The Amt signature sequences (SS1 and SS2) are highlighted as are the region showing homology to the MSF motif and the highly conserved glycine residue (G415 in E. coli AmtB) in the C-terminal tail.

map the topology. This analysis concluded that the protein contained 12 TM helices with both the N and C-termini located in the cytoplasm (Thomas et al. 2000b) (see Fig. 1). Multiple sequence alignment was then used to map this model onto other Amt proteins leading to the conclusion that many Amt proteins e.g. *S. cerevisiae* MEP2 and *A. thaliana AMT1;1*, would have only 11 TM helices with the N-terminus being extracytoplasmic. The N-terminus of *S. cerevisiae* MEP2 was confirmed as being extracytoplasmic by mapping of an N-glycosylation site to residue N4 (Marini and Andre 2000). One analysis arrived at a different model with ten TM helices and both the N and C-termini extracellular (Howitt and Udvardi 2000). These authors also proposed that the proteins exhibit internal symmetry and could have evolved from an ancestral "six plus six" topology and that certain eukaryotic Amt proteins had "lost" one or more helices. There is currently no empirical data to support this model.

The apparent difference in topology between certain Amt proteins (11 or 12 TM helices) was resolved by purification of the *E. coli* AmtB protein. Amino-terminal sequencing of this protein revealed that the first 22 residues of AmtB are cleaved off in the mature protein such that the N-terminus would then be predicted to be extracellular (Blakey and Merrick, unpublished). It would therefore appear that the putative first TM helix of *E. coli* AmtB is actually a signal sequence and indeed it has all the characteristics of such a signal (van Dommelen et al. 2001). Furthermore, multiple sequence alignments suggest that this may be a common feature of all Amt proteins from Gram negative bacteria. Hence, the currently available data are consistent with a common membrane topology for Amt proteins with 11 TM helices, an extracytoplasmic N-terminus and a cytoplasmic C-terminus (Fig. 1). This C-terminal region is minimally 30 residues but is predicted to comprise more than 150 residues in some cases.

It is worth noting that one aspect in which the Rh proteins are distinguished from other Amt proteins is that the topology of Rh proteins has been empirically determined to fit a twelve TM helix model with both the N and C termini being cytoplasmic (Avent et al. 1996). Indeed multiple sequence alignments suggest that the homology between Amt and Rh is not conserved at their N-termini and that TM helices 3 to 11 of the Amt proteins are homologous to TM helices 4 to12 of the Rh proteins (M. Merrick, unpublished).

2.3 Structure-function relationships

The models derived by bioinformatic and empirical analyses provide a starting point from which to develop hypotheses and experimental approaches to an understanding of the structure/function relationships within the Amt family. Bioinformatic analysis shows that the Amt proteins constitute a novel family of transport proteins with two distinctive signature sequences located in TM helix 5 and TM helix 10 respectively (see Fig. 1) (Saier Jr et al. 1999).

SS#1 D (F Y W S) A G (G S C) X2 (L I V) (E H)X2 (G A S) (G A) X2 (G A S) (L F)
SS#2 D D X (L I V M F C) (E D G A) (L I V AC) X3 H (G A L I V) X2 (G S) X (L I V A W) G

In a study of a large number of inactive mutants of one of the two *Aspergillus nidulans* Amt proteins (MeaA), two independent mutations affected glycine167, which is located in the cytoplasmic loop between TM helices 3 and 4. This led to the recognition of a motif (GAVAERxK/R) in this region of Amt proteins that shows similarity to part of the G-X-X-X-D/E-R/K-X-G-R/K-R/K motif found in the major facilitator superfamily (MSF) of secondary transporters, where it is located in the cytoplasmic loop between TM helices 2 and 3 (Monahan et al. 2002b). This motif has been studied in detail in the *E. coli* LacY permease and shown to be important for facilitating conformational changes necessary for substrate translocation across the membrane. Apart from this particular motif, the distinctive nature of the proteins means that predictions concerning their structures and mode of action cannot obviously be derived from other transporters. Hence, insights are most likely to be derived from *in vivo* analysis of wild type and mutant proteins and from *in vitro* studies of purified proteins.

Relatively few mutant forms of Amt proteins have yet been analysed in detail. The most detailed analysis relates to a highly conserved glycine residue in the C-terminal tail of the protein. Following the recognition that *S. cerevisiae* encodes three Amt proteins the genotype of the mutant strain initially used to isolate *S. cerevisiae MEP1* and *A. thaliana AMT1-1* was determined. The strain contains a wild type copy of *MEP3*, a complete deletion of *MEP2* (*mep2Δ*) and a single point mutation in *MEP1* (*mep1-1*) that converts glycine 412 to aspartate. This mutation is trans-dominant such that its presence inhibits the activity of Mep3 causing a defect in ammonium transport (Marini et al. 2000b). The equivalent mutation was introduced into the *A. nidulans meaA* gene and the MeaA G447D protein was also transdominant to either a wild type copy of MeaA or to the second *A. nidulans* Amt protein, MepA (Monahan et al. 2002b). These data strongly suggest interactions between Amt monomers in both *S. cerevisiae* and *A. nidulans* and indicate that such interactions may involve the C-terminal tail. Although hetero-oligomers may form naturally in species such as *S. cerevisiae* they are clearly not a prerequisite for ammonium transport as strains expressing just a single gene are transport proficient (Marini et al. 1997a, 2000b; Monahan et al. 2002a).

The equivalent amino acid exchange, G458D, has also been analysed in LeAMT1;1 from tomato. It resulted in loss of function when expressed in either oocytes or yeast, even though the G458D-GFP fusion protein still localised to the plasma membrane when expressed in yeast or plant cells (Ludewig et al. 2003). Coexpression of G458D with wild type LeAMT1;1 in oocytes provoked a dominant negative inhibition of ammonium transport suggesting homo-oligomerisation. A physical interaction between LeAMT1;1 proteins was further supported using a split-ubiquitin yeast two-hybrid system. Surprisingly, coexpression of G458D with wild type LeAMT1;2 also inhibited ammonium transport in a dominant negative manner whilst a coexpressed amino acid permease was functionally not affected in the presence of G458D (Ludewig et al. 2003). These data provide evidence for the formation of heteromeric complexes between both LeAMTs in oocytes but whether AMTs also interact in plants remains to be shown.

Recent studies suggest that oligomerisation is a general feature of secondary transport proteins although the relationship between oligomeric structure and function has only been established in a few cases (Veenhoff et al. 2002). The *E. coli* AmtB protein is the only Amt protein to have been purified so far and this protein purifies as a stable trimer that does not dissociate in SDS but is denatured by boiling (Blakey et al. 2002). Immunoblots of all three of the *S. cerevisiae* Mep proteins also show signals compatible with the existence of homomultimers. It may therefore be the case that the native state of all Amt proteins is oligomeric and indeed the Rh proteins have been proposed to associate in tetrameric complexes in the erythrocyte membrane (Eyers et al. 1994).

The C-terminal cytoplasmic region of Amt is only highly conserved within the first 30 residues and any extensions beyond this appear to be species-specific. This region has been deleted from the Amt proteins of both *E. coli* and *Aquifex aeolicus* and in both cases the proteins retain a methylammonium transport activity that is reduced to around 25% of the wild type value (Coutts et al. 2002; Soupene et al. 2002b). Hence, the C-terminus is not absolutely required for transport activity.

Of the three *S. cerevisiae* Mep proteins, only one, Mep2, is glycosylated but mutation of the glycosylation site did not inactivate the protein and only increased the K_m for methyl-ammonium by a factor of 2 (Marini and Andre 2000). A number of other point mutations that affect Amt activity have been reported in *A. nidulans* MeaA and in *Lotus japonicus* AMT1;1 (Monahan et al. 2002b; Salvemini et al. 2001) though in each case there was no evidence that the mutant proteins were actually expressed and/or stable in the membrane.

2.4 Mode of action

As ammonium exists as a mixture of the two species, NH_3 and NH_4^+, the ratio of which is dependent on pH there has been an intensive controversy on the molecular species transported by AMT/MEP/Rh proteins. Initially, uptake studies with ^{14}C-labelled methylammonium in *C. glutamicum* and *AtAMT1*-transformed yeast showed that inhibitors of the plasma membrane ATPase and protonophores strongly inhibited methylammonium transport, arguing in favour of NH_4^+ being transported (Meier-Wagner et al. 2001; Ninnemann et al. 1994; Siewe et al. 1996). Furthermore the K_m of AmtB for methylammonium transport in *C. glutamicum* (Meier-Wagner et al. 2001; Siewe et al. 1996) and in *E. coli* (Thomas and Merrick, unpublished) is essentially independent of pH, which again favours $CH_3NH_3^+$ transport or CH_3NH_2/H^+ cotransport. The conclusions drawn from these observations have been questioned on the basis that rapid assimilation of methylammonium to methylglutamine by glutamine synthetase, or compartmentation to the vacuole in the case of fungi, might create an internal sink, which subsequently would accelerate inward-directed diffusion of CH_3NH_2 (Soupene et al. 1998, 2001). However, the K_m of AmtB for methylammonium transport in *C. glutamicum* and in *E. coli*, is around 100-fold lower than that of glutamine synthetase for methylammonium assimilation (Meier-Wagner et al. 2001; Thomas and Merrick, unpublished) suggesting that in bacteria glutamine synthetase is not the driving force for (methyl)ammonium uptake by AmtB.

At concentrations of less than 1mM ammonium, growth of AmtB- or Mep-defective *E. coli* and yeast, respectively, is impaired at low but not at neutral pH. This has been taken as evidence that AmtB or MEPs are only required if the pH-dependent concentration of uncharged NH_3 drops below a critical value (Soupene et al. 1998, 2001). This rationale implies that AmtB- or Mep-mediated transport will increase with pH as the concentration of NH_3 increases, whilst that of NH_4^+ remains largely unchanged. However, electrophysiological studies on the tomato transporter LeAMT1;1 do not support this view. The properties of LeAMT1;1 were examined by expressing the protein in *Xenopus laevis* oocytes and using two-electrode voltage clamp to study the transport mechanism. Micromolar concentrations of external ammonium induced concentration- and voltage-dependent currents that remained constant over a pH range of 5.5 to 8.5 (Ludewig et al. 2002). This is in agreement with NH_4^+ being the transported species because if NH_3 was the transported species the K_m should increase by tenfold for each unit increase in pH.

A phenomenon, which has not received appropriate attention by Soupene et al. (Soupene et al. 1998, 2001, 2002a), is that ammonium transport at neutral and acid pH is accompanied by membrane depolarisation. In intact plants, membrane depolarisation was immediately induced after ammonium supply and then reversed slowly at continuing ammonium supply (Ayling 1993). This reversal is likely to be due to an enhanced activity of the H^+-ATPase or membrane fluxes of other ions to compensate for the influx of positive charge and to reinstate the negative membrane potential. In oocytes expressing LeAMT1;1, membrane depolarisation was immediately induced after addition of ammonium to the bath solution and quickly reversed after ammonium withdrawal (Ludewig et al. 2002), which is again clearly indicative of the charged species being transported.

The situation may differ for Rhesus proteins in mammalian cells, where membrane potentials are smaller and controlled by Na^+ and K^+ gradients. Based on uptake studies with ^{14}C-labelled methylammonium in oocytes expressing the Rhesus protein RhAG, transport increased with external pH but decreased with increasing internal pH of the oocyte (Westhoff et al. 2002). It was therefore suggested that transport of methylammonium might be coupled with an antiport of H^+. However, as the transporter was not characterised by electrophysiology, and as the use of methylammonium as a substrate analogue only allows monitoring of substrate accumulation in minutes rather than seconds, further experiments are definitely required to substantiate this exceptional transport mechanism.

Functional expression of RhAG and RhAK in the yeast triple *mep* background permitted not only import of ammonium but also export (Marini et al. 2000a). Export could only be observed when yeast cells were fed with arginine, in which case internal degradation of this amino acid enhances internal ammonium pools. This experiment was reproduced in *Salmonella typhimurium*, where AmtB enhanced ammonium leakage upon arginine supply; the ammonium being detected by growth of a co-cultivated strain defective in arginine catabolism (Soupene et al. 2002a). Although neither approach could define the transported ammonium species, these feeding studies indicated that Amt and Rh proteins can act bidirectionally. If NH_3 was transported, the transport direction would simply follow the concentration gradient. If NH_4^+ was transported, the transport direction would be determined by the electrochemical gradient (Ludewig et al. 2002), which in plant cells is certainly oriented outwards when cytosolic ammonium concentrations are in the millimolar range and external concentrations are approximately 100-fold lower (Britto et al. 2001b).

In conclusion, as experienced from the characterisation of transporters and channels for other substrates, complementation and tracer uptake studies in model organisms alone cannot provide conclusive information on a transport mechanism (Ward 1997). This is of particular importance if the real substrate must be substituted by a traceable analogue and if different substrate species occur at the same time, as in case of NH_3 and NH_4^+. In this case, only electrophysiological studies allow clear differentiation of the charged and non-charged species, which at least for LeAMT1;1 clearly indicated NH_4^+ uniport as the transport mechanism. Future electrophysiological studies of Amt and Mep proteins should include the use of mutant derivatives as controls to provide better assessments of the degree to which alternative routes for ammonium transport may contribute to overall transport (see 6.6).

3 Bacterial ammonium transporters

The extensive sequencing of eubacterial and archaebacterial genomes (over 100 completed and more than 300 in progress) allows the distribution of Amt proteins to be assessed. From this it is apparent that *amt* genes are present in nearly all prokaryotic genomes and there are quite often multiple paralogues e.g. three in *Archaeoglobus fulgidus* and two in *C. glutamicum*. The exceptions that have no *amt* genes include a number of intracellular and extracellular pathogens such as *Rickettsia prowazekii*, *Chlamydia trachomatis*, *Borrelia burgdorferi*, *Helicobacter pylori*, and *Haemophilus influenzae*. These organisms either rely on their hosts for the provision of nitrogenous compounds or they utilise very specific nitrogen compounds e.g. urea in the case of *Helicobacter* and glutamate in the case of *Haemophilus* (Thomas et al. 2000a).

3.1 Regulation of Amt activity in bacteria

The prokaryotic *amt* genes are remarkable in one particular respect, namely that within the eubacteria (other than the cyanobacteria) and the archaea they are almost invariably found associated with a second gene (*glnK*) that encodes a small signal transduction protein belonging to the P_{II} protein family (Thomas et al. 2000a).

P_{II} proteins act as transducers of the cellular nitrogen status in prokaryotes and are also present in plants where they are located in the chloroplast (Arcondéguy et al. 2001). The P_{II} proteins have been studied in most detail in the proteobacteria, especially in *E. coli* which expresses two P_{II} proteins, GlnK and GlnB. GlnK is highly homologous to GlnB and both proteins adopt very similar tertiary and quaternary structures (Xu et al. 1998). They are trimers and take the form of a squat barrel approximately 50 Å in diameter and 30 Å high, above the surface of which three loops (the T-loops) protrude. In response to nitrogen deprivation, the P_{II} proteins of proteobacteria are covalently modified by uridylylation of a tyrosine residue (Y51) at the apex of the T-loop and this process is reversed in nitrogen sufficiency. A similar situation occurs in cyanobacteria excepting that in this case the proteins are modified by phosphorylation of a serine residue (S49), again at the apex of the T-loop. Hence, the modification state of the P_{II} protein is an indicator of the intracellular nitrogen status. P_{II} proteins typically regulate the activities of other proteins by protein-protein interaction as exemplified by the role of GlnB in modulating the activity of the transcriptional activator NtrC by interaction with the histidine protein kinase NtrB.

Conservation of gene linkage such as that found with *amtB* and *glnK* has often been found to reflect situations where the gene products physically interact (Dandekar et al. 1998) and studies in both *Escherichia coli* and *Azotobacter vinelandii* suggest that this is the case for AmtB. GlnK binds to the cell membrane in an AmtB-dependent manner and this interaction is dependent on the nitrogen status of the cell such that binding is maximal in nitrogen-sufficient conditions (Coutts et al. 2002). This response suggests that the sequestration of GlnK is

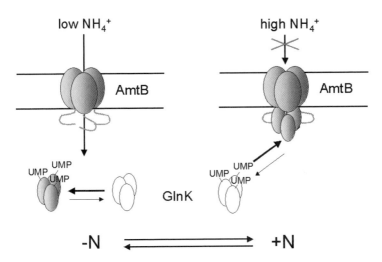

Fig. 2. Model for regulation of AmtB activity by GlnK. In conditions of nitrogen limitation (-N) GlnK is predominantly in its fully uridylylated state and is not strongly membrane-associated (non-shaded GlnK indicates the minority species in each situation). AmtB is active and will effectively scavenge ammonium from the surrounding medium. In the event of a marked rise in the availability of extracellular ammonium (+N), GlnK is rapidly deuridylylated and the unmodified form of the protein associates tightly with AmtB in the inner membrane. This association inhibits the activity of AmtB and ammonium is no longer transported into the cell until the cellular N-status drops when the inhibition is rapidly reversed.

regulated by its uridylylation state and indeed a non-uridylylatable mutant of GlnK (GlnKY51F) is bound to the membrane even in nitrogen-limiting conditions (Javelle and Merrick, unpublished). Detailed studies have shown that GlnK se-questration is rapid (being detectable after only 30 seconds exposure to elevated ammonium in the medium) and reversible. It can also be induced by as little as $50\mu M$ ammonium conditions (Javelle and Merrick, unpublished). The role of GlnK appears to be to provide rapid and sensitive regulation of AmtB activity (Fig. 2) and this is supported by experiments that demonstrate that GlnKY51F completely inhibits the transport activity of AmtB. In the event that the cellular ni-trogen status remains high then transcription of the *glnKamtB* operon ceases be-cause in *E. coli*, and probably in many other organisms, expression of the genes is controlled by the global nitrogen regulation system.

It seems likely that this mechanism will be found to be conserved throughout prokaryotes and indeed the membrane-association of GlnK has also been reported in *Azoarcus* (Martin and Reinhold-Hurek 2002) and in *Klebsiella pneumoniae* (Klopprogge et al. 2002). Of particular interest in this context are the archaea, where although *amtB* and *glnK* are linked there is presently no evidence for cova-lent modification of GlnK. In a broader context, the control of AmtB activity by GlnK raises the question as to whether similar post-translational control of Amt proteins might occur in eukaryotes and if so which proteins might effect that regulation.

The conservation of the *glnK amtB* operon also suggests that P_{II} proteins might have originally evolved together with the Amt proteins and that the primary role of the P_{II} protein was to regulate the activity of the ammonium transporter. Based on this model, the P_{II} paralogues that are now found in the bacteria and the archaea, namely the GlnB proteins of the proteobacteria and the NifI proteins in the nitrogen fixation gene clusters of some archaea (Arcondéguy et al. 2001), are likely to be the result of subsequent gene duplication and specialisation. Of particular interest in this context is the fact that both the GlnK and AmtB proteins are trimers. To date, the reason for the trimeric nature of P_{II} proteins has not been apparent, particularly as none of their known targets has trigonal symmetry. However, if GlnK evolved in concert with AmtB its trimeric structure may reflect a symmetry required for optimal interaction between the two proteins.

3.2 The role of Amt in ammonium sensing

In prokaryotes, there is relatively little evidence for ammonium sensing involving Amt proteins. However, the sequestration of GlnK does imply that elevation of external ammonium could lead to a rapid depletion of the cytoplasmic GlnK pool and consequently for any cytoplasmic system that is regulated by GlnK the AmtB protein would essentially comprise a part of an ammonium sensing mechanism. Two possible examples of this have been described.

In *Klebsiella pneumoniae*, GlnK is required to relieve the inhibitory effects of NifL on the activator protein NifA and, somewhat surprisingly, uridylylation of GlnK is not required to regulate this process (He et al. 1998; Jack et al. 1999). This leads to the question as to how NifL-mediated inhibition is restored when ammonium is added back to a nitrogen-limited medium and it was originally suggested that GlnK might be proteolysed or covalently modified by a mechanism other than uridylylation upon replenishment of ammonium (He et al. 1998). An alternative explanation is that the sequestration of deuridylylated GlnK by AmtB would rapidly lower the cytoplasmic GlnK pool and thereby release NifL to inhibit NifA activity.

In *Rhodobacter capsulatus* (and a number of other diazotrophs), the nitrogenase enzyme is rapidly covalently modified and inactivated by ADP-ribosylation in response to an extracellular ammonium shock. The process is reversible and the signal transduction pathway involved has yet to be completely elucidated but it appears to involve AmtB. *R. capsulatus* encodes two *amt* genes, *amtB* (which is linked to *glnK*) and *amtY* (which is monocistronic). An *amtB* mutant (but not an *amtY* mutant) is totally defective in ADP-ribosylation in response to ammonium addition (Yakunin and Hallenbeck 2002). *R. capsulatus* encodes two P_{II} proteins, GlnB and GlnK, and both of them appear to be involved in regulating ADP-ribosylation. A *glnB, glnK* double mutant is completely defective in nitrogenase inactivation in reponse to ammonium (Hallenbeck et al. 2002). Hence, it is certainly possible that the effects of AmtB could be explained in terms of GlnK sequestration.

4 Fungal ammonium transporters

Ammonium transporters appear to be ubiquitous in fungi and are found in all fungal genomes sequenced to date. The biological roles of the fungal Amt proteins have yet to be fully determined but in the case of the ectomycorrhizal fungi studies have suggested that they can make a significant contribution to the nitrogen nutrition of their plant hosts (Smith and Read 1997). Amt genes have been cloned and characterised from *S. cerevisiae, A. nidulans, Hebeloma cylindrosporum,* and *Tuber borchii* (Javelle et al. 2001; Marini et al. 1994, 1997a; Monahan et al. 2002a; Montanini et al. 2002). Fungi frequently encode multiple Amt proteins, there being three in *S. cerevisiae* and *H. cylindrosporum* and four in *Neurospora crassa.* Studies in *S. cerevisiae* and *H. cylindrosporum* suggest that they can be divided into two groups according to their affinity for ammonium. The high-affinity transporters ScMep1, ScMep2, HcAmt1, and HcAmt2 have a K_m for NH_4^+ in the range of <1 to 10μM whilst ScMep3 and HcAmt3 have values of >10μM to 2mM (Javelle et al. 2001, 2003b, Marini et al. 1997a). These groupings appear to be reflected in a phylogenetic analysis of fungal Amt proteins suggesting that transporters in the same affinity class may have common features (Javelle et al. 2003a).

4.1 Regulation of Amt/Mep protein activity in fungi

Regulation of protein activity may occur because of interactions with other proteins or may be caused by post-translational modification. In *S. cerevisiae*, one of the three Mep proteins has been shown to be post-translationally modified but the functional significance of this remains to be elucidated. By a combination of Western analysis, protein deglycosylation and site-directed mutagenesis Mep2 was shown to be N-glycosylated at residue N4 whereas Mep1 and Mep3 were not modified (Marini and Andre 2000). Time-dependent ammonium depletion from the external medium by a glycosylation-defective variant (Mep2 N4Q) was not significantly affected, nor was the Mep2-dependent signalling that triggers pseudohyphal growth (see Section 4.2).

As described earlier, studies of a trans-dominant mutation of *mep1* that results in a single amino acid substitution (G413D) located in the cytoplasmic C-terminus of Mep1 indicated that Mep1 can interact with Mep3 (Marini et al. 1997a, 2000b). This interaction is not necessary for transport activity of either protein as they are active individually when expressed alone. However the sum of the individual activities is higher than the transport activity of wild type cells suggesting cross-regulation between the Mep proteins (Marini et al. 2000b).

4.2 Amt proteins as ammonium sensors in fungi

In *S. cerevisiae*, nitrogen nutrition regulates pseudohyphal differentiation, a fila-mentous growth form in diploid cells induced upon nitrogen starvation or low ammonium supply. Pseudohyphal cells are more elongated than vegetative cells, employ a bipolar budding pattern and thus can efficiently invade a growth sub-strate. This morphological response is controlled by two interconnected signalling pathways, one of which involves a pheromone-responsive MAP kinase cascade, while the second involves Mep2 and the Gα-protein GPA2 (Lorenz and Heitman 1997). Since pseudohyphal differentiation and activation of GPA2 under nitrogen limitation were absent in a *mep2Δ* strain and mutational activation of GPA2 could rescue pseudohyphal growth in this strain, Mep2 was proposed to function up-stream of GPA2 (Lorenz and Heitman 1998). Thus, Mep2 apparently fulfils a sensing function by linking low ammonium availability with an adaptive growth response. The ability of ectomycorrhizal Amt proteins to carry out this sensing function was investigated by complementing a *S. cerevisiae mep2Δ* strain and complementation was found to be restricted to the high-affinity transporters HcAmt1, HcAmt2, and TbAmt1 (Javelle et al. 2003a). It was proposed that the sensing function of fungal Amt proteins is a distinct property of the high-affinity transporters.

By the construction of chimeric proteins, it was concluded that the pseudohy-phal regulatory function of *S. cerevisiae* Mep2 could be mapped to the N-terminal region of the protein (Lorenz and Heitman 1998). The authors used a topology model with only 10 TM helices and an incorrectly located N-terminus but on the current 11 TM topology model (Marini and Andre 2000) the region defined equates to the first three TM helices. Attempts were made to refine the region fur-ther, but although the authors reported that the introduction into Mep1 of a region from Mep2 encompassing TM helices 2 and 3 and the intervening extracytosolic loop was sufficient to confer the pseudohyphal regulatory function. The control reciprocal swap did not have the expected phenotype, consequently the precise features of Mep2 that determine its regulatory function remain to be defined (Lorenz and Heitman 1998).

Ammonia is also involved in long-distance signalling between neighbouring yeast colonies. Neighbouring colonies exhibit growth inhibition of the facing parts of both colonies, a phenomenon that is presumed to orient growth so as to mini-mise competition for nutrients. Colonies produce pulses of ammonia that are ap-parently perceived by neighbouring colonies which produce reciprocal pulses in response (Palková et al. 1997). However in this case, the Mep permeases are not involved in sensing of the signal and extracellular amino acids are believed to serve as the source for volatile ammonia production (Zikánová et al. 2002). Three membrane proteins, designated Ato1, Ato2, and Ato3, that are members of the YaaH family have been reported to be involved in ammonia production in *S. cere-visiae* and have been suggested to be ammonium/H^+ antiporters that extrude NH_4^+ from yeast cells and import protons (Palková et al. 2002).

5 Plant ammonium transporters

Ammonium transporters are widely distributed in plants being found in both mono- and di-cotyledonous plants and being expressed in most tissues. Most detailed studies have been carried out in two model systems, namely *Arabidopsis* and tomato but symbiotic legumes (*Lotus japonicus* and *Glycine max*) and rice are also being analysed. Plants encode multiple Amt proteins with six being encoded in the *Arabidopsis* genome and ten in the rice genome. With this level of complexity, there is clearly scope for specific physiological roles to be carried out by distinct homologues e.g. being located in specific cell types or tissues.

5.1 The function of Amt proteins in plants

A T-DNA insertion in the 5'-region of *AtAMT1;1* led to a complete loss of *AtAMT1;1* mRNA as determined by Northern analysis. Short-term uptake studies using ^{13}N-labelled ammonium showed that at micromolar concentrations of external ammonium, influx into roots of *amt1;1-1 (AtAMT1;1:T-DNA)* was decreased by 30% compared to the wild type, when plants were precultured under nitrogen deficiency (Kaiser et al. 2002). By contrast, at molar concentrations of ammonium influx by *atamt1;1-1* was even higher suggesting that low-affinity uptake was enhanced as a result of a compensatory overexpression of other AMTs in the absence of *AtAMT1;1* expression. In addition, the insertion line was lethal under exclusive ammonium nutrition and showed an altered leaf morphology (Kaiser et al. 2002). Although these observations might point to a role of *AtAMT1;1* in ammonium-dependent growth, they certainly require substantiation by investigations on the recomplemented insertion line or on further allelic insertion lines before firm conclusions can be drawn.

Similar to the *atamt1;1-1* insertion line, RNAi inhibition of *AtAMT2*, which also led to a loss of *AtAMT2* gene expression, did not provoke an altered phenotype of soil-grown plants (Sohlenkamp et al. 2002). An *AtAMT2* promoter-reporter gene fusion showed that *AtAMT2* expression was confined to the vascular system in leaves, to root tips and flowers, which at least partially explains the low expression levels detected by Northern analysis (Sohlenkamp et al. 2000). Determination of the substrate affinity in yeast and transient expression of GFP-fused *AtAMT2* indicated that *AtAMT2* encodes a high-affinity transporter that resides in the plasma membrane. Its physiological function, however, remains unclear.

Isolation of a closely related AMT2 homologue from *Lotus japonicus* showed expression throughout the whole plant, including all major tissues of root nodules. Transient expression of LjAMT2-GFP fusion proteins in plant cells also indicated plasma membrane localisation. Thus, a role of LjAMT2 in ammonium retrieval from nodules and other plant cells has been suggested (Simon-Rosin et al. 2003).

Taken together, studies on *Arabidopsis* lines with repressed expression of *AMT* genes indicate so far that there is considerable redundancy among the AMTs caused by at least partially overlapping physiological functions of the individual transporters. Thus, the generation of multiple *AMT* insertion lines or multiallelic

RNAi lines might be important strategies in future to uncover the importance of individual AMTs for ammonium transport in plants.

5.2 Biochemical transport properties of AMTs

Heterologous expression of AtAMT1;1 in yeast and short-term uptake studies of ^{14}C-labelled methylammonium as a substrate analogue showed high sensitivity against proton uncouplers (Ninnemann et al. 1994). This can be explained by the strong dependency of AMT1;1-mediated ammonium transport on the membrane potential gradient (Ludewig et al. 2002). Changes in external pH between 5.5 and 8.5 had little or no effect on the K_m for ammonium further supporting the concept that NH_4^+ and not NH_3 is the substrate for AMT1;1-like transporters (Ludewig et al. 2002; Sohlenkamp et al. 2002). AtAMT1;1 exhibited high selectivity towards ammonium, while potassium, despite its similar ion radius, could not compete for uptake by AMT1;1 paralogues (Ludewig et al. 2002; Ninnemann et al. 1994).

In yeast AtAMT1;1, AtAMT1;2, and AtAMT1;3 transported methylammonium at half-maximal velocity at external concentrations between 8 and 24 µM (Gazzarrini et al. 1999). The determination of inhibition constants further suggested that AtAMT1;1 has the highest affinity for ammonium ($K_i < 0.5$ µM) compared to AtAMT1;2 and 1;3 ($K_i = 25$ and 40 µM, respectively). This high affinity of AtAMT1;1 was not confirmed when substrate affinity was determined directly using ^{13}N-labelled ammonium for uptake studies in yeast (Sohlenkamp et al. 2002). On the other hand, a discrimination between ammonium and methylammonium transport was also observed for LeAMT1;1 when expressed in oocytes, where ammonium-induced currents were approximately fivefold higher than those of methylammonium (Ludewig et al. 2002). Such a prominent substrate discrimination at the level of AMT1;1-like transporters might explain the lower substrate affinity for methylammonium generally measured in nitrogen-deficient plant roots (Kosola and Bloom 1994; Shelden et al. 2001).

AtAMT2, the most distantly related AMT member, was initially described as having a unique ability to discriminate sharply between ammonium and methylammonium because it allowed ammonium uptake in yeast but did not confer resistance to methylammonium (Sohlenkamp et al. 2000). Uptake of ^{13}N-labelled ammonium by AtAMT2 in yeast revealed a similar K_m (approx. 20 µM) to that of AtAMT1;1 (Sohlenkamp et al. 2002) suggesting that both the AMT1 and AMT2 transporters, at least in *Arabidopsis*, encode high-affinity transporters with similar K_m values. These substrate affinities in the lower micromolar range closely match the relatively low ammonium concentrations in the soil solution, which rarely exceed 50 µM (Marschner 1995).

5.3 Transcriptional regulation of AMT transporters in plants

Given the large number of AMT proteins encoded by plants and the potential for complex tissue-specific expression it is not surprising that transcriptional

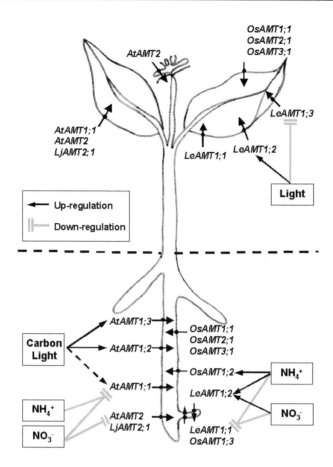

Fig. 3. Model summarising the transcriptional regulation of AMT genes in plants. Arrows indicate promoting effects on transcript levels, while capped lines represent repressive effects. The model is based on data cited in the text. Transcriptional regulation of OsAMT genes from rice is based on Suenaga et al. (2003) and Sonoda et al. (2003). Another report on OsAMT gene expression based on RT-PCR analysis found that *OsAMT1;1*, *1;2*, and *1;3* were commonly upregulated after transfer from high to low ammonium concentrations in the medium or vice versa (Kumar et al. 2003).

regulation of *AMT* genes in plants is highly complex. The complexity of the regulation revealed in studies to date is summarised in Figure 3. The plant N nutritional status and the external availability of ammonium or nitrate are two major components in the transcriptional regulation of AMT genes. In *Arabidopsis* and in tomato, transcript levels of *AMT1;1* steeply increased already after several hours of N deficiency coinciding with a concomitant increase in ammonium influx into roots (Gazzarrini et al. 1999; Rawat et al. 1999; von Wirén et al. 2000). *AtAMT1;1* expression was predominantely dependent on the local nitrogen status of the roots, because it was upregulated mainly in the root part that directly experienced nitrogen starvation when plants were grown in a split-root system (Gansel et al. 2001). Following resupply of nitrogen, root glutamine concentrations correlated best with decreasing influx, suggesting that glutamine might be the metabolic trigger down-regulating *AtAMT1;1* at the transcriptional level (Rawat et al. 1999; von Wirén et al. 2000). However, *AtAMT1;1* also responded rapidly to nitrate and was even downregulated when supplying nitrate to a nitrate reductase-deficient mutant, indicating that nitrate *per se* might act as a signal repressing *AtAMT1;1* (Gansel et al. 2001; Wang et al. 2000).

By contrast, transcript levels of *LeAMT1;2* were low under nitrogen-deficient growth conditions, but sharply increased with resupply of ammonium (Lauter et al. 1996; von Wirén et al. 2000). *LeAMT1;2* induction was highest at 5 – 50 µM ammonium but absent when ammonium assimilation was blocked by methionine sulphoxamine (MSX) (Becker et al. 2002). Besides ammonium, nitrate also induced *LeAMT1;2* (Lauter et al. 1996), which might reflect the requirement for ammonium retrieval after nitrate-stimulated efflux of ammonium. Interestingly, *LeAMT1;2* was also induced in the presence of wild type cells of the root-associated bacterium *Azospirillium brasiliense*. Induction was lower in presence an nitrogen fixation-deficient *Azospirillium* mutant, which points to a possible contribution of LeAMT1;2 in the uptake of ammonium from root-associated N_2-fixing bacteria (Becker et al. 2002).

Functional ammonium transporters have also been isolated from rice. While two of these root-expressed transporter genes were constitutively expressed (*OsAMT1;1*) or upregulated under N deficiency (*OsAMT1;3*), *OsAMT1;2* was induced under ammonium supply, even if external concentrations were lower than 200 nM (Sonoda et al. 2003). *In-situ* RNA detection confined *OsAMT1;2* gene expression to the cell surface of root tips as well as to the central cylinder, implying that one transporter might participate in different physiological functions, such as ammonium uptake from the soil solution and long distance transport. Using RT-PCR Kumar et al. observed that all three genes, *OsAMT1;1, 1;2*, and *1;3*, were negatively regulated by external ammonium and only differed in their expression levels (Kumar et al. 2003).

Three ammonium transporters in *A. thaliana* roots showed diurnal regulation with highest transcript levels at the end of the light period followed by a sharp decline in the dark (Gazzarrini et al. 1999). *AMT* transcript levels increased with light intensity and were relieved from repression after the supply of sucrose during the dark period, indicating that sugars also participate in transcriptional regulation of *AMT* genes (Lejay et al. 2003). By contrast, malate or 2-oxoglutarate could not stimulate *AMT* gene expression. Parallel investigations on the nitrate transporter gene *AtNRT2;1*, which showed a similar dependency on photosynthesis as *AtAMT1;2* and *AtAMT1;3*, suggested that hexokinase or its downstream metabolites from glycolysis might trigger induction of *AtNRT2;1* (Lejay et al. 2003). Whether this also applies to *AMT*s, however, remains to be demonstrated.

Under natural growth conditions, sugars regulating *AMT* expression are expected to mainly derive from photoassimilates. In fact, growing plants under elevated atmospheric CO_2 concentrations increased the uptake capacity of tobacco roots for ammonium versus nitrate (Matt et al. 2001). In wheat, nitrate reduction was impaired under elevated CO_2, favouring nitrogen assimilation and plant growth when ammonium nitrogen was supplied (Bloom et al. 2002). This might well be due to a CO_2-dependent regulation of AMT transporters. Transcriptional regulation of *AMT*s by CO_2 is of additional importance in leaves. Transcript levels of *LeAMT1;2* and *LeAMT1;3* decreased under elevated atmospheric CO_2, but showed an almost inverse diurnal expression pattern with highest mRNA levels for *LeAMT1;2* after onset of light and for *LeAMT1;3* during the dark. This might point to an involvement of LeAMT1;2 in ammonium retrieval, which serves to compensate for ammonium losses due to photorespiration at low CO_2 (von Wirén et al. 2000).

6 Ammonium/ammonia transport by other (than AMT-type) transporters in plants

Since the yeast triple *mep* mutant still exhibits a low ammonium uptake capacity in particular at high external supply (Marini et al. 1997a), it has been concluded that transporters other than AMT/MEPs should contribute to overall ammonium transport. Electrophysiological studies on root plasma membranes from *Arabidopsis* allowed the characterisation of so-called non-selective cation channels (NSCC) as transporters with a remarkably high selectivity for ammonium (Demidchik and Tester 2002). NSCCs can account for a major component of Na^+ influx at a relatively low dependence on membrane voltage and are inhibited by high Ca concentrations (Tyerman et al. 2002). However, the molecular identity of these transporters, that also exist in yeast (Bihler et al. 1998), has not yet been uncovered.

The chemical similarity and identical ionic radii of NH_4^+ and K^+ suggest that K^+ channels might permeate NH_4^+. Indeed, root-expressed AtKAT1 permitted ammonium-inducible inward currents (Moroni et al. 1998) and single amino acid substitutions in AtKAT1 could even increase ammonium permeability dramatically in oocytes (Uozumi et al. 1995), indicating that the transport of both substrates is not easily discriminated by this type of K^+ channel. However, ammonium uptake studies in *akt1* knockout plants must be awaited before a more precise idea is obtained about the contribution of AKT1 to overall ammonium uptake.

Ammonium transport is also of central importance at the symbiosome membrane, which surrounds N_2-fixing bacteroids and mediates the import of ammonium into the host plant cytoplasm in exchange for the export of dicarboxylates. Initially, it was believed that simple diffusion of uncharged NH_3 across the bacteroid and symbiosome membranes provided the main pathways for symbiotic nitrogen delivery to the host (Streeter 1989). Then, patch-clamp recordings with soybean symbiosome membranes proposed a voltage-activated cation channel to transport ammonium from the symbiosome space into the cytosol (Roberts and Tyerman 2002; Tyerman et al. 1995), again without knowing the molecular identity of these NH_4^+-permeable channels.

The mutant *S. cerevisiae* strain originally used to clone *S. cerevisiae MEP2* was used to isolate potential *AMT* genes from a soybean cDNA library by complementation of the yeast growth phenotype on limiting ammonium. This led to the isolation of *GmSAT1* (*Glycine max* symbiotic ammonium transporter), which was proposed to encode an ammonium transporter located on the soybean peribacteroid membrane that could be involved in the transfer of fixed nitrogen from the bacteroid to the host (Kaiser et al. 1998). However, later studies revealed that *GmSAT1* could not complement the triple *MEP* deletion strain, that *GmSAT1* caused overexpression of Mep3 and that GmSAT1 has features of a transcriptional regulatory protein including a DNA-binding domain. It was subsequently concluded that GmSAT1 is not an ammonium transporter but rather a protein that interferes with the Mep1(G413D)-dependent inhibition of Mep3 (Marini et al. 2000b). Although interference between *MEP1* and *MEP3* at the transcriptional level could not be ruled out, a direct interaction between the two Mep proteins was suggested, reminiscent of the hetero-oligomers formed between Rhesus proteins (Hartel-Schenk and Agre 1992).

The view of passive NH_3 diffusion across the symbiosome bilayer was further revised by the observation that facilitated ammonium diffusion could be inhibited by mercurials (Niemietz and Tyerman 2000) and that Nodulin 26, which represents a major intrinsic protein from the symbiosome membrane, is a mercurial-sensitive water and solute channel (Dean et al. 1999; Rivers et al. 1997). In parallel, expression of the mammalian aquaporin AQP1 in oocytes provoked an internal acidification of the oocytes in response to ammonium supply at high pH, indicating that NH_3 might be a transported substrate by aquaporins (Nakhoul et al. 2001). Thus, due to their high density in any membrane, major intrinsic proteins (MIPs) could in general provide an efficient pathway for a pH-dependent equilibration of ammonium concentrations between cellular compartments.

References

Arcondéguy T, Jack R, Merrick M (2001) P_{II} signal transduction proteins: pivotal players in microbial nitrogen control. Microbiol Mol Biol Rev 65:80-105

Avent ND, Liu W, Warner KM, Mawby WJ, Jones JW, Ridgwell K, Tanner MJ (1996) Immunochemical analysis of the human erythrocyte Rh polypeptides. J Biol Chem 271:14233-14239

Ayling SM (1993) The effect of ammonium-ions on membrane-potential and anion flux in roots of barley and tomato. Plant Cell Environ16:297-303

Bai G, Rama Rao KV, Murthy CR, Panickar KS, Jayakumar AR, Norenberg MD (2001) Ammonia induces the mitochondrial permeability transition in primary cultures of rat astrocytes. J Neurosci Res 66:981-991

Becker D, Stanke R, Fendrik I, Frommer WB, Vanderleyden J, Kaiser WM, Hedrich R (2002) Expression of the NH_4^+-transporter gene *LeAMT1;2* is induced in tomato roots upon association with N_2-fixing bacteria. Planta 215:424-429

Bihler H, Slayman CL, Bertl A (1998) NSC1: a novel high-current inward rectifier for cations in the plasma membrane of *Saccharomyces cerevisiae*. FEBS Lett 432:59-64

Blakey D, Leech A, Thomas GH, Coutts G, Findlay K, Merrick M (2002) Purification of the *Escherichia coli* ammonium transporter AmtB reveals a trimeric stoichiometry. Biochem J 364:527-535

Bloom AJ, Smart DR, Nguyen DT, Searles PS (2002) Nitrogen assimilation and growth of wheat under elevated carbon dioxide. Proc Natl Acad Sci USA 99:1730-1735

Britto DT, Siddiqi MY, Glass AD, Kronzucker HJ (2001a) Futile transmembrane NH_4^+ cycling: a cellular hypothesis to explain ammonium toxicity in plants. Proc Natl Acad Sci USA 98:4255-4258

Britto DT, Glass AD, Kronzucker HJ, Siddiqi MY (2001b) Cytosolic concentrations and transmembrane fluxes of NH_4^+/NH_3. An evaluation of recent proposals. Plant Physiol 125:523-526

Cordts ML, Gibson J (1987) Ammonium and methylammonium transport in *Rhodobacter sphaeroides*. J Bacteriol 169:1632-1638

Coutts G, Thomas G, Blakey D, Merrick M (2002) Membrane sequestration of the signal transduction protein GlnK by the ammonium transporter AmtB. EMBO J 21:1-10

Dandekar T, Snel B, Huynen M, Bork P (1998) Conservation of gene order: a fingerprint of proteins that physically interact. Trends Biochem Sci 23:324-328

Dean RM, Rivers RL, Zeidel ML, Roberts DM (1999) Purification and functional reconstitution of soybean nodulin 26. An aquaporin with water and glycerol transport properties. Biochemistry 38:347-353

Demidchik V, Tester M (2002) Sodium fluxes through nonselective cation channels in the plasma membrane of protoplasts from *Arabidopsis* roots. Plant Physiol 128:379-387

Eyers SA, Ridgwell K, Mawby WJ, Tanner MJ (1994) Topology and organization of human Rh (rhesus) blood group-related polypeptides. J Biol Chem 269:6417-6423

Fabiny JM, Jayakumar A, Chinault AC, Barnes EM (1991) Ammonium transport in *Escherichia coli*: localisation and nucleotide sequence of the *amtA* gene. J Gen Microbiol 137:983-989

Gansel X, Munos S, Tillard P, Gojon A (2001) Differential regulation of the NO_3^- and NH_4^+ transporter genes *AtNrt2.1* and *AtAmt1.1* in *Arabidopsis*: relation with long-distance and local controls by N status of the plant. Plant J 26:143-155

Gazzarrini S, Lejay L, Gojon A, Ninnemann O, Frommer WB, von Wirén N (1999) Three functional transporters for constitutive, diurnally regulated, and starvation-induced uptake of ammonium into *Arabidopsis* roots. Plant Cell 11:937-948

Gerendas J, Zhu ZJ, Bendixen R, Ratcliffe RG, Sattelmacher B (1997) Physiological and biochemical processes related to ammonium toxicity in higher plants. Zeitschrift Pflanzenernaehr und Bodenk 160:239-251

Hallenbeck PC, Yakunin AF, Drepper T, Gross S, Masepohl B, Klipp W (2002) Regulation of nitrogenase in the photosynthetic bacterium *Rhodobacer capsulatus*. In: Finan TM, O'brian MR, Layzell DB, Vessey JK, Newton WE (eds) Nitrogen Frixation: Global Perspectives. Proceedings of the 13th International Congress on Nitrogen Fixation. CABI Publishing, New York, pp 223-227

Hartel-Schenk S, Agre P (1992) Mammalian red cell membrane Rh polypeptides are selectively palmitoylated subunits of a macromolecular complex. J Biol Chem 267:5569-5574

He L, Soupene E, Ninfa AJ, Kustu S (1998) Physiological role for the GlnK protein of enteric bacteria: relief of NifL inhibition under nitrogen-limiting conditions. J Bacteriol 180:6661-6667

Howitt S, Udvardi M (2000) Structure, function, and regulation of ammonium transporters in plants. Biochem Biophys Acta 1465:152-170

Jack R, de Zamaroczy M, Merrick M (1999) The signal transduction protein GlnK is required for NifL-dependent nitrogen control of *nif* expression in *Klebsiella pneumoniae*. J Bacteriol 181:1156-1162

Javelle A, Rodriguez-Pastrana BR, Jacob C, Botton B, Brun A, Andre B, Marini AM, Chalot M (2001) Molecular characterization of two ammonium transporters from the ectomycorrhizal fungus *Hebeloma cylindrosporum*. FEBS Lett 505:393-398

Javelle A, Andre B, Marini AM, Chalot M (2003a) High-affinity ammonium transporters and nitrogen sensing in mycorrhizas. Trends Microbiol 11:53-55

Javelle A, Morel M, Rodriguez-Pastrana BR, Botton B, Andre B, Marini AM, Brun A, Chalot M (2003b) Molecular characterization, function and regulation of ammonium transporters (Amt) and ammonium-metabolizing enzymes (GS, NADP-GDH) in the ectomycorrhizal fungus *Hebeloma cylindrosporum*. Mol Microbiol 47:411-430

Kaiser BN, Finnegan PM, Tyerman SD, Whitehead LF, Bergersen FJ, Day DA, Udvardi MK (1998) Characterization of an ammonium transport protein from the peribacteroid membrane of soybean nodules. Science 281:1202-1206

Kaiser BN, Rawat SR, Siddiqi MY, Masle J, Glass AD (2002) Functional analysis of an *Arabidopsis* T-DNA "knockout" of the high-affinity NH_4^+ transporter AtAMT1;1. Plant Physiol 130:1263-1275

Klopprogge K, Grabbe R, Hoppert M, Schmitz RA (2002) Membrane association of *Klebsiella pneumoniae* NifL is affected by molecular oxygen and combined nitrogen. Arch Microbiol 177:223-234

Kosola KR, Bloom AJ (1994) Methylammonium as a transport analog for ammonium in tomato (*Lycopersicon esculentum* L.). Plant Physiol 105:435-442

Kumar A, Silim SN, Okamoto M, Siddiqi MY, Glass AD (2003) Differential expression of three members of the AMT1 gene family encoding putative high-affinity NH_4^+ transporters in roots of *Oryza sativa* subspecies *indica*. Plant Cell Environ 26:907-914

Lauter FR, Ninnemann O, Bucher M, Riesmeier JW, Frommer WB (1996) Preferential expression of an ammonium transporter and of two putative nitrate transporters in root hairs of tomato. Proc Natl Acad Sci USA 93:8139-8144

Lejay L, Gansel X, Cerezo M, Tillard P, Muller C, Krapp A, von Wiren N, Daniel-Vedele F, Gojon A (2003) Regulation of root ion transporters by photosynthesis: functional importance and relation with hexokinase. Plant Cell 15:2218-2232

Liu Z, Peng J, Mo R, Hui C-C, Huang C-H (2000) Rh type B glycoprotein is a new member of the Rh superfamily and a putative ammonia transporter in mammals. J Biol Chem 276:1424-1433

Lorenz MC, Heitman J (1997) Yeast pseudohyphal growth is regulated by GPA2, a G protein alpha homolog. EMBO J 16:7008-7018

Lorenz MC, Heitman J (1998) The MEP2 ammonium permease regulates pseudohyphal differentiation in *Saccharomyces cerevisiae*. EMBO J 17:1236-1247

Ludewig U, von Wiren N, Rentsch D, Frommer WB (2001) Rhesus factors and ammonium: a function in efflux? Genome Biol 2:1-5

Ludewig U, von Wirén N, Frommer WB (2002) Uniport of NH_4^+ by the root hair plasma membrane ammonium transporter LeAMT1;1. J Biol Chem 277:13548-13555

Ludewig U, Wilken S, Wu B, Jost W, Obrdlik P, El Bakkoury M, Marini AM, Andre B, Hamacher T, Boles E, von Wiren N, Frommer WB (2003) Homo- and heterooligomerization of AMT1 NH_4^+-uniporters. J Biol Chem. 278:45603-45610

Marini A-M, Vissers S, Urrestarazu A, Andre B (1994) Cloning and expression of the MEP1 gene encoding an ammonium transporter in *Saccharomyces cerevisiae*. EMBO J 13:3456-3463

Marini A-M, Soussi-Boudekou S, Vissers S, Andre B (1997a) A family of ammonium transporters in *Saccharomyces cerevisae*. Mol Cell Biol 17:4282-4293

Marini A-M, Urrestarazu A, Beauwens R, Andre B (1997b) The Rh (Rhesus) blood group polypeptides are related to NH_4^+ transporters. Trends Biochem Sci 22:460-461

Marini A-M, Andre B (2000) *In vivo* N-glycosylation of the Mep2 high-affinity ammonium transporter of *Saccharomyces cerevisiae* reveals an extracytosolic N-terminus. Mol Microbiol 38:552-564

Marini A-M, Matassi G, Raynal V, Andre B, Cartron JP, Cherif-Zahar B (2000a) The human Rhesus-associated RhAG protein and a kidney homologue promote ammonium transport in yeast. Nat Genet 26:341-344

Marini A-M, Springael JY, Frommer WB, Andre B (2000b) Cross-talk between ammonium transporters in yeast and interference by the soybean SAT1 protein. Mol Microbiol 35:378-385

Marschner H (1995) Mineral nutrition in higher plants. Academic Press, London, UK

Martin DE, Reinhold-Hurek B (2002) Distinct roles of P_{II}-like signal transmitter proteins and *amtB* in regulation of *nif* gene expression, nitrogenase activity, and posttranslational modification of NifH in *Azoarcus* sp. strain BH72. J Bacteriol 184:2251-2259

Matt P, Geiger M, Walch-Liu P, Engels C, Krapp A, Stitt M (2001) Elevated carbon dioxide increases nitrate uptake and nitrate reductase activity when tobacco is growing on nitrate, but increases ammonium uptake and inhibits nitrate reductase activity when tobacco is growing on ammonium nitrate. Plant Cell Environ. 24:1119-1137

Meier-Wagner J, Nolden L, Jakoby M, Siewe R, Krämer R, Burkovski A (2001) Multiplicity of ammonium uptake systems in *Corynebacterium glutamicum*: role of Amt and AmtB. Microbiol 147:135-143

Monahan BJ, Fraser JA, Hynes MJ, Davis MA (2002a) Isolation and characterization of two ammonium permease genes, meaA and mepA, from *Aspergillus nidulans*. Eukaryot Cell 1:85-94

Monahan BJ, Unkles SE, Tsing IT, Kinghorn JR, Hynes MJ, Davis MA (2002b) Mutation and functional analysis of the *Aspergillus nidulans* ammonium permease MeaA and evidence for interaction with itself and MepA. Fungal Genet Biol 36:35-46

Montanini B, Moretto N, Soragni E, Percudani R, Ottonello S (2002) A high-affinity ammonium transporter from the mycorrhizal ascomycete *Tuber borchii*. Fungal Genet Biol 36:22-34

Moroni A, Bardella L, Thiel G (1998) The impermeant ion methylammonium blocks K^+ and NH_4^+ currents through KAT1 channel differently: evidence for ion interaction in channel permeation. J Membr Biol 163:25-35

Nakhoul NL, Hering-Smith KS, Abdulnour-Nakhoul SM, Hamm LL (2001) Transport of NH_3/NH in oocytes expressing aquaporin-1. Am J Physiol Renal Physiol 281:F255-F263

Neuwald AF, Krishnan BR, Brikun I, Kulakauskas S, Suziedelis K, Tomcsanyi T, Leyh TS, Berg DE (1992) *cysQ*, a gene needed for cysteine synthesis in *Escherichia coli* K-12 only during aerobic growth. J Bacteriol 174:415-425

Niemietz CM, Tyerman SD (2000) Channel-mediated permeation of ammonia gas through the peribacteroid membrane of soybean nodules. FEBS Lett 465:110-114

Ninnemann O, Jauniaux J-C, Frommer WB (1994) Identification of a high affinity NH_4^+ transporter from plants. EMBO J 13:3464-3471

Palková Z, Janderová B, Gabriel J, Zikanová B, Pospisek M, Forstová J (1997) Ammonia mediates communication between yeast colonies. Nature 390:532-536

Palková Z, Devaux F, Ricicová M, Mináriková L, Le Crom S, Jacq C (2002) Ammonia pulses and metabolic oscillations guide yeast colony development. Mol Biol Cell 13:3901-3914

Plant PJ, Manolson MF, Grinstein S, Demaurex N (1999) Alternative mechanisms of vacuolar acidification in H^+-ATPase-deficient yeast. J Biol Chem 274:37270-37279

Rawat SR, Silim SN, Kronzucker HJ, Siddiqi MY, Glass AD (1999) *AtAMT1* gene expression and NH_4^+ uptake in roots of *Arabidopsis thaliana*: evidence for regulation by root glutamine levels. Plant J 19:143-152

Rivers RL, Dean RM, Chandy G, Hall JE, Roberts DM, Zeidel ML (1997) Functional analysis of nodulin 26, an aquaporin in soybean root nodule symbiosomes. J Biol Chem 272:16256-16261

Roberts DM, Tyerman SD (2002) Voltage-dependent cation channels permeable to NH_4^+, K^+, and Ca^{2+} in the symbiosome membrane of the model legume *Lotus japonicus*. Plant Physiol 128:370-378

Saier MH Jr, Eng BH, Fard S, Garg J, Haggerty DA, Hutchinson WJ, Jack DL, Lai EC, Liu HJ, Nussinew DP, Omar AM, Pao SS, Paulsen IT, Quan JA, Sliwinski M, Tseng T-T, Wachi S, Young GB (1999) Phylogenetic characterisation of novel transport protein families revealed by genome analysis. Biochem Biophys Acta 1422:1-56

Salvemini F, Marini A, Riccio A, Patriarca EJ, Chiurazzi M (2001) Functional characterization of an ammonium transporter gene from *Lotus japonicus*. Gene 270:237-243

Shelden MC, Dong B, de Bruxelles GL, Trevaskis B, Whelan J, Ryan PR, Howitt SM, Udvardi MK (2001) *Arabidopsis* ammonium transporters, AtAMT1;1 and AtAMT1;2, have different biochemical properties and functional roles. Plant Soil 231:151-160

Siewe RM, Weil B, Burkovski A, Eikmanns BJ, Eikmanns M, Krämer R (1996) Functional and genetic characterisation of the (methyl)ammonium uptake carrier of *Corynebacterium glutamicum*. J Biol Chem 271:5398-5403

Simon-Rosin U, Wood C, Udvardi MK (2003) Molecular and cellular characterisation of LjAMT2;1, an ammonium transporter from the model legume *Lotus japonicus*. Plant Mol Biol 51:99-108

Smith SE, Read DJ (1997) Mycorrhizal symbiosis. Academic Press, London

Sohlenkamp C, Shelden M, Howitt S, Udvardi M (2000) Characterization of *Arabidopsis* AtAMT2, a novel ammonium transporter in plants. FEBS Lett 467:273-278

Sohlenkamp C, Wood CC, Roeb GW, Udvardi MK (2002) Characterization of *Arabidopsis* AtAMT2, a high-affinity ammonium transporter of the plasma membrane. Plant Physiol 130:1788-1796

Sonoda Y, Ikeda A, Saiki S, von Wiren N, Yamaya T, Yamaguchi J (2003) Distinct expression and function of three ammonium transporter genes (OsAMT1;1-1;3) in rice. Plant Cell Physiol 44:726-734

Soupene E, He L, Yan D, Kustu S (1998) Ammonia acquisition in enteric bacteria: physiological role of the ammonium/methylammonium transport B (AmtB) protein. Proc Natl Acad Sci USA 95:7030-7034

Soupene E, Ramirez RM, Kustu S (2001) Evidence that fungal MEP proteins mediate diffusion of the uncharged species NH_3 across the cytoplasmic membrane. Mol Cell Biol 21:5733-5741

Soupene E, Lee H, Kustu S (2002a) Ammonium/methylammonium transport (Amt) proteins facilitate diffusion of NH₃ bidirectionally. Proc Natl Acad Sci USA 99:3926-3931

Soupene E, Chu T, Corbin RW, Hunt DF, Kustu S (2002b) Gas channels for NH₃: proteins from hyperthermophiles complement an *Escherichia coli* mutant. J Bacteriol 184:3396-3400

Streeter JG (1989) Estimation of ammonium concentration in the cytosol of soybean nodules. Plant Physiol 90:779-782

Suenaga A, Moriya K, Sonoda Y, Ikeda A, von Wiren N, Hayakawa T, Yamaguchi J, Yamaya T (2003) Constitutive expression of a novel-type ammonium transporter OsAMT2 in rice plants. Plant Cell Physiol 44:206-211

Thomas G, Coutts G, Merrick M (2000a) The *glnKamtB* operon: a conserved gene pair in prokaryotes. Trends Genet 16:11-14

Thomas GH, Mullins JG, Merrick M (2000b) Membrane topology of the Mep/Amt family of ammonium transporters. Mol Microbiol 37:331-344

Tyerman SD, Whitehead LF, Day DA (1995) A channel-like transporter for NH₄⁺ on the symbiotic interface of N₂-fixing plants. Nature 378:629-632

Tyerman SD, Niemietz CM, Bramley H (2002) Plant aquaporins: multifunctional water and solute channels with expanding roles. Plant Cell Environ 25:173-194

Ullrich WR, Larsson M, Larsson CM, Lesch S, Novacky A (1984) Ammonium uptake in *Lemna gibba* G-1, related membrane-potential changes, and inhibition of anion uptake. Physiol. Plant. 61:369-376

Uozumi N, Gassmann W, Cao Y, Schroeder JI (1995) Identification of strong modifications in cation selectivity in an *Arabidopsis* inward rectifying potassium channel by mutant selection in yeast. J Biol Chem 270:24276-24281

van Dommelen A, de Mot R, Vanderleyden J (2001) Ammonium transport: unifying concepts and unique aspects. Aust J Pl Physiol 28:959-967

Veenhoff LM, Heuberger EH, Poolman B (2002) Quaternary structure and function of transport proteins. Trends Biochem Sci 27:242-249

Vermeiren H, Keijers V, Vanderleyden J (2002) Isolation and sequence analysis of the *glnKamtB1amtB2* gene cluster, encoding a P$_{II}$ homologue and two putative ammonium transporters, from *Pseudomonas stutzeri* A15. DNA Seq 13:67-74

von Wirén N, Lauter FR, Ninnemann O, Gillissen B, Walch-Liu P, Engels C, Jost W, Frommer WB (2000) Differential regulation of three functional ammonium transporter genes by nitrogen in root hairs and by light in leaves of tomato. Plant J 21:167-175

Wang MY, Siddiqi MY, Ruth TJ, Glass A (1993) Ammonium uptake by rice roots (II. Kinetics of ¹³NH₄⁺ influx across the plasmalemma). Plant Physiol 103:1259-1267

Wang RC, Guegler K, Labrie ST, Crawford NM (2000) Genomic analysis of a nutrient response in *Arabidopsis* reveals diverse expression patterns and novel metabolic and potential regulatory genes induced by nitrate. Plant Cell 12:1491-1509

Ward JM (1997) Patch-clamping and other molecular approaches for the study of plasma membrane transporters demystified. Plant Physiol 114:1151-1159

Westhoff CM, Ferreri-Jacobia M, Mak DO, Foskett JK (2002) Identification of the erythrocyte Rh-blood group glycoprotein as a mammalian ammonium transporter. J Biol Chem 277:12499-12502

Xu Y, Cheah E, Carr PD, van Heeswijk WC, Westerhoff HV, Vasudevan SG, Ollis DL (1998) GlnK, a PII-homologue: structure reveals ATP binding site and indicates how the T-loops may be involved in molecular recognition. J Mol Biol 282:149-165

Yakunin AF, Hallenbeck PC (2002) AmtB is necessary for NH_4^+-induced nitrogenase switch-off and ADP-ribosylation in *Rhodobacter capsulatus*. J Bacteriol 184:4081-4088

Zikánová B, Kuthan M, Ricicová M, Forstová J, Palková Z (2002) Amino acids control ammonia pulses in yeast colonies. Biochem Biophys Res Commun 294:962-967

Merrick, Mike
Department of Molecular Microbiology, John Innes Centre, Norwich NR4 7UH, UK
mike.merrick@bbsrc.ac.uk

von Wirén, Nicolaus
Institut für Pflanzenernährung, Fachbereich "Mineralstoffwechsel und - transport", Universität Hohenheim, D-70593 Stuttgart, Germany

Role of transporter-like sensors in glucose and amino acid signalling in yeast

Eckhard Boles and Bruno André

Abstract

Nutrient sensing is crucial to enabling cells to respond properly to changing food supplies. Recent studies suggest that specific members of large transporter families have no detectable transport activity and serve, rather, as nutrient sensors. Transporter-like sensors were first described in the yeast *Saccharomyces cerevisiae*, but there is mounting evidence that similar systems exist in higher eukaryotes. The glucose sensors Snf3 and Rgt2 and the amino acid sensor Ssy1 are the prototypes of this new category of nutrient sensors in yeast. These proteins activate signalling pathways in response to detection of glucose or amino acids in the external medium, and are involved in the transcriptional regulation of genes encoding transporters for these nutrients. In this paper, we review current knowledge about the two yeast nutrient signalling pathways, with emphasis on mechanisms of nutrient perception by the sensors.

1 Introduction

Transmembrane transport proteins (channels, pumps, and uni-, sym-, and antiporters) of the plasma membrane play a crucial role in communication between cells and their immediate environment. This, of course, is because these proteins ensure selective exchange of a wide variety of compounds between the cytosol and the extracellular medium. Recent studies, however, have revealed that specific members of large transporter families have a regulatory instead of a catalytic function, i.e. these proteins have the capability to activate signalling pathways in response to detection of specific compounds in the external medium. Although these proteins seemingly act as receptors, they are usually referred to as sensors because their exact mode of functioning remains uncertain. Probably the best-studied representatives of this new category of proteins are the Snf3 and Rgt2 glucose sensors and the Ssy1 amino acid sensor of the yeast *S. cerevisiae*. Typically, these sensors strongly resemble transporters, but they differ from transporters in three important ways: they seem devoid of transport activity, they contain extended cytoplasmic domains assumed to interact with components of the signal transduction pathway, and in keeping with their regulatory role, they are expressed to much lower levels than actual nutrient transporters. Furthermore, the signalling pathways connected

Topics in Current Genetics, Vol. 9
E. Boles, R. Krämer (Eds.): Molecular Mechanisms Controlling Transmembrane Transport
DOI 10.1007/b95773 / Published online: 9 March 2004
© Springer-Verlag Berlin Heidelberg 2004

to these glucose and amino acid sensors play a central role in transcriptional regulation of genes encoding glucose and amino acid transporters, respectively. Studies performed over the past few years have unveiled several molecular aspects of these nutrient signalling pathways. In this review, we present a comparative description of these sensing systems and we also briefly mention similar situations encountered in other microorganisms and higher eukaryotes.

2 The Snf3, Rgt2, and Ssy1 sensors are members of large transporter families

2.1 The yeast hexose transporter family

The yeast hexose transporter family consists of 19 proteins, Hxt1 to -11, Hxt13 to -17, Gal2, Snf3, and Rgt2 (an *HXT12* pseudogene has also been found) (Kruckeberg 1996). The main hexose transporters are Hxt1-7 and Gal2. They act as facilitated diffusion transporters (Maier et al. 2002). Not only do these proteins differ as regards their kinetic properties, but they are also expressed differentially in response to glucose. *S. cerevisiae* cells express only those glucose transporters, which are appropriate for the amount of extracellular glucose available (Boles 2002). The expression pattern of each Hxt protein is consistent with its function as a low-, intermediate-, or high-affinity transporter. Five different kinds of transcriptional regulation can be distinguished: repression in the absence of glucose, induction by low levels of glucose, induction by high levels of glucose, induction by glucose but independently of the glucose concentration, and repression by high levels of glucose.

The properties of the yeast glucose transporters have recently been reviewed in Boles (2002). Hxt1 is a transporter with an extremely low affinity for glucose (Km $\sim$ 100 mM). In keeping with this, Hxt1 is expressed only when the extracellular glucose level is very high. Hxt2 and Hxt4 have a moderately low affinity for glucose (Km $\sim$ 10 mM). The corresponding genes are induced by low levels of glucose and strongly repressed in the presence of high concentrations of glucose. Hxt3 is a low-affinity glucose transporter with a Km of about 60 mM. *HXT3* expression is dependent on glucose but independent of its concentration. *HXT5* is expressed only in glucose-deprived cells or under a variety of starvation and stress conditions. Hxt5 has an intermediate affinity for glucose (Km $\sim$ 7 mM) (Buziol et al. 2002). Hxt6 and Hxt7 are a pair of closely related, high-affinity glucose transporters (Km = 1-2 mM). They are induced by low concentrations of glucose and strongly repressed at higher concentrations, but in contrast to all other Hxt proteins, they exhibit high basal expression also during growth on alternative carbon sources like ethanol. Gal2 is the yeast high-affinity galactose permease, and its expression is induced by galactose and repressed by glucose. Yet, it can transport both galactose and glucose. Galactose-triggered induction is under the control of the Gal1/3-Gal80-Gal4 signalling pathway, which is activated directly by binding of intracellular galactose to Gal1 or Gal3 (Bhat and Murthy 2001). The genes

HXT8 to *HXT17* (except for *HXT12* - a pseudogene) encode hexose transporters that can mediate uptake of hexoses into yeast cells only when overproduced. Under normal physiological conditions, they are very poorly transcribed. It is therefore reasonable to assume that they are involved in only very specific functions or that hexoses are not their true substrates.

The Snf3 and Rgt2 proteins lie at the boundary of the hexose transporter family, as they show only limited sequence similarity to the other members. *SNF3* and *RGT2* encode proteins of 884 and 763 amino acids respectively, whereas the other hexose transporters vary in length from 541 to 592 amino-acid residues (Kruckeberg 1996). All hexose transporters, including the sensors, contain 12 putative membrane-spanning domains. However, Snf3 and Rgt2 contain very long cytoplasmatic C-terminal extensions (~300 amino acids in Snf3, ~220 in Rgt2). *SNF3* and *RGT2* are expressed only to very low levels (Ozcan et al. 1996a) and they do not seem able to transport glucose. Even when overproduced, both proteins appear unable to restore the ability of an *hxt*-null strain (devoid of any hexose transport activity) to grow on a glucose medium (Ozcan et al. 1998; Dlugai et al. 2001). Attempts to convert them to glucose transporters by (even extensive) mutagenesis have failed (Dlugai et al. 2001). Nevertheless, these experiments do not exclude the possibility that the sensors might be able to transport glucose only in small and regulatory amounts that are too low for catabolic purposes. As detailed below, experiments have shown that, rather than being glucose transporters, Snf3 and Rgt2 are glucose sensors that control transcription of *HXT* genes in response to external glucose (Ozcan and Johnston 1999).

2.2 The yeast amino acid transporters

Yeast possesses about twenty distinct amino acid transporters likely acting as H^+ symporters. Most of these proteins belong to a single family (named AAP or YAT), which is conserved in fungi and bacteria (Grenson 1992) and is part of the APC superfamily of transporters (Saier Jr 2000). This family includes the histidine permease (Hip1), the first membrane transport protein molecularly characterised in yeast (Tanaka and Fink 1985). These amino-acid permeases of the AAP/YAT family consist of a hydrophobic core of 12 transmembrane domains flanked by hydrophilic tails facing the cytosol (Pi and Pittard 1996; Hu and King 1998; Gilstring and Ljungdahl 2000). It has long been thought that yeast has a general amino-acid permease (Gap1) regulated at both gene and protein levels according to the general nitrogen status of the cell, and several specific amino-acid permeases insensitive to these controls (e.g. Hip1, Lyp1) (Grenson 1992; Magasanik & Kaiser 2002). This view has recently been revised, however, since several broad-range-specificity amino-acid permeases have since been characterised. Moreover, several specific permeases are now known to display wider-range specificities than originally reported (Regenberg et al. 1999; Iraqui et al. 1999).

The general amino-acid permease (Gap1) (Grenson et al. 1970; Jauniaux and Grenson 1990) can recognise most if not all amino acids, including ones not present in proteins (e.g. citrulline, ornithine, γ-aminobutyric acid), D-isomers, β-alanine, and many toxic amino-acid analogues. Contrary to a widely held belief, Gap1 displays a very high affinity for most of its natural substrates (apparent Km in the micromolar range). Its affinity for citrulline (an amino acid incorporated into yeast solely via Gap1) is lower (Km ~ 0.08 mM). This permease is most active under limiting nitrogen supply conditions, e.g., when the sole nitrogen source is ammonium at low concentration or urea. The main function of Gap1 under these conditions is probably to scavenge traces of amino acids to be used as a source of nitrogen. The Gap1 permease is also highly active in cells growing on proline. This amino acid is considered to be a main source of nitrogen in the natural environment of yeast. Proline uptake is also mediated by Put4, another AAP/YAT-family protein also able to recognise GABA (Vandenbol et al. 1989), alanine, and glycine (Regenberg et al. 1999). Like Gap1, Put4 is subject to tight nitrogen control at both the transcriptional and membrane-trafficking levels (Magasanik and Kaiser 2002). When amino acids (other than proline) are present at relatively high concentrations, Gap1 is inactive and other broad-range-specificity amino-acid permeases (mainly those under Ssy1 control) are induced. Among the latter is Agp1, an amino-acid permease also subject to nitrogen regulation and able to recognise most neutral amino acids. The affinity of Agp1 for these amino acids is lower (Km ~ 100-200 μM) than that of Gap1 (Iraqui et al. 1999). Other amino-acid permeases first believed to recognise only a single or a few amino acids (Bap2, Bap3, Gnp1, Tat1, Tat2, Dip5) actually recognise larger sets of amino acids, although their specificity ranges do seem narrow as compared to Agp1 (Regenberg et al. 1999).

Besides Gap1 and the above-described broad-range-specificity amino-acid permeases, yeast cells possess a number of specific permeases apparently not under Ssy1 or nitrogen control, mediating high-affinity uptake of lysine (Lyp1) (Sychrova and Chevallier 1993), arginine (Can1, Alp1) (Hoffmann 1985; Regenberg et al. 1999), or histidine (Hip1) (Tanaka and Fink 1985). The AAP/YAT-family proteins of *S. cerevisiae* also include high-affinity permeases for S-adenosylmethionine (Sam3) and S-methylmethionine (Mmp1) (Rouillon et al. 1999) and yet another member (Agp2) that is not an amino-acid permease but a carnitine permease (van Roermund et al. 1999). Finally, at least three other amino-acid permeases of yeast are members of the APC superfamily but not of the AAP/YAT family: the specific methionine permeases (Mup1 and Mup3) (Isnard et al. 1996) and the inducible γ-aminobutyric acid (GABA) permease (Uga4) (André et al. 1993).

The Ssy1 sensor is a distant member of the AAP/YAT family. It also differs from the seventeen other yeast proteins of this family by the presence of extra peptide sequences between transmembrane (TM) domains 5-6 and 7-8 and by its N-terminal tail, much longer (~ 280 amino acids) than the tails of the other AAP/YAT proteins (~ 60 - 100 amino acids). Furthermore, codon bias index values and experiments suggest that the *SSY1* gene is expressed to very low levels (Didion et al. 1998; Iraqui et al. 1999). Several experiments have provided convincing evidence that Ssy1 is devoid of any amino-acid transport activity. First, the *SSY1* gene present on a high-copy plasmid is unable to complement the branched-chain amino-acid uptake defect displayed by *gap1 bap2 tat1 bap3* mutant cells (Didion et al. 1998). Similarly, a *gap1 agp1 SSY1$^+$* mutant is unable to grow on low concentrations (1 mM) of several amino acids as sole nitrogen source (Iraqui et al. 1999). Finally, a leucine auxotrophy introduced into a *gap1 agp1 bap2 SSY1$^+$* strain cannot be compensated by external leucine present at 0.1 mM concentration, even though Ssy1 is known to respond to extremely low concentrations of this amino acid (< 1 μM) (Bernard and André 2001a). Although these observations do not absolutely rule out the possibility that Ssy1 might have intrinsic amino-acid transport activity, they indicate that this putative function is negligible in terms of amino-acid supply. As detailed below, experiments have shown that the role of Ssy1 is not to mediate amino-acid uptake but rather to control transcription of several genes in response to the presence of external amino acids.

3 The yeast glucose and amino acid sensors: initial description and functions

3.1 The Snf3 and Rgt2 glucose sensors

Yeast *snf3* mutants (*sucrose nonfermenting*) were first isolated in a genetic screen for mutant cells unable to use raffinose and sucrose (Neigeborn and Carlson 1984). Kinetic analysis of glucose uptake showed that *snf3* mutants lack high-affinity uptake, but exhibit normal low-affinity uptake (Bisson et al. 1987). The defect in high-affinity glucose transport results in the inability to grow fermentatively on the low concentrations of glucose and fructose produced by hydrolysis of raffinose and sucrose. It was later shown that Snf3 is involved in the transcriptional regulation of the yeast high- (Hxt6 and Hxt7), intermediate- (Hxt2 and Hxt4), and low-affinity (Hxt1 and Hxt3) glucose transporters (Ozcan and Johnston 1995; Ozcan et al. 1998; Schulte et al. 2000). *RGT2* was first cloned as a dominant allele, *RGT2-1*, that enables cells to grow on low levels of glucose when Snf3 is absent (Ozcan et al. 1996a). *RGT2* is located 100 kb downstream from *SNF3* on chromosome IV, and each gene is part of a duplicated block of six genes (Van Belle and André 2001). The *RGT2* gene is involved in transcriptional induction of the low-affinity glucose transporters Hxt1 and Hxt3 (Ozcan et al. 1996a; Schmidt et al. 1999).

In contrast to the other hexose transporters, Snf3 and Rgt2 contain large C-terminal extensions. These extensions are rather dissimilar except for a 25-amino-acid peptide that occurs twice in Snf3 and only once in Rgt2. Several genes encoding Snf3/Rgt2 homologues have been found in the completely sequenced genomes of other yeast and fungal species and all these proteins contain this conserved 25-amino-acid motif (Souciet et al. 2000; Cliften et al. 2003; Kellis et al. 2003). These repeats are important for the signalling function, since deletions that remove them but leave the rest of the tail intact abolish signalling (Ozcan et al. 1998). The membrane-spanning domains of Snf3 and Rgt2 are also required for their glucose sensor function and appear responsible for the fact that these proteins respond differently to external glucose concentrations (Dlugai et al. 2001). The glucose sensors, thus, appear to function as two interacting domains: the membrane-spanning domain necessary for glucose recognition and the C-terminal extension involved in transmission of the glucose signal. In contrast to a recent speculation (Kruckeberg et al. 1998), we have found no indication that the tails of the sensors are cleaved from the rest of the proteins in response to the presence or absence of glucose (C. Cappellaro and E. Boles, unpublished results).

The key experiment confirming that Snf3 and Rgt2 perform regulatory instead of catalytic functions was the isolation of dominant constitutively signalling mutant forms of both proteins. These mutant forms cause high-level expression of *HXT* genes even in the absence of the inducer glucose. The mutations alter an arginine residue (Rgt2: Arg-231 $\rightarrow$ Lys; Snf3: Arg-229 $\rightarrow$ Lys) predicted to be located in a cytoplasmic loop just preceding the fifth transmembrane domain (Ozcan et al. 1996a). This arginine residue is highly conserved among sugar transporters of various organisms. Its role in glucose transport and transporter conformation has been investigated by site-directed mutagenesis of the mammalian GLUT4 glucose transporter (Schurmann et al. 1997). Substitution of this arginine seems to produce a profound disruption of the tertiary structure of the transporter, affecting both transport activity and extra- and intracellular ligand binding. It is reasonable to assume that the corresponding mutation in the glucose sensors locks them in a form that resembles the glucose-bound conformation.

Overexpression of either *SNF3* or *RGT2* causes constitutive *HXT* gene expression. This has been interpreted as resulting from an increase of the small fraction of the receptor naturally present in the ligand-bound form (Ozcan et al. 1998). However, the effect might also be due to sequestration of a negative factor of the glucose signalling pathway. Accordingly, the two homologous proteins Mth1 and Std1 are negative regulators of *HXT* gene expression (see below) and have been shown to bind to the C-terminal tails of Snf3 and Rgt2 (Schmidt et al. 1999; Lafuente et al. 2000). Furthermore, overexpression of the C-terminal tail of Snf3 alone has been found to produce a weak constitutive glucose signal. This signal became much stronger when the tail was attached to the plasma membrane by farnesylation and palmitoylation (Dlugai et al. 2001).

As induction of the high-affinity glucose transporters by low concentrations of glucose is absent in *snf3* mutants, Snf3 is proposed to be a sensor of low levels of glucose. On the other hand, Rgt2 appears as a sensor of high concentrations of glucose, its role being to regulate the expression of *HXT* genes encoding low-affinity glucose transporters (Ozcan and Johnston 1999). Accordingly, induction of the low-affinity transporter Hxt1 by high concentrations of glucose is strongly reduced in *rgt2* mutants. This situation differs from that described in the related yeast *Kluyveromyces lactis*. In this organism, the Rag4 glucose sensor seems to exhibit a dual sensing function for high and low glucose (Betina et al. 2001). Rag4 is highly similar to Snf3 and Rgt2 (72-74% similarity) and contains one copy of the 25-amino-acid sequence motif in its ~250-residue C-terminal tail. In a *rag4* mutant strain, both high-glucose-induced transcription of the low-affinity Rag1 and Kht1 glucose transporters and low-glucose-induced transcription of the high-affinity transporter Kht2 are absent.

The low-affinity glucose sensor Rgt2 seems to be involved not only in regulating gene expression but also in ubiquitin-dependent proteolysis of the yeast maltose transporter Mal61 (Jiang et al. 1997). Degradation of Mal61 is triggered upon arrest of cytosolic protein synthesis in combination with the presence of high concentrations of glucose. This degradation is (partially) dependent on Rgt2 and is operative even in the absence of glucose when cells harbour the dominant *RGT2-1* allele.

3.2 The Ssy1 amino acid sensor

Three laboratories have independently identified *SSY1* (= *YDR160w, SHR5, -10, APF7*) as a gene involved in sensing external amino acids. *SSY1* was first isolated by Per Ljungdahl in G. Fink's lab, in a search for mutants resisting toxic concentrations of histidine (Ljungdahl et al. 1992). Cloning of this gene (first named *SHR5*) revealed that it encodes a distant member of the AAP/YAT amino-acid permease family displaying unusual topological features (see 1.2.2). Since histidine is present at very high concentrations in the vacuole, it was first envisaged that Ssy1 might be a vacuolar basic-amino-acid permease and that its extended N-terminus encoded vacuolar targeting sequences (P. Ljungdahl, personal communication). Studies on Ssy1 were independently initiated in the group of Bruno André after a systematic computer search for yeast genes encoding membrane transporters (André 1995) revealed that a gene of chromosome IV (*YDR160w*) encodes a member of the AAP/YAT family with properties reminiscent of those distinguishing the Snf3 and Rgt2 proteins from the other proteins of the hexose transporter family (see 3.1). Thus, it was postulated that Ydr160/Ssy1 might be a sensor of extracellular amino acids (André 1996). Independently, *ssy1* mutants were isolated by the group of Morten Kielland-Brandt in a search for mutants defective in uptake of branched-chain amino acids. These strains were selected for their hypersensitivity to sulfonylurea herbicides when grown on YPD medium (SSY = <u>s</u>ensitive to <u>s</u>ulfonylurea on <u>Y</u>PD) (Jorgensen et al. 1998). These herbicides inhibit the biosynthesis of branched-chain amino acids, making cells dependent on uptake of an external source of these amino acids for growth. This mutagenesis screen led to the isolation of five classes of mutants (*SSY1* to *-5*) defective in uptake of several amino acids. The *SSY1* gene was cloned and the authors noted the striking features of Ssy1 resembling those of the Snf3 and Rgt2 sensors. Furthermore, *ssy1* mutations turned out to have a regulatory effect, since they prevented a hyper-active mutant form of Bap2 from being active. As the *BAP2* gene was known to be induced by leucine (Didion et al. 1996), it was proposed that Ssy1 might be a sensor of external leucine (Jorgensen et al. 1998).

Didion et al. (1998) reported the first data supporting a role of Ssy1 as a sensor of external amino acids. These authors showed that transcriptional induction by L-leucine of the *BAP2, BAP3, TAT1,* and *PTR2* permease genes is defective in the *ssy1* mutant. Furthermore, *BAP2* was shown to be induced by D-leucine in an Ssy1-dependent manner, even in a strain defective in D-leucine uptake, indicating that the inducer is likely external rather than internal amino acid. Thus, Ssy1 was proposed to be involved in sensing leucine in the medium (Didion et al. 1998). Later, Iraqui et al. (1999) reported that Ssy1 is essential to transcriptional induction by most neutral amino acids (but not proline) of the *AGP1* gene encoding a wide-range-specificity amino-acid permease. Other experiments confirmed that external rather than internal amino acids are the inducers, since: (i) *ssy1* mutants are totally defective in *AGP1* induction but incorporate normal levels of inducer amino acids (Gap1 being active under the conditions used); (ii) *AGP1* is still strongly induced by tryptophan in a mutant defective in tryptophan uptake; (iii) *AGP1* remains unexpressed in a *trp2* mutant strain in which the internal pool of tryptophan is ~70-fold higher than in the wild type. Finally, the same study revealed that Ssy1 is also required for induction by several amino acids of the *BAP2*, *BAP3*, *TAT1*, *GNP1* and *TAT2* genes. In the same year, Klasson and Ljungdahl (1999) reported the independent isolation of *ssy1* mutants selected for hyper-resistance to toxic concentrations of histidine. These authors showed the mutants to be defective in the uptake of several amino acids. In accordance with previous data, Ssy1 also proved to be required for normal levels of *GNP1* and *PTR2* gene transcripts in media containing amino acids. Importantly, these authors showed for the first time that Ssy1 localises to the plasma membrane and that insertion of an epitope tag in its long N-terminal tail impairs the function of Ssy1 in control of amino acid permeability (Klasson et al. 1999).

Although direct binding of amino acids to Ssy1 has not yet been shown, two observations further support the model according to which Ssy1 is involved in recognition of external amino acids. First, a mutant form of Ssy1 (*ssy1-23*) with a single substitution in the eighth predicted transmembrane domain (T639I) has been found to be defective in induction of *AGP1* by several amino acids and yet still able to respond efficiently to leucine (Bernard and André 2001a). Second, using an elegant genetic screen in which K^+ uptake is dependent on Ssy1 activation by external amino acids, Gaber et al. (2003) recently isolated an *ssy1* mutant allele (*SSY1-102*, T382K) causing amino-acid-independent transcriptional activation of the *AGP1* and *BAP2* genes. Furthermore, two other mutants (T382H and T382L) show higher apparent affinity for amino acids. These data are entirely consistent with the model that Ssy1 binds to external amino acids and that this event triggers activation of a pathway leading to transcriptional induction of several permease genes. The long N-terminal tail of Ssy1 likely plays a crucial role in signalling. Accordingly, overproduction of the Ssy1 tail alone complements the growth defects caused by the *ssy1Δ* mutation (Bernard and André 2001a) and also exerts a dominant negative effect on transcription of Ssy1-regulated genes in *SSY1+* cells (Forsberg and Ljungdahl 2001; Bernard and André 2001a).

What is the range of amino acids recognised by Ssy1? de Boer et al (1998) have isolated from the *BAP3* upstream region a minimal sequence (UAS$_{AA}$) sufficient to confer transcriptional induction by external amino acids. This UAS$_{AA}$ responds equally well to all proteinaceous amino acids but glutamate, aspartate, histidine, lysine, arginine, proline, glycine, and cysteine. Iraqui et al. (1999) similarly reported that an *AGP1-lacZ* reporter gene is induced in an Ssy1-dependent manner by all proteinaceous amino acids but proline and arginine, induction by the charged amino acids, glycine, asparagine, and glutamine being much weaker than induction by other amino acids. Similar conclusions were drawn on the basis of growth tests indicative of *AGP1* expression, except that cysteine was not an inducer in this experiment (Gaber et al. 2003). Citrulline is a good inducer of *BAP2* (de Boer et al. 1998) and *AGP1* (Iraqui et al. 1999; Gaber et al. 2003), even in a *gap1Δ* strain defective in citrulline uptake.

What are the target genes of Ssy1? Initial studies showed that they include the amino-acid permease genes *AGP1, BAP2, BAP3, TAT1, TAT2,* and *GNP1,* and the di- and tripeptide permease gene *PTR2,* all subject to transcription-level induction in response to external amino acids (Jorgensen et al. 1998; Didion et al. 1998; de Boer et al. 1998; Iraqui et al. 1999; Klasson et al. 1999; Forsberg et al. 2001b). The impact of the *ssy1Δ* mutation on levels of all gene transcripts was also examined (Forsberg et al. 2001a; Kodama et al. 2002). These experiments revealed that a lack of Ssy1 in cells grown in the presence of amino acids results in improper expression of many other genes, including some involved in amino-acid biosynthesis or nitrogen utilisation. Yet, whether these effects are indirect and due to lower amino-acid uptake or whether they result from control of these genes by Ssy1 remains to be investigated.

The sequencing of the genome of other yeast species allowed the identification of several *SSY1* orthologs (Souciet et al. 2000; Kellis et al. 2003; Cliften et al. 2003). Interestingly, the large N-terminal tails and the extra peptides between TM5 and TM6 are poorly similar while the extra peptide regions between TM7 and TM8 are highly conserved. Recently, the *SSY1* ortholog of *Candida albicans* has been functionally characterized (Brega et al. 2004). This protein (Cys1p) is required for induction at transcriptional level of several amino-acid permease genes as well as for filamentation in serum and amino-acid-based media. Interestingly, the range of amino acids to which Csy1 responds markedly differs from that of Ssy1.

3.3 Mechanisms of sensing: models

The detailed mechanisms of nutrient sensing and signal initiation by the glucose and amino acid sensors are not yet known. We here present two possible models of these processes: i) the sensors act like receptors, ii) the sensors have minor transport activity, which is crucial to signal initiation (see Fig. 1). Furthermore, we discuss several other aspects of the possible mode of functioning of the sensors.

3.3.1 Snf3, Rgt2 and Ssy1 function as receptors

According to this model (Fig. 1B), the sensors act like cell-surface receptors, which upon binding of an extracellular nutrient to their membrane-embedded domain undergo a conformational change that affects their interaction with proteins involved in transduction of the nutrient signal. According to this model, the nutrient sensors are totally devoid of transport activity, i.e. the proteins bind the inducers but these need not cross the lipid bilayer in order to activate the signalling pathway. Nevertheless, the sensors have conserved high sequence similarity to

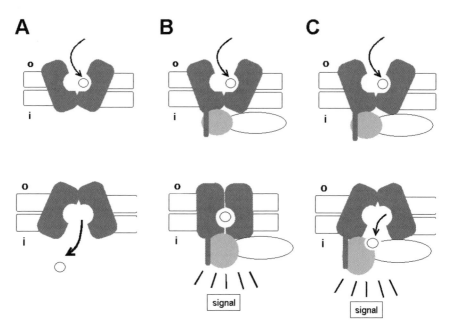

Fig. 1. Schematic presentation of two models of nutrient sensing by glucose and amino acid sensors in S. cerevisiae. A. Classical model of transporter-mediated solute translocation across the membrane. The binding of the compound induces a conformational change of the protein that transfers the solute across the membrane. B. A model of signalling by a transporter-like protein acting as a receptor. The protein has an extended cytoplasmic tail allowing it to associate with factors involved in signal transduction. The binding of the external compound induces a conformational change that is detected by factors associated with the sensor and/or by the extented tail thus leading to signal initiation. C. A model of transport-coupled signalling by a transporter-like sensor. As in model A, the binding of the external compound induces a conformational change of the protein that transfers the solute across the membrane. This conformational change and/or the incorporated solute are detected by associated transduction factors thus leading to signal initiation.

transporters and share similar modes of substrate recognition. They might also undergo similar conformational changes upon binding of the inducer. It is assumed that many transporters with 12 membrane-spanning helices oscillate between an outward- and an inward-facing conformation for transmembrane solute translocation (Fig. 1A; Holman and Rees 1987). In the receptor model, it is tempting to speculate that the sensors might have evolved from transporters in such a way that their shift to the inward-facing conformation is no longer possible (Fig. 1B). Substrate binding would rather induce a conformational change in the sensor, transmitted to a receiver molecule present at the inner face of the plasma membrane. The extended cytoplasmic tails of the sensors may play the role of signal receivers, or be involved in anchoring the proteins receiving the signal.

3.3.2 Snf3, Rgt2 and Ssy1 mediate transport-coupled signalling

The second model implies that signalling is coupled to Snf3/Rgt2/Ssy1-mediated transport of substrates across the lipid bilayer (Fig. 1B). The signal would be initiated, however, by conformational changes undergone by the proteins during the transport cycles and not by the incorporated small molecules themselves. Substrate transport *per se* is indeed not sufficient for signalling, as illustrated by the fact that overexpression of "normal" transporters does not restore normal signalling in *snf3*, *rgt2*, or *ssy1* mutants (Ozcan et al. 1998; Iraqui et al. 1999). Nor is transport of sensor substrates necessary for signalling, as induction of gene expression is normal in mutants defective in transport of these compounds (Didion et al. 1998; Iraqui et al. 1999; Ozcan 2002). Thus, this model resembles the receptor model in that what triggers signalling is the conformational change undergone by the sensor, but it differs from that model in that these conformational changes are due to transport and not just binding of the substrates.

This model is likewise compatible with all the experimental data. According to this model, the constitutive mutant forms of Snf3/Rgt2 and Ssy1 might be sensors trapped in the active "transporting" conformation. The fact that glucose sensors exist as high- and low-affinity systems and the observation that the apparent affinity of Ssy1 can be altered by single amino acid substitutions (see 1.3.2) can also be explained by the same mechanisms of substrate recognition as are attributed to transporters. This model would also nicely explain why the sensors have conserved close sequence similarity to nutrient transporters. In this respect it is also interesting to note that, e.g. the *Amanita muscaria* monosaccharide transporter (AmMST1) shown to transport glucose (Nehls et al. 1998) is much more closely related to Snf3 and Rgt2 than to the yeast hexose transporters.

Yet, another observation might be explained on the basis of this model: the strong Snf3-dependent effect on gene expression observed in an *hxt*-null strain growing on ethanol, i.e. in the absence of external inducer (Reifenberger et al. 1997). Under such conditions, low amounts of intracellular glucose are produced by hydrolysis of storage carbohydrates or gluconeogenic glucose-6-phosphate. As an *hxt*-null strain is supposed to be unable to secrete glucose, this leads to increased intracellular glucose concentrations (Jansen et al. 2002). The Snf3-dependent effects on gene transcription observed under these conditions could be due to the ability of Snf3 to sense intracellular glucose. This interpretation would argue against Snf3 acting as a glucose receptor (restricted to an outward-facing conformation) and rather be in favour of glucose sensors oscillating between outward- and inward-facing "transporting" conformations.

The observation that the hydrophobic part of an Hxt protein (mediating transmembrane transport) can replace that of the glucose sensors for glucose signalling (Ozcan et al. 1998) might be another argument in favour of the transport-coupled model. This observation suggests that the hydrophobic regions of glucose sensors function in a manner very similar to those of transporters. Yet, this idea is challenged by the recent demonstration that attachment of the C-terminal tail of Snf3 alone to the plasma membrane via farnesylation and palmitoylation produces a very strong and constitutive glucose signal (Dlugai et al. 2001). This finding suggests that in the earlier experiments the role played by the hydrophobic part of an Hxt protein was merely to target the C-terminal tail of Snf3 to the plasma membrane.

As this second model cannot be ruled out on the basis of the data currently available, the terms "glucose receptor" and "amino acid receptor" should be avoided when referring to these proteins until conclusive experimental data are provided.

3.3.3 Other aspects on the function of the Snf3, Rgt2, and Ssy1 proteins

According to the above-described models, the sensors would initiate regulatory signals only upon binding or transport of their specific ligands. An alternative view is that these proteins are also able to transmit signals in their unbound forms, the resulting effects on gene transcription being eventually different from those induced when ligands are present. What inspires this view is the unexpected observation that *snf3Δ* or *ssy1Δ* mutations also have an impact on gene transcription when the substrates of the sensors are not present in the medium (Dlugai et al. 2001; Forsberg et al. 2001a). In an *snf3Δ* mutant the activity of the *HXT7* promoter was abolished not only under inducing conditions but also on ethanol, i.e. in the complete absence of glucose (Dlugai et al. 2001). The complete loss of even basal *HXT7* transcription in an *snf3Δ* mutant suggests that Snf3 participates in signal transduction even in the absence of external glucose. The same is apparently true of Ssy1, as multiple genes have been identified that require a normal Ssy1 protein for proper expression even in amino-acid free media. Specifically, in an *ssy1Δ* mutant grown in the absence of amino acids, some genes are derepressed (e.g. *MUP3*, *ALD6*) while others are expressed to a lower level (e.g. *CAN1*, *DIP5*, *LYS9*, *LYS20*). These observations raise the question of whether Snf3, Rgt2 and Ssy1 must not be regarded as proteins having the capability to modulate gene transcription both in the presence and in the absence of external nutrients. According to this hypothesis, external nutrients (through their corresponding sensors) must be considered rather as modulators of the function of these sensors than as molecules absolutely essential to the control of gene expression by these sensors. This would open the question of whether other factors might modulate the signalling properties of the sensors. For instance, it was recently observed that the function of the Ssy1 sensor is negatively controlled by Gap1, the general amino acid permease (El Bakkoury *et al.* in preparation). Furthermore, co-immunoprecipitation and split-ubiquitin two-hybrid experiments indicate that several Hxt proteins are able to interact with Snf3 (C. Cappellaro and E. Boles, unpublished results). This raises the interesting possibility that Snf3's function might be directly influenced by hexose transporters.

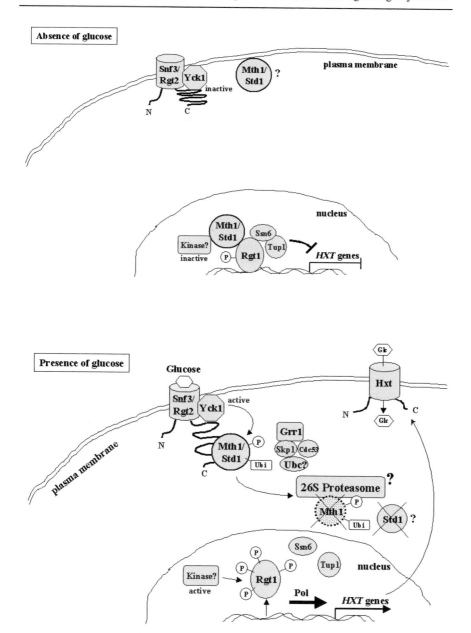

Fig. 2. The current model of the glucose-sensing pathway in *S. cerevisiae*.

4 Signal transduction pathways

4.1 The glucose-sensing pathway

4.1.1 Current view

The current model of the glucose-sensing pathway is schematically depicted in Fig. 2. An essential component of glucose signalling downstream from the glucose sensors is the SCF (Skp1 - cullin/Cdc53 - F-box protein) E3 ubiquitin ligase complex containing Grr1 as its cognate F-box protein (SCF^{Grr1}) (Li and Johnston 1997). SCF designates a family of protein complexes, each with a different F-box protein component. The nature of the F-box protein determines the specificity of the interaction between SCF and its targets (Deshaies 1999). SCF complexes associate with E2 ubiquitin-conjugating enzymes (Ubcs) to add ubiquitin to specific substrate proteins. For instance, the SCF^{Grr1} complex was originally described for its essential role in ubiquitination and degradation of G1 cyclins (Barral et al. 1995). The E2 ubiquitin-conjugating enzyme required for SCF-mediated ubiquitination of nearly all established SCF^{Grr1} targets is Cdc34 (Deshaies 1999). In the case of *HXT* gene expression, transcriptional derepression in response to glucose has been shown to require Skp1, Cdc53, and Grr1 (Li and Johnston 1997). Cdc34 does not appear necessary, however, for *HXT* gene induction. Instead, Ubc8 has been proposed as a likely E2 enzyme (Flick et al. 2003), as it is also involved in Grr1-dependent proteolysis of fructose-1,6-bisphosphatase by glucose (Schule et al. 2000; Horak et al. 2002).

Ubiquitination of substrate proteins by SCF complexes targets them for degradation by the proteasome or can alter their function (Deshaies 1999). As possible targets for glucose-induced inactivation by SCF^{Grr1}, two closely related proteins, Mth1 (=Htr1, Dgt1, Bpc1, Gsf1) and Std1 (=Msn3), have recently been identified (Flick et al. 2003) (Fig. 2). Originally, Std1 was identified as a multicopy suppressor of defects caused by overexpression of a truncated TATA-binding protein (Ganster et al. 1993). The Mth1 protein shares 61% identity with Std1 and its gene was discovered as a dominant negative mutant allele (*HTR1*) that blocks Snf3- and Rgt2-dependent glucose signalling (Schulte et al. 2000). Mth1, but not Std1, is rapidly degraded in response to glucose via a mechanism that requires SCF^{Grr1} (Flick et al. 2003). Mth1 and Std1 interact with the cytoplasmic C-terminal tails of the glucose sensors (Schmidt et al. 1999; Lafuente et al. 2000). This interaction is partially dependent on the carbon source. Both proteins seem to be involved in transduction of the glucose signal from the plasma membrane to the nucleus where they interact with the Rgt1 transcription factor, the ultimate target of the glucose induction signal (Tomas-Cobos and Sanz 2002; Lakshmanan et al. 2003).

Generally, phosphorylation of the target protein is a prerequisite to recognition by SCF complexes (Deshaies 1999). Intriguingly, the yeast membrane-bound casein kinase 1 (Yck1) has recently been found to be involved in Snf3/Rgt2-dependent glucose signalling (M. Johnston, personal communication). Preliminary experiments suggest that Yck1 is activated by the glucose sensors upon binding of glucose. Yck1 might then phosphorylate Mth1 and Std1, marking them for ubiquitination by SCF^{Grr1} and, thus, triggering their degradation/inactivation. Interestingly, the *K. lactis* equivalent of casein kinase I, encoded by *RAG8*, is required in *K. lactis* for transcriptional induction of the *RAG1* low-affinity glucose transporter gene (Blaisonneau et al. 1997).

In the absence of glucose, Mth1 and Std1 interact with the transcriptional repressor Rgt1 and prevent its phosphorylation by an as yet unknown protein kinase (Lakshmanan et al. 2003; Flick et al. 2003) (Fig. 2). Rgt1 is a Zn(II)2-Cys6-cluster-type transcription factor (Ozcan et al. 1996b). It can bind directly to the *HXT1,-2, -3,* and *-4* promoters (Flick et al. 2003; Mosley et al. 2003; Kim et al. 2003), and genetic data suggest that it also acts on the *HXT6* and *-7* promoters (Schulte et al. 2000). Phosphorylation of Rgt1 in the presence of glucose leads to its dissociation from the *HXT* promoters. It has been suggested that Rgt1 might be specifically required for maximal activation of *HXT1* gene expression at high glucose concentrations (Ozcan and Johnston 1999), but recent data show that Rgt1 is not bound to the *HXT1* promoter under these conditions (Flick et al. 2003).

The current model (Fig. 2), thus, suggests that glucose binds to the glucose sensors, thereby inducing a conformational change in the proteins that finally activates casein kinase 1 and possibly strengthens the binding of Mth1/Std1 to the C-terminal tail. Casein kinase 1 phosphorylates Mth1, thereby targeting it for ubiquitination by SCF^{Grr1} and its ultimate degradation. SCF^{Grr1} is also required to inactivate the function of the stable Std1 protein by an unknown mechanism. Blocking of the functions of Mth1 and Std1 leads to phosphorylation of the Rgt1 repressor (by an as yet unknown kinase) and its dissociation from the *HXT* promoters, resulting in derepression of *HXT* transcription. Low concentrations of glucose activate Snf3, finally inducing expression of *HXT2, -3, -6,* and *-7*; high glucose concentrations activate Rgt2, leading to the induction of *HXT1* and *-3*. The differential regulation of individual *HXT* genes by the Snf3 high-affinity and Rgt2 low-affinity sensors and by other signalling pathways is depicted in Fig. 3.

4.1.2 Data supporting this model

Each *HXT* promoter has multiple Rgt1 binding sites (Kim et al. 2003). Rgt1 binds synergistically to the multiple binding sites and mediates synergistic repression of transcription. It does so by recruiting the transcriptional corepressor complex Ssn6-Tup1. Rgt1 is not degraded after addition of glucose to yeast cells, nor is its subcellular localisation affected by glucose (Flick et al. 2003; Kim et al. 2003; Mosley et al. 2003). Instead, its DNA-binding ability is regulated by glucose. The affinity of Rgt1 for *HXT* promoters depends on the amount of glucose available, its affinity being highest in cells grown in the absence of glucose, moderate in cells grown with low levels of glucose, and practically nil in cells grown with high levels of glucose. Glucose acts by inducing hyperphosphorylation of Rgt1, thereby, relieving repression of *HXT* gene expression. The kinase, which phosphorylates Rgt1, is not yet known. Some kinases, including Snf1/AMP-activated protein kinase, protein kinase A, and Sks1/Vhs1, have already been excluded (Flick et al. 2003). Yet, as Rgt1 is always in the nucleus, the kinase is assumed to be nuclear.

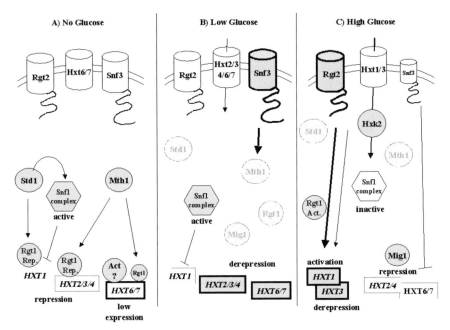

Fig. 3. Model for differential regulation of individual *HXT* genes by various signalling pathways. In the absence of glucose (A), the glucose sensors Rgt2 and Snf3 do not become activated. Therefore, Rgt1 can repress transcription of *HXT* genes as its phosphorylation is blocked by Std1 and Mth1. Furthermore, Std1 activates Snf1 kinase that exerts an inhibitory effect on *HXT1* expression. On the other hand, *HXT6* and *7* do not become completely repressed by Rgt1. In the presence of low levels of glucose (B), the high-affinity sensor Snf3 becomes activated. This leads to inactivation of Std1 and Mth1, dissociation of phosphorylated Rgt1 from *HXT* promoters, and finally derepression of *HXT* transcription. Only the *HXT1* promoter is still repressed via a Snf1-dependent mechanism. In the presence of high concentrations of glucose (C), Rgt2 becomes activated while expression of the *SNF3* gene is repressed. Snf1 kinase becomes inactivated, leading to repression of *HXT2, 4, 6* and *7* by the Mig1 repressor. *HXT1* is specifically induced by two different mechanisms, one that is dependent on Rgt1 and another one that is independent from Rgt1.

Mth1 and (apparently to a lesser extent) Std1 likely play a central role in glucose-dependent hyperphosphorylation and dissociation of Rgt1 from the promoter (Flick et al. 2003). Binding of Rgt1 to *HXT* promoters is fully abolished in *mth1 std1* double mutants under both repressing and non-repressing conditions, and this correlates with Rgt1 hyperphosphorylation. In yeast two-hybrid and co-immunoprecipitation assays, furthermore, Mth1 and Std1 have been shown to interact with Rgt1, but only in the absence of glucose (Lakshmanan et al. 2003). Finally, deletion of either *RGT1* or *MTH1/STD1* causes constitutive expression of *HXT* genes. These results suggest that Mth1 and Std1 somehow inhibit phosphorylation of Rgt1 by an as yet unknown protein kinase. They might do so by preventing the kinase from interacting with Rgt1.

Genetic data suggest that Mth1 and Std1 act upstream from Rgt1 but downstream from Grr1. Deletion of *GRR1* completely blocks glucose-induced *HXT* gene expression (Ozcan and Johnston 1995), and this defect is fully suppressed by *rgt1* mutations or *mth1 std1* double mutations (Ozcan and Johnston 1995; Flick et al. 2003). In a *grr1* mutant, Rgt1 is not hyperphosphorylated and does not dissociate from the *HXT* gene promoters in response to glucose (Flick et al 2003). It, thus, seems that Mth1 and Std1 become inactivated by glucose via a Grr1-dependent mechanism. It has been shown that Mth1 is rapidly degraded after addition of glucose to galactose-grown cells (Flick et al. 2003). This degradation depends on Grr1, suggesting that SCF^{Grr1} catalyses ubiquitination of Mth1, thus, triggering its degradation. In contrast, Std1 is not degraded in response to glucose, but its inactivation is also dependent on Grr1 (Flick et al. 2003).

The function of Std1 in the glucose induction pathway remains obscure (Fig. 3). This protein seems to be involved only modestly in regulating the *HXT2, -3,* and *-4* promoters, as inactivation of Mth1 is sufficient for full derepression of these genes (Schmidt et al. 1999; Flick et al. 2003). Moreover, Std1 seems to act predominantly on *HXT1*, whilst Mth1 has a more pronounced effect on the other *HXT* genes (Schmidt et al. 1999). Overexpression of Std1 – but not of Mth1 – causes repression of *HXT1* expression under high glucose conditions (Schmidt et al. 1999), but both Mth1 and Std1 must be inactivated for full induction of Rgt1 hyperphosphorylation and complete dissociation of Rgt1 from the *HXT* promoters in the absence of glucose. Both genes must be eliminated to bypass the requirement for Grr1 in these processes. Altogether, the data suggest that Mth1 is the primary target of Grr1 for regulation of *HXT* gene expression by glucose, with Std1 playing a minor or different role.

Interestingly, Mth1 and Std1 interact also with the C-terminal tails of Snf3 and Rgt2 (Schmidt et al. 1999; Lafuente et al. 2000). Std1, furthermore, can interact directly with the Snf1/AMP-activated protein kinase and TATA-binding protein (Hubbard et al. 1994; Tillman et al. 1995). The Snf1 kinase has a primary role in the adaptation of cells to glucose limitation, regulating both the transcription of metabolic genes and the activity of metabolic enzymes (Gancedo 1998). In keeping with the interaction results, Std1-GFP and Mth1-GFP proteins show both a nuclear localisation and punctate staining at the cytoplasmic periphery (Schmidt et al. 1999; Schulte F 1997). These localisations are not affected by the glucose concentration or by the absence of both glucose sensors. Yet, in these experiments, the Std1-GFP and Mth1-GFP proteins were expressed from high-copy-number plasmids. Overproduction of the proteins might result in partial mislocalisation and glucose-insensitivity. It, thus, remains reasonable to assume that Mth1 and Std1 shuttle between the plasma membrane and the nucleus in a manner regulated by glucose. In the presence of glucose, most of the Mth1 and Std1 proteins would be bound to the glucose sensors and become inactivated via Grr1. In the absence of glucose, they would be transferred to the nucleus where they would inhibit an unknown protein kinase or prevent it from phosphorylating Rgt1 (Lakshmanan et al. 2003). The dominant negative mutant alleles of Mth1 (Htr1-23, Dgt1-1 and Bpc1-1) that severely inhibit *HXT* gene expression obviously have a decreased affinity for the Snf3 and Rgt2 tails (Lafuente et al. 2000). Moreover, it might be speculated that their stabilities are increased. Again, the different function of Std1 is reflected in the observation that a mutant allele corresponding to the dominant negative *HTR1-23* allele had no influence on *HXT* gene expression (Schulte et al. 2000).

4.1.3 Specificity and regulation of individual HXT promoters

As detailed above (see 2.1), the individual *HXT* promoters are differentially regulated by the nature and concentration of the carbon source (see Fig. 3). How is this specific regulation achieved? A partial answer is that the Rgt2 low-affinity sensor specifically induces expression of the low-affinity glucose transporter genes *HXT1* and *HXT3*, whereas activation of Snf3 affects expression of all *HXT* genes. It is not yet clear what kind of mechanism ensures that Rgt2 regulates only the *HXT1* and *HXT3* promoters and not the others. Clearly, the different glucose signals must be modulated by other factors: In the presence of 0.2% glucose, expression of *HXT1* and *HXT3* is only partially induced and Rgt1 can associate with the promoters of both of these genes. The situation is different for *HXT4* under these conditions, since expression of the gene is fully induced and Rgt1 cannot bind to its promoter (Flick et al. 2003). Obviously, there must be ancillary proteins that act specifically at various *HXT* promoters. On the other hand, the number of Rgt1-binding sites in the different promoters might regulate the response to the glucose signal.

Moreover, various other signalling pathways converge at the individual *HXT* promoters. *HXT2, -4, -6,* and *-7* are subject to glucose repression (Gancedo 1998; Ozcan and Johnston 1999) (Fig. 3). This regulatory pathway ensures that at high glucose concentrations the levels of the high- and intermediate-affinity transporters are decreased. Surprisingly, repression of *HXT6* by high concentrations of glucose is mediated by an Snf3-dependent signal transduction pathway that is distinct from but overlaps with the general glucose repression pathway (Liang and Gaber 1996). *HXT1* is induced not only by high levels of glucose but also by hyperosmotic stress (Hirayama et al. 1995). Hence, high concentrations of glucose serve not only as a glucose signal for *HXT1* gene expression but also as a stress signal transduced by the Sln1-Hog1 osmosensing signal transduction pathway. Indeed, regulation of the *HXT1* promoter seems to be rather complex. *HXT1* expression is strongly induced only by high levels of glucose but not by low levels. This results from the fact that in the presence of low levels of glucose, activated Snf1 kinase exerts an inhibitory effect; this kinase is also controlled by Std1 (Fig. 3) (Tomas-Cobos and Sanz 2002; Kuchin et al. 2003; Schmidt et al. 1999). Moreover, Rgt1 seems to be required for full induction of *HXT1* expression at high glucose levels. It was initially proposed that Rgt1 might be converted into a transcriptional activator in the presence of high glucose concentrations (Ozcan et al. 1996b), but under such conditions Rgt1 does not bind to the *HXT1* promoter (Flick et al. 2003; Mosley et al. 2003). Its positive regulatory effect seems to be indirect. Additionally, there seems to be an Rgt1-independent activation pathway for *HXT1*. This is supported by the observation that in the absence of Rgt1, *HXT1* expression is still inducible by high glucose concentrations (Ozcan and Johnston 1995).

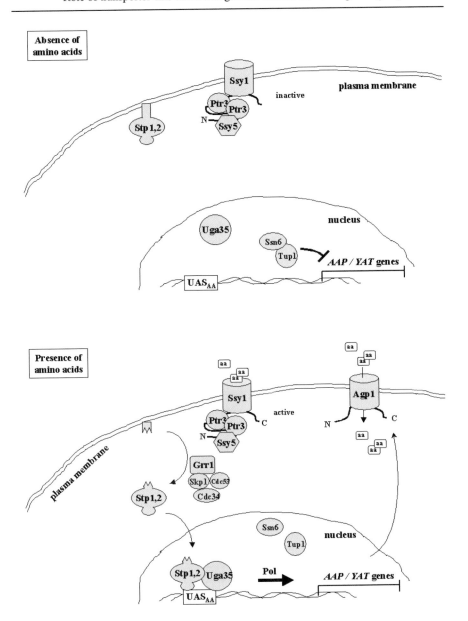

Fig. 4. The current model of the amino-acid-sensing pathway in *S. cerevisiae*.

4.2 The amino-acid-sensing pathway

4.2.1 Current view

The current model of the amino-acid-sensing pathway is schematically depicted in Fig. 4. Ssy1 is part of a plasma-membrane-associated complex named SPS (Forsberg and Ljungdahl 2001) that includes at least two other proteins: Ptr3 (Barnes et al. 1998) and Ssy5 (Jorgensen et al. 1998). In the SPS complex, Ssy1 and Ptr3 are apparently more directly involved in amino acid sensing and signal initiation. The signal generated by Ssy1 and Ptr3 would then be somehow transmitted to Ssy5, which appears to interact specifically with Ptr3 (Bernard and André 2001a). A next step in the signalling pathway is the endoproteolytic processing of transcription factors Stp1 and Stp2 (Andreasson and Ljungdahl 2002). These transcription factors belong to the Kruppel family of zinc-finger proteins, are partially redundant, and play an important role in transcriptional induction of Ssy1-target genes in response to external amino acids (Jorgensen et al. 1997; Jorgensen et al. 1998; de Boer et al. 1998; de Boer et al. 2000; Nielsen et al. 2001; Abdel-Sater et al. 2004). The endoproteolytic processing of Stp factors has been more deeply investigated in the case of Stp1 (Andreasson and Ljungdahl 2002). The cleavage induced takes place within its N-terminus, likely between residues 67 and 129, a region highly conserved between Stp1 and orthologs found in other fungal species. In addition to Ssy1, Ptr3, and Ssy5 (Andreasson and Ljungdahl 2002), the endoproteolytic processing of Stp1 requires the SCF^{Grr1} ubiquitin-ligase complex (Abdel-Sater & André, in preparation), which accordingly also plays an essential role in the response of cells to external amino acids (Iraqui et al. 1999; Bernard and André 2001b). However, cleavage of Stp1 does not require the proteasome (Andreasson and Ljungdahl 2002). The molecular details of the endoproteolytic processing of Stp1 and Stp2 remain unknown, but they likely involve tight cooperation between Ssy5 and SCF^{Grr1}.

In its unprocessed form, Stp1 seems to interact via its N-terminal domain with cell-surface components, likely factors of the SPS complex. After processing, truncated Stp1 is translocated into the nucleus (Andreasson and Ljungdahl 2002). Yet, it seems that full-length Stp1 can also potentially be translocated into the nucleus and activate gene transcription, but this process is somehow limited in a manner dependent on three proteins (Asi1, Asi2, and Asi3) located at the nuclear membrane (Forsberg et al. 2001; Zargani et al. 2003). Once in the nucleus, Stp1 and Stp2 act via an upstream sequence (UAS$_{AA}$) to activate gene transcription. Such UAS$_{AA}$ typically consist of at least two copies of the 5'-CGGC-3' tetranucleotide separated by four to nine nucleotides (de Boer et al. 1998; Abdel-Sater et al. 2004). Another transcription factor important for induction of Ssy1-regulated genes is the Uga35/Dal81 protein containing a Zn(II)$_2$-Cys$_6$-cluster-type DNA-binding domain (Iraqui et al. 1999). This transcription factor is pleiotropic. It is involved in at least two other transcriptional induction circuits, i.e. one activated by GABA (for activation of the *UGA* genes involved in GABA utilisation) and one by allophanate (for activation of the *DUR* and *DAL* genes involved in urea and allantoin utilisation). In each induction process, Uga35/Dal81 acts in conjunction with an inducer-specific transcription factor, namely, Uga3 (GABA) or Dal82/DurM (allophanate). Why these transcription induction systems and that activated by external amino acids all require a common Uga35/Dal81 factor remains unknown. Interestingly, it seems that Uga35/Dal81 is somehow activated in response to external amino acids (Abdel-Sater et al. 2004). Finally, induction of Ssy1-target genes by Stp-Uga35/Dal81 factors likely involves relieving repression exerted by the Tup1-Ssn6 corepressor complex (Forsberg et al. 2001b).

4.2.2 Data in support of the model

PTR3 was originally characterised in the group of Jeffrey Becker (Barnes et al. 1998) as a gene essential to induction of the *PTR2* gene by micromolar levels of amino acids (Island et al. 1987). Independently, the genetic screens having led to the isolation of *ssy1* mutants also identified *PTR3* (= *SSY3, SHR6, AFP3*) and *SSY5* (= *SHR4, APF8*) as two other genes absolutely essential to transcriptional induction of permease genes in response to external amino acids (Jorgensen et al. 1998; Klasson et al. 1999; Forsberg and Ljungdahl 2001; Bernard and André 2001a). The exact biochemical role of the corresponding gene products remains undetermined. Neither Ptr3 nor Ssy5 contains any typical domain indicative of a particular biochemical function, and these proteins are conserved only in fungi. Ptr3 and Ssy5 are peripherally associated membrane proteins (Klasson et al. 1999; Forsberg and Ljungdahl 2001) that likely form a complex with Ssy1 (SPS). In support to this view, two-hybrid experiments detected Ssy1-Ptr3 (Uetz et al. 2000) and Ptr3-Ssy5 interactions, and Ptr3 also interacts with itself (Bernard and André 2001a). Furthermore, HA-epitope-tagged forms of Ssy1, Ptr3, and Ssy5 exhibit abnormal electrophoretic mobilities in mutants lacking one of the other factors of the complex (Forsberg and Ljungdahl 2001). Whether other proteins, e.g. the Stp1 and Stp2 transcription factors and/or the SCF^{Grr1} ubiquitin-ligase complex, are also associated with the SPS complex at the cell surface remains uncertain. Yet, in accordance with the view that the Stp factors might be somehow complexed with SPS, the electrophoretic properties of Ssy1 are altered in the *stp1* mutant. Furthermore, Stp1 can target a heterologous protein (human SOS, a Cdc25 homologue) to the cell surface, and this targeting is ineffective in an *ssy1* mutant. Models of a possible association between Stp1 and the SPS complex must, nevertheless, also take into account the observation that the N-terminal domain of Stp1 targets human SOS to the cell surface in both wild type and *ssy1* mutant strains (Andreasson and Ljungdahl 2002).

In the SPS complex, Ssy1 and Ptr3 appear to be more involved in the detection of external amino acids and in subsequent signal initiation, since mutations in *SSY1* (Bernard and André 2001a; Gaber et al. 2003) and *PTR3* (Abdel-Sater and André, in preparation) can result in an altered response of cells to external amino acids. It is noteworthy that a sequence in Ptr3 appears significantly similar to a region shared by AAP/YAT permeases (Klasson et al. 1999). Perhaps this region enables Ptr3 to intercept amino acids crossing the lipid bilayer through Ssy1, thus, triggering a conformational change that would then be transmitted to Ssy5 (see 3.3). The long cytosolic N-terminal tail of Ssy1 must also play a key role in transmitting the amino-acid signal, as overproduction of this domain alone is sufficient to complement the growth defects caused by the *ssy1Δ* mutation (Bernard and André 2001a). When overproduced in wild type cells, furthermore, this domain interferes negatively and in a dominant manner with induction of amino-acid permease genes by external amino acids (Forsberg and Ljungdahl 2001; Bernard and André 2001a). Yet, the factors interacting with the cytosolic tail of Ssy1 remain to be identified. Ssy5 is thought to act downstream from Ssy1 and Ptr3. In support of this view, overproduction of Ssy5 relieves the growth defects caused by the *ssy1Δ* and *ptr3Δ* deletions. This overproduction leads in fact to high-level constitutive activation of the genes normally induced by external amino acids. Neither Ssy1 nor Ptr3 is required at all for this activation (Bernard and André 2001a; Abdel-Sater and André, in preparation). Furthermore, the electrophoretic mobility of Ptr3 extracted from cells growing in the presence of amino acids is altered in *ssy1Δ* but not in *ssy5Δ* mutants (Forsberg and Ljungdahl 2001).

The likely next step in the pathway is endoproteolytic processing of the latent membrane-associated Stp1 and Stp2 transcription factors, followed by translocation of their N-terminally truncated forms into the nucleus. In support of the model postulating that the N-terminal domain of Stp1 anchors the transcription factor to the cell surface, thus, preventing it from being translocated into the nucleus, deletion of residues 9 to 66 of Stp1 (Stp1Δ131) restores high-level expression of the *AGP1*, *GNP1*, and *BAP2* genes and a wild type growth phenotype in an *ssy1Δ* mutant strain growing in the presence of amino acids (Forsberg et al. 2001b; Andreasson and Ljungdahl 2002). Furthermore, full-length Stp1 or its N-terminal domain alone can target a heterologous protein (hSOS) to the cell surface, whereas the N-terminally truncated form of Stp1 cannot. Finally, Stp1 is mainly located in the nucleus only when amino acids are present in the medium (Andreasson and Ljungdahl 2002). It was also proposed initially that the endoproteolytic processing of Stp1 might not require any factors other than the SPS complex components (Andreasson and Ljungdahl 2002), but it is now established that the SCF^{Grr1} ubiquitin-ligase complex is also essential to cleavage of Stp1 in response to amino acids (Abdel-Sater & André, in preparation). Further experiments are needed to clarify the mechanism of Stp1 endoprotelytic processing (e.g. the protease involved remains unknown) and to determine the exact role in this process of SCF^{Grr1} (e.g. does Stp1 processing require its prior SCF^{Grr1}-catalysed ubiquitination or does SCF^{Grr1} mediate ubiquitin-dependent maturation of one of the SPS complex components, e.g. Ssy5?). Interestingly, mutations in genes encoding subunits of the proteasome do not impair the endoproteolytic processing of Stp1 (Andreasson and Ljungdahl 2002). The apparent lack of involvement of the proteasome in Stp1 cleavage contrasts with all reported cases of activation of transcription factors by ubiquitin-dependent processing (Rape and Jentsch 2002).

The role of the *ASI1*, *ASI2*, and *ASI3* genes in the signalling pathway also remains to be clarified (Forsberg et al. 2001b). Each of the proteins Asi1 and Asi3 contains an ubiquitin-ligase RING domain at its extreme C-terminus, and Asi2 has three predicted trans-membrane domains. Loss-of-function mutations in the *ASI* genes restore high-level expression of the Ssy1-target genes in an *ssy1Δ* strain growing in the presence of amino acids. Thus, the Asi proteins would seem normally to inhibit the expression of Ssy1-regulated genes by preventing the full-length Stp factors from being active (Forsberg et al. 2001b).

After their translocation into the nucleus, Stp1 and Stp2 act via UAS_{AA} elements present in the control regions of Ssy1-target genes to activate their transcription. A detailed analysis of the UAS_{AA} regions of the *BAP2*, *BAP3*, and *AGP1* genes has been published (de Boer et al. 1998; de Boer et al. 2000; Abdel-Sater et al. 2004), and similar sequences are present in the upstream regions of the other AAP/YAT genes under Ssy1 control. The UAS_{AA} of the *BAP3* and *AGP1* genes is sufficient to drive transcriptional induction of a reporter gene by multiple amino acids (see 2.2). Results of band-shift and one-hybrid assays suggest that Stp1 and Stp2 bind to UAS_{AA}-containing DNA fragments (Nielsen et al. 2001), although these experiments are based on Stp1 proteins containing an N-terminal extrapeptide of 58 amino acids due to incorrect use of the ATG initiation codon (Wang et al. 1992; Andreasson and Ljungdahl 2002). The Uga35/Dal81 protein containing a $Zn(II)_2$-Cys_6-cluster-type DNA-binding domain (Coornaert et al. 1991) is required for normal induction of at least two Ssy1 target genes, i.e. *AGP1* and *BAP2* (Iraqui et al. 1999; Bernard and André 2001a) and is likely involved as well in induction of the other genes under Ssy1 control. Studies focused on *AGP1* showed that Stp1 and Uga35/Dal81 act synergistically through UAS_{AA} to induce high-level transcription while Stp2 does not significantly contribute. Remarkably, an Uga35/Dal81-dependent residual induction of *AGP1* subsists in the *stp1* and *stp1 stp2* mutants suggesting that Uga35/Dal81 is also activated by the external amino-acid signalling pathway (Abdel-Sater et al. 2004). Further experiments are needed to specify the mechanism of this activation and the exact physiological role of this pleiotropic factor in Ssy1-dependent transcriptional induction. Finally, several reports indicate that even more general transcription factors like Tup1, Ssn6, and possibly Abf1 are also involved in transcription of Ssy1-regulated genes (de Boer et al. 2000; Nielsen et al. 2001; Forsberg et al. 2001b). In particular, mutations in *TUP1* and *SSN6* restore high-level expression of Ssy1-regulated genes in *ssy1Δ* and *ptr3Δ* mutants grown in the presence of amino acids (Forsberg et al. 2001b), suggesting that transcriptional induction of these permease genes involves relief from a general repression exerted by the Tup1-Ssn6 complex, possibly via chromatin remodelling.

4.2.3 Specificity and regulation of individual amino-acid permease genes

The amino-acid permease genes under Ssy1 control (Agp1, Gnp1, Bap2, Bap2, Tat1, Tat2, Dip5) display broad-range specificities for amino acids (Iraqui et al. 1999; Regenberg et al. 1999). Largely unexplored, however, are the affinities of these proteins for each of their substrates and their respective roles in amino-acid uptake according to the amino acid content of the medium.

Although the above-mentioned permease genes are all under Ssy1 control, their transcription appears not to respond similarly to amino acid availability. In at least some cases, it seems that these differences are due to the permease genes being subject to other transcriptional control mechanisms adding to or interfering with the Ssy1-dependent induction circuit. For instance, although asparagine is an efficient inducer of UAS$_{AA}$-driven transcription (de Boer et al. 1998), it is a very poor inducer of the *AGP1* and *PTR2* genes (Iraqui et al. 1999; Bernard and André 2001a). This is probably due to the fact that *AGP1* and *PTR2* are also subject to nitrogen catabolite repression (NCR) and that the latter control is fully operative when asparagine (a preferential nitrogen source) is used as a nitrogen source by the cells. Consistently, *BAP2* is insensitive to NCR (Regenberg et al. 1999) and is induced to a high level by asparagine (Bernard and André 2001a). Similarly, histidine and cysteine are good inducers of *BAP2* but poor inducers of *AGP1* and *PTR2* and of a reporter gene under UAS$_{AA}$ control. In this case, it was shown that these inductions do not depend on Ssy1, i.e. *BAP2* is induced by these amino acids (as well as by methionine) via one or several other transcriptional pathways whose *cis*- and *trans*-components remain unknown (Bernard and André 2001a). It was further shown that *BAP2* is under the positive control of Leu3. *BAP2* expression is higher if the internal leucine pool is reduced. This control requires a Leu3-binding site in the *BAP2* upstream region, a sequence not found in the upstream region of any other gene under Ssy1 control (Nielsen et al. 2001). Transcription of the *PTR2* gene is also induced by dipeptides. This regulation involves activation by dipeptides of the Ubr1 ubiquitin-ligase enzyme, which then promotes degradation of the Cup9 repressor (Turner et al. 2000), a condition sufficient for high-level activation of *PTR2*. Interestingly, lack of Ubr1 also prevents *PTR2* from being induced by amino acids via the Ssy1 pathway (Bernard and André 2001a), likely because Cup9 is abundant under these conditions. This shows that other transcriptional controls can interfere negatively with Stp-Uga35/Dal81-mediated induction of the Ssy1-regulated genes by external amino acids.

5 Are there transporter-like sensors in higher eukaryotes?

The presence of unusually long cytosolic tails in specific members of transporter families has been used to predict the existence in yeast of other potential sensors of extracellular compounds (Van Belle and André 2001). For at least one of these proteins, namely the Pho87 low-affinity transporter (Wykoff and O'Shea 2001), data consistent with an additional role in external phosphate sensing have recently been reported (Giots et al. 2003; Auesukaree et al. 2003).

In plants, a lowly expressed protein of the sucrose transporter family (SUT2), with an extended cytoplasmic domain, is proposed to be involved in sucrose signalling (Barker et al. 2000), although this hypothesis is a matter of discussion (Barth et al. 2003). Another plant protein, EIN2, is a member of the Nramp family of metal transporters, but differs from the other proteins of this family by the presence of a long cytosolic C-terminal tail. This protein plays an essential role in ethylene signalling and its C-terminal domain is crucial to this function. For instance, mutations in the tail of EIN2 lead to complete ethylene insensitivity, and overexpression of the tail alone is sufficient to constitutively activate responses to ethylene (Alonso et al. 1999). Intriguingly, however, EIN2 is not the sensor of ethylene, as it acts downstream from several ethylene receptors similar to bacterial two-component histidine kinases (Hua and Meyerowitz 1998) and from a Raf-like protein kinase (Alonso et al. 1999; Hall and Bleecker 2003).

Membrane transporters of animal origin usually have long cytoplasmic tails and it is, thus, difficult to predict a role for some of them in signalling on the basis of this criterion. No protein of any transporter family in animal cells has thus far been shown to lack a transport-catalysing function and to be involved in sensing. A possible exception is a human member of the sodium/glucose cotransporter family, hSGLT3, recently proposed to be a glucose sensor in the plasma membrane of cholinergic neurons and cells at the neuromuscular junction in skeletal muscle (Diez-Sampedro et al. 2003). When expressed in oocytes, hSGLT3 did not mediate the uptake of glucose, but glucose did produce a significant inward current, thereby modulating membrane potential. Thus, it was suggested that hSGLT3 might be a glucose sensor involved in the regulation of muscle activity. It is noteworthy that the sequence of hSGLT3 does not display any particularities that might account for the differences in function between the sensor and other known SGLT transporter proteins.

On the other hand, there is mounting evidence that some membrane transporters of animal cells in fact play a dual role, i.e. a role in both transport catalysis and signalling. For instance, there is evidence of a possible direct role of the GLUT2 glucose transporter in glucose-regulated gene expression (Guillemain et al. 2000). Yet, most glucose-sensing mechanisms in mammalian cells require intense glucose metabolism, and the triggering events are related to the phosphorylation of glucose or to intracellular signals (Rolland et al. 2001). In another study, calmodulin has been shown to be tethered to the C-terminal tail of the L-type voltage-gated calcium channel (L-VGCC) and to serve as a sensor for "local" calcium entry through L-VGCC (Dolmetsch et al. 2001). In response to L-VGCC-mediated calcium flow, calmodulin promotes feed-back inhibition of the channel and activates the Ras/MAPK cascade, which in turn causes prolonged phosphorylation of the cAMP response element binding protein (CREB), a nuclear transcription factor crucial to neuronal development and plasticity (Dolmetsch et al. 2001). Finally, studies based on the use of non-metabolisable substrates or antagonists have provided evidence that the System A amino-acid transporter SAT2 (Ling et al. 2001; Gazzola et al. 2001) and the glial glutamate transporter EAAT1 (Duan et al. 1999) are involved directly in the initiation of intracellular signalling in response to amino acids (Hyde et al. 2003). These studies are of high interest and raise the possibility that membrane transporters might play a more important role in signalling than anticipated. Further investigation is needed to unravel the molecular details of these signalling processes.

6 Prospects and future directions

Many aspects of the glucose and amino acid detection and signalling pathways remain unclear. For instance, whether the Snf3/Rgt2/Ssy1 proteins just bind the external nutrients or whether they can catalyse nutrient uptake (e.g. for subsequent trapping by tethered transduction factors) remains to be determined. This justifies, for now, the use of the term "sensor" rather than "receptor" to qualify them. Also unknown are the molecular details of the association of these sensors with signal-transducing proteins into membrane-associated complexes, and also the conformational changes these complexes likely undergo upon binding to external nutrients. The subsequent steps of nutrient signalling seem to differ between the glucose and amino-acid pathways. Induction of the *HXT* genes by glucose requires relief from Rgt1-mediated repression. In contrast, induction of *AAP/YAT* genes by external amino acids involves activation by endoproteolytic processing of Stp1 and Stp2, which then act together with Uga35/Dal81 to activate genes. Although the SCF^{Grr1} ubiquitin ligase complex plays a central role in both glucose and amino acid signalling, we do not know whether this is achieved through similar mechanisms. Nor do we know whether SCF^{Grr1} is itself activated by nutrients or whether its role is limited to recognition and ubiquitination of a pathway component (for instance, once the latter has been phosphorylated in response to nutrients).

Other unanswered questions specifically concern the glucose-sensing pathway: what are the properties that make Snf3 a high-affinity sensor and Rgt2 a low-affinity sensor, and how are the different signals distinguished? Which kinase phosphorylates Rgt1 in the presence of glucose and is this kinase regulated by Mth1 and Std1? How is specific regulation of individual *HXT* promoters by Snf3 and Rgt2 achieved? Regarding the amino-acid-sensing pathway there also remain questions to be answered, such as: what is the protease responsible for Stp1 and Stp2 endoproteolytic processing? What roles do the Asi proteins play in Stp-dependent transcription? How is the specific regulation of individual AAP/YAT genes achieved according to amino-acid availability? Finally, it will be most interesting to determine to what extent these nutrient sensing systems are conserved in other fungi, including pathogens.

Acknowledgement

We thank Mark Johnston for sharing information prior to publication. We also thank Per Ljungdahl and Morten Kielland-Brandt for many fruitful discussions. This work was supported by grant FRSM 3.4597.00 F for Medical Scientific Research, Belgium, to B.A.

References

Abdel-Sater F, Iraqui I, Urrestarazu A, André B (2004). The external amino acid signalling pathway promotes activation of Stp1 and Uga35/Dal81 transcription factors for induction of the *AGP1* gene in *Saccharomyces cerevisiae*. Genetics: *in press*

Alonso JM, Hirayama T, Roman G, Nourizadeh S, Ecker JR (1999) EIN2, a bifunctional transducer of ethylene and stress responses in *Arabidopsis*. Science 284:2148-2152

Andreasson C, Ljungdahl PO (2002) Receptor-mediated endoproteolytic activation of two transcription factors in yeast. Genes Dev 16:3158-3172

André B (1995) An overview of membrane transport proteins in *Saccharomyces cerevisiae*. Yeast 11:1575-1611

André, B (1996). Whole genome in silico analysis of *Saccharomyces cerevisiae* membrane transport proteins. Proceedings of the 14th Small Meeting on Yeast Transport and Energetics, Bonn, Germany.

André B, Hein C, Grenson M, Jauniaux JC (1993) Cloning and expression of the *UGA4* gene coding for the inducible GABA-specific transport protein of *Saccharomyces cerevisiae*. Mol Gen Genet 237:17-25

Auesukaree C, Homma T, Kaneko Y, Harashima S (2003) Transcriptional regulation of phosphate-responsive genes in low-affinity phosphate-transporter-defective mutants in *Saccharomyces cerevisiae*. Biochem Biophys Res Commun 306:843-850

Barker L, Kuhn C, Weise A, Schulz A, Gebhardt C, Hirner B, Hellmann H, Schulze W, Ward JM, Frommer WB (2000) SUT2, a putative sucrose sensor in sieve elements. Plant Cell 12:1153-1164

Barnes D, Lai W, Breslav M, Naider F, Becker JM (1998) *PTR3*, a novel gene mediating amino acid-inducible regulation of peptide transport in *Saccharomyces cerevisiae*. Mol Microbiol 29:297-310

Barral Y, Jentsch S, Mann C (1995) G1 cyclin turnover and nutrient uptake are controlled by a common pathway in yeast. Genes Dev 9:399-409

Barth I, Meyer S, Sauer N (2003) PmSUC3: characterization of a SUT2/SUC3-type sucrose transporter from Plantago major. Plant Cell 15:1375-1385

Bernard F, André B (2001a) Genetic analysis of the signalling pathway activated by external amino acids in *Saccharomyces cerevisiae*. Mol Microbiol 41:489-502

Bernard F, André B (2001b) Ubiquitin and the SCF(Grr1) ubiquitin ligase complex are involved in the signalling pathway activated by external amino acids in *Saccharomyces cerevisiae*. FEBS Lett 496:81-85

Betina S, Goffrini P, Ferrero I, Wesolowski-Louvel M (2001) *RAG4* gene encodes a glucose sensor in *Kluyveromyces lactis*. Genetics 158:541-548

Bhat PJ, Murthy TV (2001) Transcriptional control of the *GAL/MEL* regulon of yeast *Saccharomyces cerevisiae*: mechanism of galactose-mediated signal transduction. Mol Microbiol 40:1059-1066

Bisson LF, Neigeborn L, Carlson M, Fraenkel DG (1987) The *SNF3* gene is required for high-affinity glucose transport in *Saccharomyces cerevisiae*. J Bacteriol 169:1656-1662

Blaisonneau J, Fukuhara H, Wesolowski-Louvel M (1997) The *Kluyveromyces lactis* equivalent of casein kinase I is required for the transcription of the gene encoding the low-affinity glucose permease. Mol Gen Genet 253:469-477

Boles E (2002) Yeast as a model system for studying glucose transport. In Quick MW, ed. Transmembrane transporters. Wiley, Inc. 19-36.

Brega E, Zufferey R, Ben Manoun C (2004) Cys1p of *Candida albicans* is a nutrient sensor important for activation of amino acid uptake and hyphal morphogenesis. Eukaryot Cell: in press

Buziol S, Becker J, Baumeister A, Jung S, Mauch K, Reuss M, Boles E (2002) Determination of *in vivo* kinetics of the starvation-induced Hxt5 glucose transporter of *Saccharomyces cerevisiae*. FEM Yeast Res 2:283-291

Cliften P, Sudarsanam P, Desikan A, Fulton L, Fulton B, Majors J, Waterston R, Cohen BA, Johnston M (2003) Finding functional features in *Saccharomyces* genomes by phylogenetic footprinting. Science 301:71-76

Coornaert D, Vissers S, André B (1991) The pleiotropic *UGA35(DURL)* regulatory gene of *Saccharomyces cerevisiae*: cloning, sequence and identity with the *DAL81* gene. Gene 97:163-171

de Boer M, Bebelman JP, Goncalves PM, Maat J, Van Heerikhuizen H, Planta RJ (1998) Regulation of expression of the amino acid transporter gene BAP3 in *Saccharomyces cerevisiae*. Mol Microbiol 30:603-613

de Boer M, Nielsen PS, Bebelman JP, Heerikhuizen H, Andersen HA, Planta RJ (2000) Stp1p, Stp2p and Abf1p are involved in regulation of expression of the amino acid transporter gene *BAP3* of *Saccharomyces cerevisiae*. Nucleic Acids Res 28:974-981

Deshaies RJ (1999) SCF and Cullin/Ring H2-based ubiquitin ligases. Annu Rev Cell Dev Biol 15:435-467

Didion T, Grauslund M, Kielland-Brandt MC, Andersen HA (1996) Import of branched-chain amino acids in *Saccharomyces cerevisiae*. Folia Microbiol (Praha) 41:87

Didion T, Regenberg B, Jorgensen MU, Kielland-Brandt MC, Andersen HA (1998) The permease homologue Ssy1p controls the expression of amino acid and peptide transporter genes in *Saccharomyces cerevisiae*. Mol Microbiol 27:643-650

Diez-Sampedro A, Hirayama BA, Osswald C, Gorboulev V, Baumgarten K, Volk C, Wright EM, Koepsell H (2003) A glucose sensor hiding in a family of transporters. Proc Natl Acad Sci USA 100:11753-11758

Dlugai S, Hippler S, Wieczorke R, Boles E (2001) Glucose-dependent and -independent signalling functions of the yeast glucose sensor Snf3. FEBS Lett 505:389-392

Dolmetsch RE, Pajvani U, Fife K, Spotts JM, Greenberg ME (2001) Signaling to the nucleus by an L-type calcium channel-calmodulin complex through the MAP kinase pathway. Science 294:333-339

Duan S, Anderson CM, Stein BA, Swanson RA (1999) Glutamate induces rapid upregulation of astrocyte glutamate transport and cell-surface expression of GLAST. J Neurosci 19:10193-10200

Flick KM, Spielewoy N, Kalashnikova TI, Guaderrama M, Zhu Q, Chang HC, Wittenberg C (2003) Grr1-dependent inactivation of Mth1 mediates glucose-induced dissociation of Rgt1 from HXT gene promoters. Mol Biol Cell 14:3230-3241

Forsberg H, Gilstring CF, Zargari A, Martinez P, Ljungdahl PO (2001a) The role of the yeast plasma membrane SPS nutrient sensor in the metabolic response to extracellular amino acids. Mol Microbiol 42:215-228

Forsberg H, Hammar M, Andreasson C, Moliner A, Ljungdahl PO (2001b) Suppressors of *ssy1* and *ptr3* null mutations define novel amino acid sensor-independent genes in *Saccharomyces cerevisiae*. Genetics 158:973-988

Forsberg H, Ljungdahl PO (2001) Genetic and biochemical analysis of the yeast plasma membrane Ssy1p-Ptr3p-Ssy5p sensor of extracellular amino acids. Mol Cell Biol 21:814-826

Gaber RF, Ottow K, Andersen HA, Kielland-Brandt MC (2003) Constitutive and hyperresponsive signaling by mutant forms of *Saccharomyces cerevisiae* amino acid sensor Ssy1. Eukaryot Cell 2:922-929

Gancedo JM (1998) Yeast carbon catabolite repression. Microbiol Mol Biol Rev 62:334-361

Ganster RW, Shen W, Schmidt MC (1993) Isolation of STD1, a high-copy-number suppressor of a dominant negative mutation in the yeast TATA-binding protein. Mol Cell Biol 13:3650-3659

Gazzola RF, Sala R, Bussolati O, Visigalli R, Dall'Asta V, Ganapathy V, Gazzola GC (2001) The adaptive regulation of amino acid transport system A is associated to changes in ATA2 expression. FEBS Lett 490:11-14

Gilstring CF, Ljungdahl PO (2000) A method for determining the *in vivo* topology of yeast polytopic membrane proteins demonstrates that Gap1p fully integrates into the membrane independently of Shr3p. J Biol Chem 275:31488-31495

Giots F, Donaton MC, Thevelein JM (2003) Inorganic phosphate is sensed by specific phosphate carriers and acts in concert with glucose as a nutrient signal for activation of the protein kinase A pathway in the yeast *Saccharomyces cerevisiae*. Mol Microbiol 47:1163-1181

Grenson M (1992) Amino acid transporters in yeast: structure, function and regulation 291. In J. J. L. L. M. De Pont, ed. Molecular aspects of transport proteins. Amsterdam: Elsevier Science. 219-245.

Grenson M, Hou C, Crabeel M (1970) Multiplicity of the amino acid permeases in *Saccharomyces cerevisiae*. IV. Evidence for a general amino acid permease. J Bacteriol 103:770-777

Guillemain G, Loizeau M, Pincon-Raymond M, Girard J, Leturque A (2000) The large intracytoplasmic loop of the glucose transporter GLUT2 is involved in glucose signaling in hepatic cells. J Cell Sci 113 (Pt 5):841-847

Hall AE, Bleecker AB (2003) Analysis of combinatorial loss-of-function mutants in the *Arabidopsis* ethylene receptors reveals that the ers1 etr1 double mutant has severe developmental defects that are EIN2 dependent. Plant Cell 15:2032-2041

Hirayama T, Maeda T, Saito H, Shinozaki K (1995) Cloning and characterization of seven cDNAs for hyperosmolarity-responsive (HOR) genes of *Saccharomyces cerevisiae*. Mol Gen Genet 249:127-138

Hoffmann W (1985) Molecular characterization of the *CAN1* locus in *Saccharomyces cerevisiae*. A transmembrane protein without N-terminal hydrophobic signal sequence. J Biol Chem 260:11831-11837

Holman GD, Rees WD (1987) Photolabelling of the hexose transporter at external and internal sites: fragmentation patterns and evidence for a conformational change. Biochim Biophys Acta 897:395-405

Horak J, Regelmann J, Wolf DH (2002) Two distinct proteolytic systems responsible for glucose-induced degradation of fructose-1,6-bisphosphatase and the Gal2p transporter in the yeast *Saccharomyces cerevisiae* share the same protein components of the glucose signaling pathway. J Biol Chem 277:8248-8254

Hu LA, King SC (1998) Membrane topology of the *Escherichia coli* gamma-aminobutyrate transporter: implications on the topography and mechanism of prokaryotic and eukaryotic transporters from the APC superfamily. Biochem J 336 (Pt 1):69-76

Hua J, Meyerowitz EM (1998) Ethylene responses are negatively regulated by a receptor gene family in *Arabidopsis thaliana*. Cell 94:261-271

Hubbard EJ, Jiang R, Carlson M (1994) Dosage-dependent modulation of glucose repression by MSN3 (STD1) in *Saccharomyces cerevisiae*. Mol Cell Biol 14:1972-1978

Hyde R, Taylor PM, Hundal HS (2003) Amino acid transporters: roles in amino acid sensing and signalling in animal cells. Biochem J 373:1-18

Iraqui I, Vissers S, Bernard F, De Craene JO, Boles E, Urrestarazu A, André B (1999) Amino acid signaling in *Saccharomyces cerevisiae*: a permease-like sensor of external amino acids and F-Box protein Grr1p are required for transcriptional induction of the AGP1 gene, which encodes a broad-specificity amino acid permease. Mol Cell Biol 19:989-1001

Island MD, Naider F, Becker JM (1987) Regulation of dipeptide transport in *Saccharomyces cerevisiae* by micromolar amino acid concentrations. J Bacteriol 169:2132-2136

Isnard AD, Thomas D, Surdin-Kerjan Y (1996) The study of methionine uptake in *Saccharomyces cerevisiae* reveals a new family of amino acid permeases. J Mol Biol 262:473-484

Jansen ML, de Winde JH, Pronk JT (2002) Hxt-carrier-mediated glucose efflux upon exposure of *Saccharomyces cerevisiae* to excess maltose. Appl Environ Microbiol 68:4259-4265

Jauniaux JC, Grenson M (1990) *GAP1*, the general amino acid permease gene of *Saccharomyces cerevisiae*. Nucleotide sequence, protein similarity with the other bakers' yeast amino acid permeases, and nitrogen catabolite repression. Eur J Biochem 190:39-44

Jiang H, Medintz I, Michels CA (1997) Two glucose sensing/signaling pathways stimulate glucose-induced inactivation of maltose permease in Saccharomyces. Mol Biol Cell 8:1293-1304

Jorgensen MU, Bruun MB, Didion T, Kielland-Brandt MC (1998) Mutations in five loci affecting GAP1-independent uptake of neutral amino acids in yeast. Yeast 14:103-114

Jorgensen MU, Gjermansen C, Andersen HA, Kielland-Brandt MC (1997) *STP1*, a gene involved in pre-tRNA processing in yeast, is important for amino-acid uptake and transcription of the permease gene BAP2. Curr Genet 31:241-247

Kellis M, Patterson N, Endrizzi M, Birren B, Lander ES (2003) Sequencing and comparison of yeast species to identify genes and regulatory elements. Nature 423:241-254

Kim JH, Polish J, Johnston M (2003) Specificity and regulation of DNA binding by the yeast glucose transporter gene repressor Rgt1. Mol Cell Biol 23:5208-5216

Klasson H, Fink GR, Ljungdahl PO (1999) Ssy1p and Ptr3p are plasma membrane components of a yeast system that senses extracellular amino acids. Mol Cell Biol 19:5405-5416

Kodama Y, Omura F, Takahashi K, Shirahige K, Ashikari T (2002) Genome-wide expression analysis of genes affected by amino acid sensor Ssy1p in *Saccharomyces cerevisiae*. Curr Genet 41:63-72

Kruckeberg AL (1996) The hexose transporter family of *Saccharomyces cerevisiae*. Arch Microbiol 166:283-292

Kruckeberg AL, Walsh MC, van Dam K (1998) How do yeast cells sense glucose? Bioessays 20:972-976

Kuchin S, Vyas VK, Kanter E, Hong SP, Carlson M (2003) Std1p (Msn3p) positively regulates the Snf1 kinase in *Saccharomyces cerevisiae*. Genetics 163:507-514

Lafuente MJ, Gancedo C, Jauniaux JC, Gancedo JM (2000) Mth1 receives the signal given by the glucose sensors Snf3 and Rgt2 in *Saccharomyces cerevisiae*. Mol Microbiol 35:161-172

Lakshmanan J, Mosley AL, Ozcan S (2003) Repression of transcription by Rgt1 in the absence of glucose requires Std1 and Mth1. Curr Genet 44:19-25

Li FN, Johnston M (1997) Grr1 of *Saccharomyces cerevisiae* is connected to the ubiquitin proteolysis machinery through Skp1: coupling glucose sensing to gene expression and the cell cycle. EMBO J 16:5629-5638

Liang H, Gaber RF (1996) A novel signal transduction pathway in *Saccharomyces cerevisiae* defined by Snf3-regulated expression of HXT6. Mol Biol Cell 7:1953-1966

Ling R, Bridges CC, Sugawara M, Fujita T, Leibach FH, Prasad PD, Ganapathy V (2001) Involvement of transporter recruitment as well as gene expression in the substrate-induced adaptive regulation of amino acid transport system A. Biochim Biophys Acta 1512:15-21

Ljungdahl PO, Gimeno CJ, Styles CA, Fink GR (1992) SHR3: a novel component of the secretory pathway specifically required for localization of amino acid permeases in yeast. Cell 71:463-478

Magasanik B, Kaiser CA (2002) Nitrogen regulation in *Saccharomyces cerevisiae* 475. Gene 290:1-18

Maier A, Volker B, Boles E, Fuhrmann GF (2002) Characterisation of glucose transport in *Saccharomyces cerevisiae* with plasma membrane vesicles (countertransport) and intact cells (initial uptake) with single Hxt1, Hxt2, Hxt3, Hxt4, Hxt6, Hxt7 or Gal2 transporters. FEM Yeast Res 2:539-550

Mosley AL, Lakshmanan J, Aryal BK, Ozcan S (2003) Glucose-mediated phosphorylation converts the transcription factor Rgt1 from a repressor to an activator. J Biol Chem 278:10322-10327

Nehls U, Wiese J, Guttenberger M, Hampp R (1998) Carbon allocation in ectomycorrhizas: identification and expression analysis of an *Amanita muscaria* monosaccharide transporter. Mol Plant Microbe Interact 11:167-176

Neigeborn L, Carlson M (1984) Genes affecting the regulation of SUC2 gene expression by glucose repression in *Saccharomyces cerevisiae*. Genetics 108:845-858

Nielsen PS, van den HB, Didion T, de Boer M, Jorgensen M, Planta RJ, Kielland-Brandt MC, Andersen HA (2001) Transcriptional regulation of the *Saccharomyces cerevisiae* amino acid permease gene BAP2. Mol Gen Genet 264:613-622

Ozcan S (2002) Two different signals regulate repression and induction of gene expression by glucose. J Biol Chem 277:46993-46997

Ozcan S, Dover J, Johnston M (1998) Glucose sensing and signaling by two glucose receptors in the yeast *Saccharomyces cerevisiae*. EMBO J 17:2566-2573

Ozcan S, Dover J, Rosenwald AG, Wolfl S, Johnston M (1996a) Two glucose transporters in *Saccharomyces cerevisiae* are glucose sensors that generate a signal for induction of gene expression. Proc Natl Acad Sci USA 93:12428-12432

Ozcan S, Johnston M (1995) Three different regulatory mechanisms enable yeast hexose transporter (HXT) genes to be induced by different levels of glucose. Mol Cell Biol 15:1564-1572

Ozcan S, Johnston M (1999) Function and regulation of yeast hexose transporters. Microbiol Mol Biol Rev 63:554-569

Ozcan S, Leong T, Johnston M (1996b) Rgt1p of *Saccharomyces cerevisiae*, a key regulator of glucose-induced genes, is both an activator and a repressor of transcription. Mol Cell Biol 16:6419-6426

Pi J, Pittard AJ (1996) Topology of the phenylalanine-specific permease of *Escherichia coli*. J Bacteriol 178:2650-2655

Rape M. Jentsch S (2002) Taking a bite: proteasomal protein processing. Nat Cell Biol 4:E113-E116

Regenberg B, During-Olsen L, Kielland-Brandt MC, Holmberg S (1999) Substrate specificity and gene expression of the amino-acid permeases in *Saccharomyces cerevisiae*. Curr Genet 36:317-328

Reifenberger E, Boles E, Ciriacy M (1997) Kinetic characterization of individual hexose transporters of *Saccharomyces cerevisiae* and their relation to the triggering mechanisms of glucose repression. Eur J Biochem 245:324-333

Rolland F, Wanke V, Cauwenberg L, Ma P, Boles E, Vanoni M, de Winde JH, Thevelein JM, Winderickx J (2001) The role of hexose transport and phosphorylation in cAMP signalling in the yeast *Saccharomyces cerevisiae*. FEMS Yeast Res 1:33-45

Rouillon A, Surdin-Kerjan Y, Thomas D (1999) Transport of sulfonium compounds. Characterization of the S-adenosylmethionine and S-methylmethionine permeases from the yeast *Saccharomyces cerevisiae*. J Biol Chem 274:28096-28105

Saier MH Jr (2000) Families of transmembrane transporters selective for amino acids and their derivatives. Microbiology 146 (Pt 8):1775-1795

Schmidt MC, McCartney RR, Zhang X, Tillman TS, Solimeo H, Wolfl S, Almonte C, Watkins SC (1999) Std1 and Mth1 proteins interact with the glucose sensors to control glucose-regulated gene expression in *Saccharomyces cerevisiae*. Mol Cell Biol 19:4561-4571

Schule T, Rose M, Entian KD, Thumm M, Wolf DH (2000) Ubc8p functions in catabolite degradation of fructose-1, 6-bisphosphatase in yeast. EMBO J 19:2161-2167

Schulte F (1997). Molekulargenetische und physiologische Untersuchungen zur Funktion von *HTR1* beim Hexosetransport in der Hefe *Saccharomyces cervisiae*. Thesis, Heinrich-Heine-Universität Düsseldorf.

Schulte F, Wieczorke R, Hollenberg CP, Boles E (2000) The *HTR1* gene is a dominant negative mutant allele of *MTH1* and blocks Snf3- and Rgt2-dependent glucose signaling in yeast. J Bacteriol 182:540-542

Schurmann A, Doege H, Ohnimus H, Monser V, Buchs A, Joost HG (1997) Role of conserved arginine and glutamate residues on the cytosolic surface of glucose transporters for transporter function. Biochemistry 36:12897-12902

Souciet J, Aigle M, Artiguenave F, Blandin G, Bolotin-Fukuhara M, Bon E, Brottier P, Casaregola S, de Montigny J, Dujon B, Durrens P, Gaillardin C, Lepingle A, Llorente B, Malpertuy A, Neuveglise C, Ozier-Kalogeropoulos O, Potier S, Saurin W, Tekaia F, Toffano-Nioche C, Wesolowski-Louvel M, Wincker P, Weissenbach J (2000) Genomic exploration of the hemiascomycetous yeasts: 1. A set of yeast species for molecular evolution studies. FEBS Lett 487:3-12

Sychrova H, Chevallier MR (1993) Cloning and sequencing of the *Saccharomyces cerevisiae* gene LYP1 coding for a lysine-specific permease. Yeast 9:771-782

Tanaka J, Fink GR (1985) The histidine permease gene (*HIP1*) of *Saccharomyces cerevisiae*. Gene 38:205-214

Tillman TS, Ganster RW, Jiang R, Carlson M, Schmidt MC (1995) STD1 (MSN3) interacts directly with the TATA-binding protein and modulates transcription of the *SUC2* gene of *Saccharomyces cerevisiae*. Nucleic Acids Res 23:3174-3180

Tomas-Cobos L, Sanz P (2002) Active Snf1 protein kinase inhibits expression of the *Saccharomyces cerevisiae HXT1* glucose transporter gene. Biochem J 368:657-663

Turner GC, Du F, Varshavsky A (2000) Peptides accelerate their uptake by activating a ubiquitin-dependent proteolytic pathway. Nature 405:579-583

Uetz P, Giot L, Cagney G, Mansfield TA, Judson RS, Knight JR, Lockshon D, Narayan V, Srinivasan M, Pochart P, Qureshi-Emili A, Li Y, Godwin B, Conover D, Kalbfleisch

T, Vijayadamodar G, Yang M, Johnston M, Fields S, Rothberg JM (2000) A comprehensive analysis of protein-protein interactions in *Saccharomyces cerevisiae*. Nature 403:623-627

Van Belle D, André B (2001) A genomic view of yeast membrane transporters. Curr Opin Cell Biol 13:389-398

van Roermund CW, Hettema EH, van den BM, Tabak HF, Wanders RJ (1999) Molecular characterization of carnitine-dependent transport of acetyl-CoA from peroxisomes to mitochondria in *Saccharomyces cerevisiae* and identification of a plasma membrane carnitine transporter, Agp2p. EMBO J 18:5843-5852

Vandenbol M, Jauniaux JC, Grenson M (1989) Nucleotide sequence of the *Saccharomyces cerevisiae* PUT4 proline-permease-encoding gene: similarities between CAN1, HIP1 and PUT4 permeases. Gene 83:153-159

Wang SS, Stanford DR, Silvers CD, Hopper AK (1992) *STP1*, a gene involved in pre-tRNA processing, encodes a nuclear protein containing zinc finger motifs. Mol Cell Biol 12:2633-2643

Wykoff DD, O'Shea EK (2001) Phosphate transport and sensing in *Saccharomyces cerevisiae*. Genetics 159:1491-1499

Boles, Eckhard
    Institut fuer Mikrobiologie, Goethe-Universitaet Frankfurt, Marie-Curie-Str. 9, D-60439 Frankfurt am Main, Germany
    E.Boles@em.uni-frankfurt.de

André, Bruno
    Université Libre de Bruxelles CP300, Institut de Biologie et de Médecine Moléculaires (IBMM), Laboratoire de Physiologie Moléculaire de la Cellule, Rue des Pr. Jeener et Brachet, 12 6041 Gosselies Belgium
    bran@ulb.ac.be

Osmoregulation and osmosensing by uptake carriers for compatible solutes in bacteria

Susanne Morbach and Reinhard Krämer

Abstract

In order to circumvent deleterious effects of hypo- and hyperosmotic conditions in their environment, bacterial cells have developed a number of mechanisms to counteract osmotic stress. The first response to an osmotic upshift is the activation of uptake systems for so-called compatible solutes, e.g. betaine, proline, or ectoine. Compatible solutes are taken up at the expense of metabolic energy and accumulated in the cytoplasm until osmotic compensation is reached. These uptake systems respond to osmotic stress by regulation at the level of both protein activity and gene expression. Activity regulation of uptake systems includes elaborate signal transduction mechanisms, involving sensing of appropriate stimuli and signal transduction to the catalytic (transport) domains of the carrier protein. Mechanisms of osmosensing and osmoregulation will be discussed mainly for BetP, the betaine uptake carrier of *Corynebacterium glutamicum*, as well as for ProP from *Escherichia coli*, OpuA from *Lactococcus lactis,* and a few other examples.

1 Introduction

Osmotic stress is one of the most common environmental stress factors encountered by living cells including bacteria. A highly important aspect of cellular homeostasis is to keep the water activity in the cell's cytoplasm at a constant level, which guarantees proper physiological function of cellular systems, viability, and growth. The first and inevitable reaction following a change in external osmolality is water flux, directed into the cytoplasm upon hypoosmotic stress, i.e. a decrease in external osmolality, and, *vice versa*, directed out of the cytoplasm under conditions of hyperosmotic stress. The ultimate consequence of these events, namely cell rupture in the former and plasmolysis in the latter case, would be deleterious or even lethal. Thus, bacteria have developed efficient strategies to overcome osmotic challenge. Under hypoosmotic stress, the concentration of internal solutes is instantly lowered to decrease the driving force for water entry. This is managed by the action of mechanosensitive channels located in the plasma membrane. Upon a hypoosmotic shock, these channels open in a millisecond time scale and cause fast efflux of small solutes. On the other hand, as a response to hyperosmotic shock, the cell increases its internal osmolality, which may lead to extremely high cyto-

Topics in Current Genetics, Vol. 9
E. Boles, R. Krämer (Eds.): Molecular Mechanisms Controlling Transmembrane Transport
DOI 10.1007/b95846 / Published online: 9 March 2004

plasmic solute concentrations. For this purpose, particular solutes are used by the cell, which are compatible with normal physiological functions when present at high concentrations, so-called compatible solutes. These are highly soluble, polar, and often zwitterionic substances, such as betaines (glycine betaine and proline betaine), amino acids (proline, glutamate, and glutamine), amino acid derivatives (ectoine, taurine), and polyols (glycerol, trehalose, and glucosylglycerol).

Compatible solutes can be accumulated in the cell by two different mechanisms, namely biosynthesis and uptake. In general, the latter process is faster and more favorable in terms of energy and carbon cost for the cell, whereas the former renders the cell independent of the availability of this solute in the environment. The systems responsible for solute accumulation, i.e. biosynthetic enzymes or carrier proteins, must be effectively regulated in order to guarantee a well-adapted response to varying hyperosmotic stress (osmoregulation). For the instant response to hyperosmotic stress, enzymes and transporters are regulated at the level of protein activity, whereas a long-term adaptation to conditions of high or low external osmolality is achieved by regulation at the level of gene expression and protein synthesis.

Osmoregulation depends on the ability of cells to properly sense stimuli related to osmotic stress (osmosensing) and to transduce the corresponding signals to the targets of osmoregulatory networks. This article deals with a typical signal transduction system including carrier systems as a core part. In osmosensing, the change in external osmolality, which causes alterations of internal conditions (solute concentration) because of related water flux, acts as a stimulus for so far largely unknown receptor proteins. A number of possible stimuli are currently discussed, including changes in external and internal composition and concentration of solutes, cell turgor, membrane strain, molecular crowding etc. (Wood 1999). Due to this complexity, the understanding of osmosensing is still very limited. Osmorelated stimuli are then transduced via two different kinds of signal pathways. On the one hand, signals are transferred to proteins responding to osmotic stress at the level of activity, and, on the other hand, to osmoresponsive operons, leading to regulated expression of certain genes. Only the former aspect of osmoregulation will be discussed in more detail in this review. We will focus on the understanding how the catalytic (transport) activity of carrier proteins in bacterial plasma membranes responds to changes of osmotic conditions in their direct environment. Closely connected to this issue is the question how the information of a change in osmolality is perceived by these carrier proteins or by receptors connected to them and how the signal is transduced ultimately leading to a regulation of the transporter's catalytic activity.

2 Osmodependent transport systems in bacteria

A number of transport systems in bacteria are known whose activity and/or synthesis is influenced by the osmolality of the surrounding medium. Several of them have been studied in great detail with respect to activity, specificity, and

Table 1. Selected bacterial osmoregulated transporters.

Transporter family	Transport protein	Source (organism)	Preferred substrate
ABC	ProU	*E. coli*	betaine,
	OpuA, Opu C	*B. subtilis*	betaine
	OpuB	*B. subtilis*	coline
	OpuA (BusA)	*L. lactis*	betaine
MFS	ProP	*E. coli*	proline (?)
	ProP	*C. glutamicum*	proline, ectoine
SSS	OpuE	*B. subtilis*	proline
BCCT	BetT	*E. coli*	choline
	OpuD	*B. subtilis*	betaine
	BetP	*C. glutamicum*	betaine
	EctP	*C. glutamicum*	ectoine > betaine > proline

physiological function, in *Escherichia coli*, *Salmonella typhimurium*, *Bacillus subtilis*, *Lactococcus lactis*, *Listeria monocytogenes*, and *Corynebacterium glutamicum* (for reviews see Csonka and Hanson 1991; Wood 1999; Bremer and Krämer 2000; Poolman et al. 2002; Morbach and Krämer 2002). Although expression regulation of genes coding for osmoregulated transport proteins is highly relevant and still not well understood, it is not the main topic of this article. Consequently, for the following overview, transport systems have been selected which are known to be strongly regulated at the level of activity in response to osmotic challenge (Table 1).

Secondary carriers are transport systems in which solute transport is coupled to the (electro)chemical gradient of another solute, i.e. the cosubstrate, which provides the driving force for the transport reaction. The secondary systems studied in most detail and discussed here are ProP of *E. coli* and BetP of *C. glutamicum*.

In primary carriers, transport is directly coupled to a chemical reaction, mostly ATP hydrolysis, to provide the driving force. The best studied ABC transporter in terms of activity regulation is OpuA, also named BusA from *L. lactis*. Among osmoregulated primary transport systems, the multicomponent high-affinity K^+ uptake system KdpFABC of *E. coli* is frequently discussed, too. KdpD, the sensor kinase of the KdpDE two component regulatory system located within the *kdpFABCDE* operon, is one of the best-characterized osmosensors.

Mechanosensitive (MS) channels have been found in many bacterial cells, and their rapid action upon hypoosmotic shock is essential for cell viability. In *E. coli*, at least three types of channels, MscL, MscS, and MscM have been discriminated based on their conductance in the open state by patch clamp experiments (Berrier et al. 1996; Martinac 2001). In difference to osmosensitive carrier systems, high resolution 3D structures of both MscL of *Mycobacterium tuberculosis* (Chang et al. 1998) and MscS of *E. coli* (Bass et al. 2002) are available.

3 Osmoregulated solute carriers of *C. glutamicum*: osmosensing and osmoregulation

The Gram-positive soil bacterium *Corynebacterium glutamicum* possesses five different uptake systems for compatible solutes (Peter et al. 1996, 1998b, unpublished observation). Four of them, namely BetP, EctP, ProP, and LcoP respond to hyperosmotic stress by an instant increase in transport activity, whereas PutP is a proline carrier responsible for proline uptake for anabolic purposes. BetP is highly specific for its substrate glycine betaine, whereas the other carriers have a broader range of substrates, including also proline and ectoine (Peter et al. 1998b). BetP, ProP, and LcoP, as well as EctP to some extent, are regulated at the level of gene expression, too, being induced to higher levels after a hyperosmotic shock. In difference to other well-studied organisms, such as *E. coli* or *B. subtilis*, *C. glutamicum* uses only secondary systems for uptake of compatible solutes, in spite of the fact that a large number of ABC transporters are present in its genome. *C. glutamicum* uses mainly proline, glycine betaine, and ectoine as compatible solutes, as well as glutamate and glutamine to some extent. Under conditions of nitrogen limitation, these solutes are replaced by trehalose (Wolf et al. 2003). *C. glutamicum* is not able to synthesize betaine and ectoine, uptake of compatible solutes is thus of high significance, and is preferred for synthesis if external solutes are available. Within these five uptake carriers, BetP is by far the best studied, both in terms of catalytic as well as regulatory properties.

An osmoresponsive uptake system is characterized by a number of different functional properties related to the response to osmotic challenge. Three different functional levels can be discriminated. (i) The level of catalytic activity, i.e. transport function, includes kinetic properties (affinity, activity, and specificity), the kind of transport mechanism, and the type of energetic coupling. (ii) The level of osmoregulation describes the property of the transport system to correctly adapt its catalytic activity to the extent of osmotic stress. Finally (iii), the property of osmosensitivity indicates that a system is able to perceive stimuli related to osmotic stress. In order to function properly, such a system must be able to transduce the perceived stimulus into a signal, which is recognized by other elements of the system leading to a transport activity that is correctly adapted to both the extent of stress and the physiological situation of the cell. In the case of BetP, similar to ProP of *E. coli* (Racher et al. 1999) and OpuA of *L. lactis* (van der Heide and Poolman 2000), it was proven by functional reconstitution in proteoliposomes, that the transporter harbors all three different functional levels (Rübenhagen et al. 2000). In addition, BetP was shown to correctly function when heterologously expressed in *E. coli*, which, besides successful reconstitution, argues against additional proteinaceous factors being involved in any of these functions.

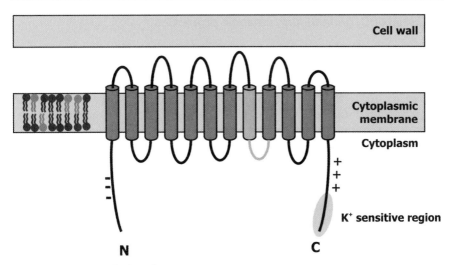

Fig. 1. Topology prediction of BetP.

3.1 BetP of *C. glutamicum*: structural properties

BetP of *C. glutamicum* is a polytopic membrane protein consisting of 595 amino acids. It is predicted to span the membrane 12 times, and it contains hydrophilic domains of 55 – 62 amino acids at both the N- and the C-terminal end. Both terminal domains face the cytoplasm (Rübenhagen et al. 2001) (Fig. 1). Like EctP and LcoP, BetP is a member of the BCCT family of transporters. Carrier proteins of this family are known to transport solutes with a positively charged, quaternary N-group, such as betaines, choline, carnitine, and ectoine. A conserved motif, which is located in the cytoplasmic loop between the 8th and 9th transmembrane segment, is supposed to be involved in substrate binding.

Recently, strep-tagged BetP was isolated and purified, and two-dimensional crystals have been obtained (Ziegler et al. 2004). The crystals were analyzed by electron cryo-microscopy and projection maps were calculated to 7.5 Å. In the crystal, BetP monomers are associated forming a dimer of trimers. In each monomer 10-12 transmembrane α-helices can be distinguished as well as two pore like features suggesting potential transport pathways. The projection map of BetP does not resemble projection structures of other secondary carriers. The trimeric arrangement of BetP subunits was verified by analytical ultracentrifugation of BetP solubilized in detergent. The fact that both solubilized BetP, which was proven to be functionally active after reconstitution into proteoliposomes, as well as BetP in 2D crystals form trimers, strongly suggests that BetP is also trimeric in the native membrane.

3.2 BetP of *C. glutamicum*: catalytic properties

BetP couples the electrochemical Na^+ potential to betaine flux via co-transport of 2 Na^+ ions (Farwick et al. 1995). Extremely high steady state accumulation ratios of up to $4 \cdot 10^6$ (internal/external concentration) have been determined. When measuring steady-state internal pools at low external betaine concentrations and under variation of the electrochemical Na^+ potential, the chemical potential (betaine concentration gradient) and the electrochemical Na^+ potential were found to be in equilibrium (Farwick et al. 1995). These observations argue for the following properties of the system, (i) a strict coupling of substrate and cosubstrate fluxes (no internal leak), (ii) virtual impermeability of the membrane to betaine, and (iii), at least under the experimental conditions tested, absence of (external) leaks, i.e. betaine efflux channels or carriers. It is important to note that *C. glutamicum* cannot metabolize betaine.

The affinity of the external binding site of BetP for betaine in *C. glutamicum* is relatively high (Km of 8 µM), whereas the Na^+ affinity is low (Km of 4 mM). The affinity of the other carriers in *C. glutamicum*, namely EctP, ProP, and LcoP, for their substrates is significantly lower in the high µM to mM range (Peter et al. 1998b). By testing substances structurally related to betaine as competitors for BetP-mediated betaine uptake, e.g. N,N-dimethyl- and N-methylglycine, butyrobetaine, carnitine and related substances, BetP was shown to be strictly specific for glycine betaine (Burger 1999).

When induced at the level of expression and activated at the level of protein activity, BetP is, together with acetate (Ebbighausen et al. 1991) and glucose uptake (Marx et al. 1996), one of the fastest uptake systems in *C. glutamicum*, characterized by V_{max} values up to 110 µMol $\cdot$ g cdm^{-1} $\cdot$ min^{-1} (Farwick et al. 1995). The high activity of BetP in comparison to the other osmoregulated carriers in *C. glutamicum* is one of the reasons why betaine uptake in general dominates the uptake of other compatible solutes, provided that it is present in the environment.

3.3 BetP of *C. glutamicum*: regulatory properties

Accumulation of compatible solutes must be adapted to the actual situation of the cell, depending on internal and external osmolality and consequently on turgor pressure. Besides by variation of the biosynthesis rate, solute accumulation can be regulated by modifying the rate of uptake, the rate of efflux or both. Furthermore, regulation may take place, for an instant response, at the level of protein activity and, for long-term adaptation, at the level of gene expression. BetP is probably one of the best studied osmoregulated carrier proteins in terms of regulation of protein activity, whereas this is not yet the case for transcriptional regulation.

3.3.1 Regulation of BetP at the level of activity

BetP is able to instantly respond to osmotic stress, and its activity is adapted to the actual extent of stress. BetP thus represents a well-suited example for studying

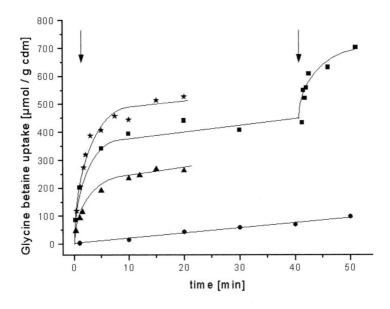

Fig. 2. Betaine uptake by BetP after osmotic upshift. Cells are suspended in buffer at an osmolality of 125 mOsmol/kg. At time zero, NaCl is added to reach final osmolalities of 125 mOsmol/kg (no addition, circles), 625 mOsmol/kg (triangles), 925 mOsmol/kg (squares), or 1325 mOsmol/kg (stars). At 40 min, the osmolality is increased to 1600 mOsmol/kg.

control mechanisms of solute transport. Two different features of osmoregulation, as far as it concerns hyperosmotic stress, are recognized (Fig. 2). As an osmoregulated uptake system, it is more or less silent in the absence of hyperosmotic or in the presence of hypoosmotic stress. Upon an osmotic upshift, BetP becomes activated. Once the hyperosmotic stress has been compensated by accumulation of compatible solutes, uptake activity is reduced in order to prevent an overshoot of internal solute concentration. The decrease of uptake activity upon stress compensation is not likely to resemble the pre-activated state, since the conditions of compensated stress are different from those before activation occurred. In the following, the first regulation event will be called activation and the second activity adaptation.

The time course of activation and activity adaptation under *in vivo* conditions is shown in Figure 2 for BetP mediated betaine uptake. The observed response time of BetP to an osmotic upshift is shorter than the time resolution of conventional kinetic experiments, i.e. less than one second. If the initial activity, measured immediately after the onset of osmotic stress, is plotted in dependence of the extent of osmotic stress, a kind of dose – response curve is obtained (Fig. 3). Without osmotic stress, BetP is virtually inactive. The threshold of activation is observed at

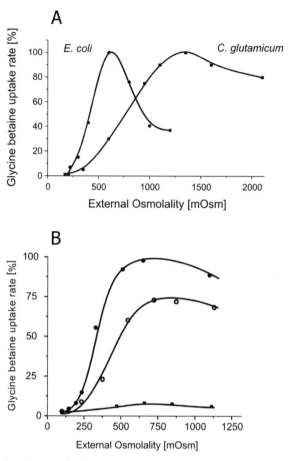

Fig. 3. Profile of BetP activation by osmotic stress. A, The external osmolality was adjusted by adding NaCl to cells of *C. glutamicum* and *E. coli* and betaine uptake activity was determined. In *E. coli* cells (strain MKH13, devoid of osmoregulated carriers), BetP was heterologously expressed. B, Activation of betaine uptake by BetP reconstituted in proteoliposomes. The external osmolality was adjusted by NaCl (full circles), sorbitol (open circles), or glycerol (squares). Only non-permeable osmolytes induce an osmotic gradient and thus osmotic activation of BetP.

about 300 - 400 mOsmol/kg, and BetP reaches an optimum activity around 1300 mOsmol/kg. This activation profile is very similar *in vivo*, i.e. in intact cells of *C. glutamicum* (Fig. 3A) and *in vitro* when BetP is reconstituted in proteoliposomes (Fig. 3B) (Rübenhagen et al. 2000). We do not routinely observe a decrease of transport activity after reaching an optimum of activation in the liposomal system where the driving force is the K^+ diffusion potential together with a Na^+ gradient. It may be argued that the decrease of activity observed in intact cells at high osmotic stress originates from a decrease of the driving force, the electrochemical Na^+ potential, under these unfavorable physiological conditions, however, this has not been proven. When heterologously expressed in *E. coli*, activation of BetP by hyperosmotic stress is characterized by a basically similar regulation pattern (Fig. 3A) (Peter et al. 1996). Interestingly, although showing a similar shape, the optimum of stimulation is drastically shifted to lower values of osmotic stress. It was later shown on the basis of experiments in proteoliposomes that this shift is likely to be due to the different phospholipid composition of the membrane and not to the different value of turgor pressure of the Gram-negative *E. coli* and the Gram-positive *C. glutamicum* (Rübenhagen et al. 2000).

Several conclusions can be drawn from these experiments. First, BetP does not need additional effector components for its regulatory function. Its full functional competence observed both in *E. coli* and in proteoliposomes argues for BetP harboring all three properties, i.e. that of a transporter, an osmoregulator and an osmosensor. Second, the stimulus responsible for osmoregulation must be functional in all systems tested, *C. glutamicum* and *E. coli* cells, as well as proteoliposomes. Third, a different membrane environment of BetP leads to modulation of the activation profile. This may be due to a modification in the stimulus being responsible for activation or to a change of the surrounding hydrophilic or hydrophobic phases interacting with the carrier protein.

It is interesting to think about a possible model, which explains intermediate activities observed during activation. This is not a trivial question, since it is of basic significance for interpreting molecular mechanisms of osmoregulation whether an intermediate activity is the consequence of different activity levels of single BetP molecules or whether it is the result of averaging over a changing share of active and inactive BetP molecules. A plausible model has been proposed for ProP of *E. coli*, where an equilibrium between active and inactive forms of the carrier is assumed (Culham et al. 2003). This equilibrium is then supposed to be shifted by osmotic stress towards the active form and *vice versa*.

After betaine accumulation, the cell reaches osmotic compensation, and the uptake of compatible solutes decreases (Fig. 2). The apparent steady state reached is not a steady state typically observed in uptake systems, independent whether it is explained on thermodynamic (balance of electrochemical potentials) or kinetic reasons (e.g. trans inhibition). The fact, that the steady state levels of solute accumulation depend on the extent of osmotic stress to be compensated (Farwick et al. 1995), argues that the mechanism by which the steady state is defined is an intrinsic aspect of osmoregulation. At least three possible models could explain this observation, (i) a controlled shift in BetP activity from unidirectional uptake to counter exchange, (ii) downregulation of betaine uptake at the level of BetP activity, and (iii) onset of efflux activity via export mechanisms (channels or carriers) which counterbalance betaine uptake. The major contribution in *C. glutamicum* seems to be downregulation of BetP activity (model ii) combined with a functional shift to a counter exchange (model i) (unpublished observation). In other bacteria, osmostress-dependent product inhibition due to high internal substrate accumulation has been discussed as a possible explanation for the observed decrease of activity (model ii). This has been suggested for *Staphyolococcus aureus* (Pourkomailian and Booth 1994; Stimeling et al. 1994), *Lactobacillus plantarum* (Glaasker et al. 1996), and *Listeria monocytogenes* (Verheul et al. 1997). As an alternative mechanism (model iii), in *Lactobacillus plantarum*, separate systems for (primary) uptake and efflux of betaine were discriminated and the efflux system was postulated to be inactive under hyperosmotic conditions (Glaasker et al. 1996).

3.3.2 Regulation of BetP at the level of expression

Since transcription regulation is not the focus of this article, it will only briefly be discussed here. Analysis of transport activity after growth under varying external osmolality revealed that BetP and ProP are regulated at the level of gene expression, whereas EctP seemed to be constitutively expressed. Northern blot and RNA hybridization analysis showed that all four carrier proteins, i.e. BetP, EctP, ProP, and LcoP respond to hyperosmotic stress to a different extent by an increase in mRNA synthesis, and the mechanosensitive channel MscL by a corresponding decrease (Möker 2003). Interestingly, transcription of *betP*, *proP*, and *mscL* was found to strictly depend on the presence of one out of the 13 two-component regulatory systems present in *C. glutamicum*. Based on these results, the following arguments indicate the presence of several input pathways into osmoregulatory signal transduction in *C. glutamicum*. First, the carrier proteins are regulated at the level of activity responding to osmotic stimuli. Second, the transcription of *betP*, *proP*, and *mscL* is regulated via signal input through a two-component system and signal transduction to the corresponding genes. Third, there must be further pathways of signal input since *ectP*, which is not regulated via a two-component system, is transcriptionally regulated in dependence of osmotic stress, too. Moreover, the same holds for transcription regulation of osmoregulated enzymes catalyzing the biosynthesis of compatible solutes, e.g. proline or trehalose (Ley 2001; Wolf et al. 2003).

3.4 BetP of *C. glutamicum*: sensory properties

Processes of osmoregulation are integrated into signal transduction pathways in which a sensory element, measuring the extent of osmotic stress, is located at the beginning. When activity regulation is concerned, the osmosensor will relay a signal to the transport protein leading to osmotic activation or activity adaptation. Since BetP was shown to possess all three functions relevant for signal transduction, it should harbor sensor domain(s) and must be able to relay an appropriate signal to the catalytic (transport) domain regulating its activity.

A long list of possible stimuli is considered as being sensed by bacterial cells (Wood 1999) (Fig. 4). Mainly four different categories are discussed: (i) Stimuli directly originating from the environment, e.g. external osmolality, ionic strength, or concentration of particular solutes. (ii) The same parameters may be relevant at the cytoplasmic site, too, since a change in internal water activity is the consequence of a change in external osmolality. Molecular crowding of cytoplasmic macromolecules may be relevant too. (iii) Membrane related parameters such as cell turgor and membrane strain may be important for carrier proteins.

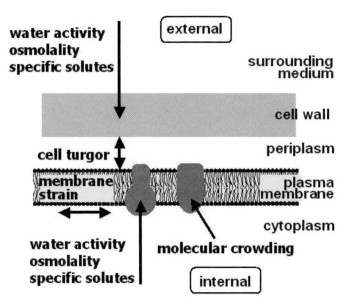

Fig. 4. Possible stimuli related to hyperosmotic stress in bacteria.

(iv) Changes in the surrounding osmolality may also directly influence soluble and membrane-embedded proteins by altering their surface hydration and thus their conformation.

The true nature of the stimuli perceived is not yet clear for most of the osmo-regulated systems. As one of few exceptions, it has been shown for reconstituted BetP of *C. glutamicum* that an increase in the lumenal K^+ concentration, i.e. at the side where the hydrophilic domains are located, is sufficient to activate BetP. In these experiments, several possible triggers were excluded, namely, changes of external solutes, as well as changes of internal solutes except K^+, e.g., choline$^+$, NH_4^+ (Rübenhagen et al. 2001) and Na^+ (unpublished observations), furthermore membrane strain and cell turgor, since proteoliposomes are lacking turgor pressure. Consequently, by identifying an increase in lumenal K^+ as an activating stimulus, BetP has literally been converted from an osmosensor to a chemosensor. The internal threshold concentration of K^+ necessary for activation of BetP in *E. coli* phospholipid liposomes was around 220 mM (Rübenhagen et al. 2001). Interestingly, a strong influence of the membrane phospholipid composition on activation was observed (Rübenhagen et al. 2000). The higher the share of negatively charged phospholipids, the higher was the threshold of K^+ necessary for activation. It has to be pointed out that this result identifying K^+ as a major stimulus for BetP activation, strictly speaking, holds true only for the system in which it has been measured, namely in proteoliposomes made from *E. coli* phospholipids. Also in intact cells, however, a rise in internal K^+ concentration immediately following hyperosmotic stress has been determined in *E. coli* (Dinnbier et al. 1988) as well as in *C. glutamicum* (unpublished observations).

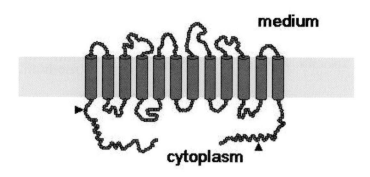

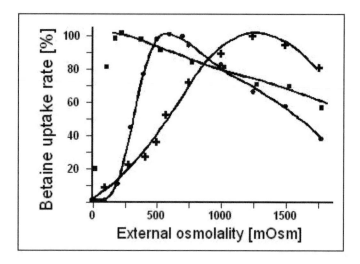

Fig. 5. Consequence of N- and C-terminal truncations of BetP on transport activity. Betaine uptake activity of wild type BetP in *C. glutamicum* cells (crosses) increases in response to increasing hyperosmotic stress. After truncation of 60 amino acids at the N-terminal hydrophilic domain of BetP, as indicated in the upper panel, the activation profile is shifted to higher osmolalities (circles). Truncation of 25 amino acids at the C-terminal domain leads to deregulation of BetP. The increase at low Na$^+$ concentration is caused by the dependence on Na$^+$ as a cosubstrate.

So far, sensing by and activation of BetP has been described on a phenomenological level only, however, data are available, which give a closer insight at the molecular level. The terminal hydrophilic domains of BetP, both at the N- and the C-terminal end, have been shown to strongly influence activation in intact *C. glutamicum* cells (Peter et al. 1998a) (Fig. 5). Truncation of the N-terminal domain did not significantly change the catalytic activity of the carrier protein, however, the activation profile was shifted to higher osmolalities, i.e. higher osmotic stress was required to activate this mutant form of BetP. Truncations at the C-terminal domain had a more drastic effect. When truncating 25 or 45 amino acids at the end of this domain, which consists of about 56 amino acids, deregulation of the protein was observed, whereas truncation of the terminal 12 amino acids led to partial deregulation only. These mutant forms were found to be constantly active in intact *C. glutamicum* cells even in the absence of osmotic stress, although at a somewhat reduced V_{max}. These results argue for the C-terminal domain being involved in osmosensing and may thus be interpreted in terms of locating the sensory input at this hydrophilic domain. Interestingly, they also indicate that the C-terminal domain acts as an inhibitory element. In other words, this domain seems to be required to keep BetP in an inactive state in the absence of osmotic stress. If the C-terminal domain would function by activating BetP, a truncation should result in inactivation of the protein. Notably, these results have recently been confirmed using isolated wild type and mutant forms of BetP reconstituted in proteoliposomes. BetP was activated in dependence of the lumenal K^+ concentration, and the terminal truncations led to the same changes in the regulation pattern as observed in intact cells (unpublished observations).

It has not been elucidated on a mechanistic level, how the increase in internal K^+ concentration causes BetP activation. Several observations can be put together to a hypothetical model. The structure of the heterologously expressed C-terminal domain of BetP was found to be influenced by the presence of amphiphilic surfaces. CD-spectroscopy of this domain consisting of 56 amino acids revealed a random structure, which was only slightly shifted to a more structured conformation by addition of detergents below their critical micellar concentration. The peptide, however, adopted a 90% α-helical structure in the presence of detergent micelles (Burger 2002). It was furthermore shown by resonant mirror spectroscopy that the C-terminal domain binds strongly to lipid monolayer surfaces. As expected, the binding capacity depended on the ionic strength of the buffer used. Surprisingly, the affinity of the C-terminal domain to the lipid bilayer was increased with increasing K^+ concentrations in the assay buffer (concentration range 0-0.4M). When exceeding this threshold level of K^+ the binding affinity of the peptide was strongly reduced. If cations other than K^+ were used the binding affinity decreased more or less linearly with increasing osmotic stress (Burger 2003). On the basis of these observations, a model can be suggested where (i) the state of binding of the C-terminal domain of BetP to the membrane surface determines the state of activity of the carrier, and (ii) the terminal domain changes its conformation and thus its binding state in dependence of the surrounding K^+ concentration. The K^+ and membrane-surface dependent conformational change is then supposed to be responsible for activation of BetP. This model would also explain the observed dependence of the activation threshold on the kind and charge of phospholipids present in the membrane.

In contrast to the BetP activation after an osmotic upshift, not much is known concerning stimuli and signal transduction responsible for downregulation of BetP activity after osmotic compensation (activity adaptation). In accordance with the view that the adapted state is not equivalent to the 'ground state' before osmotic activation (see above), K^+ could be ruled out as a stimulus for this regulation in intact *C. glutamicum* cells, since the internal K^+ concentration was found not to change during activity adaptation (unpublished observations). In addition, a change in external parameters could be excluded. For this kind of regulatory process, stimuli related to membrane properties (turgor or membrane strain) could be possible candidates. Consequently, these studies in intact cells have to be extended to proteoliposomal systems in order to define the responsible stimuli.

Besides BetP, *C. glutamicum* possesses three other osmoregulated uptake carriers for compatible solutes. In this context, EctP is of particular interest, since it belongs to the same carrier family as BetP, the BCCT family. The *ectP* gene encodes a protein of 615 amino acids, which is predicted to contain 12 transmembrane segments as well as N- and C-terminal domains with a predicted length of 23 and 108 amino acids, respectively. The response of EctP activity to osmotic stress is very similar to that of BetP (cf. Fig. 3). Interestingly, truncation of the N- and the C-terminal domain of EctP led to identical results as compared to BetP, i.e. modulation of stress response in N-terminally truncated mutants and deregulation of catalytic and regulatory activity in C-terminally truncated constructs, respectively (Steger 2002). In spite of this functional similarity, sequence similarity between BetP and EctP is found only in the central part, but not in the terminal domains, neither in sequence, in length, or in charge distribution. Consequently, the question was addressed whether the terminal domains of BetP and EctP are able to replace each other in terms of osmosening and/or osmoregulation. A full set of chimeras of BetP and EctP (each N- and each C-terminal domain fused to the membrane part of either BetP and EctP) revealed that these domains, in spite of having similar functions in the respective intact proteins, are not able to functionally replace each other (Steger 2002). These results may be interpreted in two ways. Either the similar function in terms of a conformational change does not require structure similarity, or the function arises from an additional interaction of the C-terminal domains with a so far unidentified part of the rest of the carrier protein thereby generating the specific response.

Finally, we would like to point out, that sensory events are certainly of crucial importance for expression regulation of the *betP* gene, too. The membrane-embedded sensor elements of the two-component regulatory system must be able to perceive stimuli related to osmotic stress and must convert them into appropriate signals. It is, however, unclear whether these stimuli are related to those, which are perceived by BetP in course of regulation at the level of protein activity.

4 Other osmoregulated carrier systems

A significant number of transport systems with osmoregulatory properties are available in the literature, however, only few of them have been studied on the molecular level with respect to regulation of transport and sensing osmotic stimuli. In particular, this holds true as far as the level of protein activity is concerned. For this reason, only two further transporters will be discussed, namely ProP from *E. coli*, and OpuA/BusA from *L. lactis*. Some others will be briefly mentioned, and, in addition, osmoregulated solute efflux channels will be included.

4.1 ProP of *E. coli*

ProP of *E. coli*, a member of the major facilitator family (MFS), is a secondary, H^+ symport carrier accepting a broad range of compatible solutes related to glycine betaine and proline (Cairney et al. 1985a; Wood 1999). Together with the primary ABC-type system ProU and the secondary choline transport system BetT, it mediates the uptake of most compatible solutes in *E. coli*. In contrast to ProP, the ProU system responds to osmotic stress mainly at the level of gene expression (Cairney et al. 1985b). ProP was the first carrier protein, which has been shown to function both as osmosensor and as osmoregulator by analysis in proteoliposomes (Racher et al. 1999). ProP is activated upon hyperosmotic stress within about 1 min (Milner et al. 1988). External K^+ was found to be required for activation of ProP in *E. coli* and in *S. typhimurium* (Koo et al. 1991; MacMillan et al. 1999). ProP consists of 500 amino acids and is predicted to possess 12 transmembrane segments as well as N- and C-terminal hydrophilic domains presumably facing the cytoplasm. As a particularly interesting feature, the C-terminal domain of ProP contains so-called heptad repeats, a characteristic of α-helical coiled-coil forming peptides (Culham et al. 1993, 2000). Experimental evidence, both *in vivo* and *in vitro*, has been accumulated indicating that this structure has regulatory functions in ProP. (i) Truncation of 26 C-terminal amino acids rendered the protein inactive, (ii) stabilization of the coiled-coil structure by specific amino acid replacements led to higher activation thresholds of ProP in intact cells, and (iii) the synthesized C-terminal domain of ProP was shown to form a dimeric, antiparallel coiled-coil structure (Culham et al. 2000). Based on these results, it has been suggested that the propensity of the C-terminal domain of ProP to interact with other peptides by forming homodimeric (ProP-ProP) or heterodimeric (ProP-X) coiled-coil structures plays a central role in osmosensing. On the other hand, formation of coiled-coil structures does not seem to be a general feature related to osmosensing, since other osmoregulated transport systems lack this structure, e.g. the uptake carriers BetP, ProP, and EctP of *C. glutamicum*.

Recently, ProP was shown to respond in proteoliposomes to changes in external osmolality by sensing stimuli related to the cytoplasmic concentration of solutes (Culham et al. 2003). When reconstituted in *E. coli* lipid liposomes, the activity of ProP depended on a complex mixture of parameters including concentration, chemistry, and molecular size of osmolytes in the lumen. In difference to BetP of *C. glutamicum* and OpuA of *L. lactis*, ProP was found to respond not only to ionic but also to nonionic solutes in the liposomal lumen. A major contribution to activation was related to the total internal concentration of monovalent cations, of which K^+ is certainly the most relevant in physiological terms. Interestingly, the lumenal cation concentration necessary for half-maximal activation decreased in the presence of increasing amounts of large polyethylene glycols. This could in principle mimic the effect of macromolecular crowding and thus an activation of ProP by steric exclusion from the surface of the protein. A number of possible reasons other than the latter phenomenon, however, can also be made responsible for the activating influence of PEGs on ProP, namely direct influences on the lumenal K^+ activity as well as on the membrane.

The osmoresponsive action of *E. coli* ProP seems to be more complicated as, for example, that of BetP of *C. glutamicum*, since ProQ, a soluble cytoplasmic protein consisting of 232 amino acids, was shown to influence ProP activity, too (Kunte et al. 1999). In a ProQ defective *E. coli* strain, ProP activity was fivefold lower, and ProQ was found to be necessary for maintaining ProP in an active state for prolonged periods of time after osmotic stimulation. Since ProP alone is fully functional in osmosensing, ProQ was supposed to function in fine tuning the osmotic response (Wood 1999).

4.2 OpuA / BusA of *L. lactis*

Because of their limited capacity to synthesize compatible solutes, lactic acid bacteria depend on uptake systems for these solutes in the presence of elevated osmolalities. An osmoregulated primary ABC-type transport system has been characterized, which was named BusA (Obis et al. 1999) or OpuA (van der Heide and Poolman 2000b). It belongs to a new type of ABC transporters in which the binding protein is fused to the membrane components of the carrier system. OpuA consists of 573 amino acid residues and contains 8 α-helical transmembrane domains as well as the binding protein domain facing the external side. Functional reconstitution has proven that this transporter, like ProP of *E. coli* and BetP of *C. glutamicum*, harbors both osmosensing and osmoregulatory activity. In contrast to the other osmosensing systems discussed here, it has not been studied so far which parts of this multidomain transporter may be responsible for or related to the sensing function.

Again, the proteoliposomal system offered the possibility to discriminate various stimuli in their significance for triggering the carrier's activity. Several parameters, such as turgor pressure or absolute external and internal osmolality, were experimentally ruled out. The results obtained suggested that the activity of OpuA is controlled by interaction with its lipid surrounding, and in particular, with the surface charges, whose impact on the carrier protein is influenced by the lumenal ionic strength (van der Heide et al. 2001). Thus, a major contribution of the lumenal ionic strength (or K^+ concentration (BetP), or concentration of monovalent cations (ProP), respectively) to activity regulation has been found for all three carriers studied. Moreover, an influence of lipid headgroup charge has been reported for both OpuA and BetP (van der Heide et al. 2001; Rübenhagen et al. 2001), although the contribution of membrane properties to osmosensing, and in particular that of membrane surface charges, has been much more elaborated for OpuA (van der Heide et al. 2001). The fraction of anionic (charged) lipids seems to be of major importance, whereas variations in acyl chain length, position, and configuration of the double bond, and the fraction of so-called non-bilayer lipids had relatively minor effects. These results indicate that it is the local surface charge in dependence of changes in the lumenal ionic strength and not a perturbation in the lateral pressure profile, which triggers OpuA activation. Consistent with this idea was the observation that low concentrations of small amphiphilic molecules with different net charges, known as local anesthetics, were able to mimic the osmotic activation of OpuA (van der Heide and Poolman 2000a). Again, similar observations had been reported for BetP of *C. glutamicum* (Rübenhagen et al. 2000).

OpuA or BusA from *L. lactis* is interesting also from another point of view. Like other osmoregulated systems, it is also regulated at the level of gene expression. Recently, by using a genetic screen the transcription factor BusR was identified in close neighborhood of the *opuA* (or *busA*) gene (Romeo et al. 2003). BusR is responsible for transcription regulation of *busA* (*opuA*), and represents the first known transcription factor for osmoregulated uptake systems. It seems to be interesting to study how the information of an elevated external osmolality is transferred to BusR, in other words, whether osmosensing on the basis of controlling expression of a carrier gene is similar to osmosensing relevant for regulating the catalytic activity of the corresponding carrier protein.

4.3 Kdp of *E. coli*

Osmoresponsive uptake of K^+ in *E. coli* is mediated by a number of uptake systems; most important are Trk and the primary Kdp system. The mechanisms of osmoregulation of the Trk system is not well studied, and this is also true for the P-type ATPase KdpFABC, and KdpA, the component catalyzing K^+ uptake. On the other hand, the two-component system KdpDE, responsible for osmodependent expression regulation of the *kdpFABC* operon, is one of the best studied systems involved in controlling the synthesis of an osmoresponsive transport system (KdpFABC). KdpD, however, is not part of a transport system and, thus, controls the activity of KdpA, the K^+ pump, only indirectly via regulating its synthesis.

Signal perception induces autophosphorylation of the membrane-bound sensor kinase KdpD. The phosphate group is transferred to the soluble response regulator KdpE, which binds to the *kdpFABC* promoter and stimulates transcription. KdpD is also responsible for repression of the *kdpFABC* operon, since it possesses phosphatase activity resulting in deactivation of KdpE (Nakashima et al. 1993a, 1993b). The kind of stimulus perceived by KdpD is not fully clear, both turgor, low external K^+ concentration, as well as a shift in internal K^+ have been suggested, based on experiments in cells and in reconstituted systems (Laimins et al. 1981; Gowrishankar 1985; Sutherland at al. 1986). Mutant forms of KdpD have been generated, which proved to be sensitive to different stimuli, indicating that KdpD can possibly be activated in different ways. Analysis of various regions within the KdpD protein, which consists of 894 amino acids and functions as a homodimer, with the help of a long list of site specific amino acid replacements led to the identification of at least two regions being important for activity regulation and sensing, as monitored by the ratio of kinase and phosphatase activity (Zimmann et al. 1995). A long N-terminal domain facing the cytoplasm as well as the last two (out of four) transmembrane segments of the central domain proved to be important for the sensory function of KdpD. In experiments using right-side-out vesicles (Jung et al. 2000) and intact cells (Roe et al. 2000), K^+ was shown to directly influence the activity of KdpD.

4.4 Osmoregulation and osmosensing by mechanosensitive channels

Mechanosensitive efflux channels (MSC) have not very much in common with osmoregulated uptake carriers, except that they are membrane-embedded and respond to osmotic changes. There are, however, two aspects, which contribute to the topic of this article, in spite of the fact that a channel is not a carrier. There is strong evidence that the activity of MSC is directly controlled via changes in the lateral tension of the membrane because of changes in external osmolality. Thus, they are models for stimulus perception via the lipid phase. Furthermore, studies on the osmoreactivity of these proteins are favored by the fact that crystallographic 3D structures of both MscL and MscS are available, which is not the case for osmoregulated uptake carrier proteins. Bacteria possess several MSC, which have first been functionally differentiated by patch-clamp techniques. Later on, the responsible genes have been identified (Sukharev et al 1994; Levina et al 1999) and recently, the 3D structures of MscL from *M. tuberculosis* (Chang et al. 1998) and MscS from *E. coli* (Bass et al. 2002) have been solved by X-ray crystallography.

MscL of *M. tuberculosis* consists of 151 amino acids and the functional unit is a pentamer (Chang et al. 1998). Each monomer contributes two transmembrane helices to both an outer and an inner, pore-lining ring. The inner helices are tilted and form an 'inverted teepee' structure with an opening of 2 Å in the closed state, representing the crystallized form. Experimental evidence suggests that MscL has a pore diameter of 30-40 Å in the open form. An iris-like movement of the inner ring of helices sliding one along the other was suggested to lead to their separation at the constriction side and to pore opening (Sukharev et al. 2001). The proposed conformational change of the inner and the outer ring is a complex and highly coordinated process, which has been modeled in detail (Gullingsrud et al. 2001). The trigger for these conformational changes leading to opening and closing of the channel is supposed to be a change in lateral membrane tension.

MscS is a symmetric homoheptamer with three helical segments contributed by each subunit (Bass et al. 2002). Its structure is different from MscL and the crystallized 3D-structure, characterized by a pore opening with a diameter of about 10 Å, probably corresponds to the open form, in difference to MscL crystals. Whereas gating of MscL is thought to occur by changes in helical tilting of the transmembrane segments, pore opening and closing is not fully understood in the case of MscS. Notably, the membrane tension necessary for MscS opening is much smaller (about 4 dyne/cm) than that which triggers MscL gating (about 12 dyne/cm), the latter being close to the tension needed to rupture the membrane (Martinac 2001; Bezanilla and Perozo 2002). As a consequence, upon hypoosmotic stress, MscS opens first and MscL acts as the ultimate safety valve. Both MscS and MscL are thus obvious examples for controlling the activity of a transport protein via a direct stimulus from the membrane phase, i.e. they are supposed to open and close in response to mechanical stress applied directly to the membrane and their gating function depends on the coupling mechanism between protein conformation and membrane stretch.

5 General concepts and conclusions

Transport systems, which are controlled by environmental factors, either from the external, the cytoplasmic, or the membrane phase, and either at the level of protein activity or of gene expression, are frequently involved in signal transduction systems. In a simple case, an effector molecule directly influences the activity of the carrier, in elaborate cases, sensory and regulatory elements are combined to lead to an adapted control of transport activity.

In systems essential for stress protection, regulation is in general observed at both levels, and both are important, either for the emergency response or for long term adaptation to stress. Modulation of the activity of transporters for compatible solutes upon changes in the external osmolality consists of complicated mechanisms involving stimulus sensing, signal transduction, and activity regulation. Mechanisms regulating gene expression may include a number of elements interacting with promoter sites at the DNA for the purpose of controlling transporter synthesis. In both cases, at the beginning of the signal transduction pathway, basically similar processes are expected for signal perception by sensory elements. Understanding the control of transport systems requires elucidation of both the mechanism of control itself (regulation of transport activity), as well as the mechanism how a stimulus is translated into a relevant signal (stimulus sensing and signal transduction). Osmoregulated carriers are well suited examples to study these aspects.

A long list of possible osmotic stimuli has been made responsible for influencing the activity of carriers for compatible solutes. Recent success based on the application of reconstituted systems in identifying the change in cytoplasmic K^+ concentration (BetP of *C. glutamicum*), ionic strength, and its impact on membrane lipids (OpuA of *L. lactis*), or both concentration of lumenal cations and macromolecules (ProP of *E. coli*), represents a step forward in unraveling the mechanism of osmosensing and osmoregulation.

The simplicity of liposomal systems was the main reason for their successful use in these studies. However, the results obtained have to be taken with some caution with respect to their general significance. First, they have been obtained in proteoliposomes, which differ from the situation in intact cells by many aspects. Under *in vivo* conditions, a high concentration of macromolecules (proteins) in the surrounding hydrophilic spaces is present as well as membrane proteins in the lipid phase, the composition of lipids is more complex, and, last not least, the cell wall is present which gives rise to turgor pressure. Second, the influence of the physical state of the membrane and/or the membrane surface seems to be more important than has been considered so far. On the one hand, there is still no clear experimental evidence that the concept of intrinsic stress or alterations of lateral pressure in the lipid bilayer, which have been suggested as a trigger related to osmotic stress based on studies using protein-free liposomes (Hallett et al. 1993), may be relevant for osmosensing and/or osmoregulation. Studies carried out mainly with OpuA of *L. lactis*, on the other hand, indicate the involvement of the lipid phase on regulatory processes of the embedded transporter protein. Moreover, the fact that turgor is absent in proteoliposomes does not necessarily rule out its influence on regulating osmoreactive carriers in intact cells. It should also be taken into account that mechanosensitive efflux channels in fact seem to be regulated by a change in the lateral membrane pressure. Third, there is direct evidence that a multiplicity of stimuli may be relevant for controlling osmoregulated carrier proteins. Several examples have been described above. It should also be noted that only activation upon osmotic stress has been studied in detail so far, but not activity adaptation of osmoregulated carrier proteins after stress compensation. Consequently, more detailed investigations, in particular in intact cells, are required to elucidate possible additional stimuli, which may play a role in the control of osmoregulated transport proteins. Finally, it should be emphasized that another aspect has also not been studied in much detail so far, namely the question whether the oligomeric state of these transporters may be relevant for osmosensing and/or osmoregulation. This issue has been brought up by studies of *E. coli* ProP, which contains a coiled-coil domain and thus a putative protein-protein interaction site. The recent finding that BetP seems to be a trimer in its native state, adds further ideas in this direction, since monomer interaction may be an element of osmoregulation too.

Future studies are required to solve this series of interesting questions concerning osmosensing and osmoregulation by transporters for compatible solutes. Nevertheless, it seems convincing that those factors which have recently been identified for BetP of *C. glutamicum*, ProP of *E. coli*, and OpuA of *L. lactis* to trigger activation, are at least of major importance for the instant activation of bacterial osmoregulated carrier proteins upon hyperosmotic stress.

References

Bass RB, Strop P, Barclay M, Rees DC (2002) Crystal structure of *Escherichia coli* MscS, a voltage-modulated and mechanosensitive channel. Science 298:1582-1587

Berrier CA, Besnard M, Ajouz B, Coulombe A, Ghazi A (1996) Multiple mechanosensitive ion channels from *Escherichia coli*, activated at different thresholds of applied pressure. J Membr Biol 151:175-187

Bezanilla F, Perozo E (2002) Force and voltage sensors in one structure. Science 298:1562-1563

Bremer E, Krämer R (2000) Coping with osmotic challenges: osmoregulation through accumulation and release of compatible solutes, p.79-97. In: Storz G, Hengge-Aronis R (eds), Bacterial stress responses. ASM Press, Washington DC

Burger U (1999) Untersuchungen zu Substratspezifität an zwei osmoregulierten Carrierproteinen (BetP und EctP) in *Corynebacterium glutamicum*. Diploma thesis, University of Köln, Germany

Burger U (2002) Struktur- und Funktionsanalysen am osmotisch regulierten Transporter BetP aus *Corynebacterium glutamicum*. PhD thesis, University of Köln, Germany

Cairney J, Booth IR, Higgins CF (1985a) *Salmonella typhimurium proP* gene encodes a transport system for the osmoprotectant betaine. J Bacteriol 164:1218-1223

Cairney J, Booth IR, Higgins CF (1985b) Osmoregulation of gene expression in *Salmonella typhimurium: proU* encodes an osmotically induced betaine transport system. J Bacteriol 164:1224-1232

Chang G, Spencer RH, Lee AT, Barclay MT, Rees DC (1998) Structure of the MscL homolog from *Mycobacterium tuberculosis*: a gated mechanosensitive channel. Science 282:2220-2226

Csonka LN, Hanson AD (1991) Prokaryotic osmoregulation: genetics and physiology. Annu Rev Microbiol 45:569-606

Culham DE, Lasby B, Marangoni AG, Milner JL, Steer BA, van Nues RW, Wood JM (1993) Isolation and sequencing of *Escherichia coli* gene *proP* reveals unusual structural features of the osmoregulatory proline/betaine transporter, ProP. J Mol Biol 229:268-276

Culham DE, Tripet B, Racher KI, Voegele RT, Hodges RS, Wood JM (2000) The role of the carboxyl terminal alpha-helical coiled-coil domain in osmosensing by transporter ProP of *Escherichia coli*. J Mol Recognit 13:309-322

Culham DE, Henderson J, Crane R, Wood JM (2003) Osmosensor ProP of *Escherichia coli* responds to the concentration, chemistry, and molecular size of osmolytes in the proteoliposome lumen. Biochemistry 42:410-420

Dinnbier U, Limpinsel E, Schmid R, Bakker EP (1988) Transient accumulation of potassium glutamate and its replacement by trehalose during adaptation of growing cells of *Escherichia coli* K-12 to elevated sodium chloride concentrations. Arch Microbiol 150:348-357

Ebbighausen H, Weil B, Krämer R (1991) Carrier mediated acetate uptake in *Corynebacterium glutamicum*. Arch Microbiol 155:505-510

Farwick M, Siewe RM, Krämer R (1995) Glycine betaine uptake after hyperosmotic shift in *Corynebacterium glutamicum*. J Bacteriol 177:4690-4695

Glaasker E, Konings WN, Poolman B (1996) Glycine betaine fluxes in *Lactobacillus plantarum* during osmostasis and hyper- and hypo-osmotic shock. J Biol Chem 271:10060-10065

Gowrishankar J (1985) Identification of osmoresponsive genes in *Escherichia coli*: evidence of participation of potassium and praline transport systems in osmoregulation. J Bacteriol 164:434-445

Gullingsrud J, Kosztin D, Schulten K (2001) Structural determinants of MscL gating structure studied by molecular dynamics simulation. Biophys J 80:2074-2081

Hallett FR, Marsh J, Nickel BG, Wood JM (1993) Mechanical properties of vesicles. II. A model for osmotic swelling and lysis. Biophys J 64:435-442

Jung K, Veen M, Altendorf K (2000) K^+ and ionic strength directly influence the autophosphorylation activity of the putative turgor sensor KdpD of *Escherichia coli*. J Biol Chem 275:40142-40147

Koo S-P, Higgins CF, Booth IR (1991) Regulation of compatible solute accumulation in *Salmonella typhimurium*: evidence for a glycine betaine efflux system. J Gen Microbiol 137:2617-2625

Kunte HJ, Crane RA, Culham DE, Richmond D, Wood JM (1999) Protein ProQ influences osmotic activation of compatible solute transporter ProP in *Escherichia coli* K-12. J Bacteriol 181:1537-1543

Laimins LA, Rhoads DB, Epstein W (1981) Osmotic control of *kdp* operon expression in *Escherichia coli*. Proc Natl Acad Sci USA 78:464-468

Levina N, Tötemeyer S, Stokes NR, Louis P, Jones MA, Booth IR (1999) Protection of *Escherichia coli* cells against extreme turgor by activation of MscS and MscL mechanosensitive channels: identification of genes required for MscS activity EMBO J 18:1730-1737

Ley O (2001) Bedeutung der Prolin-Biosynthese bei der Osmoregulation von *Corynebacterium glutamicum*. Diploma thesis, University of Köln, Germany

MacMillan SV, Alexander DA, Culham DE, Kunte HJ, Marshall EV, Rochon D, Wood JM (1999) The ion coupling and organic substrate specificities of osmoregulatory transporter ProP in *Escherichia coli*. Biochim Biophys Acta 1420:30-44

Martinac B (2001) Mechanosensitive channels in prokaryotes. Cell Physiol Biochem 11:61-76

Marx A, de Graaf AA, Wiechert W, Eggeling L, Sahm H (1996) Determination of fluxes in the central metabolism of *Corynebacterium glutamicum* by nuclear magnetic resonance spectroscopy combined with metabolite balancing. Biotech Bioeng 49:111-129

Milner JL, Grothe S, Wood JM (1988) Proline porter II is activated by a hyperosmotic shift in both whole cells and membrane vesicles of *Escherichia coli* K12. J Biol Chem 263:14900-14905

Möker N (2002) Einfluss von Zwei-Komponenten-Systemen auf die Osmoregulation in *Corynebacterium glutamicum*. Diploma thesis, University of Köln, Germany

Morbach W, Krämer R (2002) Body shaping under water stress: osmosensing and osmoregulation of solute transport in bacteria. ChemBioChem 3:384-397

Nakashima K, Sugiura A, Kanamaru K, Mizuno T (1993a) Signal transduction between the two regulatory components involved in the regulation of the *kdpABC* operon in *Escherichia coli*: phosphorylation-dependent functioning of the positive regulator, KdpE. Mol Microbiol 7:109-116

Nakashima K, Sugiura A, Mizuno T (1993b) Functional reconstitution of the putative *Escherichia coli* osmosensor, KdpD, in liposomes. J Biochem 114:615-621

Obis D, Guillot A, Gripon JC, Renault P, Bolotin A, Mistou MY (1999) Genetic and biochemical characterization of a high-affinity betaine uptake system (BusA) in *Lactococcus lactis* reveals a new functional organization within bacterial ABC transporters. J Bacteriol 181: 6238-6246

Peter H, Burkovski A, Krämer R (1996) Isolation, characterization and expression of the *Corynebacterium glutamicum* betP gene, encoding the transport system for the compatible solute glycine betaine. J Bacteriol 178:5229-5234

Peter H, Burkovski A, Krämer R (1998a) Osmosensing by N- and C-terminal extensions of the glycine betaine uptake system BetP of *Corynebacterium glutamicum*. J Biol Chem 273:2567-2574

Peter H, Weil B, Burkovski A, Krämer R, Morbach S (1998b) *Corynebacterium glutamicum* is equipped with four secondary carriers for compatible solutes: identification, sequencing and characterization of the proline/ectoine uptake system ProP, and the ectoine/proline/glycine betaine carrier EctP. J Bacteriol 180:6005-6012

Poolman B, Blount P, Folgering JHA, Friesen RHE, Moe PC, van der Heide T (2002) How do proteins sense water stress? Mol Microbiol 44:889-902

Pourkomailian B, Booth IR (1994) Glycine betaine transport by *Staphylococcus aureus*: evidence for feedback regulation of the activity of the two transport systems. Microbiology 140:3131-3138

Racher KI, Voegele RT, Marshall EV, Culham DE, Wood JM, Jung H, Bacon M, Cairns MT, Ferguson SM, Liang W-J, Henderson PJF, White G, Hallet FR (1999) Purification and reconstitution of an osmosensor: transporter ProP of *Escherichia coli* senses and responds to osmotic shifts. Biochemistry 38:1676-1684

Roe AJ, McLaggan D, O'Byrne CP, Booth IR (2000) Rapid inactivation of the *Escherichia coli* KdpK$^+$ uptake system by high potassium concentrations. Mol Microbiol 35:1235-1243

Romeo Y, Obis D, Bouvier J, Guillot A, Fourcans A, Bouvier I, Gutierrez C, Mistou M-Y (2003) Osmoregulation in *Lactococcus lactis*: BusR, a transcriptional repressor of the glycine betaine uptake system BusA. Mol Microbiol 47:1135-1147

Rübenhagen R, Rönsch H, Jung H, Krämer R, Morbach S (2000) Osmosensor and osmoregulator properties of the betaine carrier BetP from *Corynebacterium glutamicum* in proteoliposomes. J Biol Chem 275:735-741

Rübenhagen R, Morbach S, Krämer R (2001) The osmoreactive betaine carrier BetP from *Corynebacterium glutamicum* is a sensor for cytoplasmic K$^+$. EMBO J 20:5412-5420

Steger R (2002) Vergleichende Studien zur Aktivitätsregulation osmosensitiver Transporter aus *Corynebacterium glutamicum*. PhD thesis, University of Köln, Germany

Stimeling KW, Graham JE, Kaenjak A, Wilkinson BJ (1994) Evidence for feedback (trans) regulation of, and two systems for, glycine betaine transport by *Staphylococcus aureus*. Microbiology 140:3139-3144

Sukharev SI, Blount P, Martinac B, Blattner FR, Kung C (1994) A large-conductance mechanosensitive channel in *E. coli* encoded by *mscL* alone. Nature 368:265-268

Sukharev SI, Betanzos M, Chiang CS, Guy HR (2001) The gating mechanism of the large mechanosensitive channel MscL. Nature 409:720-724

Sutherland L, Carney J, Elmore MJ, Booth IR, Higgins CF (1986) Osmotic regulation of transcription: induction of the *proU* betaine transport gene is dependent on accumulation of intracellular potassium. J Bacteriol 168:805-814

van der Heide T, Poolman B (2000) Osmoregulated ABC transport system of *Lactococcus lactis* senses water stress via changes in the physical state of the membrane. Proc Natl Acad Sci USA 97:7102-7106

van der Heide T, Stuart MC, Poolman B (2001) On the osmotic signal and osmosensing mechanism of an ABC transport system for glycine betaine. EMBO J 20:7022-7032

Verheul A, Glaasker E, Poolman B, Abee T (1997) Betaine and L-carnitine transport by *Listeria monocytogenes* Scott A in response to osmotic signals. J Bacteriol 179:6979-6985

Wolf A, Krämer R, Morbach S (2003) Three pathways for trehalose metabolism in *Corynebacterium glutamicum* ATCC 13032 and their significance in response to osmotic stress. Mol Microbiol 49:1119-1134

Wood JM (1999) Osmosensing by bacteria: signals and membrane-based sensors. Microbiol Mol Biol Rev 63:230-262

Ziegler C, Morbach S, Schiller D, Krämer, R, Tziatzios C, Schubert D, Kühlbrnadt W (2004) Propjection structure and oligomeric state of the osmoregulated sodium/glycine betaine symporter BetP from *Corynebacterium glutamicum*. J Mol Biol, in press

Zimmann P, Puppe W, Altendorf K (1995) Membrane topology analysis of the sensor kinase KdpD of *Escherichia coli*. J Biol Chem 270:28282-28288

Krämer, Reinhard
  Institute of Biochemistry, Universität Köln, Zülpicher Str. 47, 50674 Köln,
  Germany
  r.kraemer@uni-koeln.de

Morbach, Susanne
  Institute of Biochemistry, Universität Köln, Zülpicher Str. 47, 50674 Köln,
  Germany
  s.morbach@uni-koeln.de

The bacterial phosphotransferase system: a perfect link of sugar transport and signal transduction

Jörg Stülke and Matthias H. Schmalisch

Abstract

Bacteria transport and concomitantly phosphorylate several sugars via the multi-component phosphoenolpyruvate:sugar phosphotransferase system (PTS). This allows the introduction of these sugars into the glycolytic pathway at low cost and provides, thus, an advantage as compared to the use of other sugar transport systems. Four of the PTS proteins are involved in phosphotransfer from phosphoenolpyruvate to the incoming sugar. The phosphorylation state of these proteins can, thus, indicate the sugar supply, and different PTS proteins are indeed involved in different signal transduction strategies that link the availability of sugars and the physiological state of the cell to the activity of transport proteins, metabolic enzymes, and transcriptional regulators. Moreover, there are PTS-related proteins that are exclusively devoted to regulatory purposes. The general physiological scenarios (input information, regulatory output) of many regulatory events caused by PTS proteins are conserved in a broad range of bacteria. However, their actual molecular mechanisms may differ substantially.

1 Introduction

Carbon is the main chemical constituent of any life known to us. Thus, the efficient acquisition of carbon by the cells is crucial and the invention of more effective systems for carbon source transport and utilisation provides any species with a selective advantage in the competition for scarce resources. Thus, it is not surprising that a multitude of transport systems for carbon sources emerged in the course of evolution. As other substances, carbon sources can be transported by primary or secondary transporters (for review see Saier 2000). In addition, bacteria invented a system devoted not only to carbohydrate transport but also to carbohydrate sensing and signal transduction, the phosphoenolpyruvate:sugar phosphotransferase system (PTS). The PTS offers the advantage of directly linking sugar transport to its phosphorylation, thus, reducing the cost for the introduction of sugar into the glycolytic pathway. Moreover, the PTS fulfills a variety of regulatory functions, which will be the scope of this review. In addition to its role in the regulation of metabolism, the PTS is also involved in the control of bacterial chemotaxis. This

Topics in Current Genetics, Vol. 9
E. Boles, R. Krämer (Eds.): Molecular Mechanisms Controlling Transmembrane Transport
DOI 10.1007/b95776 / Published online: 9 March 2004
© Springer-Verlag Berlin Heidelberg 2004

subject is beyond the scope of this work, the reader is referred to the papers by Lux et al. (1995, 1999) and Garrity et al. (1998). The combination of a highly efficient transport system with the capability to transduce signals may be one of the reasons for the evolutionary success of the bacteria that possess a PTS. The small size of a prokaryotic cell and the resulting high surface/volume ratio demands rapid and appropriate responses to the changes in the environmental conditions and the PTS allows the bacteria to meet this demand.

The PTS is made up of two general components, Enzyme I (EI) and HPr, and one to four sugar specific proteins that are collectively termed Enzyme II (Postma et al. 1993). The Enzyme II consists of two soluble proteins (or domains), EIIA and EIIB, that are involved in phosphate transfer and one (sometimes two) membrane-bound protein(s), EIIC (or EIID). The phosphate is transferred from phosphoenolpyruvate via EI, HPr, EIIA, and EIIB to the incoming sugar (see Fig. 1). The *Escherichia coli* EIIA specific for glucose is of general regulatory importance and therefore, is called EIIA^{Crr} (Saier and Reizer 1992).

The progress of the bacterial genome sequencing projects allows us to get an idea about the emergence of the PTS during bacterial evolution. As can be seen in Figure 2, PTS components appeared early in bacterial evolution before the radiation of the major bacterial phyla. However, in the lineages that contain representatives with a PTS, some species lack the system. Moreover, the PTS seems to be lost in the cyanobacteria. This is not surprising given the autotrophic lifestyle of these bacteria.

In addition to the PTS proteins involved in transport, some bacteria possess regulatory paralogues of PTS components and proteins devoted to the control of PTS activity. In *Bacillus subtilis* and closely related bacteria, the Crh protein fulfills regulatory functions in carbon catabolite repression. In contrast to the similar HPr protein, Crh is not phosphorylated by EI and, thus, not part of the phosphotransferase system (Galinier et al. 1997). In *E. coli* and many other proteobacteria, a PTS not involved in sugar transport but rather in nitrogen regulation was discovered. The corresponding enzymes are termed EI^{Ntr}, NPr, and EIIA^{Ntr} (Powell et al. 1995; Rabus et al. 1999; Reizer et al. 1992). Finally, a protein kinase (HPrK/P) that controls the activity of HPr by reversible phosphorylation was discovered and studied in *B. subtilis* and a variety of other Gram-positive and Gram-negative bacteria (Galinier et al. 1998; Reizer et al. 1998; Kravanja et al. 1999; Boël et al. 2003).

During the last few years, the structures of many PTS and PTS-related proteins and their complexes were solved in an attempt to understand the biochemical and regulatory interactions between these proteins (for reviews see Waygood 1998; Peterkofsky et al. 2001; Nessler et al. 2003). Some of the PTS proteins with determined structures are listed in Table 1.

Table 1. Structures of PTS and PTS-related proteins[1]

Protein(s)	Function	Organism	PDB access
EI (N-terminal fragment)	Phosphotransferase	E. coli	1ZYM
EI (N-terminal fragment)/ HPr	Phosphotransferase complex	E. coli	3EZA
HPr	Phosphotransferase	E. coli, B. subtilis, E. faecalis	1HDN, 1JEM, 1QFR
HPr H15D	Phosphotransferase	E. coli	1CM3
HPr S46D	Phosphotransferase	E. coli	1OPD
HPr(His~P)	Phosphotransferase	E. coli	1PFH
HPr(Ser-P)	Phosphotransferase	E. coli	1FU0
IIA^{Crr}/ HPr	Phosphotransferase complex	E. faecalis	1GGR
IIA^{Glc}	Phosphotransferase	E. coli, B. subtilis	1F3G, 1GPR
IIA^{Crr}/Zn^{2+}	Phosphotransferase	E. coli	1F3Z
IIA^{Crr} / IIB^{Glc}	Phosphotransferase complex	E. coli	1O2F
IIB^{Glc}	Phosphotransferase	E. coli	1IBA
HPrK/P	Kinase/Phosphorylase	L. casei, M. pneumoniae, S. xylosus	1JB1, 1KNX, 1KO7
HPrK/P	Kinase/Phosphorylase/ Phosphotransferase complex	L. casei/ B. subtilis	1KKL
HPrK/P / HPr	Kinase/ Phosphorylase/ Phosphotransferase complex	L. casei/ B. subtilis	1KKM
HPrK/P / HPr(Ser46)-P	Phosphotransferase complex	L. casei/ B. subtilis	1KKM
Crh	Regulatory protein	B. subtilis	1K1C
IIA^{Crr}/ glycerol kinase	Phosphotransferase/enzyme complex	E. coli	1GLA
LicT (RNA-binding domain)/ RNA	Antiterminator/RNA complex	B. subtilis	1L1C
LicT (PRDs)	Antiterminator	B. subtilis	1H99

[1] PTS and PTS-related proteins mentioned in this review are indicated. The accession numbers for the protein database PDB (http://www.rcsb.org/pdb/index.html) are shown.

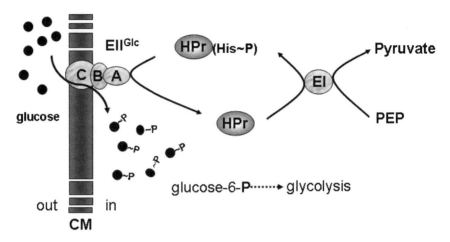

Fig. 1. Schematic overview of the glucose-specific phosphotransferase system (PTS) of *Bacillus subtilis*. First, Enzyme I (EI) is autophosphorylated in the presence of phosphoenolpyruvate (PEP). The phosphate residue (-P) is then transferred to a conserved histidine residue (His15) of HPr. HPr phosphorylates the glucose-specific Enzyme IIA domain (A). After intramolecular transport to the EIIB domain (B), the phosphate is transferred to the incoming glucose (•). The sugar enters the cell *via* the cytoplasmic membrane (CM) spanning IIC domain (C). The phosphorylated glucose (•-P) is then further metabolised by glycolysis.

2 The phosphorylation state of PTS proteins is tightly controlled

In the presence of PTS sugars, the phosphate residues of the PTS proteins are drained to the incoming sugars and the PTS proteins are subsequently thought to be non-phosphorylated under these conditions. Experimental studies concerning the phosphorylation of PTS proteins have focussed on EIIA^{Crr} from *E. coli* and HPr from Gram-positive bacteria since these proteins are most relevant for signal transduction in the respective bacteria. The regulatory consequences of the different forms of EIIA^{Crr} and HPr will be discussed below.

2.1 Control of EIIA^{Crr} phosphorylation in *E. coli*

As expected, EIIA^{Crr} is completely dephosphorylated if *E. coli* is grown on glucose whereas full phosphorylation was detected after growth in the presence of succinate (Hogema et al. 1998b). However, a partial dephosphorylation of EIIA^{Crr} was also observed after growth of the bacteria on a variety of other sugars including mannose, fructose, lactose, galactose, gluconate, and arabinose. Even lactate

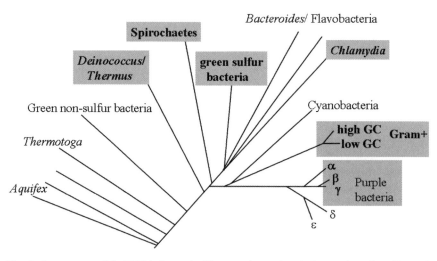

Fig. 2. Occurrence of the PTS in bacteria. The tree shows the phylogenetic order of bacteria (Woese 1987). Groups, which possess the general PTS components are highlighted in gray. The evidence is based on the PEDANT database as of August, 2003 (http://pedant.gsf.de/#Complete; Frishman et al. 2003). The presence of general PTS components in the *Deinococcus/Thermus* phylum is based on experimental evidence (Darbon et al. 1999).

induces a slight but significant dephosphorylation of EIIA^{Crr}. These carbon sources are transported by different systems, and they are catabolised by different pathways (glycolysis, pentose phosphate pathway, citric acid cycle). It was, thus, astonishing that they affect the phosphorylation state of EIIA^{Crr}. Detailed studies with glycolytic mutants revealed that metabolic activity is crucial for EIIA^{Crr} dephosphorylation. Thus, both PTS-dependent sugar transport and glycolytic activity control the phosphorylation state of EIIA^{Crr}. It was demonstrated that changes in EIIA^{Crr} phosphorylation were paralleled by altered ratios of the PEP and pyruvate concentrations. If the PEP concentration is high (as compared to the pyruvate concentration), EIIA^{Crr} is found predominantly in the phosphorylated form. High PEP concentrations and accumulation of phosphorylated EIIA^{Crr} correspond to metabolically inactive cells. In contrast, a low concentration of PEP (and a high pyruvate concentration) will result in EIIA^{Crr} dephosphorylation, thus, indicating high metabolic activity (Hogema et al. 1998b). This observation is supported by the kinetic parameters of the phosphotransfer reactions (Meadow and Roseman 1996; Weigel et al. 1982).

Phosphorylation of EIIA^{Crr} does not only affect its ability to donate the phosphate residue but has also implications for the conformation of the protein, which is important for the regulatory interactions of the two forms of EIIA^{Crr} (see below). The phosphorylation site of EIIA^{Crr}, His-90, is located in a depression of the protein. Upon phosphorylation, a net of hydrogen bonds is established that links Thr-73, His-75, the phosphate residue, Asp-94, and *via* His-90, Gly-92. The formation of this network affects the interactions with other proteins and, thus, the regulatory properties of EIIA^{Crr} (Pelton et al. 1993, 1996).

2.2 Control of HPr phosphorylation in Gram-positive bacteria

In Gram-positive bacteria, HPr, rather than EIIA^{Glc} is important for signal trans-duction (Stülke and Hillen 1998; Brückner and Titgemeyer 2002). In these bacteria, HPr can be phosphorylated on two distinct sites: His-15 is the target of EI-dependent phosphorylation and Ser-46 is phosphorylated by the metabolite-regulated HPrK/P (Deutscher and Saier 1983; Deutscher et al. 1986). The former phosphorylation is reversible, and HPr(His~P) can donate the phosphate to all the EIIA proteins and to some enzymes and regulators (see below). In contrast, HPr(Ser-P) cannot phosphorylate any other protein, and dephosphorylation re-quires the phosphorylase activity of HPrK/P. The two phosphorylation events are more or less mutually exclusive, phosphorylation at one site was shown to inhibit the second phosphorylation event about 600-fold (Deutscher et al. 1984). Consid-ering the phosphorylation/dephosphorylation reactions, it is obvious that the activ-ity of HPrK/P decides the phosphorylation state of HPr. Kinase activity of *Bacil-lus subtilis* HPrK/P is triggered by high ATP concentrations. Another effector of HPrK/P is fructose-1.6-bisphosphate, an indicator of high metabolic activity. At low ATP concentrations, fructose-1.6-bisphosphate acts to stimulate kinase activ-ity in a co-operative manner with ATP. In contrast, phosphorylase activity prevails at low ATP and fructose-1.6-bisphosphate concentrations and is stimulated by in-organic phosphate (Jault et al. 2000; Hanson et al. 2002, Mijakovic et al. 2002; Ramström et al. 2003). However, the *in vivo* phosphorylation state of HPr became only recently the subject of experimental analysis. In the model organism *B. sub-tilis*, HPr is present as HPr(His~P) and in the non-phosphorylated form if cells grow in the absence of glucose. In contrast, about 50% of HPr are phosphorylated on Ser-46 and 50% are non-phosphorylated during growth in the presence of glu-cose. A very small fraction of HPr is doubly phosphorylated, whereas HPr(His~P) is not detectable under these conditions (Monedero et al. 2001b; Ludwig et al. 2002). Similar results have been obtained with *Streptococcus mutans* and *Lacto-coccus lactis* (Thevenot et al. 1995; Monedero et al. 2001a). Thus, non-phosphorylated HPr as a target for both phosphorylation events is available in the absence and in the presence of glucose, whereas the phosphorylated forms are ex-clusively found either in the presence (HPr(Ser-P)) or in the absence (HPr(His~P)) of glucose.

The structural consequences of HPr phosphorylation have been studied to some detail. Upon phosphorylation of His-15 by EI, local conformational adjustments occur around the active center of *B. subtilis* HPr. In contrast, phosphorylation of *B. subtilis* HPr on Ser-46, which is located on the cap of α-helix B, results in sta-bilisation of this helix rather than in a conformational change. For HPr from *En-terococcus faecalis*, loss of hydrophobic interaction with EI was described as the major structural effect of Ser-46 phosphorylation (Pullen et al. 1995; Jones et al. 1997b; Audette et al. 2000).

2.3 Regulation of HPr kinase/phosphorylase activity

As discussed above, kinase activity of HPrK/P is triggered by ATP and fructose-1.6-bisphosphate. However, HPr is not completely phosphorylated on Ser-46 even if the cells grow in a rich medium in the presence of glucose. Free HPr is needed under these conditions as a target for EI-dependent phosphorylation and as a part of the sugar transport system. This raises the question how HPrK/P kinase activity may be controlled to prevent phosphorylation of the complete HPr pool. An answer to this question comes from studies with *B. subtilis* and *E. faecalis ccpA* mutants (Ludwig et al. 2002; Leboeuf et al. 2000). CcpA is the main regulatory protein of carbon metabolism in Gram-positive bacteria. In the presence of glucose or other readily metabolisable carbon sources, it binds to specific sites on the DNA and represses or activates large sets of genes (Henkin 1996; Moreno et al. 2001; Yoshida et al. 2001; Blencke et al. 2003). In *ccpA* mutants, the portion of HPr(Ser-P) is considerably increased as compared to the wild type strains. Even the doubly phosphorylated form of HPr is easily detectable in *ccpA* mutants. In *B. subtilis*, HPr is nearly completely phosphorylated on Ser-46 in the presence of glucose in the *ccpA* mutant strain and this mutant is not able to transport glucose and other PTS sugars. Therefore, it was concluded that a negative feedback mechanism, which normally prevents excessive HPr(Ser-P) formation is not operative in the *ccpA* mutant (Ludwig et al. 2002). It is, however, unknown how CcpA may affect the activity of HPrK/P. Four possible explanations were discussed: (i) the amount of HPrK/P may be increased in the *ccpA* mutant, thus, resulting in higher kinase activity. (ii) The concentrations of the effectors of kinase activity (ATP, fructose-1.6-bisphosphate) may be increased in the *ccpA* mutant. (iii) The concentrations of inorganic phosphate, the positive effector of HPrK/P phosphorylase activity may be decreased in the *ccpA* mutant. (iv) There might exist a negative feedback regulation of HPrK/P kinase activity. The experimental evidence provided so far excludes the first three possibilities and suggests the operation of a novel feedback control of HPrK/P activity (Ludwig et al. 2002). This feedback control might be exerted by CcpA directly or it may depend on effectors, which are formed in a CcpA-dependent manner. Since CcpA has no effect on *in vitro* HPr phosphorylation by HPrK/P and all known effects of *ccpA* mutants result from loss of DNA binding, we may conclude that CcpA controls the formation of a negative effector (a protein or a metabolic intermediate) of HPrK/P kinase activity (Ludwig and Stülke 2001; Ludwig et al. 2002) (see Fig. 3).

The activity of HPrK/P does finally determine the phosphorylation state of HPr and, thus, the availability of HPr for sugar transport and of the two phosphorylated forms of HPr for regulatory purposes (see below). As shown for *B. subtilis*, HPrK/P activity is triggered by high ATP concentrations in most other low-GC Gram-positive bacteria studied so far including the enzymes of *E. faecalis*, *Streptococcus salivarius*, and *Staphylococcus xylosus*. In addition to ATP, fructose-1.6-bisphosphate stimulates the kinase activity of HPrK/P in *B. subtilis* and to some extent in *S. xylosus*, whereas it does not affect the enzymes from *E. faecalis* and *S. salivarius* (Jault et al. 2000; Brochu and Vadeboncoeur 1999; Kravanja et al. 1999; Huynh et al. 2000). It was proposed that phosphorylase activity is the

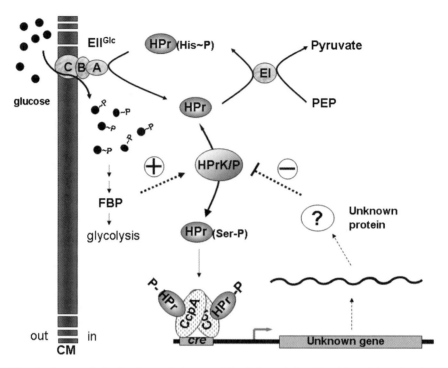

Fig. 3. Proposed feedback regulation model of *B. subtilis* HPr kinase/phosphorylase (HPrK/P). During growth on glucose, HPrK/P activity is stimulated by fructose-1.6-bisphosphate (FBP) generated in glycolysis. This model suggests that HPr(Ser-P) in complex with catabolite control protein A (CcpA) leads to expression of a yet unknown gene. This gene encodes an unknown protein factor, which is responsible for negative feedback inhibition of HPrK/P. Synthesis of this protein results in inhibition of HPrK/P kinase activity, either by direct protein-protein interaction or mediated by a generated product.

default activity of *B. subtilis* HPrK/P and this may apply to the enzymes from the other bacteria as well (Hanson et al. 2002). In contrast, the HPrK/P of *Mycoplasma pneumoniae* is not controlled by the ATP concentration. In this organism, a high concentration of inorganic phosphate and a low concentration of fructose-1.6-bisphosphate trigger the phosphorylase activity of HPrK/P. Thus, *M. pneumoniae* HPrK/P is by default a kinase (Steinhauer et al. 2002). These different modes of HPrK/P activities may reflect the different habitats and lifestyles of the different bacteria. While *M. pneumoniae* is closely adapted to its ecological niche on nutrient-rich human mucous membranes, *B. subtilis* and the other mentioned Gram-positive bacteria face typically nutrient limitations in their natural environments. Thus, the default state of HPrK/P indicates a good (*M. pneumoniae*, kinase activity) or a poor (*B. subtilis*, phosphorylase activity) nutrient supply. The additional regulation of kinase activity by fructose-1.6-bisphosphate in *B. subtilis* may be necessary to distinguish between different pathways to respond to nutrient exhaustion: under conditions of energy limitation these bacteria are capable of initiating a complex program of cell differentiation resulting in the formation of dormant endospores. The possibility to check the glycolytic activity (reflected by the pool of fructose-1.6-bisphosphate) allows the cells to choose to sporulate or to dephosphorylate HPr(Ser-P) with consequences for gene expression (see below). It will be very interesting to study the control of HPrK/P activity in Gram-negative bacteria as well.

3 Control of transporter and enzyme activities by PTS components

The PTS controls the activities of several transporters for non-PTS carbohydrates and enzymes that catalyze the initial steps of the introduction of secondary carbon sources into metabolism. Basically, the activity of these proteins is inhibited in the presence of glucose or other preferred carbon sources, and thus, the inducers of the respective genes and operons cannot be formed. This regulatory scenario is one of the major mechanisms of carbon catabolite repression and is called inducer exclusion (Saier 1989; Postma et al. 1993; Inada et al. 1996; Stülke and Hillen 2000). While the general principle of inducer exclusion is the same in Gram-negative and Gram-positive bacteria, the actual molecular mechanisms differ substantially. Therefore, we will first present a discussion of the regulation of transporter activities in Gram-negative and Gram-positive bacteria, and will then present findings concerning the control of glycerol kinase by PTS components.

3.1 Control of transporter activities by the PTS in Gram-negative bacteria

With the classical finding of a glucose-lactose diauxie in *E. coli* (Monod 1942), the reason for this phenomenon became a matter of debate. While regulation of the *lac* operon by cAMP/Crp was long considered to be the main factor for diauxic growth (Saier 1989), recent studies provided compelling evidence that the control of lactose uptake (and, thus, inducer formation) is crucial for the glucose-lactose diauxie (Inada et al. 1996). In the presence of glucose, non-phosphorylated EIIA^{Crr} may bind and thereby inactivate the lactose permease LacY (Fig. 4). Similar regulatory interactions of EIIA^{Crr} were observed with the permeases for maltose, melibiose, and raffinose (Saier 1989). Three aspects have been in the focus of research on inducer exclusion by interaction of EIIA^{Crr} with sugar transporters: (i) the biochemical and kinetic parameters of the interaction, (ii) the mechanism of specific recognition of different permeases by EIIA^{Crr}, and (iii) the regulatory relevance of inducer exclusion.

It is important to consider that simultaneous inactivation of all non-PTS sugar permeases by EIIA^{Crr} might require more of the latter protein than is actually available. Cooperativity of binding of EIIA^{Crr} and the respective sugar substrates allows to inactivate those permeases for which the substrates are actually available in the medium. This implies that no EIIA^{Crr} is wasted for the inactivation of permeases, which are inactive due to the absence of their substrate (Saier et al. 1983;

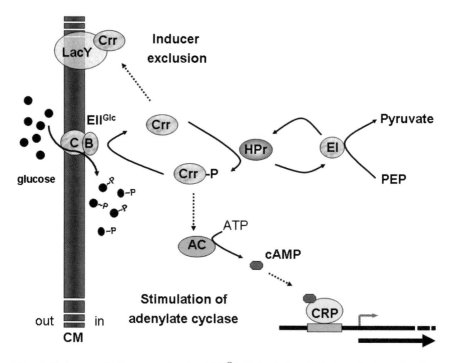

Fig. 4. Carbon catabolite repression by EIIA^{Crr} (Crr) of *E. coli*. In *E. coli*, the ratio of un-phosphorylated EIIA^{Crr} (Crr) to its phosphorylated form (Crr-P) reflects the physiological state of the cell. At high metabolic activity (presence of glucose), most of EIIA^{Crr} is un-phosphorylated due to the rapid transfer of phosphate to EIICB (C,B). In its unphosphory-lated form, EIIA^{Crr} binds to membrane-bound permeases like the lactose permease (LacY), thus, preventing the uptake of other sugars. Absence of glucose leads to accumulation of Crr-P, which stimulates the activity of adenylate cyclase (AC). Cyclic adenosine mono-phosphate (cAMP) binds to its receptor protein and enables the expression of catabolic genes.

Seok et al. 1997b). Recently, a mutant form of the *E. coli* lactose permease, which binds EIIA^{Crr} even in the absence of lactose or other galactosides, was studied. This mutation is located close to the EIIA^{Crr} binding loop and the galactoside bind-ing site in the lactose permease (Sondej et al. 2003; Abramson et al. 2003). Bind-ing of EIIA^{Crr} to the lactose permease is also important for autoregulation of lac-tose transport. In the presence of lactose, about 50% of EIIA^{Crr} is present in the non-phosphorylated form (Hogema et al. 1998b) and, thus, capable of binding and inactivating the lactose permease. Indeed, lactose transport activity is inhibited at high lactose concentrations. This inhibition was not observed in *crr* mutant defec-tive in EIIA^{Crr} and in a class of *lacY* mutants in which the lactose permease is in-sensitive to inducer exclusion (Hogema et al. 1999).

Studies of the binding sites of EIIA^{Crr} to the lactose permease revealed that it binds to two cytoplasmic loops that link the transmembrane helices IV/V and

VI/VII. It was demonstrated that substrate binding brings these loops in an arrangement that makes it competent for EIIA^{Crr} binding (Hoischen et al. 1996; Seok et al. 1997b). Moreover, binding of different substrates of the lactose permease results in a collection of similar conformations of the EIIA^{Crr} binding site that allow interaction to varying degrees (Sondej et al. 2002). Comparative studies allowed to deduce consensus sequences, which are present in the two loops of the lactose permease and also in the permeases for raffinose and maltose (Titgemeyer et al. 1994; Sondej et al. 1999). A database search revealed 37 proteins of E. coli, which contain the two consensus sequences. Therefore, it was proposed that EIIA^{Crr} might control more cellular processes than it is known today (Sondej et al. 1999). Interestingly, one of the two consensus sequences is also found in HPr in the interaction surface between EIIA^{Glc} and HPr around the active site at His-15 (Herzberg 1992; Sondej et al. 1999). It is, thus, probable that the proteins interacting with EIIA^{Crr} exhibit a similar structure around the binding site. In a complementary approach, the topography of the interaction surface was also studied in EIIA^{Crr}. Site-directed mutagenesis identified a surface covering the active site histidine to be important for binding of the lactose permease. Thus, phosphorylation of His-90 would have the same consequences as a mutation in this region and interfere with binding to the lactose permease (Sondej et al. 2000). Moreover, this surface of EIIA^{Crr} does also interact with the active site of HPr and EIIBC^{Glc} (Chen et al. 1993; Gemmecker et al. 1997). This observation supports the idea that EIIA^{Crr} contacts different proteins in a similar manner. The surface that interacts with the lactose permease is conserved in EIIA^{Crr} of *Salmonella typhimurium* and *Haemophilus influenzae*. Interestingly, this surface is not conserved in the EIIA^{Glc} from *B. subtilis* and *M. pneumoniae*. This is in good agreement with the fact that HPr rather than EIIA^{Glc} is involved in regulation in Gram-positive bacteria.

While inducer exclusion by EIIA^{Crr}-mediated inactivation of substrate-specific transporters was originally regarded as auxiliary mechanism of carbon catabolite repression, recent evidence clearly demonstrates that this view strongly underestimates the role of inducer exclusion. The glucose-lactose diauxie in *E. coli* is mainly caused by inhibition of the lactose permease and the subsequent failure to generate the intracellular inducer of the *lac* operon, allolactose. In *E.* coli, *lacI* mutants that do not express a functional lac repressor, no glucose repression of the lactose operon is observed. Moreover, artificial induction of the *lac* operon with isopropyl-β-D-thiogalactoside (IPTG), which can enter the cell independent from the lactose permease, is insensitive to glucose repression (Inada et al. 1996). Glucose transport is essential for inducer exclusion since EIIA^{Crr} is dephosphorylated only upon glucose phosphorylation (Kimata et al. 1997). Moreover, also non-PTS substrates such as glucose-6-phosphate and gluconate can exert inducer exclusion because their metabolism causes partial dephosphorylation of EIIA^{Crr} (see above, Hogema et al. 1997, 1998a, 1998b).

3.2 Control of transporter activities by the PTS in Gram-positive bacteria

In Gram-positive bacteria, inducer exclusion by the control of specific transporter activities has so far only been demonstrated in lactic acid bacteria. In contrast to the situation in *E. coli*, HPr seems to play a pivotal role in inducer exclusion in Gram-positive bacteria rather than $EIIA^{Glc}$.

In *Lactobacillus casei*, transport of the non-PTS sugar maltose is strongly inhibited in the presence of glucose. This inhibition depends on HPr phosphorylated on the regulatory site, Ser-46 (Viana et al. 2000; Dossonnet et al. 2000). Similarly, transport of ribose and maltose is inhibited in the presence of glucose in *L. lactis*. As observed for *L. casei*, this inhibition depends on HPr(Ser-P) (Monedero et al. 2001a). However, the molecular mechanism(s) causing this type of inducer exclusion is so far unknown. In *Lactobacillus brevis*, galactose transport via a sugar:H^+ symport mechanism is also inhibited by glucose. It was proposed that this inhibition results from HPr(Ser-P) binding to the galactose symporter GalP and its subsequent inactivation (see Fig. 5) (Djordjevic et al. 2001). Regulation of transporter activities by HPr in lactic acid bacteria is likely to cause diauxic growth with mixtures of glucose and non-PTS sugars. This is suggested by the observation that hierarchical utilisation of glucose and non-PTS sugars such as galactose was abolished in a *S. salivarius ptsH* mutant. The HPr protein of this mutant is affected in a site close to the regulatory Ser-46, Met-48. Interestingly, this mutant HPr is still phosphorylated by the HPrK/P. It can, thus, be concluded that it is not alone the structural stabilisation of helix B of HPr upon Ser-46 phosphorylation, which is important for regulatory effects. Thus, the phosphorylation itself and the region surrounding Ser-46 may be important for regulatory interactions with sugar permeases (Plamondon et al. 1999).

A different mechanism of transporter control by the PTS was discovered for the lactose permease from *Streptococcus thermophilus*. In this bacterium, the lactose:H^+ symporter LacS has a unique organisation. The N-terminal part of LacS is similar to transporter proteins such as the melibiose permease of *E. coli*, whereas the C-terminus exhibits similarity to $EIIA^{Glc}$. This unusual arrangement immediately suggested a regulatory function for the EIIA domain of LacS (Poolman et al. 1989). The $EIIA^{LacS}$ domain is indeed phosphorylated by HPr(His~P) on a histidyl residue. Although the $EIIA^{LacS}$ domain is not required for full transport activity, it has been shown that phosphorylation increases the transport activity of LacS about fivefold (Poolman et al. 1995; Gunnewijk and Poolman 2000) (see Fig. 5). As discussed above, the phosphorylation state of HPr is crucial for determining the phosphorylation of $EIIA^{LacS}$. In the presence of HPr(His~P), phosphorylated $EIIA^{LacS}$ accumulates in the cell. In addition, the expression of *lacS* is also controlled by the phosphorylation state of HPr since this gene is subject to CcpA/ HPr(Ser-P)-dependent carbon catabolite repression (see below) (Gunnewijk and Poolman 2000). While this mechanism of control of a sugar transporter by the PTS has so far only been described for *S. thermophilus*, it is probably more widespread among the lactic acid bacteria. A blast search with LacS (http://www.ncbi.nlm.nih.gov/sutils/blink.cgi?pid=475108&quq=97893) revealed

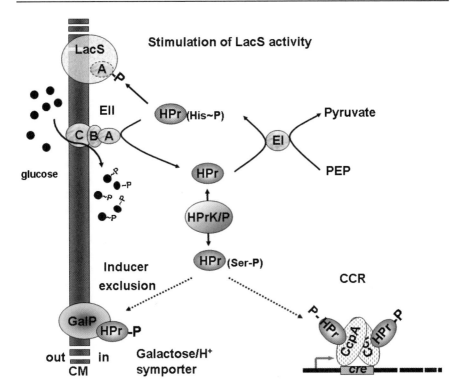

Fig. 5. Regulation of transporter activity, inducer exclusion and carbon catabolite repression (CCR) by phosphorylated HPr in Gram-positive bacteria.
HPr(Ser-P), generated by HPrK/P during growth on glucose plays a key role in inducer exclusion and carbon catabolite repression. HPr(Ser-P) binds to the galactose/H$^+$ symporter, thus leading to the inhibition of sugar uptake (inducer exclusion). In addition, HPr(Ser-P) in complex with CcpA inhibits the expression of catabolic genes (CCR). In the absence of glucose, HPr is phosphorylated at histidine 15. HPr(His~P) can transfer the phosphate to the EIIA^{LacS} domain of the lactose permease (A, LacS). This phosphorylation results in an increased lactose transport activity of LacS.

the presence of transporters with a similar domain organisation in different *Streptococcus, Lactobacillus, Leuconostoc,* and *Pediococcus* species.

3.3 Control of enzymatic activities by the PTS

In addition to the control of transporters, inducer exclusion can be obtained via the regulation of enzymatic activities by PTS components. In both *E. coli* and *B. subtilis*, the glycerol kinase is a target of PTS-dependent regulation. Moreover, the *E. coli* glycogen phosphorylase catalyzing the first step in glycogen degradation is controlled by the PTS.

As described for the transporters, a similar regulatory scenario (*i.e.* no production of inducer) is operative for the glycerol kinases in *E. coli* and *B. subtilis*, the actual regulatory mechanisms are, however, very different. While a direct protein-protein interaction with EIIA^{Crr} controls the activity of glycerol kinase in *E. coli*, HPr-dependent phosphorylation of *B. subtilis* glycerol kinase determines whether the inducer of the *glp* regulon, glycerol-3-phosphate, can be formed or not. In *E. coli*, glycerol kinase is bound by unphosphorylated EIIA^{Crr}, *i.e.* in the presence of glucose, and thereby, inactivated. Moreover, glycerol kinase is allosterically inhibited by fructose-1.6-bisphosphate, which acts also as an indicator of a high glycolytic activity (Novotny et al. 1985). As observed for the lactose permease (see above), phosphorylated EIIA^{Crr} is not capable of binding and inhibiting glycerol kinase due to a direct disruption of the protein-protein interaction by the phosphate group. Interestingly, one of the two binding motifs for EIIA^{Crr} identified in the lactose permease and other permeases is also present in the glycerol kinase suggesting the same principal arrangement of the interaction (Sondej et al. 1999). Indeed, the glycerol kinase binds to the same region of EIIA^{Crr} as do the permeases and HPr (Hurley et al. 1993; Peterkofsky et al. 2001). Interestingly, EIIA^{Crr} binds far from the enzymatically active site of glycerol kinase suggesting a long-range conformational change that mediates the inhibition. Recently, glycerol kinase of *H. influenzae*, which is not a target of regulation by EIIA^{Crr} was converted to an enzyme that is allosterically inhibited by introducing only 11 mutations (out of 117 differences to the *E. coli* enzyme) (Pawlyk and Pettigrew 2002). Thus, the mechanism of interaction and the coupling of EIIA^{Crr} binding and enzymatic activity are very well understood for the glycerol kinase and make it one of the best-characterised enzymes. Interestingly, the control of glycerol kinase activity by the PTS is not a unidirectional event: glycerol kinase may also "retroinhibit" the PTS activity, possibly by sequestering EIIA^{Crr} (Rohwer et al. 1998).

In *B. subtilis*, glycerol but not glycerol-3-phosphate utilisation depends on a functional PTS. This observation suggested that the PTS is somehow involved in the generation of the inducer of the *glp* genes, glycerol-3-phosphate (Beijer and Rutberg 1992). Mutations that bypass the need for a functional PTS for glycerol utilisation were identified in the glycerol kinase indicating that the activity of glycerol kinase is positively regulated by PTS components (Wehtje et al. 1995). Studies using the purified enzymes from *Enterococcus casseliflavus* and *B. subtilis* revealed phosphorylation by HPr(His~P). This phosphorylation stimulates the activity of the enzyme about 9-fold (Charrier et al. 1997; Darbon et al. 2002). Thus, the phosphorylation status of HPr links the availability of preferred carbon sources to the formation of glycerol-3-phosphate: In the absence of glucose, HPr is phosphorylated on His-15 (see above) and can in turn phosphorylate and stimulate glycerol kinase. In the presence of glucose, no HPr(His~P) is available for glycerol kinase phosphorylation and in its unstimulated form, the enzyme is not able to produce sufficient glycerol-3-phosphate to induce the *glp* genes. This mechanism of HPr-dependent inducer exclusion may occur even in Gram-negative bacteria: phosphorylation of glycerol kinase by HPr was also demonstrated in *Thermus flavus* (Darbon et al. 1999). However, the regulatory relevance of this phosphorylation event remains to be analysed.

While EIIA^{Crr} is the major regulatory component of the PTS in *E. coli*, HPr is also capable of performing regulatory interactions in this organism. The enzymatic activity of *E. coli* glycogen phosphorylase, an enzyme involved in glycogen degradation, is stimulated by binding to unphosphorylated HPr (Seok et al. 1997a). As demonstrated for the regulatory interactions of EIIA^{Crr}, HPr uses the same surface surrounding the active site His-15 to interact with its different partners, EI, EI-IA^{Crr}, and glycogen phosphorylase (Wang et al. 2000).

4 Regulation of transcription by PTS components

The PTS does not only control the activities of transporters and metabolic enzymes but also of transcriptional regulators. This control is mediated by three different mechanisms: (i) the cofactors of transcription regulators may be derived from the PTS, (ii) proteins of the PTS may directly control the activity of the regulators by phosphorylation or, (iii) PTS components may bind transcription factors and, thus, control their activity. This latter mechanism is discussed in detail elsewhere in this volume for the sequestration of the repressor protein Mlc by the membrane-bound EIICB^{Glc} (see chapter by Böhm and Boos, and for a recent review see also Plumbridge 2002). In any scenario, it is again the phosphorylation state of the PTS proteins that determines the regulatory output.

4.1 PTS-dependent generation of cofactors for transcription regulators

The PTS is the main sugar-sensing system in most bacteria that possess it. The PTS is directly involved in the regulation of hierarchical expression of catabolic genes and operons. In the presence of glucose and other preferred carbon sources, the genes required for the utilisation of secondary substrates are not expressed. This regulatory phenomenon is referred to as carbon catabolite repression. It is exerted at the level of direct transcription regulation by general regulator proteins (catabolite repression *sensu stricto*), by the regulation of operon-specific transcription regulators by the carbon source (induction prevention, see below) and by controlling the uptake and generation of inducers (inducer exclusion, see above) (Stülke and Hillen 1999; Brückner and Titgemeyer 2002).

In *E. coli*, the cyclic AMP receptor protein (Crp) is the global transcriptional regulator that affects gene expression in response to the carbon supply. Its DNA-binding and transcription activation activity depends on binding of its cofactor, cyclic AMP (cAMP). cAMP in turn is synthesised by adenylate cyclase, an enzyme that is stimulated by an interaction with phosphorylated EIIA^{Crr} in the absence of glucose (see Fig. 4). Although the basic fact of the stimulation has been known for more than 25 years (Peterkofsky 1977), the underlying molecular mechanisms have escaped their elucidation until now. So far, there is only genetic evidence for the proposed interaction between phosphorylated EIIA^{Crr} and adenylate cyclase but no biochemical data have been provided. However, the recent findings concerning the phosphorylation state of EIIA^{Crr} (Hogema et al. 1998b; Hogema et al. 1999) are in good agreement with analyses of the intracellular cAMP concentrations (Inada et al. 1996): EIIA^{Crr} is dephosphorylated and the cAMP concentrations are similarly low in the presence of glucose or lactose. Moreover, mutations of the EIIA^{Crr} phosphorylation site prevent stimulation of adenylate cyclase and, thus, activation of catabolic genes by Crp (Reddy and Kamireddi 1998).

In *B. subtilis* and other low-GC Gram-positive bacteria, HPr and its regulatory paralogue Crh are involved in carbon catabolite repression. As outlined above, the HPrK/P is active as a kinase in the presence of glucose and other well-metabolisable carbon sources, and HPr(Ser-P) and Crh(Ser-P) act as cofactors of the pleiotropic transcription regulator CcpA to stimulate its DNA-binding activity. The nonphosphorylated forms of HPr and Crh or HPr(His~P) do not interact with CcpA (Deutscher et al. 1995; Jones et al. 1997a; Galinier et al. 1999; Aung-Hilbrich et al. 2002) (see Fig. 5). The interaction between CcpA and either of its cofactors occurs at a surface that is specifically conserved in all CcpA proteins but not in other members of the LacI/GalR family of transcriptional regulators (Kraus et al. 1998). Binding of CcpA to promoter regions of the controlled genes results in either transcription repression or activation of genes involved in carbon catabolism, central metabolism, amino acid metabolism, and several other functions (Moreno et al. 2001; Yoshida et al. 2001; Blencke et al. 2003) (see Fig. 5).

4.2 Control of transcription regulators by direct phosphorylation

Substrate-specific induction of catabolic operons depends on regulatory proteins that somehow sense the availability of the respective substrate. Many regulators are controlled by direct binding to the substrate or a derivative of it (*e.g.* the *E. coli* lac repressor). A second class of regulators senses the substrate via a PTS-dependent phosphorylation event. These regulators possess a common domain, the PTS regulation domain (PRD) (Stülke et al. 1998; van Tilbeurgh and Declerck 2001; Greenberg et al. 2002). The PRD-containing regulators can act as transcriptional antiterminators that bind RNA to exert regulation or they can be DNA-binding transcriptional activators. The transcriptional antiterminators are all composed of a N-terminal RNA-binding domain and two reiterated PRDs. The activators contain a DNA-binding helix-turn-helix motif, two PRDs and EIIB and EIIA-like domains (Greenberg et al. 2002). Most work has been devoted to the *B. subtilis* antiterminator LicT and the activator protein LevR, and only finding related to these proteins will be discussed here (Schnetz et al. 1996; Débarbouillé et al. 1991).

The utilisation of β-glucosides such as salicin is catalyzed by the enzymes encoded in the *B. subtilis bglPH* operon. The first gene of the operon, *bglP*, encodes a β-glucoside-specific EII of the PTS (Le Coq et al. 1995). Expression of the operon is induced by the substrate, salicin, and repressed by glucose. Transcription is initiated both in the absence and in the presence of the inducer but is immediately

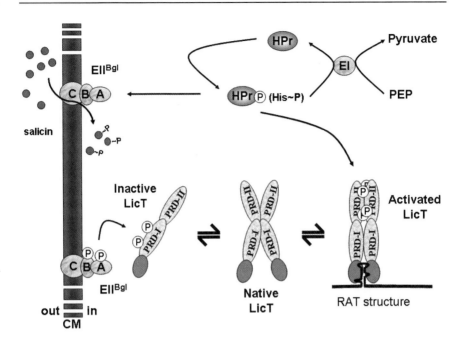

Fig. 6. Proposed model for the regulation of LicT activity. LicT is present in different forms in the cell due to presence or absence of sugars. In the presence of the inducer salicin, the phosphoryl group is transferred from HPr(His~P) to the incoming sugar via EII^Bgl. Simultaneously, HPr(His~P) also phosphorylates the LicT antiterminator protein in the PRD-II. This phosphorylation leads to a stabilisation of the dimer (Activated LicT). In its active form, LicT binds to the ribonucleic antiterminator (RAT), thus enabling transcription of the *bglPH* operon. This positive phosphorylation depends on the availability of HPr(His~P). In the presence of rapidly metabolisable sugars, *e.g.* glucose, HPr is found in its unphosphorylated form or as HPr(Ser-P), thus, no activation of LicT occurs and LicT remains in an open dimer conformation (Native LicT), which is inactive in antitermination. In the absence of salicin, EII^Bgl is phosphorylated. In this case, the phosphate is transferred to the PRD-I of LicT, thus, leading to inactivation of the LicT protein. Inactive LicT is thought to be monomeric and maybe sequestered by the EII^Bgl as described for the homologous protein (BglG) from *E. coli* (Görke and Rak 2001).

terminated if no salicin is present. In contrast, if the inducer is available, the regulatory protein LicT binds to the mRNA prevents transcription termination and allows the *bglPH* operon to be expressed. The mechanisms of protein-RNA interaction and the formation of mutually exclusive RNA structures is beyond the scope of this review (refer to Yang et al. 2002; Stülke 2002). The RNA-binding activity of LicT is controlled by two antagonistically acting phosphorylation events. In the absence of the inducer, the β-glucoside-specific EII BglP is phosphorylated, and the genetic evidence suggests that it can phosphorylate and, thereby, inactivate LicT under these conditions (Fig. 6). This negatively acting phosphorylation of LicT occurs on a histidine residue in the N-terminal PRD (also called PRD-I)

(Tortosa et al. 2001). If salicin becomes available, the phosphate in PRD-I is drained from LicT *via* BglP to the incoming sugar. This form of LicT (non-phosphorylated in PRD-I) is competent for antitermination activity. However, a second condition must be fulfilled that LicT can act as an antiterminator: glucose or other preferred carbon sources must be absent from the medium (Krüger et al. 1996). If no glucose is available, HPr is present in its unphosphorylated form or as HPr(His~P) (see above; Ludwig et al. 2002). This latter form of HPr can phosphorylate the C-terminal PRD-II of LicT thereby stimulating the activity of the protein (Lindner et al. 1999). This second phosphorylation is a mechanism of carbon catabolite repression of the *bglPH* operon that is exerted in addition to CcpA/HPr(Ser-P)-dependent catabolite repression and is referred to as induction prevention (see above) (Krüger et al. 1996; Lindner et al 2002). Dual regulation by two independent antagonistic phosphorylation events is observed for many antiterminators of this family. However, there are antiterminators whose activity does not depend on phosphorylation by HPr. The genes controlled by these antiterminators, *B. subtilis sacB* and the *ptsGHI* operon, are not subject to catabolite repression (Crutz et al. 1990; Stülke et al. 1997). To understand the mechanism by which these antagonistic phosphorylation events control the RNA-binding activity of LicT, the structure of the PRDs was investigated (van Tilbeurgh et al. 2001). They were found to form dimers with the phosphorylation sites buried at the dimer interface. It was proposed that two conformations of the LicT dimer exist: a native open conformation and an active closed conformation. This latter conformation is adopted upon phosphorylation on PRD-II by HPr. In contrast, phosphorylation on PRD-I by BglP is thought to disrupt the LicT dimer, thus, inactivating the protein (Declerck et al. 2001; van Tilbeurgh et al. 2001; van Tilbeurgh and Declerck 2001) (see Fig. 6). Dual antagonistically acting phosphorylation events were also described for the *E. coli* counterpart of LicT, BglG, as well as several other antiterminator proteins of this family (Görke and Rak 1999; Stülke et al. 1998).

The *B. subtilis* LevR protein controls the expression of the levanase operon. In the presence of the inducer fructose and in the absence of glucose it binds to a region upstream of the promoter of the operon and activates transcription by interacting with the RNA polymerase (Débarbouillé et al. 1991). The involvement of LevR in induction and carbon catabolite repression is again mediated by PTS-dependent phosphorylation (Martin-Verstraete et al. 1995; Stülke et al. 1995). In the absence of fructose, EIIB^{Lev} phosphorylates LevR at PRD-II and, thus, inactivates the protein. In contrast, the nonphosphorylated protein exhibits an intermediate transcription activation activity. If the protein is phosphorylated at its EIIA-like domain by HPr(His~P), a further increase in activity is observed (Martin-Verstraete et al. 1998). Thus, the phosphorylation state of LevR links the expression of the levanase operon to the availability of carbon sources and PTS phosphorylation.

5 Conclusion

The bacterial PTS is very well suited for the purpose of economical uptake of sugars. Moreover, it is a perfect signal transduction system that can interact with a wide variety of proteins – enzymes, transporters and transcriptional regulators – to control their activity. These interactions may occur either as protein-protein interactions as we saw it for EIIA^{Crr} with the lactose permease or the glycerol kinase from *E. coli* and for HPr(Ser-P) with CcpA in *B. subtilis*. Alternatively, the PTS can control the activity of other proteins by phosphorylation as discussed for antagonistically acting phosphorylations of antiterminator proteins by specific EII and HPr(His~P) in both *E. coli* and *B. subtilis*, and for glycerol kinase and the lactose permease LacS in Gram-positive bacteria. The different phosphorylated forms of the PTS proteins make them so versatile for such a variety of regulatory interactions.

Given this versatility and obvious high utility of the PTS for adaptation of bacteria to their ecological niches, one may ask why this system is restricted to bacteria and not more widespread in nature. The PTS evolved early in bacterial evolution before the radiation of the major phyla (see Fig. 2). It was certainly incorporated by early eukaryotes when the endosymbiontic event with α-proteobacteria, that gave rise to mitochondria, took place. However, the PTS was lost in eukaryotes. A reason for the absence of the PTS from eukaryotes may be the large difference in cell size. A recent study indicates that the bacterial cell is sufficiently small to support the regulatory chain of the PTS. In contrast, the diffusional space would be much larger in eukaryotic cells, and the PTS proteins would not be able to transduce their signals. Thus, the eukaryotes had to develop their own systems for signal transduction (Francke et al. 2003).

Acknowledgements

We are grateful to S. Bachem, K. Steinhauer, K-G Hanson, I. Langbein, H. Ludwig, H-M Blencke, C. Detsch, and M. Merzbacher for fruitful cooperation in the elucidation of regulatory functions of the PTS. Work in the authors' lab was supported by the Deutsche Forschungsgemeinschaft through SFB473 and the Fonds der Chemischen Industrie.

References

Abramson J, Smirnova I, Kasho V, Verner G, Kaback HR, Iwata S (2003) Structure and mechanism of lactose permease of *Escherichia coli*. Science 301:610-615
Audette GF, Engelmann R, Hengstenberg W, Deutscher J, Hayakawa K, Quail JW, Delbaere LTJ (2000) The 1.9 Å resolution structure of phospho-serine 46 in HPr from *Enterococcus faecalis*. J Mol Biol 303:545-553

Aung-Hilbrich LM, Seidel G, Wagner A, Hillen W (2002) Quantification of the influence of HPrSer46P on CcpA-*cre* interaction. J Mol Biol 319:77-85

Beijer L, Rutberg L (1992) Utilisation of glycerol and glycerol-3-phosphate is differently affected by the phosphotransferase system in *Bacillus subtilis*. FEMS Microbiol Lett 100:217-220

Blencke HM, Homuth G, Ludwig H, Mäder U, Hecker M, Stülke J (2003) Transcriptional profiling of gene expression in response to glucose in *Bacillus subtilis*: regulation of the central metabolic pathways. Metab Engn 5:133-149

Boël G, Mijakovic I, Mazé A, Poncet S, Taha MK, Laribe M, Darbon E, Khemiri A, Galinier A, Deutscher J (2003) Transcription regulators potentially controlled by HPr kinase/phosphorylase in Gram-negative bacteria. J Mol Microbiol Biotechnol 5:206-215

Brochu D, Vadeboncoeur C (1999) The HPr(Ser) kinase of *Streptococcus salivarius*: purification, properties, and cloning of the *hprK* gene. J Bacteriol 181:709-717

Brückner R, Titgemeyer F (2002) Carbon catabolite repression in bacteria: choice of the carbon source and autoregulatory limitation of sugar utilization. FEMS Microbiol Lett 209:141-148

Charrier V, Buckley E, Parsonage D, Galinier A, Darbon E, Jaquinod M, Forest E, Deutscher J, Claiborne A (1997) Cloning and sequencing of two enterococcal *glpK* genes and regulation of the encoded glycerol kinases by phosphoenolpyruvate-dependent, phosphotransferase system-catalyzed phosphorylation of a single histidyl residue. J Biol Chem 272:14166-14174

Chen Y, Reizer J, Saier MH, Fairbrother WJ, Wright PE (1993) Mapping of the binding interfaces of the proteins of the bacterial phosphotransferase system, HPr and IIA^{Glc}. Biochemistry 32:32-37

Crutz AM, Steinmetz M, Aymerich S, Richter R, Le Coq D (1990) Induction of levansucrase in *Bacillus subtilis*: an antitermination mechanism negatively controlled by the phosphotransferase system. J Bacteriol 172:1043-1050

Darbon E, Ito K, Huang HS, Yoshimoto T, Poncet S, Deutscher J (1999) Glycerol transport and phosphoenolpyruvate-dependent enzyme I- and HPr-catalysed phosphorylation of glycerol kinase in *Thermus flavus*. Microbiology 145:3205-3212

Darbon E, Servant P, Poncet S, Deutscher J (2002) Antitermination by GlpP, catabolite repression via CcpA and inducer exclusion triggered by P~GlpK dephosphorylation control *Bacillus subtilis glpFK* expression. Mol Microbiol 43:1039-1052

Declerck N, Dutartre H, Receveur V, Dubois V, Royer C, Aymerich S, van Tilbeurgh H (2001) Dimer stabilzation upon activation of the transcriptional antiterminator LicT. J Mol Biol 314:671-681

Débarbouillé M, Martin-Verstraete I, Klier A, Rapoport G (1991) The transcriptional regulator LevR of *Bacillus subtilis* has domains homologous to both σ^{54} - and phosphotransferase system-dependent regulators. Proc Natl Acad Sci USA 88:2212-2216

Deutscher J, Saier MH (1983) ATP-dependent protein-kinase catalyzed phosphorylation of a seryl residue in HPr, the phosphoryl carrier protein of the phosphotransferase system in *Streptococcus pyogenes*. Proc Natl Acad Sci USA 80: 6790-6794

Deutscher J, Kessler U, Alpert CA, Hengstenberg W (1984) Bacterial phosphoenolpyruvate-dependent phosphotransferase system: P-ser-HPr and its possible regulatory function. Biochemistry 23: 4455-4460

Deutscher J, Pevec B, Beyreuther K, Kiltz HH, Hengstenberg W (1986) Streptococcal phosphoenolpyruvate-sugar phosphotransferase system: amino acid sequence and site of ATP-dependent phosphorylation of HPr. Biochemistry 25:6543-6551

Deutscher J, Küster E, Bergstedt U, Charrier V, Hillen W (1995) Protein kinase-dependent HPr/CcpA interaction links glycolytic activity to carbon catabolite repression in Gram-positive bacteria. Mol Microbiol 15:1049-1053

Djordjevic GM, Tchieu JH, Saier MH (2001) Genes involved in control of galactose uptake in *Lactobacillus brevis* and reconstitution of the regulatory system in *Bacillus subtilis*. J Bacteriol 183:3224-3236

Dossonnet V, Monedero V, Zagorec M, Galinier A, Perez-Martinez G, Deutscher J (2000) Phosphorylation of HPr by the bifunctional HPr Kinase/P-ser-HPr phosphatase from *Lactobacillus casei* controls catabolite repression and inducer exclusion but not inducer expulsion. J Bacteriol 182:2582-2590

Francke C, Postma PW, Westerhoff HV, Blom JG, Petelier MA (2003) Why the phosphotransferase system of *Escherichia coli* escapes diffusion limitation. Biophys J 85:612-622

Frishman D, Mokrejs M, Kosykh D, Kastenmüller G, Kolesov G, Zubrzycki I, Gruber C, Geier B, Kaps A, Albermann K, Volz A, Wagner C, Fellenberg M, Heumann K, Mewes HW (2003) The PEDANT genome database. Nucl Acids Res 31:207-211

Galinier A, Haiech J, Kilhoffer MC, Jaquinod M, Stülke J, Deutscher J, Martin-Verstraete I (1997) The *Bacillus subtilis* crh gene encodes a HPr-like protein involved in carbon catabolite repression. Proc Natl Acad Sci USA 94:8439-8444.

Galinier A, Kravanja M, Engelmann R, Hengstenberg W, Kilhoffer MC, Deutscher J, Haiech J (1998) New protein kinase and protein phosphatase families mediate signal transduction in bacterial catabolite repression. Proc Natl Acad Sci USA 95:1823-1828

Galinier A, Deutscher J, Martin-Verstraete I (1999) Phosphorylation of either Crh or HPr mediates binding of CcpA to the *Bacillus subtilis xyn cre* and catabolite repression of the *xyn* operon. J Mol Biol 286:307-314

Garrity LF, Schiel SL, Merrill R, Saier MH Jr, Ordal GW (1998) Unique regulation of carbohydrate chemotaxis in *Bacillus subtilis* by the phosphoenolpyruvate-dependent phosphotransferase system and the methyl-accepting chemotaxis protein McpC. J Bacteriol 180:4475-4480

Gemmecker G, Eberstadt M, Buhr A, Lanz R, Grdadolnik SG, Kessler H, Erni B (1997) Glucose transporter of *Escherichia coli*: NMR characterization of the phosphocysteine form of the IIB^{Glc} domain and its binding interface with the IIA^{Glc} subunit. Biochemistry 36:7408-7417

Görke B, Rak B (1999) Catabolite control of *Escherichia coli* regulatory protein BglG activity by antagonistically acting phosphorylations. EMBO J 18:3370-3379

Görke B, Rak B (2001) Efficient transcriptional antitermination from the *Escherichia coli* cytoplasmic membrane. J Mol Biol 308:131-145

Greenberg DB, Stülke J, Saier MH (2002) Domain analysis of transcriptional regulators bearing PTS regulatory domains. Res Microbiol 153:519-526

Gunnewijk MG, Poolman B (2000) Phosphorylation state of HPr determines the level of expression and the extent of phosphorylation of the lactose transport protein of *Streptococcus thermophilus*. J Biol Chem 275:34073-34079

Hanson KG, Steinhauer K, Reizer J, Hillen W, Stülke J (2002) HPr kinase/phosphatase of *Bacillus subtilis*: expression of the gene and effects of mutations on enzyme activity, growth and carbon catabolite repression. Microbiology 148:1805-1811

Henkin TM (1996) The role of the CcpA transcriptional regulator in carbon metabolism in *Bacillus subtilis*. FEMS Microbiol L 135: 9-15

Herzberg O (1992) An atomic model for protein-protein phosphoryl group transfer. J Biol Chem 267:24819-24823

Hogema BM, Arents JC, Inada T, Aiba H, van Dam K, Postma PW (1997) Catabolite repression by glucose-6-phosphate, gluconate and lactose in *Escherichia coli*. Mol Microbiol 24:857-867

Hogema BM, Arents JC, Bader R, Eijkemans K, Inada T, Aiba H, Postma PW (1998a) Inducer exclusion by glucose-6-phosphate in *Escherichia coli*. Mol Microbiol 28:755-765

Hogema BM, Arents JC, Bader R, Eijkemans K, Yoshida H, Takahashi H, Aiba H, Postma PW (1998b) Inducer exclusion in *Escherichia coli* by non-PTS substrates: the role of the PEP to pyruvate ratio in determining the phosphorylation state of enzyme IIA^{Glc}. Mol Microbiol 30:487-498

Hogema BM, Arents JC, Bader R, Postma PW (1999) Autoregulation of lactose uptake through the LacY permease by enzyme IIA^{Glc} of the PTS in *Escherichia coli* K-12. Mol Microbiol 31:1825-1833

Hoischen C, Levin J, Pitaknarongphorn S, Reizer J, Saier MJ (1996) Involvement of the central loop of the lactose permease of *Escherichia coli* in its allosteric regulation by the glucose-specific enzyme IIA of the phosphoenolpyruvate-dependent phosphotransferase system. J Bacteriol 178:6082-6086

Hurley JH, Faber HR, Worthylake D, Meadow ND, Roseman S, Pettigrew DW, Remington SJ (1993) Structure of the regulatory complex of *Escherichia coli* III^{Glc} with glycerol kinase. Science 259:673-677

Huynh PL, Jankovic I, Schnell NF, Brückner R (2000) Characterization of an HPr kinase mutant of *Staphylococcus xylosus*. J Bacteriol 182:1895-1902

Inada T, Kimata K, Aiba H (1996) Mechanism responsible for glucose-lactose diauxie in *Escherichia coli*: challenge to the cAMP model. Genes to Cells 1:293-301

Jault J M, Fieulaine S, Nessler S, Gonzalo P, Di Pietro A, Deutscher J, Galinier A (2000) The HPr kinase from *Bacillus subtilis* is a homo-oligomeric enzyme which exhibits strong positive cooperativity for nucleotide and fructose 1,6-bisphosphate binding. J Biol Chem 275:1773-1780

Jones BE, Dossonnet V, Küster E, Hillen W, Deutscher J, Klevit RE (1997a) Binding of the catabolite repressor protein CcpA to its DNA target is regulated by phosphorylation of its corepressor HPr. J Biol Chem 272: 26530-26535

Jones BE, Rajagopal P, Klevit RE (1997b) Phosphorylation on histidine is accompanied by localized structural changes in the phosphocarrier protein, HPr from *Bacillus subtilis*. Protein Sci 6:2107-2119

Kimata K, Takahashi H, Inada T, Postma P, Aiba H (1997) cAMP receptor protein- cAMP plays a crucial role in glucose-lactose diauxie by activating the major glucose transporter gene in *Escherichia coli*. Proc Natl Acad Sci USA 94:12914-12919

Kraus A, Küster E, Wagner A, Hoffmann K, Hillen W (1998) Identification of a corepressor binding site in catabolite control protein CcpA. Mol Microbiol 30:955-963

Kravanja M, Engelmann R, Dossonnet V, Blüggel M, Meyer HE, Frank R, Galinier A, Deutscher J, Schnell N, Hengstenberg W (1999) The *hprK* gene of *Enterococcus faecalis* encodes a novel bifunctional enzyme: the HPr kinase/phosphatase. Mol Microbiol 31:59-66

Krüger S, Gertz S, Hecker M (1996) Transcriptional analysis of *bglPH* expression in *Bacillus subtilis*: Evidence for two distinct pathways mediating carbon catabolite repression. J Bacteriol 178:2637-2644

Leboeuf C, Leblanc L, Auffray Y, Hartke A (2000) Characterization of the *ccpA* gene of *Enterococcus faecalis*: identification of starvation-inducible proteins regulated by *ccpA*. J Bacteriol 182:5799-5806

Le Coq D, Lindner C, Krüger S, Steinmetz M, Stülke J (1995) New β-glucoside (*bgl*) genes in *Bacillus subtilis*: the *bglP* gene product has both transport and regulatory functions, similar to those of BglF, its *Escherichia coli* homolog. J Bacteriol 177:1527-1535

Lindner C, Galinier A, Hecker M, Deutscher J (1999) Regulation of the activity of the *Bacillus subtilis* antiterminator LicT by multiple PEP-dependent, enzyme I- and HPr-catalysed phosphorylation. Mol Microbiol 31:995-1006

Lindner C, Hecker M, Le Coq D, Deutscher J (2002) *Bacillus subtilis* mutant LicT antiterminators exhibiting enzyme I- and HPr-independent antitermination affect catabolite repression of the bglPH operon. J Bacteriol 184:4819-4828

Ludwig H, Stülke J (2001) The *Bacillus subtilis* catabolite control protein CcpA exerts all its regulatory functions by DNA-binding. FEMS Microbiol Lett 203:125-129

Ludwig H, Rebhan N, Blencke HM, Merzbacher M, Stülke J (2002) Control of the glycolytic *gapA* operon by the catabolite control protein A in *Bacillus subtilis*: a novel mechanism of CcpA-mediated regulation. Mol Microbiol 45:543-553

Lux R, Jahreis K, Bettenbrock K, Parkinson JS, Lengeler JW (1995) Coupling the phosphotransferase system and the methyl-accepting chemotaxis protein-dependent chemotaxis signaling pathways of *Escherichia coli*. Proc Natl Acad Sci USA 92:11583-11587

Lux R, Munasinghe VR, Castellano F, Lengeler JW, Corrie JE, Khan S (1999) Elucidation of a PTS-carbohydrate chemotactic signal pathway in *Escherichia coli* using a time-resolved behavioral assay. Mol Biol Cell 10:1133-1146

Martin-Verstraete I, Stülke J, Klier A, Rapoport G (1995) Two different mechanisms mediate catabolite repression of the *Bacillus subtilis* levanase operon. J Bacteriol 177:6919-6927

Martin-Verstraete I, Charrier V, Stülke J, Galinier A, Erni B, Rapoport G, Deutscher J (1998) Antagonistic effects of dual PTS catalyzed phosphorylation on the *Bacillus subtilis* transcriptional activator LevR. Mol Microbiol 28:293-303

Meadow ND, Roseman S (1996) Rate and equilibrium constants for phosphoryl transfer between active site histidines of *Escherichia coli* HPr and the signal transducing protein III^{Glc}. J Biol Chem 271:33440-33445

Mijakovic I, Poncet S, Galinier A, Monedero V, Fieulaine S, Janin J, Nessler S, Marquez JA, Scheffzek K, Hasenbein S, Deutscher J (2002) Pyrophosphate-producing protein dephosphorylation by HPr kinase/phosphorylase: A relic of early life? Proc Natl Acad Sci USA 99:13442-13447

Monedero V, Kuipers OP, Jamet E, Deutscher J (2001a) Regulatory functions of serine-46 phosphorylated HPr in *Lactococcus lactis*. J Bacteriol 183:3391-3398

Monedero V, Poncet S, Mijakovic I, Fieulaine S, Dossonnet V, Martin-Verstraete I, Nessler S, Deutscher J (2001b) Mutations lowering the phosphatase activity of HPr kinase/phosphatase switch off carbon metabolism. EMBO J 20:3928-3937

Monod J (1942) Recherches sur la croissance des cultures bactériennes. Hermann et Cie, Paris

Moreno MS, Schneider BL, Maile RR, Weyler W, Saier MH Jr (2001) Catabolite repression mediated by CcpA protein in *Bacillus subtilis*: novel modes of regulation revealed by whole-genome analyses. Mol Microbiol 39:1366-1381

Nessler S, Fieulaine S, Poncet S, Galinier A, Deutscher J, Janin J (2003) HPr kinase/phosphorylase, the sensor enzyme of catabolite repression in Gram-positive bacteria: structural aspects of the enzyme and the complex with its protein substrate. J Bacteriol 185:4003-4010

Novotny MJ, Frederickson WL, Waygood EB, Saier MH (1985) Allosteric regulation of glycerol kinase by enzyme III^{glc} of the phosphotransferase system in *Escherichia coli* and *Salmonella typhimurium*. J Bacteriol 162:810-816

Pawlyk AC, Pettigrew DW (2002) Transplanting allosteric control of enzyme activity by protein-protein interactions: coupling a regulatory site to the conserved catalytic core. Proc Natl Acad Sci USA 99:11115-11120

Pelton JG, Torchia DA, Meadow ND, Roseman S (1993) Tautomeric states of the active-site histidines of phosphorylated and unphosphorylated III^{Glc}, a signal-transducing protein from *Escherichia coli*, using two-dimensional heteronuclear NMR techniques. Protein Sci 2: 543-558

Pelton JG, Torchia DA, Remington SJ, Murphy KP, Meadow ND, Roseman S (1996) Structures of active site histidine mutants of III^{Glc}, a major signal-transducing protein in *Escherichia coli*. J Biol Chem 271:33446-33456

Peterkofsky A (1977) Regulation of *Escherichia coli* adenylate cyclase by phosphorylation-dephosphorylation. Trends Bioochem Sci 2:12-14

Peterkofsky A, Wang G, Garrett DS, Lee BR, Seok YJ, Clore GM (2001) Three-dimensional structures of protein-protein complexes in the *E. coli* PTS. J Mol Microbiol Biotechnol 3:347-354

Plamondon P, Brochu D, Thomas S, Fradette, Gauhier L, Vaillancourt K, Buckley N, Frenette M, Vadeboncoeur C (1999) Phenotypic conequences resulting from a methionine-to-valine substitution at position 48 in the HPr protein of *Streptococcus salivarius*. J Bacteriol 181:6914-6921

Plumbridge J (2002) Regulation of gene expression in the PTS in *Escherichia coli*: the role and interactions of Mlc. Curr Opin Microbiol 5:187-193

Poolman B, Royer TJ, Mainzer SE, Schmidt BF (1989) Lactose transport system of *Streptococcus thermophilus*: a hybrid protein with homology to the melibiose carrier and enzyme III of phosphoenolpyruvate-dependent phosphotransferase systems. J Bacteriol 171:244-253

Poolman B, Knol J, Mollet B, Nieuwenhuis B, Sulter G (1995) Regulation of bacterial sugar-H$^+$ symport by phosphoenolpyruvate-dependent enzyme I/HPr-mediated phosphorylation. Proc Natl Acad Sci USA 92:778-782

Postma PW, Lengeler JW, Jacobson GR (1993) Phosphoenolpyruvate:carbohydrate phosphotransferase systems of bacteria. Microbiol Rev 57:543-594

Powell BS, Court DL, Inada T, Nakamura Y, Michotey V, Cui X, Reizer A, Saier MH, Reizer J (1995) Novel proteins of the phosphotransferase system encoded within the *rpoN* operon of *Escherichia coli*. Enzyme IIA^{Ntr} affects growth on organic nitrogen and the conditional lethality of an *era*^{ts} mutant. J Biol Chem 270:4822-4839

Pullen K, Rajagopal P, Branchini BR, Huffine ME, Reizer J, Saier MH, Scholtz JM, Klevit RE (1995) Phosphorylation of serine-46 in HPr, a key regulatory protein in bacteria, results in stabilization of its solution structure. Protein Sci 4:2478-2486

Rabus R, Reizer J, Paulsen IT, Saier MH (1999) Enzyme I^{Ntr} from Escherichia coli: A novel enzyme of the phosphoenolpyruvate-dependent phosphotransferase system exhibiting strict specificity for its phosphoryl acceptor, NPr. J Biol Chem 274:26185-26191

Ramström H, Sanglier S, Leize-Wagner E, Philippe C, van Dorsselaer A, Haiech J (2003) Properties and regulation of the bifunctional enzyme HPr kinase/phosphatase in *Bacillus subtilis*. J Biol Chem 278:1174-1185

Reddy P, Kamireddi M (1998) Modulation of *Escherichia coli* adenylyl cyclase activity by catalytic-site mutants of protein IIA^{Glc} of the phosphoenolpyruvate:sugar phosphotransferase system. J Bacteriol 180: 732-736

Reizer J, Reizer A, Saier MH, Jacobson GR (1992) A proposed link between nitrogen and carbon metabolism involving protein phosphorylation in bacteria. Protein Sci 1:722-726

Reizer J, Hoischen C, Titgemeyer F, Rivolta C, Rabus R, Stülke J, Karamata D, Saier MH, Hillen W (1998) A novel bacterial protein kinase that controls carbon catabolite repression. Mol Microbiol 27:1157-1170

Rohwer JM, Bader R, Weserhoff HV, Postma PW (1998) Limits to inducer exclusion: inhibition of the bacterial phosphotransferase system by glycerol kinase. Mol Microbiol 29:641-652

Saier, MH (1989) Protein phosphorylation and allosteric control of inducer exclusion and catabolite repression by the bacterial phosphoenolpyruvate:sugar phosphotransferase system. Microbiol Rev 53:109-120

Saier MH (2000) A functional-phylogenetic classification system for transmembrane solute transporters. Microbiol Mol Biol Rev 64:354-411

Saier MH, Novotny MJ, Comeau-Fuhrman D, Osumi T, Desai JD (1983) Cooperative binding of the sugar substrates and allosteric regulatory protein (enzyme III^{Glc} of the phosphotransferase system) to the lactose and melibiose permeases in *Escherichia coli* and *Salmonella typhimurium*. J Bacteriol 155:1351-1357

Saier MH, Reizer J (1992) Proposed uniform nomenclature for the proteins and protein domains of the bacterial phospoenolpyruvate:sugar phosphotransferase system. J Bacteriol 174:1433-1438

Schnetz K, Stülke J, Gertz S, Krüger S, Krieg M, Hecker M, Rak B (1996) LicT, a *Bacillus subtilis* transcriptional antiterminator of the BglG family. J Bacteriol 178:1971-1978

Seok YJ, Sondej M, Badawi P, Lewis MS, Briggs MC, Jaffe Y, Peterkofsky A (1997a) High affinity binding and allosteric regulation of *Escherichia coli* glycogen phosphorylase by the histidine phosphocarrier protein, HPr. J Biol Chem 272:26511-26521

Seok YJ, Sun J, Kaback HR, Peterkofsky A (1997b) Topology of allosteric regulation of lactose permease. Proc Natl Acad Sci USA 94:13515-13519

Sondej M, Sun J, Seok YJ, Kaback HR, Peterkofsky A (1999) Deduction of consensus binding sequences on proteins that bind IIA^{Glc} of the phosphoenolpyruvate:sugar phosphotransferase system by cysteine scanning mutagenesis of *Escherichia coli* lactose permease. Proc Natl Acad Sci USA 96:3525-3530

Sondej M, Seok YJ, Badawi P, Koo BM, Nam TW, Peterkofsky A (2000) Topography of the surface of the *Escherichia coli* phosphotransferase system enzyme IIA^{Glc} that interacts with lactose permease. Biochemistry 39:2931-2939

Sondej M, Weinglass AB, Peterkofsky A, Kaback HR (2002) Binding of enzyme IIA^{Glc}, a component of the phosphoenolpyruvate:sugar phosphotransferase system, to the *Escherichia coli* lactose permease. Biochemistry 41:5556-5565

Sondej M, Vázquez-Ibar JL, Farshidi A, Peterkofsky A, Kaback HR (2003) Characterization of a lactose permease mutant that binds IIA^{Glc} in the absence of ligand. Biochemistry 42:9153-9159

Steinhauer K, Jepp T, Hillen W, Stülke J (2002) A novel mode of control of *Mycoplasma pneumoniae* HPr kinase/phosphatase reflects its parasitic lifestyle. Microbiology 148:3277-3284

Stülke J (2002) Control of transcription termination in bacteria by RNA-binding proteins that modulate RNA structures. Arch Microbiol 177:433-440

Stülke J, Martin-Verstraete I, Charrier V, Klier A, Deutscher J, Rapoport G (1995) The HPr protein of the phosphotransferase system links induction and catabolite repression of the *Bacillus subtilis* levanase operon. J Bacteriol 177:6928-6936

Stülke J, Martin-Verstraete I, Zagorec M, Rose M, Klier A, Rapoport G (1997) Induction of the *Bacillus subtilis ptsGHI* operon by glucose is controlled by a novel antiterminator, GlcT. Mol Microbiol 25: 65-78

Stülke J, Arnaud M, Rapoport G, Martin-Verstraete I (1998) PRD - A protein domain involved in PTS-dependent induction and carbon catabolite repression of catabolic operons in bacteria. Mol Microbiol 28: 865-874

Stülke J, Hillen W (1998) Coupling physiology and gene regulation in bacteria: the phosphotransferase sugar uptake system delivers the signals. Naturwissenschaften 85:583-592

Stülke J, Hillen W (1999) Carbon catabolite repression in bacteria. Curr Opin Microbiol 2:195-201

Stülke J, Hillen, W (2000) Regulation of carbon catabolism in *Bacillus* species. Annu Rev Microbiol 54:849-880

Thevenot T, Brochu D, Vadeboncoeur C, Hamilton IR (1995) Regulation of ATP-dependent P-(Ser)-HPr formation in *Streptococcus mutans* and *Streptococcus salivarius*. J Bacteriol 177:2751-2759

Titgemeyer F, Mason RE, Saier MH (1994) Regulation of the raffinose permease of *Escherichia coli* by the glucose-specific enzyme IIA of the phosphoenolpyruvate:sugar phosphotransferase system. J Bacteriol 176:543-546

Tortosa P, Declerck N, Dutartre H, Lindner C, Deutscher J, Le Coq D (2001) Sites of positive and negative regulation in the *Bacillus subtilis* antiterminators LicT and SacT. Mol Microbiol 41:1381-1393

van Tilbeurgh H, Le Coq D, Declerck N (2001) Crystal structure of an activated form of the PTS regulation domain from the LicT transcriptional antiterminator. EMBO J 20:3789-3799

van Tilbeurgh H, Declerck N (2001) Structural insights into the regulation of bacterial signalling proteins containing PRDs. Curr Opin Struct Biol 11:685-693

Viana R, Monedero V, Dossonnet V, Vadeboncoeur C, Pérez-Martinez G, Deutscher J (2000) Enzyme I and HPr from *Lactobacillus casei*: their role in sugar transport, carbon catabolite repression and inducer exclusion. Mol Microbiol 36:570-584

Wang G, Sondej M, Garrett DS, Peterkofsky A, Clore GM (2000) A common interface on histidine-containing phosphocarrier protein for interaction with its partner proteins. J Biol Chem 275:16401-16403

Waygood EB (1998) The structure and function of HPr. Biochem Cell Biol 76:359-367

Wehtje C, Beijer L, Nilsson RP, Rutberg B (1995) Mutations in the glycerol kinase gene restore the ability of a *ptsGHI* mutant of *Bacillus subtilis* to grow on glycerol. Microbiology 141:1193-1198

Weigel N, Kukuruzinska MA, Nakazawa A, Waygood EB, Roseman S (1982) Sugar transport by the bacterial phosphotransferase system. Phosphoryl transfer reactions catalyzed by enzyme I of *Salmonella typhimurium*. J Biol Chem 257:14477-14491

Woese CR (1987) Bacterial evolution. Microbiol Rev 51:221-271

Yang Y, Declerck N, Manival X, Aymerich S, Kochoyan M (2002) Solution structure of the LicT-RNA antitermination complex: CAT clamping RAT. EMBO J 21:1987-1997

Yoshida KI, Kobayashi K, Miwa Y, Kang CM, Matsunaga M, Yamaguchi Y, Tojo S, Yamamoto M, Nishi R, Ogasawara N, Nakayama T, Fujita Y (2001) Combined transcriptome and proteome analysis as a powerful approach to study genes under glucose repression in *Bacillus subtilis*. Nucl Acids Res 29:6683-6692

Schmalisch, Matthias H.
Department of General Microbiology, Institute for Microbiology and Genetics, Georg-August University Göttingen, Grisebachstr. 8, D-37077 Göttingen, Germany

Stülke, Jörg
Department of General Microbiology, Institute for Microbiology and Genetics, Georg-August University Göttingen, Grisebachstr. 8, D-37077 Göttingen, Germany
jstuelk@gwdg.de

Ancillary proteins in membrane targeting of transporters

Tomas Nyman, Jhansi Kota, and Per O. Ljungdahl

Abstract

The integral protein components of the eukaryotic plasma membrane (PM) are targeted to the PM via the secretory pathway. Movement between the compartments of the secretory pathway occurs via small transport vesicles that form and bud from donor compartments, and that specifically target to and fuse with distinct acceptor compartments. The sorting and concentration of cargo proteins, including the proteins required for vesicle targeting, occurs concomitantly with the formation of transport vesicles. A detailed knowledge of the events facilitating the formation of transport vesicles is key to understanding the vectorial transport of proteins through the secretory pathway. A class of ancillary proteins, all integral components of the endoplasmic reticulum (ER) membrane, is required for the incorporation of discrete sets of metabolite transporters into ER-derived COPII coated transport vesicles. These ancillary proteins, also called packaging chaperones, exert highly specific effects and function in a manner that only contributes to the packaging of a very limited set of related transport proteins - their cognate substrates. In cells lacking a particular packaging chaperone only its cognate cargo accumulates within the ER. Here we review the current state of understanding of how packaging chaperones facilitate the specific selection of metabolite transporters as cargo during the formation of COPII vesicles.

1 Introduction

The plasma membrane (PM) is a dynamic structure comprised of lipids and proteins that establishes and maintains the integrity of the cell. Newly synthesized secreted and integral PM proteins are inserted into the endoplasmic reticulum (ER) and subsequently targeted to the PM via the secretory pathway. The secretory pathway, first defined in mammalian cells (Palade 1975), consists of a number of discrete membrane-bounded compartments that are connected by a network of vesicular carriers. In addition to enabling cells to secrete soluble proteins, this pathway is required for the assembly, post-translational modification, and delivery of newly synthesized membrane proteins and lipids to several distinct intracellular destinations. One of the established tenets of cell biology is that sequence motifs present within proteins function as entry tickets into the secretory pathway, and as labels for proper progression through the secretory pathway (Blobel and Dobber-

Topics in Current Genetics, Vol. 9
E. Boles, R. Krämer (Eds.): Molecular Mechanisms Controlling Transmembrane Transport
DOI 10.1007/b96974 / Published online: 9 March 2004
© Springer-Verlag Berlin Heidelberg 2004

stein 1975 a, b; Blobel 1980). During the last 25 years, the dynamic and successful integration of genetic and biochemical approaches have yielded substantial insights into how the various components within the secretory pathway recognize and interpret these targeting sequences (Rothman 2002; Schekman 2002). One of the fundamental conclusions that can be drawn from these studies is that the cellular processes ensuring proper intracellular protein trafficking are highly conserved throughout the eukaryotic world (Mellman and Warren 2000).

Due to their distinct protein and lipid compositions, each secretory compartment has unique structural features and biochemical profiles. A central question critical to understanding intracellular protein trafficking is how the proteins that transit the secretory pathway, the cargo, are sorted and separated away from the resident proteins present in each distinct compartment. This question has to be placed in the context of the dynamic processes underlying the continual anterograde and retrograde vesicle mediated membrane traffic occurring between each of the secretory compartments. Clearly, the processes underlying the sorting of proteins lie at the heart of organelle identity, and the quest to elucidate the molecular mechanisms that enable proteins to be identified as cargo represents one of the most active areas of study in the field of cell biology.

An early paradigm of cargo sorting was established in seminal studies examining the clathrin mediated endocytosis of the LDL receptor (Davis et al. 1986). It was found that the LDL receptor has a short cytoplasmic sequence required for the receptor to localize to clathrin coated pits. Subsequent analysis revealed that this sorting sequence binds directly to clathrin adaptor complexes (AP2) (Hirst and Robinson 1998). These findings demonstrated that cargo recruitment into transport vesicles can be driven by direct interactions between a sorting signal and a coat forming complex. Additional sorting signals have been found that are important for transport of cargo through distinct stages of the secretory pathway, many of these sequence motifs interact directly with vesicle coat protein complexes. Examples include signals promoting anterograde transport between the ER and Golgi (Barlowe 2003), retrograde transport between the Golgi and the ER (Letourneur et al. 1994), and transport from the Golgi to vacuole (Darsow et al. 1998). These findings clearly indicate that the processes of cargo selection and transport vesicle formation are linked.

In the yeast *Saccharomyces cerevisiae*, several metabolite transport proteins require the assistance of ancillary proteins to become incorporated into ER-derived transport vesicles. Each of these ancillary proteins exert highly specific effects and appear to function in a manner that only contributes to the packaging of a very limited set of related transport proteins - their cognate substrates. The metabolite transporters that depend upon these ancillary proteins, termed packaging chaperones, include the members of the amino acid permease (AAP) protein family, the phosphate transporter (Pho84p) and the sugar transporters Hxt1p and Gal2p (Ljungdahl et al. 1992; Sherwood and Carlson 1999; Lau et al. 2000). Advances in our understanding of how these transport proteins are selected as cargo are illuminating novel aspects of the molecular mechanisms associated with the formation of ER-derived transport vesicles. This review provides a status report regarding the role of packaging chaperones in the cargo selection process. To understand the function of these specialized ancillary proteins, and to put their role in context with core secretory components, it is first necessary to briefly review the general secretory pathway and vesicular mediated transport.

2 The secretory pathway

The yeast secretory pathway is schematically presented in Figure 1 (for a comprehensive review, see Kaiser et al. 1997). Secreted and membrane proteins containing appropriate signal sequences are targeted to enter the secretory pathway, and are translocated across or inserted into the ER membrane, respectively (Fig. 1, Step 1). Targeting occurs by either of two parallel pathways, the co-translational signal recognition particle (SRP) or post-translational Sec63 complex mediated pathways (Ng et al. 1996). Both targeting pathways deliver proteins to the ER membrane translocon that is composed of oligomers of the Sec61 complex (Meacock et al. 2000). Once inserted, membrane proteins fold and attain native conformations, a process that often requires specific processing events (Step 2). ER quality control mechanisms facilitate proper processing and function to ensure that only correctly assembled proteins enter the subsequent stages of the secretory pathway and exit the ER (Ellgaard et al. 1999).

Properly processed cargo proteins are incorporated into small transport vesicles (Step 3). The formation of transport vesicles originating from the ER and the Golgi requires the transient assembly of protein coats on the cytoplasmic face of donor membranes; two distinct coat complexes COPI (Waters et al. 1991) and COPII (Barlowe et al. 1994) have been identified. Anterograde transport out of the ER is obligatorily coupled with the formation of COPII-coated vesicles (Antonny and Schekman 2001). Retrograde transport from the Golgi to the ER is mediated by COPI-coated vesicles (Cosson and Letourneur 1997; Gaynor and Emr 1997).

The Golgi is comprised of a stack of membrane cisternae, each with a defined enzymatic profile. A high resolution tomographic analysis has recently shown that there are no connections between the Golgi cisternae in the yeast *Pichia pastoris* (Mogelsvang et al. 2003); this is likely to be the case in most other eukaryotes as well. The anterograde movement of proteins through the Golgi (Step 4) has been the subject of rigorous debate; at issue is whether COPI-coated vesicles significantly contribute to the forward transport of proteins (Storrie and Nilsson 2002). Recent studies provide strong support that COPI mediated transport within the Golgi is restricted to the retrograde transport of resident Golgi proteins (Pelham 2001). Large protein aggregates, too large to be packaged into transport vesicles, translocate through the Golgi stack without leaving the lumen in a process termed cisternal progression, i.e. the sequential maturation of early-Golgi into late-Golgi cisternae. The maintenance of cisternal identity, is dependent upon the balanced retrograde flow of resident Golgi proteins that is mediated by COPI coated vesicles. In two independent and sophisticated microscopic analyses, it could be determined that even small cargo proteins are transported through the Golgi without

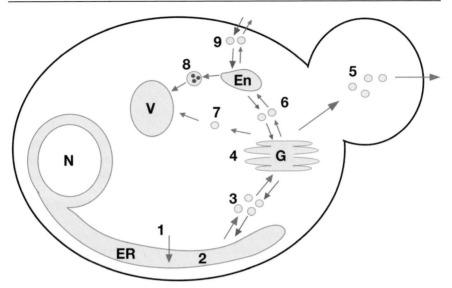

Fig. 1. The yeast secretory pathway. Newly synthesized proteins and lipids move in an an-terograde direction (green arrows) through the secretory pathway that initiates in the endo-plasmic reticulum (ER). The ER membrane is contiguous with the membrane surrounding the nucleus (N). The Golgi complex (G) is depicted as a stack of membrane-bounded com-partments, each with a defined profile of enzymatic activity. In *S. cerevisiae* these stacks are rarely observed, none-the-less, processes occurring in (cis) early- and (trans) late-Golgi are separated within discrete vesicular structures. In the late-Golgi, soluble and membrane proteins destined for the vacuole (V) are separated from secreted and plasma membrane proteins. Vacuolar proteins exit the Golgi and are either directly targeted to the vacuole, or enter the multivesicular body (MVB-PVC) pathway that passages proteins through the en-dosomal compartment (En). The endocytic and major retrograde routes of transport are in-dicated with red arrows. Small transport vesicles mediate movement between the individual membrane-bounded compartments. Details regarding the major steps, numbered 1-9, are provided in the text.

leaving the lumen and without entering COPI vesicles (Martínez-Menargue et al. 2001; Mironov et al. 2001). In contrast, resident Golgi proteins presumably mov-ing in the retrograde direction were found at significant levels in COPI-coated vesicles. These findings, entirely consistent with cisternal progression, indicate that a common mechanism independent of forward vesicular traffic is responsible for the translocation of small and large secretory cargo through the Golgi.

At least three anterograde pathways (Steps 5-7) facilitate the exit of proteins out of the late Golgi (Harsay and Bretscher 1995; Gurunathan et al. 2002). In a rather ill characterized pathway, low-density vesicles form at the Golgi and subse-quently target to the bud, the major site of exocytosis in yeast (Step 5). Although coat-forming proteins most certainly function to facilitate the formation of these low-density vesicles, their identity has not been clearly established. In a second pathway, clathrin coated vesicles bud from the Golgi. These vesicles direct trans-port to the endosomal membrane system (Step 6) (Gurunathan et al. 2002). This step represents the initial stages of the well-studied PVC-MVB (prevacuolar compartment – multivesicular body) pathway (Bryant and Stevens 1998). In the third route, clathrin-like proteins facilitate the formation of transport vesicles that target the vacuole in the ALP (alkaline phosphate) pathway (Step 7) (Cowles et al. 1997; Nothwehr et al. 2000; Darsow et al. 2001). The latter stages of the PVC-MVB pathway (Step 8) results in formation of endosomal structures that typically have a multivesicular appearance (Katzmann et al. 2002). Endocytic vesicle form at the PM membrane (Step 9), these vesicles primarily target to the endosomal compartment. For a more detailed discussion regarding post-Golgi and endocytic transport, the reader is referred to two additional reviews in this volume (see contributions by M. Prado, B. André, and R. Haguenauer-Tsapis, respectively). Also, the important role of lipids in ensuring the proper progression of proteins through the secretory pathway and their enzymatic activity is addressed in this volume in a review by M. Opekarova.

3 Vesicle-mediated transport

The guiding principles of vesicle-mediated transport, schematically depicted in Figure 2, can be summarized as follows. Vesicle budding requires the ordered recruitment of soluble protein coat complexes to the cytoplasmic face of donor membranes, as previously described. These coats can be defined both by their unique protein composition and the membranes at which they function (Kirchhausen 2000). The process of oligomerization and assembly of coat complexes is thought to provide the necessary force to deform the plane of the membrane and the formation of a vesicle bud. It is important to stress that in each case, cargo proteins destined for transport are separated from resident proteins concomitantly with the formation of vesicles.

The assembly of vesicle coats is initiated by priming complexes that form when small G-proteins are recruited to donor membranes (Springer et al. 1999). These G-proteins are soluble when complexed with GDP. They interact with membrane bound guanine nucleotide exchange factors (GEFs) that exchange GDP to GTP, an event that induces a conformational change, or switch, that exposes membrane-anchoring motifs (Goldberg 1998; Bi et al. 2002). Consequently, the GTP bound forms of these G-proteins are able to directly bind membranes. Their membrane association facilitates the recruitment of inner coat proteins via direct protein-protein interactions resulting in the formation of pre-budding complexes. The inner coat proteins possess cargo-binding sites that bind cytoplasmically oriented sorting signals present within cargo proteins. Pre-budding complexes with bound cargo serve as nucleation sites for the subsequent recruitment and oligomerization of outer coat proteins. Successive rounds of coat protein recruitment lead to the formation of a transport vesicle. The intrinsic GTPase activity of the G-proteins, eventually leads to the hydrolysis of GTP, an event that triggers the disassembly of vesicle coats.

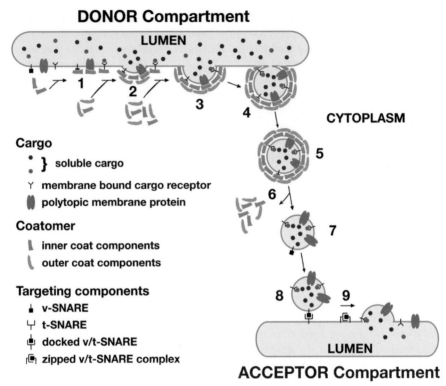

Fig. 2. General scheme of vesicle mediated transport between the membrane-bounded compartments of the secretory pathway. Proteins destined to exit a donor compartment are selected as cargo via interactions with specific sets of inner coat proteins (blue) resulting in the formation of pre-budding complexes (stage 1). Membrane protein cargo including polytopic proteins (green), v-SNAREs, and cargo receptor proteins (**Y**) are selected via direct interactions with inner coat proteins. Soluble proteins (red) can be selected for packaging by binding to cargo receptor proteins. Prebudding complexes function as nucleation sites (stage 2) for the recruitment of additional coatomer. The assembly of outer coat proteins (yellow) is thought to provide the actual force to deform the plane of the membrane and to create a vesicle bud (stage 3). A mature vesicle forms after successive rounds of coat protein assembly, a process that culminates in a membrane fusion event (stage 4) that releases a coated vesicle (stage 5). Some proteins (dark blue) are incorporated during the process of vesicle formation without the apparent involvement of other proteins by mechanisms commonly referred to as bulk flow. Disassembly of the vesicle coat (stage 6) is prerequisite for uncovering and exposing v-SNAREs that are needed for correct targeting (stage 7) and docking (stage 8) with the membrane of an appropriate acceptor compartment carrying cognate t-SNAREs. Cargo proteins within correctly docked vesicles are released into the acceptor compartment subsequent to membrane fusion catalyzed by appropriately paired v-SNARE/t-SNARE complexes (stage 9).

Coat disassembly is required to expose SNARE proteins, receptor-like entities that must be located on both the vesicle (v-SNARE) and on the target membrane (t-SNARE complexes). These proteins are thought to provide the molecular basis for both vesicle targeting and fusion to acceptor membranes (Söllner et al. 1993; McNew et al. 2000). Directionality is conferred by combinatorial mechanisms involving specific interactions between cognate v-SNARE and t-SNARE pairs. Small GTP binding proteins (Rab Family GTPases) function at every transport vesicle docking and fusion step (Novick and Zerial 1997). One of the more critical functions of Rab proteins is to ensure that only interactions between appropriate combinations of v-SNAREs and t-SNAREs lead to productive membrane fusion events (Chen and Scheller 2001).

Finally, bidirectional flow must occur to balance the distribution of membrane associated components of transport machinery. Retrograde transport has been found to occur at every stage of the secretory pathway and serves to recycle components needed for subsequent rounds of anterograde vesicle formation, and to retrieve proteins that are inappropriately transported (Pelham 1991; Spang and Schekman 1998). Importantly, the sorting mechanisms occurring during retrograde transport ensures that secretory cargo proceed in a forward manner, whereas resident proteins are prevented from making significant net advances.

4 Formation of COPII-coated vesicles

Transport from the ER to the Golgi requires the formation of COPII coated transport vesicles (Fig. 3A). The formation of COPII vesicles has been studied in great detail, and it has been shown that the COPII coat is comprised of five proteins, Sar1p, Sec23p, Sec24p, Sec13p, and Sec31p (Barlowe et al. 1994). Sar1p is a small G-protein that in its GTP-bound form binds directly to ER membranes; Sec23p and Sec24p form a heterodimeric complex; and Sec13p and Sec31p form a heterotetrameric complex comprised of two molecules of each subunit (Lederkremer et al. 2001). Reconstitution experiments have demonstrated that this set of COPII proteins is sufficient to drive vesicle budding from synthetic liposomes in the absence of any membrane proteins (Matsuoka et al. 1998), and can self assemble in solution to form spherical structures that are 60-120 nm in diameter (Antonny et al. 2003). In these latter minimalistic experiments, the assembly and formation of COPII coated vesicles requires the use of non-hydrolysable forms of GTP, demonstrating that additional factors regulate the COPII assembly-disassembly cycle during export from ER membranes.

The COPII coatomer components function in concert in three stages. First, the soluble G-protein Sar1p associates with the ER membrane to initiate the process of vesicle formation. Sec12p (Nakano et al. 1988) has been shown to function as a GTP exchange factor for Sar1p (Oka et al. 1991; Barlowe and Schekman 1993). The GTP-bound form of Sar1p has an exposed N-terminal membrane anchor sequence, the first ten residues not exposed in the GDP-bound state, that facilitates its binding to the ER membrane (Bi et al. 2002). Second, the conformational

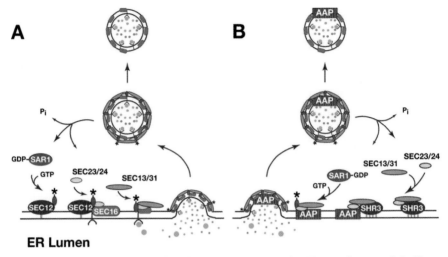

ER Lumen

Fig. 3. Multiple pathways recruit COPII components to the ER membrane and facilitate coatomer assembly. Panel A depicts the best studied pathway leading to the formation of ER transport vesicles. Panel B depicts the pathway facilitated by the ancillary protein Shr3p. See text for details.

change induced by GTP binding enhances its affinity for the Sec23/Sec24p heterodimer (inner coatomer) (Bi et al. 2002). Binding of Sec23/Sec24p leads to the formation of a pre-budding complex that incorporates SNAREs and cargo. Third, the Sec13/Sec31p heterotetramer (outer coatomer) binds and incorporates prebudding complexes by bridging them together, an interaction that nucleates multiple rounds of coatomer recruitment. Ultimately the coatomer proteins polymerize to form a fully coated COPII vesicle. EM studies examining COPII coated vesicles have provided evidence of a layered coat structure with an inner shell of Sec23/Sec24p and an outer shell of Sec13/Sec31p (Matsuoka et al. 2001). Presumably due to restrictions regarding the possible coatomer subunit-subunit interactions, the resulting vesicles are remarkably uniform in size, and are roughly 60 nm in diameter (Lederkremer et al. 2001).

GTP is hydrolyzed by the intrinsic GTPase activity of Sar1p, an event that destabilizes the vesicle coat (Barlowe et al. 1994). Consistently, inhibition of GTP hydrolysis by Sar1p results in the accumulation of COPII-coated vesicles (Barlowe et al. 1994; Oka and Nakano 1994; Saito et al. 1998). The GTP hydrolyzing activity of Sar1p is enhanced by the GTPase activating protein (GAP) activity of Sec23p (Yoshihisa et al. 1993). The relatively low GAP activity of Sec23p is significantly enhanced in the presence of Sec13/Sec31p (Antonny et al. 2001). This latter finding suggests that GTPase hydrolysis occurs only after prebudding complexes become linked together. The fact that a component of the vesicle coat functions as a GAP and that coatomer assembly stimulates this activity plainly demonstrates the dynamic nature of vesicle formation. Clearly, the nucleotide exchange and GTP hydrolysis reactions must be regulated within a framework of spatial and temporal constraints. Consistent with this notion, other factors have been implicated in regulating the COPII assembly-disassembly cycle during export from ER membranes.

Sec16p is an essential and large peripherally associated ER membrane protein to which both the Sec23p/24p and Sec13p/31p complexes bind, and that consequently gets incorporated into transport vesicles (Gimeno et al. 1995). Based on these results, it has been proposed that Sec16p functions to help organize coatomer assembly onto the membrane (Espenshade et al. 1995; Gimeno et al. 1996; Shaywitz et al. 1997). The presence of Sec16p in the natural membranes used *in vitro* budding assays allows vesicle formation to occur in the presence of GTP as the sole nucleotide. It is, thus, possible that Sec16p negatively affects the GTPase activity of Sar1p, either directly or by altering the GAP activity of Sec23p. The integral ER protein Sed4p, a close homolog of Sec12p, has also been shown to inhibit the Sar1p GTPase (Saito-Nakano and Nakano 2000).

The mechanisms governing ER-vesicle budding appear to be conserved. Homologs of all COPII components have been reported in mammals. There are two identified isoforms of both mammalian Sar1p (Kuge et al. 1994) and Sec23p (Paccaud et al. 1996). The cloning of four isoforms of mammalian Sec24 has been reported (Pagano et al. 1999; Tang et al. 1999; Tani et al. 1999). A single form of mammalian Sec13 has been found (Shaywitz et al. 1995), and the existence of two mammalian proteins that are homologous to yeast Sec31p has been reported (Tang et al. 2000). Several of these mammalian proteins function interchangeably when heterologously expressed in yeast, a fact that underscores the high degree of conservation.

5 Cargo selection – sorting motifs bind Sec24p

A quantum leap in understanding the process of COPII vesicle formation was recently made possible when the structure of the Sar1p/Sec23/Sec24p pre-budding complex was deduced by crystallographic analysis (Bi et al. 2002). In a stunningly direct manner, this structural approach was pushed to an even greater level of sophistication when the binding of ER-to-Golgi SNAREs to this pre-budding complex was analyzed (Mossessova et al. 2003). The SNAREs investigated included Sed5p, Bos1p, and Sec22p, which together form a t-SNARE, and Bet1p the cognate v-SNARE. With the exception of Bos1p, all of these SNARE proteins were found to bind directly to Sec24p. The precise peptide sequences in Bet1p and Sed5p that mediated the binding were successfully mapped (see Table 1) and the corresponding peptides were co-crystallized with Sec24p. This analysis revealed that these peptides bind to Sec24p at two distinct sites, designated A and B. The A site was defined by its binding the hydrophobic motif of Sed5p. The crystal structure suggested that this site relies on shape-based recognition with hydrophobic interactions playing an important role in peptide binding. Site B was defined by the binding of acidic motifs present in peptide sequences within Bet1p, Sed5p, and Sys1p. Sys1p had earlier been found to bind Sec24p (Votsmeier and Gallwitz 2001). Analysis of site B crystal structures suggested that peptide binding relies on

Table 1. ER exit sorting motifs

Motif	Protein	Reference
Di-acidic		
NT localized		
YSQSTLASLESQ*	Bet1p	(Mossessova et al. 2003)
QLMLMEEGQ	Sed5p	(Mossessova et al. 2003)
SDDIEK	Yor5p	(Epping and Moye-Rowley 2002)
CT localized		
AEKMDIDTGR	Gap1p	(Malkus et al. 2002)
NSFCYENEVAL	Kir2.0, Kir2.1	(Stockklausner et al. 2001) (Ma et al. 2001)
QLKDLESQI	Sys1p	(Votsmeier and Gallwitz 2001)
IYTDIEMNRLGK	VSV-G	(Nishimura and Balch 1997; Sevier et al. 2000)
NDFENRS	Yor5p	(Epping and Moye-Rowley 2002)
Hydrophobic		
NT localized		
YNNSNPF	Sed5p	(Mossessova et al. 2003)
CT localized		
YYMFRINQDIKKVKLL	Emp47p	(Sato and Nakano 2002)
RRFFEVTSLV	Emp24p	(Nakamura et al. 1998)
YIMYRSQQEAAAKKFF	ERGIC53	(Kappeler et al. 1997)
YQPDDKTKGILDR	Erv41p	(Otte and Barlowe 2002)
KLFYKAQRSIWGKKSQ	Erv46p	(Otte and Barlowe 2002)
YLRRFFKAKKLIE	p24δ1	(Fiedler et al. 1996; Dominguez et al. 1998)
YWRQEYPGVDEFF	Prm8p	(Miller et al. 2003)

*mutations that alter the underlined residues reduce ER export

side-chain based chemistry recognition. Importantly, the B site accommodated the different acidic motifs using overlapping but distinct pattern of contacts. This suggests that there is flexibility in the substrate binding, an attribute that may be important in binding other related motifs (Table 1).

In a parallel study, the crystal structure of Sec24p was used to plan a directed alanine scanning approach that generated mutations affecting equatorial regions of Sec24p, i.e. surfaces not membrane proximal or distal (Miller et al. 2003). These mutations were initially screened for their inability to complement a null allele of *sec24*. One mutant allele was found that altered two amino acid residues within the site B binding pocket. This prompted further rounds of mutagenesis that individually modified other site B residues known to provide contacts for peptide binding (Mossessova et al. 2003). When these mutant versions of Sec24p were used in budding assays, Bet1p and several other membrane protein cargo were not efficiently incorporated into vesicles. Importantly, the ability of the mutant forms of Sec24p to form vesicles was not impaired, nor did they diminish the efficiency of packaging soluble secretory pro-α-factor, indicating that the mutations were not inducing dramatic alterations in the overall structure of Sec24p.

The ability of these individual site B mutants to complement *sec24* null alleles was tested, and the degree to which they complemented was found to be quite variable; full (wildtype), partial (temperature sensitive), and no complementation (inviability) was observed. In most instances the ability to partially or fully complement *sec24* null alleles was dependent upon the presence of either of two non-essential genes, *ISS1* and *LST1*, which encode Sec24-related proteins. This finding encouraged Miller et al. (2003) to survey additional alanine scanning mutations and identified a novel mutation that exhibited synthetic lethality with an *iss1* null allele. The mutation giving rise to the synthetic lethality mapped to a new site on Sec24p (Arg342). In budding assays, this Sec24p mutant specifically failed to incorporate Sec22p, whereas, the other cargo proteins monitored were packaged at normal levels. Together these results are consistent with Sec24p having at least three distinct cargo binding sites, and with previous findings that cargo-containing pre-budding complexes can be isolated from ER membrane extracts after incubation in the presence of Sar1p and Sec23/Sec24p (Kuehn and Schekman 1997; Aridor et al. 1998; Kuehn et al. 1998). As will be discussed in a following section, the existence of Sec24p homologs (Lst1p and Iss1p) may provide additional means to increase the repertoire of binding sites available for direct cargo interactions.

6 Sorting signals

Based on the discussion in the preceding section, it is clear that inner COPII coatomer complexes directly participate in the sorting of membrane cargo. Consistently, sorting signals facilitating ER exit have been defined in a diverse set of cargo proteins (Table 1). Based upon the identity of the residues required to facilitate ER export, these peptide motifs can be coarsely grouped into two classes, acidic and hydrophobic. Although in most instances it remains to be proven, these sorting motifs are likely to bind cargo-binding sites on Sec24p (Miller et al. 2003; Mossessova et al. 2003). Sorting motifs have been found in both cytoplasmically oriented N-terminal and C-terminal domains of cargo proteins. Sed5p has two motifs - one of each of the two classes (Mossessova et al. 2003). The hydrophobic motif within Sed5p exhibits higher affinity binding than the di-acidic motif, but both motifs have been shown to participate in binding Sec24p. Two motifs have also been defined in Yor5p (Epping and Moye-Rowley 2002). These motifs may function synergistically. Similarly, sorting motifs may be comprised of sequences present on more than one subunit of multimeric protein complexes (Otte and Barlowe 2002; Sato and Nakano 2002).

In certain instances, it has been shown that sorting sequences can be transferred, and accelerate the rate of ER export when fused to neutral cargo. For example, the amino acid permease Gap1p has a di-acidic sorting motif within its C-terminal domain (Malkus et al. 2002). Truncated versions of Gap1p that lack this motif are inefficiently exported from the ER. Appending the di-acidic sorting motif of Sys1p to C-terminal truncated Gap1p restores ER export, indicating that the Sys1p sorting signal can replace the intrinsic di-acidic motif of Gap1p (Miller et al. 2003). However, in the inverse situation, appending the Gap1p sorting sequence to C-terminally truncated Sys1p did not facilitate the resulting fusion protein to exit the ER (Miller et al. 2003). These finding indicate that sorting motifs may be interpreted in a context dependent manner.

It should be noted that the binding of cargo to Sec24p may be dependent upon the participation of the other subunits of the pre-budding complex, as well as higher order interactions that occur as coatomer complexes coalesce to form a polymerized coat structure. For example, Sed5p binds to Sec24p independently of both Sec23p and Sar1p (Peng et al. 1999), but Bet1p has been shown to bind to Sec24p in a Sar1p-dependent manner (Springer and Schekman 1998). Finally, the interactions between sorting motifs in cargo and prebudding complexes are transient in nature, COPII disassembly occurs efficiently after a vesicle buds from the membrane. Thus, the avidity of cargo-coatomer binding is likely to be determined by the GDP/GTP switching property of Sar1p. The GTPase activity of Sar1p is influenced by coatomer-coatomer interactions, between both prebudding complexes (inner-inner) and prebudding complexes and Sec13/Sec31p (inner-outer) (Antonny et al. 2001).

7 Sec24p homologs – cargo selection by combinatorial mechanisms

Iss1p and Lst1p are homologous to Sec24p sharing 56% and 23% identity, respectively. In similarity to Sec24p, both Lst1p and Iss1p are able to form heterodimers with Sec23p (Roberg et al. 1999; Kurihara et al. 2000; Shimoni et al. 2000). *iss1* null mutants are viable, and do not exhibit growth or cargo specific export defects. The overexpression of *ISS1* is able to suppress the temperature sensitive growth phenotype of *sec24* mutants grown under restrictive conditions (Peng et al. 2000). Additionally, temperature sensitive *sec24* mutations are inviable when introduced into *iss1Δ* null mutant strains (Miller et al. 2003). Together these results indicate that Iss1p functions similarly to Sec24p. Consistently, Iss1p can substitute for Sec24p in *in vitro* budding assays (Kurihara et al. 2000).

Mutations in *LST1* were identified based on being synthetically lethal with the *sec13-1* allele (Roberg et al. 1999). Null mutations of *LST1* specifically affect the export of the plasma membrane ATPase (Pma1p) from the ER; Gas1p, invertase, pro-α-factor, and CPY exit the ER as in wildtype cells. The overexpression of *LST1* does not rescue the loss of Sec24p function; however, conversely, the overexpression of *SEC24* partially suppresses the Pma1p secretion defect of *lst1* null mutants. These results suggest that a specialized form of the vesicle coat is responsible for incorporating Pma1p into COPII vesicles. Consistent with this model, vesicles formed *in vitro* using Sec23/Lst1p were found to contain only a limited set of cargo proteins, distinct from that incorporated in the presence of Sec23/Sec24p. Specifically, the SNAREs Bos1p, Bet1p, and Sec22p were not incorporated into vesicles when Sec23/Lst1p was used. Consequently, vesicles generated with Sec23/Lst1p are unable to fuse with acceptor Golgi membranes (Miller et al. 2002, 2003).

In vitro budding assays have shown that Pma1p packaging is optimal when a mixture of Sec23/Sec24p and Sec23/Lst1p complexes is used. In these experiments Pma1p could be co-immunoprecipitated with either anti-Lst1p or anti-Sec24p antibodies. The vesicles formed were found to be competent to fuse with the Golgi, indicating that both these inner coatomer complexes were polymerized into the same vesicle coat. The Sec23/Lst1p and Sec23/Sec24p COPII vesicles were found to be morphologically similar to Sec23/Sec24p vesicles, however, they were slightly larger, suggesting the possibility that incorporation of Lst1p may facilitate the entry of bulky cargo (Shimoni et al. 2000).

8 Ancillary proteins – packaging chaperones

The findings regarding sorting signals and their capacity to bind to pre-budding complexes (inner coatomer) provide convincing evidence that cargo loading is directly coupled to vesicle formation. However, there are proteins that possess established sorting signals that are not recognized as cargo without the assistance of ancillary proteins. Depending on how they mediate assistance, these ancillary proteins can be divided into three categories: outfitters, escorts, and guides (Herrmann et al. 1999). Outfitters and escorts function similarly, although outfitters, in contrast to escorts, do not accompany their cognate cargo out of the ER. These proteins are thought to function by enabling sorting signals present within cargo to be presented in a context that is recognized by the coat components, e.g. Sec24p. Guides, the third class of accessory proteins, function as cargo receptors that bind cargo that do not themselves possess signals recognized by coat components.

Several metabolite transporters have been found to require ancillary proteins belonging to the outfitter class of accessory proteins to be included into ER-derived COPII transport vesicles. When these accessory proteins are mutated or deleted, their cognate cargo do not enter into COPII transport vesicles and specifically accumulate within the ER. These proteins, termed packaging chaperones, are integral components of the ER membrane that share no significant sequence or structural homology with one another (Fig. 4). The characterized packaging chaperones function as outfitters in that they do not exit the ER at appreciable rates (Kuehn et al. 1996; Lau et al. 2000). This is in contrast to Erv14p that guides Axl2p out of the ER (Powers and Barlowe 1998), and the receptor-like proteins

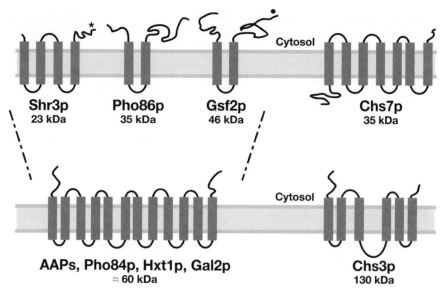

Fig. 4. Schematic illustration of the membrane topology of ER packaging chaperones (upper panel) and their cognate cargo (lower panel). The topologies of Shr3p (J. Kota and P.O. Ljungdahl unpublished results) and Gsf2p (T. Hamacher and E. Boles, personal communication) have been experimentally determined by the method of Gilstring and Ljungdahl (2003). The topology of Pho86p and Chs7p are based upon hydropathy analysis. The asterisk (*) and the solid circle (•) indicate the positions of a putative ER retention signal and retrieval signal, respectively. The topology of the amino acid permease Gap1p has been experimentally determined (Gilstring and Ljungdahl 2000). The other metabolite transporters Hxt1p, Gal2p and Pho84p are predicted to have 12 membrane spanning segments, and Chs3p is predicted to have 6 transmembrane domains. Note that the proteins are not drawn to scale, although Chs3p (~130 kDa) has fewer membrane spanning segments it has twice the molecular mass compared to the metabolite transporters (~60 kDa).

ERGIC53 (Appenzeller et al. 1999; Hauri et al. 2000) and the p24 family (Muniz et al. 2000) that co-transport along with cargo between the ER and Golgi.

9 Shr3p - the original packaging chaperone

The first indication that ancillary proteins facilitate ER to Golgi transport was obtained in studies examining amino acid uptake in yeast. A family of eighteen structurally related permeases mediate amino acid uptake into yeast. The AAPs are comprised of 12 membrane-spanning segments, with the N- and C-terminal domains oriented towards the cytoplasm (Gilstring and Ljungdahl 2000). The individual members of the AAP family exhibit unique substrate affinities. However, many permeases exhibit extensive functional overlap and can transport several different amino acids (Iraqui et al. 1999; Regenberg et al. 1999; Andréasson et al.

2004). Over the course of the last forty years, single mutations that block the up-take of a diverse spectrum of amino acids into yeast have been isolated (Sorsoli et al. 1964; Surdin et al. 1965; Grenson and Hennaut 1971; Lasko and Brandriss 1981; McCusker and Haber 1990; Ljungdahl et al. 1992). The existence of viable strains exhibiting the pleiotropic loss of multiple amino acid transport proteins, each encoded by distinct genes that are differentially regulated, could not readily be understood. However, their existence indicated that AAPs share a common re-quirement for function. Clarity regarding how these pleiotropic mutations exert their affects was achieved when one of the corresponding genes, *SHR3*, was cloned (Ljungdahl et al. 1992).

Shr3p is an integral membrane component of the ER with four membrane-spanning segments and a hydrophilic, cytoplasmically oriented carboxy-terminal domain (Fig. 4). Mutations in *SHR3* specifically impede the transport of the mem-bers of the AAP family from the ER (Ljungdahl et al. 1992; Horák and Kotyk 1993; Kuehn et al. 1996; Gilstring et al. 1999), including Ssy1p, the primary re-ceptor of the SPS sensor of extracellular amino acids (Klasson et al. 1999)(see contribution in this volume by B. André and E. Boles). In detailed studies using Gap1p as a model, it was found that each of the 12 membrane spanning segments were integrated into the ER membrane in the correct orientation independently of Shr3p (Gilstring et al. 1999; Gilstring and Ljungdahl 2000). In contrast to muta-tions that result in Gap1p misfolding, the AAPs that accumulate in the ER mem-brane of *shr3* mutants do not activate the ER stress response pathway (Gilstring et al. 1999). These results are consistent with Shr3p acting after AAPs are correctly integrated and folded in the ER membrane.

In vitro budding assays using purified COPII components and yeast membrane preparations showed that Shr3p is required for packaging of AAPs into vesicles (Kuehn et al. 1996). In these experiments, 35-45% of the total Gap1p and the his-tidine AAP (Hip1p) present in *SHR3* wildtype membranes were incorporated into vesicles. In contrast only 3% of Gap1p and Hip1p were incorporated when mem-brane preparations from *shr3* null mutants were used. Other cargo (i.e. pro-α-factor, Pma1p, CPY, invertase) was packaged normally independently of Shr3p function (Kuehn et al. 1996). Importantly, these *in vitro* experiments clearly showed that Shr3p was not itself incorporated into COPII vesicles.

The ability of prebudding complexes to bind AAPs was directly tested using a functional Sar1p-GST fusion protein (Kuehn et al. 1998). Sar1p-Sec23/Sec24p prebudding complexes were purified after incubation with detergent solubilized membranes isolated from either *SHR3* or *shr3* null mutant cells. The purifications were performed in the presence of the non-hydrolysable GTP analogue GMP-PNP to stabilize the binding of Sar1p to Sec23p. Both Gap1p and Hip1p co-purified with the prebudding complexes in a Shr3p dependent manner, no prebudding complexes containing AAPs could be recovered when solubilized membranes from *shr3* mutants were used. Consistent with its lack of being incorporated into vesicles, Shr3p was found to be excluded from these pre-budding complexes. When the Sec13/Sec31p complex was added, prebudding complexes were not re-covered, instead the AAPs were found in the vesicle-containing supernatant, suggesting that the Sar1p-GST/Sec23/Sec24p/AAP complexes were consumed during vesicle formation.

In a similar approach, a functional GST-Shr3p fusion protein was expressed in yeast (Gilstring et al. 1999). The fusion protein was purified from membrane preparations solubilized with N-dodecylmaltoside, a mild nonionic detergent. It was found that Gap1p, but not other polytopic membrane proteins, such as Sec61p, Gal2p, or Pma1p, co-purified with Shr3p. Consistent with its role as a packaging chaperone, the association of Shr3p with Gap1p was shown to be transient, a reflection of the transport of Gap1p from the ER. Further analysis indicated that the COPII coatomer components Sec13p, Sec23p, Sec24p, and Sec31p, but not Sar1p, bind Shr3p via interactions with its hydrophilic carboxy-terminal domain. Thus, the possibility exists that Sar1p is unable to bind Sec23p in the presence of Shr3p. The dissociation of Shr3p away from the AAP-coatomer complex could be a prerequisite for Sar1p binding, and the conversion of the pre-budding complex into a *bona fide* vesicle bud site. As Shr3p does itself not exit the ER, Shr3p must dissociate and diffuse away prior to completion of vesicle formation. The precise mechanisms governing the dissociation of Shr3p remain to be elucidated. In summary, the available data suggest that Shr3p functions as a packaging chaperone that directs the formation of vesicle buds in the vicinity of AAPs and, thereby, ensures their inclusion into transport vesicles (Fig. 3B).

10 Additional packaging chaperones – Pho86p, Gsf2p, and Chs7p

In media containing low levels of phosphate, the major phosphate transporter Pho84p is localized to the plasma membrane (Lau et al. 2000). Mutations in *PHO86* block the ER exit of Pho84p and consequently exhibit similar phenotypes as mutations in *PHO84*, including a defect in phosphate uptake and reduced expression of acid phosphatase (Pho5p). Pho86p is predicted to have two membrane spanning segments and a hydrophilic C-terminal domain that is oriented towards the cytoplasm (Fig. 4). *In vitro* budding assays have demonstrated that similar to Shr3p, Pho86p is not transported out of the ER. Gal2p, which shares substantial homology with Pho84p (22% identity and 37% similarity), and Pma1p are correctly targeted to the plasma membrane in the absence of Pho86p. These findings suggest that the function of Pho86p is specific for Pho84p. Interestingly, both *PHO84* and *PHO86* are among the most highly induced genes when cells are starved for phosphate indicating that sorting of Pho84p may be nutritionally regulated (Lau et al. 2000).

Hxt1p is a low-affinity hexose transporter that is active when glucose is abundant. It requires Gsf2p, to exit the ER and to be functionally expressed at the PM (Sherwood and Carlson 1999). Gsf2p contains two membrane spanning domains (T. Hamacher and E. Boles, personal communication) and a hydrophilic cytoplasmically oriented C-terminus with a putative ER retrieval motif (KKXX) (Fig. 4). The galactose transporter Gal2p, which is closely related to Hxt1p (69% identity, 83% similarity), is also dependent on Gsf2p for its proper localization to the PM. In contrast, the high-affinity hexose transporter Hxt2p (62% identical and 78% similar to Hxt1p), the maltose transporter and amino acid permeases exit the ER in the absence of Gsf2p. Thus, Gsf2p appears to specifically promote the secretion of only a subset of hexose transporters.

Chs3p is the catalytic and largest subunit of chitin synthase III (CSIII), a PM localized complex that is responsible for the synthesis of the majority of cellular chitin, including the chitin synthesized during yeast mating and sporulation. The ER resident protein Chs7p, predicted to have seven membrane spanning segments (Fig. 4), is required for CSIII activity (Trilla et al. 1999). In cells lacking Chs7p, Chs3p specifically accumulates in the ER leading to reduced rates of chitin synthesis. Overexpression of *CHS3* alone does not lead to higher levels of CSIII, however, the simultaneous overexpression of *CHS3* and *CHS7* increased the activity of CSIII, indicating that Chs7p is a limiting factor for CSIII activity.

11 Packaging chaperones - presentation of sorting motifs

Relatively little is known regarding the precise mechanisms enabling ancillary proteins to function as packaging chaperones, and it remains to be demonstrated whether all packaging chaperones function similarly. Based upon our current understanding, these ancillary proteins may facilitate the packaging of specific membrane protein cargo by the selective stabilization, or pre-assembly, of COPII components within a complex, minimally consisting of a specific packaging chaperone, membrane protein cargo and the inner coatomer Sec23/Sec24p components. Such primed-cargo complexes closely resemble pre-budding complexes (Kuehn et al. 1998), except that they include packaging chaperones. As previously discussed with respect to Shr3p, the binding of Sar1p may displace the participating packaging chaperone and, thereby, convert the primed-cargo complex into an active pre-budding complex. Active cargo containing pre-budding complexes may function as nucleation sites that recruit soluble coatomer complexes resulting in the polymerization of a vesicle coat. Alternatively, active pre-budding complexes containing bound cargo diffuse laterally within the ER membrane until they encounter other pre-budding complexes, e.g. Sar1p-Sec16p-Sed4p-Sec23/Sec24p complexes (Espenshade et al. 1995; Gimeno et al. 1995, 1996; Bednarek et al. 1996). These encounters presumably facilitate the recruitment of the Sec13/Sec31p outer coatomer (Matsuoka et al. 2001). These two post-priming possibilities are mechanistically distinct. It is presently unclear how additional coatomer complexes are incorporated into polymerizing coats during later stages of vesicle formation.

In the case of amino acid permeases, Gap1p and Hip1p have been isolated in association with Sar1p-Sec23/Sec24p pre-budding complexes (Kuehn et al. 1998). The capacity of Gap1p to bind to these complexes was found to be dependent on Shr3p and the presence of a di-acidic sorting signal in its cytoplasmically oriented C-terminal domain (Table 1) (Malkus et al. 2002; Miller et al. 2003). Similar sorting motifs are found on all AAPs. These results have raised the question - why are the sorting signals present in AAPs not recognized and able to bind Sec24p without the assistance of Shr3p?

The recent studies regarding the structural analysis of Sec24p has enabled this question to be more clearly framed. As previously discussed, appending the di-acidic sorting motif of Sys1p to a C-terminally truncated, and consequently ER retained form of Gap1p restored its ability to exit the ER (Miller et al. 2003). Unfortunately the role of Shr3p in the packaging of this fusion protein was not investigated. The converse experiment, appending the Gap1p di-acidic sorting motif to Sys1p lacking its C-terminal domain, did not facilitate ER exit of the fusion protein. The di-acidic motif of Sys1p has been shown to bind to the B site of Sec24p (Mossessova et al. 2003), and mutations affecting site B impair the binding of Sys1p (Miller et al. 2003). These B site mutations do not affect the incorporation of Gap1p, Hip1p or Can1p into vesicles, strongly suggesting that the di-acidic motifs present on AAPs bind other sites on Sec23/Sec24p (Miller et al. 2003). However, it has not been possible to detect direct interactions between the Gap1 C-terminal tail and Sec23/Sec24p (Miller et al. 2003).

These findings suggest that the C-terminal domain of Gap1p has to be presented in a context that can be recognized by Sec23/Sec24p, and clearly implicate a role for Shr3p in this process. Importantly, mutations that remove the di-acidic sorting motif of Gap1p do not affect its capacity to co-purify with Shr3p (F. Gilstring and P.O. Ljungdahl, unpublished data presented in Malkus et al. 2002). Therefore, it is unlikely that Shr3p recognizes these motifs; rather Shr3p is likely to interact with membrane domains of AAPs. The two well-documented activities of Shr3p, the capacity to interact with AAPs and COPII coatomer, provide ample basis to speculate on possible models as to how the AAP sorting motifs might be presented. First, the availability of sorting motifs within SNAREs for binding to Sec24p was found to depend upon whether they were assembled into complexes; certain sorting motifs were inaccessible, whereas, others became exposed (Mossessova et al. 2003). Similarly, Shr3p may influence the conformation of AAPs in such a manner that exposes their C-terminal domains. This could be accomplished as a consequence of direct interactions that induce a conformational change in AAPs that renders the C-terminal domain to be accessible. An equally likely possibility is that Shr3p prevents AAPs from engaging in inappropriate interactions with themselves or with other ER membrane components. Such nonproductive interactions could prevent the exposure of sorting signals and consequently block binding of AAPs to Sec23/Sec24p. Finally, the capacity of Shr3p to interact with Sec23/Sec24p in the absence of AAPs raises an additional possibility. Shr3p may influence the structural properties of the Sec23/Sec24p complex so that a latent binding site becomes available. As previously discussed, once the sorting motif has been correctly presented and an AAP binds to Sec23/Sec24p, Shr3p must dissociate, a process that may be facilitated by the action of Sar1p. Clearly, more work is needed to differentiate between these and other possible models.

12 Coatomer asymmetry may influence the differential packaging of cargo

COPII vesicles have been shown to carry different sets of secretory cargo, suggesting that not all proteins can be simultaneously incorporated into a single vesicle (Muniz et al. 2001). ER-derived vesicles were immunoisolated with antibodies recognizing the amino acid permease Gap1p. Soluble pro-α-factor and the integral membrane protein alkaline phosphatase (ALP) were efficiently recovered from these vesicles, indicating that Gap1p, ALP, and pro-α-factor are co-packaged together into the same vesicles. This finding is consistent with earlier work (Kuehn et al. 1996). In contrast, the GPI-anchored proteins Gas1p and Yps1p were not detected in these vesicles (Muniz et al. 2001). The differences in sorting were not due to different rates of packaging, and Gap1p and Gas1p vesicles could be separated by a sucrose density-gradient. Together, these results show that two GPI-anchored proteins (Gas1p and Yps1p) are packaged into distinct vesicles with clearly defined physical properties that enable them to be distinguished from vesicles carrying other types of secretory cargo.

In light of the structural studies of COPII coatomer and the current understanding of ER vesicle formation, it is possible to imagine how distinct sets of cargo are differential packaged. The crystallographic analysis of Sar1/Sec23/Sec24p revealed that this inner coatomer complex forms a flat (≈ 40 Å thick) elongated bow tie shaped structure (Bi et al. 2002). One end of the bow tie is comprised of Sar1p-Sec23p and the other of Sec24p (Bi et al. 2002). The complex is 150 Å long and has a concave positively charged face that is likely to be proximal to the membrane. The width varies along the bow tie from ≈ 100 Å at the Sec23p-Sar1p end, to ≈ 30 Å at the Sec24p-Sec23p interface, to ≈ 75 Å at the Sec24p end. The defined cargo binding sites A and B are located > 80 Å apart on opposite sides at the periphery of Sec24p, approximately 20 Å from the membrane-proximal surface (Mossessova et al. 2003). Based on an EM analysis, the Sec13/Sec31p heterotetramer (outer coatomer complex) is an extended flexible structure with well-defined globular domains that resemble a strand of five closely spaced beads (Lederkremer et al. 2001). The complex is approximately 300 Å in length and each globular domain is approximately 60 Å in diameter. Based on a variety of biochemical studies the Sec13/Sec31p complex is likely to be organized with a side-by-side arrangement of subunits. At one end, the globular domain is comprised of two Sec13p subunits; the remaining four domains are formed by two parallel Sec31p subunits that are oriented such that their carboxy-termini are located in the most distal domain with respect to Sec13p.

These structural studies clearly show that both inner and outer complexes are inherently asymmetric. The known protein-protein interactions, e.g. Sec23/Sec24p associate with the C-terminal domains of Sec31p (Shaywitz et al. 1997), restrict the possible higher order assemblies that ultimately form the vesicle coat. Furthermore, the binding of cargo is likely to influence the structural characteristics of Sar1/Sec23/Sec24p pre-budding complexes (inner coatomer) and add to the overall structural asymmetry. For example, and figuratively, depending upon what binding sites on Sec24p are used, the binding of cargo could favor the formation of "left" or "right" handed pre-budding complexes. Due to restrictions regarding coatomer subunit interactions, left-handed complexes may be structurally incompatible with right-handed complexes and, therefore, differentially gathered by outer coatomer (Sec13/Sec31p). Consequently, two populations of vesicles are formed that contain either left- or right-handed inducing cargo.

13 Concluding remarks

The current understanding of how secretory proteins are selected as cargo and included into ER-derived COPII coated vesicles is largely derived from studies using purified components, identified in reductionistic approaches aimed at elucidating the minimal set of required core components. These extremely powerful approaches, including the recent crystallographic analysis that demonstrated that sequence motifs present on cargo proteins bind directly to the inner coatomer component Sec24p, have substantially increased our level of understanding. However, the repertoire of characterized cargo known to function this way is still quite limited. Additionally, many of the identified sorting motifs fail to function as signals when transferred to other proteins, indicating that sorting motifs are likely to be presented in a context dependent manner. As more information becomes available regarding the factors that contribute to the presentation of sorting signals, we will undoubtedly be forced to make adjustments in the way we view these core processes.

The work regarding the role of packaging chaperones, in particular with Shr3p, is not entirely consistent with the accepted view of the events initiating COPII vesicle formation. For example, the finding that Sec23/Sec24p and Sec13/Sec31p co-purify with Shr3p, whereas, the G-protein Sar1p does not, suggests that there may be alternative pathways to prime vesicle formation. Consistent with this notion, the overexpression of *SHR3* or *shr3-23*, a non-functional allele of *SHR3* encoding a mutant protein that is unable to interact with AAPs but that retains the ability to bind COPII coatomer, partially suppresses *sec12-1* mutations (Gilstring et al. 1999). These findings suggest that productive cargo-coatomer interactions may occur prior to the association of Sar1p with the membrane. Thus, the precise order of events leading to the binding of Sar1p to Sec23p may not be absolutely dependent upon Sec12p. The possibility exists that primed-cargo complexes themselves stimulate GDP-GTP exchange that enables Sar1p to bind. Interestingly, a comparative analysis of just fungal sequences revealed that Sec12p is not a well-conserved core component (Martínez and Ljungdahl 2000). When the *S. cerevisiae* components known to be involved with ER-derived vesicle formation were used to query the translated *S. pombe* genome database (Sanger Centre), Sec12p is among the least conserved components, exhibiting a similar degree of homology as the corresponding *S. pombe* Shr3p homolog. These observations suggest that based on their ability to bind and, thereby, recruit coatomer to the ER membrane, packaging chaperones may act as nucleation sites that facilitate the polymerization of additional COPII coatomer.

It is likely that many large polytopic membrane proteins require the action of packaging chaperones, cognate Shr3p-like proteins, to exit the ER. Notably, the yeast genome encodes over 1000 proteins that have one to four transmembrane domains, the majority of which have not been assigned a function. Thus, there exist many possible candidates. Additionally, packaging chaperones may recognize structural motifs formed by two or more subunits. The possibility exists that packaging chaperones may govern the secretion of membrane protein complexes that assemble within the ER membrane. Incompletely assembled complexes would be unable to interact with packaging chaperones and, thus, the individual subunits would be unable to enter transport vesicles prematurely. In such cases, packaging chaperones would function as components of the ER quality control system. An example of such a phenomenon may be the ER retention of unassembled subunits of the integral membrane portion (V_0) of the vacuolar-ATPase. The integral subunits of the V_0 complex assemble in the ER; the exit of fully assembled V_0 complex from the ER depends upon the resident ER protein Vma21p (Graham et al. 1998). In vma21 mutants, the 100-kDa integral membrane subunit is retained in the ER and is rapidly degraded by non-vacuolar proteases. Furthermore, the existence of coatomer homologs, e.g. Lst1p and Iss1p, in yeast (Pagano et al. 1999; Roberg et al. 1999) and in mammalian cells (Pagano et al. 1999; Tang et al. 1999) raises the possibility that packaging chaperones participate in cargo selection by recruiting specific combinations of coatomer components.

The identification of packaging chaperones may help explain the difficulties of obtaining functional expression of heterologous membrane proteins. In yeast, heterologously expressed polytopic mammalian membrane proteins are often retained in the ER, e.g. the rat glucose transporters GLUT1 and GLUT4 (Kasahara and Kasahara 1996, 1997). The ER retention of heterologously expressed GLUT1 and GLUT4 was not due to their being misfolded; these proteins catalyzed glucose transport when reconstituted into liposomes. In a systematic genetic approach, mutations were selected that enabled the functional expression of these rat transporters (Wieczorke et al. 2003). Strains carrying a combination of genomic mutations, fgy1-1 and fgy4X, functionally expressed GLUT1 and GLUT4, which enabled their activity to be monitored in growth-based assays. Although the identity of the corresponding wildtype proteins has not been reported, the mutations appear to affect different steps of the secretory pathway. The fgy4X mutations enabled GLUT4 to exit the ER, suggesting that these mutations may alter the activity of endogenous packaging chaperones such that the mutant forms recognize the rat transporters. The fgy1-1 mutation affected a post-Golgi step and enabled PM localized GLUT1 and GLUT4 to function. The success of these studies provides encouragement for similar approaches to achieve functional expression of other heterologously expressed proteins. Interestingly, due to the high degree of conservation of the core components of the secretory pathway, it may be possible to obtain functional expression of heterologous transporters if their cognate packaging chaperones are co-expressed. Genetic approaches to identify such packaging chaperones are currently underway.

Finally, most studies to date have focused on priming and the initial stages of coatomer assembly. Events associated with the subsequent stages of vesicle formation also need to be better clarified. Clearly, a major challenge for future work will be to completely integrate all of the experimentally observed partial reactions. Ultimately a more realistic picture of the function and regulation of this fascinating and important process will emerge.

Acknowledgements

The vastness of the literature regarding transport vesicle formation and our focus on yeast as a model organism has precluded referencing many relevant papers, and we have undoubtedly inadvertently failed to cite some papers that are of equal or greater value than the ones cited. We apologize to the investigators of any uncited work. Original research in our laboratory is supported by the Ludwig Institute for Cancer Research.

References

Andréasson C, Neve EPA, Ljungdahl PO (2004) Four permeases import proline and the toxic proline analog azetidine-2-carboxylate into yeast. Yeast 21:193-199

Antonny B, Gounon P, Schekman R, Orci L (2003) Self-assembly of minimal COPII cages. EMBO Rep 4:419-424

Antonny B, Madden D, Hamamoto S, Orci L, Schekman R (2001) Dynamics of the COPII coat with GTP and stable analogues. Nat Cell Biol 3:531-537

Antonny B, Schekman R (2001) ER export: public transportation by the COPII coach. Curr Opin Cell Biol 13:438-443

Appenzeller C, Andersson H, Kappeler F, Hauri HP (1999) The lectin ERGIC-53 is a cargo transport receptor for glycoproteins. Nat Cell Biol 1:330-334

Aridor M, Weissman J, Bannykh S, Nuoffer C, Balch WE (1998) Cargo selection by the COPII budding machinery during export from the ER. J Cell Biol 141:61-70

Barlowe C (2003) Signals for COPII-dependent export from the ER: what's the ticket out? Trends Cell Biol 13:295-300

Barlowe C, Orci L, Yeung T, Hosobuchi M, Hamamoto S, Salama N, Rexach MF, Ravazzola M, Amherdt M, Schekman R (1994) COPII: a membrane coat formed by Sec proteins that drive vesicle budding from the endoplasmic reticulum. Cell 77:895-907

Barlowe C, Schekman R (1993) SEC12 encodes a guanine-nucleotide-exchange factor essential for transport vesicle budding from the ER. Nature 365:347-349

Bednarek SY, Orci L, Schekman R (1996) Traffic COPs and the formation of vesicle coats. Trends Cell Biol 6:468-473

Bi X, Corpina RA, Goldberg J (2002) Structure of the Sec23/24-Sar1 pre-budding complex of the COPII vesicle coat. Nature 419:271-277

Blobel G (1980) Intracellular protein topogenesis. Proc Natl Acad Sci USA 77:1496-1500

Blobel G, Dobberstein B (1975a) Transfer of proteins across membranes. I. Presence of proteolytically processed and unprocessed nascent immunoglobulin light chains on membrane-bound ribosomes of murine myeloma. J Cell Biol 67:835-851

Blobel G, Dobberstein B (1975b) Transfer to proteins across membranes. II Reconstitution of functional rough microsomes from heterologous components. J Cell Biol 67:852-862

Bryant NJ, Stevens TH (1998) Vacuole biogenesis in *Saccharomyces cerevisiae*: protein transport pathways to the yeast vacuole. Microbiol Mol Biol Reviews 62:230-247

Chen YA, Scheller RH (2001) SNARE-mediated membrane fusion. Nat Rev Mol Cell Biol 2:98-106

Cosson P, Letourneur F (1997) Coatomer (COPI)-coated vesicles: role in intracellular transport and protein sorting. Curr Opin Cell Biol 9:484-487

Cowles CR, Odorizzi G, Payne GS, Emr SD (1997) The AP-3 adaptor complex is essential for cargo-selective transport to the yeast vacuole. Cell 91:109-118

Darsow T, Burd CG, Emr SD (1998) Acidic di-leucine motif essential for AP-3-dependent sorting and restriction of the functional specificity of the Vam3p vacuolar t-SNARE. J Cell Biol 142:913-922

Darsow T, Katzmann DJ, Cowles CR, Emr SD (2001) Vps41p function in the alkaline phosphatase pathway requires homo-oligomerization and interaction with AP-3 through two distinct domains. Mol Biol Cell 12:37-51

Davis CG, Lehrman MA, Russell DW, Anderson RG, Brown MS, Goldstein JL (1986) The JD mutation in familial hypercholesterolemia: amino acid substitution in cytoplasmic domain impedes internalization of LDL receptors. Cell 45:15-24

Dominguez M, Dejgaard K, Fullekrug J, Dahan S, Fazel A, Paccaud JP, Thomas DY, Bergeron JJ, Nilsson T (1998) gp25L/emp24/p24 protein family members of the cis-Golgi network bind both COP I and II coatomer. J Cell Biol 140:751-765

Ellgaard L, Molinari M, Helenius A (1999) Setting the standards: quality control in the secretory pathway. Science 286:1882-1888

Epping EA, Moye-Rowley WS (2002) Identification of interdependent signals required for anterograde traffic of the ATP-binding cassette transporter protein Yor1p. J Biol Chem 277:34860-34869

Espenshade P, Gimeno RE, Holzmacher E, Teung P, Kaiser CA (1995) Yeast *SEC16* gene encodes a multidomain vesicle coat protein that interacts with Sec23p. J Cell Biol 131:311-324

Fiedler K, Veit M, Stamnes MA, Rothman JE (1996) Bimodal interaction of coatomer with the p24 family of putative cargo receptors. Science 273:1396-1399

Gaynor EC, Emr SD (1997) COPI-independent anterograde transport: cargo-selective ER to Golgi protein transport in yeast COPI mutants. J Cell Biol 136:789-802

Gilstring CF, Ljungdahl PO (2000) A method for determining the *in vivo* topology of yeast polytopic membrane proteins demonstrates that Gap1p fully integrates into the membrane independently of Shr3p. J Biol Chem 275:31488-31495

Gilstring CF, Melin-Larsson M, Ljungdahl PO (1999) Shr3p mediates specific COPII coatomer-cargo interactions required for the packaging of amino acid permeases into ER-derived transport vesicles. Mol Biol Cell 10:3549-3565

Gimeno RE, Espenshade P, Kaiser CA (1995) *SED4* encodes a yeast endoplasmic reticulum protein that binds Sec16p and participates in vesicle formation. J Cell Biol 131:325-338

Gimeno RE, Espenshade P, Kaiser CA (1996) COPII coat subunit interactions: Sec24p and Sec23p bind to adjacent regions of Sec16p. Mol Biol Cell 7:1815-1823

Goldberg J (1998) Structural basis for actibation of ARF GTPase: mechansim of quaninine nucleaotide exchange and GTP-myristol switching. Cell 95:237-248

Graham LA, Hill KJ, Stevens TH (1998) Assembly of the yeast vacuolar H+-ATPase occurs in the endoplasmic reticulum and requires a Vma12p/Vma22p assembly complex. J Cell Biol 142:39-49

Grenson M, Hennaut C (1971) Mutation affecting activity of several distinct amino acid transport systems in Saccharomyces cerevisiae. J Bacteriol 105:477-482

Gurunathan S, David D, Gerst JE (2002) Dynamin and clathrin are required for the biogenesis of a distinct class of secretory vesicles in yeast. EMBO J 21:602-614

Harsay E, Bretscher A (1995) Parallel secretory pathways to the cell surface in yeast. J Cell Biol 131:297-310

Hauri HP, Kappeler F, Andersson H, Appenzeller C (2000) ERGIC-53 and traffic in the secretory pathway. J Cell Sci 113:587-596

Herrmann JM, Malkus P, Schekman R (1999) Out of the ER--outfitters, escorts and guides. Trends Cell Biol 9:5-7

Hirst J, Robinson MS (1998) Clathrin and adaptors. Biochim Biophys Acta 1404:173-193

Horák J, Kotyk A (1993) Functional analysis of apf1 mutation causing defective amino acid transport in Saccharomyces cerevisiae. Biochem Mol Biol 29:907-912

Iraqui I, Vissers S, Bernard F, de Craene JO, Boles E, Urrestarazu A, André B (1999) Amino acid signaling in Saccharomyces cerevisiae: a permease-like sensor of external amino acids and F-Box protein Grr1p are required for transcriptional induction of the AGP1 gene, which encodes a broad-specificity amino acid permease. Mol Cell Biol 19:989-1001

Kaiser CA, Gimeno RE, Shaywitz DA (1997) Protein secretion, membrane biogenesis, and endocytosis. In: Pringle JR, Broach JR and Jones EW (eds) The Molecular and Cellular Biology of the Yeast Saccharomyces, vol. 3. Cold Spring Harbor Laboratory Press, pp 91-227

Kappeler F, Klopfenstein DR, Foguet M, Paccaud JP, Hauri HP (1997) The recycling of ERGIC-53 in the early secretory pathway ERGIC-53 carries a cytosolic endoplasmic reticulum-exit determinant interacting with COPII. J Biol Chem 272:31801-31808

Kasahara T, Kasahara M (1996) Expression of the rat GLUT1 glucose transporter in the yeast Saccharomyces cerevisiae. Biochem J 315:177-182

Kasahara T, Kasahara M (1997) Characterization of rat Glut4 glucose transporter expressed in the yeast Saccharomyces cerevisiae: comparison with Glut1 glucose transporter. Biochim Biophys Acta 1324:111-119

Katzmann DJ, Odorizzi G, Emr SD (2002) Receptor downregulation and multivesicular-body sorting. Nat Rev Mol Cell Biol 3:893-905

Kirchhausen T (2000) Three ways to make a vesicle. Nature Rev Mol Cell Biol 1:187-198

Klasson H, Fink GR, Ljungdahl PO (1999) Ssy1p and Ptr3p are plasma membrane components of a yeast system that senses extracellular amino acids. Mol Cell Biol 19:5405-5416

Kuehn MJ, Herrmann JM, Schekman R (1998) COPII-cargo interactions direct protein sorting into ER-derived transport vesicles. Nature 391:187-190

Kuehn MJ, Schekman R (1997) COPII and secretory cargo capture into transport vesicles. Curr Opin Cell Biol 9:477-483

Kuehn MJ, Schekman R, Ljungdahl PO (1996) Amino acid permeases require COPII components and the ER resident membrane protein Shr3p for packaging into transport vesicles *in vitro*. J Cell Biol 135:585-595

Kuge O, Dascher C, Orci L, Rowe T, Amherdt M, Plutner H, Ravazzola M, Tanigawa G, Rothman JE, Balch WE (1994) Sar1 promotes vesicle budding from the endoplasmic reticulum but not Golgi compartments. J Cell Biol 125:51-65

Kurihara T, Hamamoto S, Gimeno RE, Kaiser CA, Schekman R, Yoshihisa T (2000) Sec24p and Iss1p function interchangeably in transport vesicle formation from the endoplasmic reticulum in *Saccharomyces cerevisiae*. Mol Biol Cell 11:983-998

Lasko PF, Brandriss MC (1981) Proline transport in *Saccharomyces cerevisiae*. J Bacteriol 148:241-247

Lau WT, Howson RW, Malkus P, Schekman R, O'Shea EK (2000) Pho86p, an endoplasmic reticulum (ER) resident protein in *Saccharomyces cerevisiae*, is required for ER exit of the high-affinity phosphate transporter Pho84p. Proc Natl Acad Sci USA 97:1107-1112

Lederkremer GZ, Cheng Y, Petre BM, Vogan E, Springer S, Schekman R, Thomas , Walz T, Kirchhausen T (2001) Structure of the Sec23p/24p and Sec13p/31p complexes of COPII. Proc Natl Acad Sci USA 98:10707-10709

Letourneur F, Gaynor EC, Hennecke S, Demolliere C, Duden R, Emr SD, Riezman H, Cosson P (1994) Coatomer is essential for retrieval of dilysine-tagged proteins to the endoplasmic reticulum. Cell 79:1199-1207

Ljungdahl PO, Gimeno CJ, Styles CA, Fink GR (1992) SHR3: a novel component of the secretory pathway specifically required for localization of amino acid permeases in yeast. Cell 71:463-478

Ma D, Zerangue N, Lin YF, Collins A, Yu M, Jan YN, Jan LY (2001) Role of ER export signals in controlling surface potassium channel numbers. Science 291:316-319

Malkus P, Jiang F, Schekman R (2002) Concentrative sorting of secretory cargo proteins into COPII-coated vesicles. J Cell Biol 159:915-921

Martínez P, Ljungdahl PO (2000) The SHR3 homologue from *S pombe* demonstrates a conserved function of ER packaging chaperones. J Cell Sci 113:4351-4362

Martínez-Menargue JA, Prekeris R, Oorschot VMJ, Scheller R, Slot JW, Geuze HJ, Klumperman J (2001) Peri-Golgi vesicles contain retrograde but not anterograde proteins consistent with the cisternal progression model of intra-Golgi transport. J Cell Biol 155:1213-1212

Matsuoka K, Orci L, Amherdt M, Bednarek SY, Hamamoto S, Schekman R, Yeung T (1998) COPII-coated vesicle formation reconstituted with purified coat proteins and chemically defined liposomes. Cell 93:263-275

Matsuoka K, Schekman R, Orci L, Heuser JE (2001) Surface structure of the COPII-coated vesicle. Proc Natl Acad Sci USA 98:13705-13709

McCusker JH, Haber JE (1990) Mutations in *Saccharomyces cerevisiae* which confer resistance to several amino acid analogs. Mol Cell Biol 10:2941-2949

McNew JA, Parlati F, Fukuda R, Johnston RJ, Paz K, Paumet F, Söllner TH, Rothman JE (2000) Compartmental specifcity of cellular membrane fusion encoded in SNARE proteins. Nature 407:153-159

Meacock SL, Greenfield JJ, High S (2000) Protein targeting and translocation at the endoplasmic reticulum membrane--through the eye of a needle? Essays Biochem 36:1-13

Mellman I, Warren G (2000) The road taken: past and future foundations of membrane traffic. Cell 100:88-112

Miller E, Antonny B, Hamamoto S, Schekman R (2002) Cargo selection into COPII vesicles is driven by the Sec24p subunit. EMBO J 21:6105-6113

Miller EA, Beilharz TH, Malkus PN, Lee MC, Hamamoto S, Orci L, Schekman R (2003) Multiple cargo binding sites on the COPII subunit Sec24p ensure capture of diverse membrane proteins into transport vesicles. Cell 114:497-509

Mironov AA, Beznoussenko GV, Nicoziani P, Martella O, Trucco A, Kweon HS, Di Giandomenico D, Polishchuk RS, Fusella A, Lupetti P, Berger EG, Geerts WJC, Koster AJ, Burger KNJ, Luini A (2001) Small cargo proteins and large aggregates can traverse the Golgi by a common mechanism without leaving the lumen of cisternae. J Cell Biol 155:1225-1212

Mogelsvang S, Gomez-Ospina N, Soderholm J, Glick BS, Staehelin LA (2003) Tomographic evidence for continuous turnover of Golgi cisternae in *Pichia pastoris*. Mol Biol Cell 14:2277-2291

Mossessova E, Bickford LC, Goldberg J (2003) SNARE selectivity of the COPII coat. Cell 114:483-495

Muniz M, Morsomme P, Riezman H (2001) Protein sorting upon exit from the endoplasmic reticulum. Cell 104:313-320

Muniz M, Nuoffer C, Hauri HP, Riezman H (2000) The Emp24 complex recruits a specific cargo molecule into endoplasmic reticulum-derived vesicles. J Cell Biol 148:925-930

Nakamura N, Yamazaki S, Sato K, Nakano A, Sakaguchi M, Mihara K (1998) Identification of potential regulatory elements for the transport of Emp24p. Mol Biol Cell 9:3493-3503

Nakano A, Brada D, Schekman R (1988) A membrane glycoprotein, Sec12p, required for protein transport from the endoplasmic reticulum to the Golgi apparatus in yeast. J Cell Biol 109:851-863

Ng DT, Brown JD, Walter P (1996) Signal sequences specify the targeting route to the endoplasmic reticulum membrane. J Cell Biol 134:269-278

Nishimura N, Balch WE (1997) A di-acidic signal required for selective export from the endoplasmic reticulum. Science 277:556–558

Nothwehr SF, Ha SA, Bruinsma P (2000) Sorting of yeast membrane proteins into an endosome-to-Golgi pathway involves direct interaction of their cytosolic domains with Vps35p. J Cell Biol 151:297-310

Novick P, Zerial M (1997) The diversity of Rab proteins in vesicle transport. Curr Opin Cell Biol 9:496-504

Oka T, Nakano A (1994) Inhibition of GTP hydrolysis by Sar1p causes accumulation of vesicles that are a functional intermediate of the ER-to-Golgi transport in yeast. J Cell Biol 124:425-434

Oka T, Nishikawa S, Nakano A (1991) Reconstitution of GTP-binding Sar1 protein function in ER to Golgi transport. J Cell Biol 114:671-679

Otte S, Barlowe C (2002) The Erv41p-Erv46p complex: multiple export signals are required in trans for COPII-dependent trans- export from the ER. EMBO J 21:6095–6104

Paccaud JP, Reith W, Carpentier JL, Ravazzola M, Amherdt M, Schekman R, Orci L (1996) Cloning and functional characterization of mammalian homologues of the COPII component Sec23. Mol Biol Cell 7:1535-1546

Pagano A, Letourneur F, Garcia-Estefania D, Carpentier JL, Orci L, Paccaud JP (1999) Sec24 proteins and sorting at the endoplasmic reticulum. J Biol Chem 274:7833-7840

Palade G (1975) Intracellular aspects of the process of protein synthesis. Science 189:347-358

Pelham HR (1991) Recycling of proteins between the endoplasmic reticulum and Golgi complex. Curr Opin Cell Biol 3:585-591

Pelham HR (2001) Traffic through the Golgi apparatus J Cell Biol 155:1099-1101

Peng R, De Antoni A, Gallwitz D (2000) Evidence for overlapping and distinct functions in protein transport of coat protein Sec24p family members. J Biol Chem 275:11521-11528

Peng R, Grabowski R, De Antoni A, Gallwitz D (1999) Specific interaction of the yeast cis-Golgi syntaxin Sed5p and the coat protein complex II component Sec24p of endoplasmic reticulum-derived transport vesicles. Proc Natl Acad Sci USA 96:3751-3756

Powers J, Barlowe C (1998) Transport of Axl2p depends on Erv14p, an ER-vesicle protein related to the *Drosophila cornichon* gene product. J Cell Biol 142:1209-1222

Regenberg B, During-Olsen L, Kielland-Brandt MC, Holmberg S (1999) Substrate specificity and gene expression of the amino-acid permeases in *Saccharomyces cerevisiae*. Curr Genet 36:317-328

Roberg KJ, Crotwell M, Espenshade P, Gimeno R, Kaiser CA (1999) *LST1* is a *SEC24* homologue used for selective export of the plasma membrane ATPase from the endoplasmic reticulum. J Cell Biol 145:659-672

Rothman JE (2002) Lasker Basic Medical Research Award - The machinery and principles of vesicle transport in the cell. Nat Med 8:1059-1062

Saito Y, Kimura K, Oka T, Nakano A (1998) Activities of mutant Sar1 proteins in guanine nucleotide binding, GTP hydrolysis, and cell-free transport from the endoplasmic reticulum to the Golgi apparatus. J Biochem 124:816-823

Saito-Nakano Y, Nakano A (2000) Sed4p functions as a positive regulator of Sar1p probably through inhibition of the GTPase activation by Sec23p Genes. Cells 5:1039-1048

Sato K, Nakano A (2002) Emp47p and its close homolog Emp46p have a tyrosine-containing endoplasmic reticulum exit signal and function in glycoprotein secretion in *Saccharomyces cerevisiae*. Mol Biol Cell 13:2518–2532

Schekman R (2002) Lasker Basic Medical Research - Award *SEC* mutants and the secretory apparatus. Nat Med 8:1055-1058

Sevier CS, Weisz OA, Davis M, Machamer CE (2000) Efficient export of the vesicular stomatitis virus G protein from the endoplasmic reticulum requires a signal in the cytoplasmic tail that includes both tyrosine-based and di-acidic motifs. Mol Biol Cell 11:13–22

Shaywitz DA, Espenshade PJ, Gimeno RE, Kaiser CA (1997) COPII subunit interactions in the assembly of the vesicle coat. J Biol Chem 272:25413-25416

Shaywitz DA, Orci L, Ravazzola M, Swaroop A, Kaiser CA (1995) Human SEC13Rp functions in yeast and is located on transport vesicles budding from the endoplasmic reticulum. J Cell Biol 128:769-777

Sherwood PW, Carlson M (1999) Efficient export of the glucose transporter Hxt1p from the endoplasmic reticulum requires Gsf2p. Proc Natl Acad Sci USA 96:7415-7420

Shimoni Y, Kurihara T, Ravazzola M, Amherdt M, Orci L, Schekman R (2000) Lst1p and Sec24p cooperate in sorting of the plasma membrane ATPase into COPII vesicles in *Saccharomyces cerevisiae*. J Cell Biol 151:973-984

Söllner T, Whiteheart SW, Brunner M, Erdjument-Bromage H, Geromanos S, Tempst P, Rothman JE (1993) SNAP receptors implicated in vesicle targeting and fusion. Nature 362:318-324

Sorsoli WA, Spence KD, Parks LW (1964) Amino acid accumulation in ethionine-resistant *Saccharomyces cerevisiae*. J Bacteriol 88:20-24

Spang A, Schekman R (1998) Reconstitution of retrograde transport from the Golgi to the ER *in vitro*. J Cell Biol 143:589-599

Springer S, Schekman R (1998) Nucleation of COPII vesicular coat complex by endoplasmic reticulum to Golgi vesicle SNAREs. Science 281:698-700

Springer S, Spang A, Schekman R (1999) A primer on vesicle budding. Cell 97:145-148

Stockklausner C, Ludwig J, Ruppersberg JP, Klocker N (2001) A sequence motif responsible for ER export and surface expression of Kir20 inward rectifier K(+) channels. FEBS Lett 493:129-133

Storrie B, Nilsson T (2002) The Golgi apparatus: balancing new with old. Traffic 3:521-529

Surdin Y, Sly W, Sire J, Bordes AM, de Robichon-Szulmajster H (1965) Properties and genetic control of the amino acid accumulation system in *Saccharomyces cerevisiae*. Biochim Biophys Acta 107:546-566

Tang BL, Kausalya J, Low DY, Lock ML, Hong W (1999) A family of mammalian proteins homologous to yeast Sec24p. Biochem Biophys Res Commun 258:679-684

Tang BL, Zhang T, Low DY, Wong ET, Horstmann H, Hong W (2000) Mammalian homologues of yeast sec31p An ubiquitously expressed form is localized to endoplasmic reticulum (ER) exit sites and is essential for ER-Golgi transport. J Biol Chem 275:13597-13604

Tani K, Oyama Y, Hatsuzawa K, Tagaya M (1999) Hypothetical protein KIAA0079 is a mammalian homologue of yeast Sec24p. FEBS Lett 447:247-250

Trilla JA, Duran A, Roncero C (1999) Chs7p, a new protein involved in the control of protein export from the endoplasmic reticulum that is specifically engaged in the regulation of chitin synthesis in *Saccharomyces cerevisiae*. J Cell Biol 145:1153-1163

Votsmeier C, Gallwitz D (2001) An acidic sequence of a putative yeast Golgi membrane protein binds COPII and facilitates ER export. EMBO J 20:6742-6750

Waters MG, Serafini T, Rothman JE (1991) 'Coatomer': a cytosolic protein complex containing subunits of non-clathrin-coated Golgi transport vesicles. Nature 349:248-251

Wieczorke R, Dlugai S, Krampe S, Boles E (2003) Characterisation of mammalian GLUT glucose transporters in a heterologous yeast expression system. Cell Physiol Biochem 13:123-134

Yoshihisa T, Barlowe C, Schekman R (1993) Requirement for a GTPase-activating protein in vesicle budding from the endoplasmic reticulum. Science 259:1466-1468

Kota, Jhansi
Ludwig Institute for Cancer Research, Box 240, S-171 77 Stockholm

Ljungdahl, Per O.
Ludwig Institute for Cancer Research, Box 240, S-171 77 Stockholm
plju@licr.ki.se

Nyman, Tomas
Ludwig Institute for Cancer Research, Box 240, S-171 77 Stockholm

Regulation of transporter trafficking by the lipid environment

Miroslava Opekarová

Abstract

The function of lipid molecules in membrane bilayers as mere structural elements has been fundamentally revised during the last two decades. There is increasing evidence for the essential roles that specific lipids play in a great many controlling and regulatory mechanisms in cells. It is the purpose of this review to follow the trafficking of typical plasma membrane proteins, such as transporters, and to describe the current level of knowledge on the role that is played by individual lipids, the membrane lipid composition and its reorganization in this process. It will be demonstrated that, in addition to the specific lipid molecules required for stabilization, correct folding and full activity of transporter molecules, particular aggregates of sterols and sphingolipids are involved in protein sorting and trafficking.

1 Introduction

In living cells, various types of proteins residing in membranes mediate the communication between the interior and exterior, and between individual cellular compartments. The most simplified view of a biological membrane is a lipid bilayer in which the proteins are integrated, immersed, or attached to. The organization of the membrane components is, however, much more complex and is of implicit importance for all cellular functions. Both in prokaryotes and eukaryotes, the membrane composition is a dynamic entity and varies in response to numerous endogenous and exogenous stimuli. In eukaryotic cells, the lipid and protein compositions of subcellular membranes differ from that of the plasma membrane and also between the inner and outer leaflets of the bilayer. In addition, a large body of evidence supports the idea that not even the composition within the individual leaflets of the plasma membrane is uniform, but that lipids and proteins display heterogeneous lateral distribution (Simons and Ikonen 1997). Transporters (permeases, carriers) are integrated into the plasma membrane by several membrane-spanning domains with a characteristic hydrophobic profile. The hydrophobic parts of the protein molecule are in full contact with the adjacent lipids and, thus, it is not surprising that their activity is influenced by the surrounding lipid milieu. This effect is exerted both by the physicochemical properties of the lipid bilayer

Topics in Current Genetics, Vol. 9
E. Boles, R. Krämer (Eds.): Molecular Mechanisms Controlling Transmembrane Transport
DOI 10.1007/b95874 / Published online: 9 March 2004
© Springer-Verlag Berlin Heidelberg 2004

(reviewed in de Kruijff 1997; Epand 1998) and by a single or a few lipid molecules selectively affecting individual transporters (reviewed in Opekarová and Tanner 2003).

Due to the hydrophobicity of the prevailing part of these molecules, transporters have to be and, in fact, are in contact with lipids since they are synthesized at the ribosomes attached to the bacterial membrane or to the membrane of the endoplasmic reticulum (ER) in eukaryotes. The *curriculum vitae* of transporters and lipids can be seen as being closely interconnected throughout their journey from their birthplace to the place of their final engagement.

2 Itineration of transporters to the plasma membrane

The first step in the process of trafficking of a polytopic membrane protein, as a transporter, is analogous both in prokaryotic and eukaryotic cells. The newly synthesized transporter has to be targeted and inserted into the bacterial or the ER membranes of eukaryotes. After emerging from the ribosome, the first hydrophobic segment of a transporter is bound to a specific soluble protein, which delivers the protein-ribosome-mRNA complex to the bacterial or the ER membrane-bound receptor. Subsequently, the soluble protein is released from the receptor and the hydrophobic region of the newly synthesized transporter enters a channel formed by a specific set of proteins. After a certain number of hydrophobic regions have entered the channel, they can integrate into the lipid bilayer (reviewed in van Voorst and de Kruijff 2000; Dalbey et al. 2000). Many components of the protein-translocation machineries are integral membrane proteins or they at least interact with lipids during the translocation process. Understandingly, any changes in the phospholipid composition will implicitly compromise the whole insertion and/or translocation machinery.

After insertion into the membrane, the transporter has to be folded and eventually assembled. The correct folding and assembly is an absolute prerequisite for the proper function of a transporter, and it is frequently assisted by specific protein molecule(s) called chaperones. An interesting exception from this rule was reported by Bogdanov and Dowhan, who demonstrated a novel function of phosphatidyl ethanolamine (PE) in the folding of *E. coli* lactose permease (lac Y). A combination of *in vivo* and *in vitro* studies documented that this phospholipid acts as a non-protein chaperone directing the proper assembly of the permease in the membrane bilayer (Dowhan 1998). The authors showed that the critical folding step occurs after the permease is inserted into the membrane (Bogdanov et al. 1998), and its topology can be changed in a reversible manner in response to the presence or absence of PE (Bogdanov et al. 2002).

In eukaryotic cells, before reaching the plasma membrane, integral membrane proteins have to pass multiple membrane compartments of the secretory pathway (Hamamoto et al. 1994; Rothman and Wieland 1996; Schekman and Orci 1996). Trafficking of a newly synthesized protein from the ER to the Golgi apparatus, in between the Golgi networks, and towards the plasma membrane is mediated by

COPII and COPI secretory (transport) vesicles and is briefly outlined in Fig. 1. In the individual compartments of the Golgi apparatus, the proteins undergo specific modifications, and some of them associate with complex sphingolipids (see below). Coat proteins of the secretory vesicles allow the selective transfer of macromolecules from one compartment of the secretory pathway to another. In addition to these proteins, highly specific proteins were identified as essential for assorting of one or a few types of the trafficked transporter molecules (the topic is reviewed by Ljungdahl in this issue). However, during the last decade, it became evident that, in addition to these defined proteins, also lipid molecules play an indispensable and active role in the trafficking and sorting machinery. Great progress in understanding of the lipid-based membrane protein sorting has been achieved through the introduction of the "lipid raft" concept.

To gain better insight into the transporter/lipid companionship along the secretory pathway, a brief view on the topology of lipid biosynthesis in eukaryotic cells is given.

3 Topology of lipid synthesis along the secretory pathway

Phosphoglycerolipids, sterols and ceramides are synthesized in the ER (Bishop and Bell 1988; Hechtberger et al. 1994; Kent 1995; Kohlwein et al. 1996; Patton and Lester 1991; van Meer 1989, 1998). The mammalian and fungal ceramides typically contain very long saturated acyl chains consisting of 16-26 carbons (Gu 1997; Han et al. 2002). Acyl chain elongation occurs at the ER membrane and, as will be discussed later, it represents an important factor in protein sorting.

Sterol structures are characterized by a planar and rigid tetracyclic ring. A great variety of sterols were identified in living organisms but always one type is abundant in, and characteristic for members of a definite life-kingdom. The major sterol in animal cells is cholesterol; ergosterol is the typical sterol of fungal cells, plants contain a mixture of sterols, such as, e.g. sitosterol, campesterol, and stigmasterol (Hartmann 1998). Complex sphingolipids are synthesized from the ER-derived ceramides in the Golgi apparatus (Futerman et al. 1990; Hirschberg et al. 1993; Jeckel et al. 1990). Sphingomyelin and glycosphingolipids represent the basic sphingolipids of mammalian cells, whereas, in yeast, these are inositolphosphoglyceramide and its mannosylated derivatives.

Although synthesized in compartments of the early secretory pathway, both the sterols and sphingolipids are found predominantly in the plasma membrane (Daum et al. 1998; Lange et al. 1989; Levine et al. 2000). It was proposed that the trafficking of lipids from the sites of their syntheses and their sorting towards the plasma membrane can, similarly as the secreted proteins, use the same means of the secretory pathway (van Meer 1989, 1998). In addition to the secretory vesicles, COPII and COPI, it was shown that ER-Golgi direct membrane contacts play an important role in the so-called, nonvesicular transport of lipids (Funato and Riezman 2001). Both ways offer an opportunity for common trafficking of the two

Topology of synthesis and trafficking of lipids and transporters

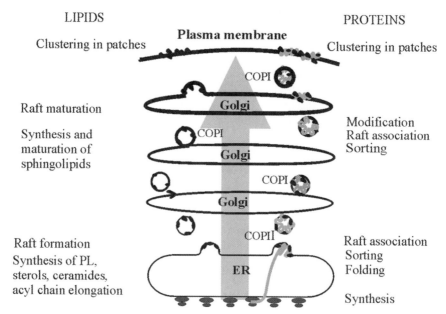

Fig. 1. The topology of lipid synthesis is depicted at the left, alongside the compartments of the secretory pathway. The black rods symbolize lipid rafts. The membrane-line thickening represents raft enrichment in the membranes of the pathway brought along by sphingolipid/sterol segregation from the membranes of the earlier Golgi cisternae. Conveying of rafts is carried out by transport vesicles either alone (left) or in association with a protein (right). Individual proteins synthesized at ribosomes of the ER are folded and sorted to COPII vesicles; some of the proteins are selectively associated with ceramide/sterol rafts. Further sorting occurs in the cisternae of the Golgi, the process being dependent on the protein modification and additional raft association. After reaching the plasma membrane, either the rafts alone or associated with a transporter (or its oligomer) can cluster into large patches. The large gray arrow depicts the flow of lipids and integral proteins towards the plasma membrane.

partners, i.e. an integral membrane protein and lipids. Of course, the lipids can also be trafficked independently of proteins (reviewed in van Meer and Holthuis 2000).

The fellow trafficking of lipids and a transporter protein along the secretory pathway is briefly summarized in Fig. 1.

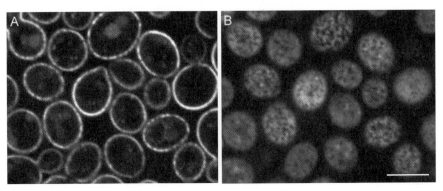

Fig. 2. Fluorescence pattern of Pma1p-GFP in living *S. cerevisiae* cells. (**A**) Individual transversal optical section; (**B**) superposition of four consecutive surface optical sections. Bar: 5 μm. (Provided by K. Malínská).

4 Lipid rafts and associated transporters

Lipid rafts (DRMs - detergent resistant membranes, DIGs - detergent insoluble glycolipid-enriched microdomains, or GEMs - glycolipid enriched membranes) are described as dynamic assemblies of sterols and sphingolipids that are laterally distributed in the plasma membrane of most, if not all, eukaryotes (Simons and Ikonen 1997; Brown and London 1998). Long saturated acyl chains of ceramides and sphingomyelins have high affinity for sterols. By restricting the acyl chain mobility on the molecule, sterols induce the formation of these compact domains. Individual lipid rafts appear to be small in size but, in total, they may constitute a large fraction of the plasma membrane. The area of the yeast plasma membrane occupied by PMA1p is shown in Fig. 2. Due to the physicochemical properties of the lipid constituents, rafts can be formed spontaneously *in vitro* (Sankaram and Thompson 1990; Schroeder et al. 1998). In living cells, they are often associated with specific membrane proteins (Brown and Rose 1992; Simons and Ikonen 1997; Muniz and Riezman 2000). A distinctive feature of lipid rafts is their insolubility in mild nonionic detergents (typically Triton X-100) at 4°C (Brown and Rose 1992; Rietveld and Simons 1998). These lipid domains can selectively incorporate or exclude particular proteins and it was suggested that, in this manner, they play an important role in protein sorting (Simons and Ikonen 1997; Ikonen 2001; Galbiati et al. 2001).

Raft-based protein sorting was first postulated for GPI anchored proteins in mammalian cells (Brown and Rose 1992). These proteins spanning one leaflet of the membrane are considered to be typical raft residents. However, within the last couple of years, an increasing number of multi-helix transmembrane proteins have been reported to associate with cholesterol/ergosterol and sphingolipid enriched domains. So far, about 20 transporters were reported to reside in rafts, and are listed in Table 1. Identification of all of them is based on their resistance to solubilization by mild detergents in the cold, and on their occurrence in the floating low-density fractions of the centrifugation gradients performed subsequently. A decrease in the content of sterols in membranes by chemical depletion or by a block in sterol biosynthesis, or by a manipulation of the ceramide/sphingolipds content results in the raft disruption and checkmates the delivery of the transporters to the cell surface.

5 Raft-dependent sorting of transporters in the secretory pathway

Sorting of different integral plasma membrane proteins into secretory vesicles was shown to require specific ER membrane integral proteins. Thus, for example, Shr3p is obligatory for COPII packing of all amino acid permeases in yeast (Ljungdahl et al. 1992), integration of Pma1 into COPII vesicles requires Lst1p (Lee et al. 2002). Inactivation of these proteins halts the secretion of the pertinent transporters. Similar effects of an arrest of a transporter secretion have been also observed under conditions preventing raft formation. This suggests that the protein-raft association represents an additional controlling and/or checking mechanism of sorting and trafficking of some transporters. It is documented that compositionally distinct lipid microdomains are assembled and may coexist within a given membrane. They can differ both in their resistance towards different detergents (Roper et al. 2000; Schuck et al. 2003) and/or associated proteins. Distinctive microdomains can be regulated, at least in part, by the content of membrane cholesterol (Brown and London 1997; Simons and Ikonen 1997; Anderson and Jacobson 2003; Brown and London 2000; Keller and Simons 1997). The different solubility might be further related to the sphingolipid composition and to the strength of the association with the hydrophobic parts of individual transporters. Gradual maturation of sphingolipids occurring in the cisternae of the Golgi (Levine et al. 2000; van Meer and Holthuis 2000; Funato and Riezman 2001) allows structural diversity of the raft components and consequently contributes to the diversity of the rafts associating selectively with individual proteins. All the above factors can result in formation of domains differing in size, charge, detergent resistance, compactness, etc. and each of the properties may participate in the diversification of the sorting process.

5.1 Location of transporter-raft association

Due to the location of sterol and ceramide syntheses, newly synthesized integral proteins can first come into the contact with rafts composed of a sterol and ceramides as early as in the ER. Coat protein complexes assemble on putative sorting signals on the ER membrane that might be cytoplasmatically exposed sequence motifs of the membrane proteins and/or local thickenings in the ER membrane

Table 1. Transporters and channels associated with rafts of the plasma membrane

Trans-porter	Organism	Remarks	Reference
Pma1p	*S. cerevisiae*	H^+-ATPase,	(Bagnat et al. 2000; Lee et al. 2002)
Fur4p	*S. cerevisiae*	Uracil transporter	(Hearn et al. 2003; Dupre and Haguenauer-Tsapis 2003)
Can1p	*S. cerevisiae*	Specific arginine permease	(Malínská et al. 2003)
Tat2p	*S. cerevisiae*	Tryptophan permease	(Umebayashi and Nakano 2003)
P-gp	Chinese hamster ovary cells, brain capillaries	P-glycoprotein, multidrug resistance transporter (colocalized with caveolin)	(Demeule et al. 2000)
GLUT1	Polarized epithelial cells	Hexose transport facilitator; GLUT3 localizes in a fluid membrane domain	(Sakyo and Kitagawa 2002)
GLUT4	Adipose cells (insulin signaling)	Raft localization of Cb1-CAP complex regulates glucose uptake. Rafts contribute to GLUT4 translocation to the cell surface	(Baumann et al. 2000) (Chiang et al. 2001) (Ros-Baro et al. 2001)
SGLT1	Renal epithelial cells	Na/glucose cotransporter (associates with vimentin)	(Runembert et al. 2002)
ATRC1	Human embryonic and canine kidney	y^+ cationic amino acid transporter Half of the population associates with actin, one fourth associates with rafts.	(Kizhatil and Albritton 2002)
ENaC A6 cell line	Lung, distal colon, kidney collecting duct	Epithelial sodium channel Recombinant ENaC is not located to rafts	(Hill et al. 2002) (Hanwell et al. 2002)
K(V,Ca)-hSlo	Canine kidney epithelial cells	Human Slowpock channel subunite of voltage/ Ca^{2+}- activated K^+ channel	(Bravo-Zehnder et al. 2000)
Kir3	Hippocampal neurons	Neuronal K^+channels Both endogenous and recombinant Kir3 in rafts	(Delling et al. 2002)
Kv2.1	Rat brain	Voltage-gated K^+-channel; Kv4.2 is not located to rafts	(Martens et al. 2000)
NHE1	Mammalian plasma membranes	Na^+/H^+ exchanger coprecipitates with caveolin	(Bullis et al. 2002)
NHE3	Opossum kidney Rabbit ileus	Brush border Na^+/H^+ exchanger	(Akhter et al. 2002; Li et al. 2001)
Na^+ channels	Renal epithelial cells	Amiloride-sensitive Na^+ channels	(Shlyonsky et al. 2003)
NCX1	Sarcolemmal vesicles	Cardiac Na^+- Ca^{2+}exchanger coprecipitates with caveolin-3	(Bossuyt et al. 2002)

caused by an enrichment of raft-protein agglomerations. In yeast, it was documented that Gas1p, a GPI anchored protein, associates with rafts in the ER (Bagnat et al. 2000), whereas Pma1p raft association was reported to occur later in the Golgi (Bagnat et al. 2001). Differential sorting of integral plasma membrane proteins into secretory vesicles has also been documented by *in vitro* studies (Muniz et al. 2001). The authors showed that at least GPI anchored proteins, like Gas1p, are sorted from other proteins as early as at their exit from the ER. Gas1p was localized to a different type of vesicles from those occupied by, e.g. a 12 membrane spanning protein, a general amino acid permease (Gap1p). On the other hand, two secretory proteins with different final destination, Gap1p (integral plasma membrane protein) and alkaline phosphatase (vacuole) were shown to leave the ER in the same type of vesicles, implying that these proteins are sorted later, most likely in the Golgi apparatus.

5.2 Lipid requirements for transporter-raft association

Since the Pma1p has been established as a typical raft resident (Bagnat et al. 2000), it became the best-characterized transporter protein as regards its intracellular trafficking. Before exiting the ER, Pma1p form large complexes of MW ~1.8 MDa that are packed into COPII vesicles (Lee et al. 2002). The oligomerization is correlated to ceramide synthesis. Pma1p isolated from *lcb1-100* cells (Zanolari et al. 2000) containing lower sphingolipid levels as compared to wild type, migrated almost exclusively as the monomeric form. Both the monomeric and oligomeric Pma1p are packed into COPII vesicles, but the monomers are rerouted to the vacuole for degradation. Lee at al. (2002) found Pma1p homododecamers associated with rafts already in the COPII vesicles, which suggests that entry into rafts and oligomerization begin before arrival at the Golgi. A very strict structural requirement concerning lipids for the Pma1p-raft association has been described in a yeast mutant defective in the ER-associated acyl-chain elongation complex. Deletion of one component of the complex, Elo3, required for the final acyl-chain elongation from C22/C24 to C26 (Oh et al. 1997) in combination with a specific defect in ergosterol biosynthesis, *ERG6*, resulted in altered plasma membrane lipid composition (sterol/sphingolipid ratio) and in a dramatic reduction of the Pma1p-raft association. In the single elo3Δ mutant, rapid degradation of newly synthesized Pma1p occurred at 37 °C. On the other hand, raft integration of Gas1p was not affected under the same conditions (Eisenkolb et al. 2002). The requirements of specifically structured lipid components for association of individual proteins further supports the idea of the existence of different types of lipid rafts along the secretory pathway.

5.3 Asymmetric distribution of transporters in polarized cells

In mammalian cells, uptake of hexoses is facilitated by GLUT 1-11 hexose transporters (Joost and Thorens 2001). Despite the high degree of sequence homology, each homologue can be subject to distinct modes of regulation concerning its cellular localization and transport activity. In polarized cells, such as Caco-2 and MDCK cells, GLUT1 is expressed on the basolateral surface, while GLUT3 is sorted to the apical surface (Harris et al. 1992; Pascoe et al. 1996). A striking difference in cellular localization of GLUT1 and GLUT3 was reported in non-polarized HeLa cells. Sakyo and Kitagawa (2002) showed that, in platelets and neuronal cells, GLUT1 distributes to cholesterol-rich DRM domains, whereas

GLUT3 localizes in fluid membrane domains. The differential distribution of GLUT1 and GLUT3 may occur irrespective of the N-glycosylation state, the tumorigenic state or cell type, implying that the raft association is mainly due to the intrinsic properties of the individual protein. The role of lipid rafts was reported in insulin-stimulated GLUT4 translocation (Baumann et al. 2000; Chiang et al. 2001). In adipose cells, GLUT4-containing vesicles fused with the plasma membrane only when the pertinent SNARE proteins were incorporated into rafts (Chamberlain and Gold 2002).

There is no longer any doubt that sphingolipid/sterol rafts are involved in the process of trafficking and sorting of a number of transporter molecules and in their stabilization within the plasma membrane. However, the detailed mechanism of the process is still based mainly on sophisticated speculations.

5.4 Molecular models for raft-based protein sorting

In principle, two non-conflicting hypotheses were postulated for the raft-based protein sorting: (i) transmembrane domain-dependent sorting takes into account the compatibility of the length of hydrophobic parts of the protein molecule with the thickness of the membrane bilayer. (ii) The lipid shell hypothesis emphasizes mutual affinity of the hydrophobic parts of an integral protein and definite lipid molecules.

(i) Transmembrane domain-dependent sorting determines the final destination of membrane proteins. The notion that the bilayer thickness increases towards the trans-Golgi network (TGN) was reported as early as in 1968 (Grove et al. 1968). The membrane thickening has been later attributed to the maturation of sphingolipids and to the sphingolipid/sterol gradient formed across the Golgi stack (Holthuis et al. 2001). The membrane of the ER has the thickness of a pure phospholipid (PL) bilayer ~3.5 nm; the PL bilayer enriched in cholesterol has a thickness of about 4 nm; sphingolipid-containing rafts may be 4.5-5.5 nm thick depending on the length of the amide-linked fatty acids (Sprong et al. 2001). The great difference between the thickness of the plasma membrane and the membranes of the early secretory compartments could lead to mechanistic problems in trafficking of cell surface proteins.

In mammalian cells, it was shown that the transmembrane domains (TMDs) of Golgi-residing proteins are, on an average, five residues shorter than those of plasma membrane residents (Bretscher and Munro 1993). To examine bilayer thickness along the secretory pathway in yeast, the TMD length of known one-membrane-spanning integral proteins was paralleled with their location in the membranes of the secretory pathway compartments. It was found that the bilayer thickness is constant until late in the Golgi apparatus, and then it is apparently more than 50% thicker in the plasma membrane (Levine et al. 2000). Based on the above observations, it has been postulated that, due to their shorter TMDs, the membrane-resident proteins of the early secretory pathway would be excluded from the thicker sphingolipid/sterol-enriched membrane regions destined for the cell surface and hence retained in the subcellular membranes (Bretscher and

Munro 1993; Munro 1998; Rayner and Pelham 1997). A short transmembrane α-helix can incorporate into a thin bilayer, but cannot incorporate into a bilayer that is too thick. A long transmembrane α-helix can incorporate into either a thick or a thin bilayer but, in the latter case, the hydrophobic domain must be tilted, which may alter the transmembrane protein conformation and function. Obviously, the effect of the bilayer thickness on protein function will be stronger in the case of proteins containing a bundle of multiple transmembrane helices.

A coupling between hydrophobic membrane thickness and transporter protein activity has been documented for, e.g. the $(Ca^{2+}-Mg^{2+})$-ATPase of skeletal muscle sarcoplasmic reticulum (Starling et al. 1993; Cornea and Thomas 1994).

(ii) Sorting based on the postulation of a lipid shell as thermodynamically stable aggregates of sterols and sphingolipids (different from boundary lipids) that have an affinity for pre-existing rafts or raft-like structures (Anderson and Jacobson 2003). It has been postulated that the lipid shells represent the smallest aggregates in a hierarchy of laterally organized lipids that exist in the bilayer of biological membranes. The dynamic process of protein sorting depends on the self-assembly of sterol-sphingolipid complexes and on the propensity of the protein to associate with these complexes. Once a protein is associated with the lipid-condensed complexes, it is targeted to specific locations in the cell.

5.5 Determination of raft association

It can be expected that both mechanisms participate in the lipid-dependent sorting process, and their interplay with specific accessory proteins and export motifs in the cargo molecules contribute to diversification of the routing of individual transporters. The following example may document the complexity of the process. All amino acid permeases in yeast are destined for cell surface delivery, and their capture to the COPII vesicles requires an ER-localized accessory protein, Shr3p (Ljungdahl et al. 1992, Gilstring et al. 1999). However, Gap1p is delivered to the plasma membrane without being incorporated into rafts (Bagnat and Simons 2002; Malínská, unpublished data), whereas the secretion of a specific arginine permease (Can1p) does require raft association (Malínská et al. 2003). What event determines, and at which step of the secretion is it decided whether the two highly homologous permeases will associate with rafts or not? Recently, it was shown that the efficiency of the cargo capture to the COPII vesicles is determined by the presence of specific signal motifs within the molecule of the individual amino acid permeases. It was shown that the Gap1p capture requires a COOH-terminal diacidic signal, whereas efficient packing of Can1p relies in addition on the ER export determinants outside the COOH-terminal domain (Malkus et al. 2002). Can the raft association be determined by analogous internal motifs within the Can1p molecule? Another striking difference was observed in the localization pattern of these two permeases within the plasma membrane. Visualization by the means of confocal fluorescence microscopy revealed that the Can1-GFP fusion protein was organized in bulky clusters, whereas Gap1p was distributed homogenously in the plasma membrane of yeast (Bagnat and Simons 2002; Malínská, unpublished results). Although not yet proven, it also may be possible that the clustering of one permease and not of the other is related to the raft association with some type of cytoskeletal elements. However, the relevance of the different distribution within the plasma membrane of the two permeases above, that are both functionally and structurally so similar, has not yet been elucidated.

6 Clustering and oligomerization of transporter molecules in the plasma membrane

When cells are examined in situ after extraction by cold Triton X-100, a large fraction of the cell surface (70-80%) remains unextracted. This is interpreted as indicating that most of the plasma membrane of cells is in a more ordered state (l_o) that has been attributed to rafts (Hao et al. 2001; Mayor and Maxfield 1995), which led to a revision of the original fluid mosaic model concept of biological membranes (Singer and Nicolson 1972).

New technologies employed in membranology allowed visualization of both rafts and raft-associated proteins, and revealed that a number of them are organized in large aggregates (clusters or patches), in which the mobility of the residing proteins is restricted. As documented by means of GFP technology in yeast, patches can occupy most of the cell surface (Malínská et al. 2003). The density of patches within the plasma membrane may also prevent the free movement of proteins not residing in rafts.

Double fluorescent-labeling experiments performed in several types of mammalian cells suggest that the patches are assembled by association of glycolipid-enriched domains with the actin cytoskeleton (Rodgers and Zavzavadjian 2001; Jacobson and Dietrich 1999 and reference within). In brush border membranes of cultured renal proximal tubular cells, sodium-glucose cotransporter, SGLT1, co-clustered with a GPI raft-anchored protein, 5'-nucleosidase, and vimentin, a cytosceleton filamental protein. The absence of vimentin in rafts selectively reduces the transporter activity through preclusion of SGLT1 localization to rafts (Runembert et al. 2002). Intracellular trafficking dynamics of human reduced foliate carrier (hRFC)-containing vesicles and hRFC delivery to the cell surface were shown to be critically dependent on intact microtubules but not microfilaments (Marchant et al. 2002). The same observation was reported for human thiamine transporter hTHTR1 (Subramanian et al. 2003).

Double fluorescent labeling of several transporters residing in the plasma membrane of *S. cerevisiae* revealed that these proteins occupied at least three distinct subcompartments (Malínská et al. 2003). Can1p and Pma1p are organized in non-overlapping patches, whereas Hxt1p (hexose uptake facilitator) is evenly distributed within the plasma membrane. The patches are stable for at least for 30 minutes, which also suggests a sort of stabilization, e.g. by an association with some elements of the cytoskeleton.

Recent findings indicate that oligomerization is a general feature of secondary transport systems (reviewed in Veenhoff et al. 2002). Questions arise as to whether the oligomerization is of functional and/or regulatory significance. It has been suggested that cooperation between subunits of sugar transporters, LacS (Veenhoff et al. 2001) and GLUT1 (Zottola et al. 1995), plays a role in reorienting the empty binding sites. Biochemical evidence suggests that neurotransmitter transporters exist as oligomeric complexes in cells (Kilic and Rudnick 2000; Schmid et al. 2001; Eskandari et al. 2000; Hastrup et al. 2001) but the relevance of this process for the transporter function is not clear. Recently, assembly of human dopamine transporter monomers was shown to play a crucial role for both the expression and function of this protein (Torres et al. 2003).

Members of primary solute-transport systems (ATP-binding cassette superfamily) have also been shown to occur in oligomeres. Formation of a multimer of 8-10 monomers was reported for the *Schizosaccharomyces pombe* H^+- ATPase (Dufour and Goffeau 1980), *Neurospora crassa* H^+- ATPase was observed to form hexamers (Chadwick et al. 1987; Auer et al. 1998), whereas Pma1p from *S. cerevisiae* capable for the surface delivery was found as a homododecamer (Lee et al. 2002). The Pma1p oligomerization was shown to be necessary for raft association and cell surface delivery (Bagnat et al. 2001; Lee et al. 2002), whereby, it might not be a prerequisite for H^+-ATPase activity, since a monomeric form of the same enzyme from *N. crassa* could be reconstituted *in vitro* as a fully active protein (Goormaghtigh et al. 1987).

7 Lipid function as molecular chaperons

Delivery of a transporter molecule to the plasma membrane is initiated by export of the properly folded and assembled molecule from the ER to the Golgi apparatus (Scheckman and Orci, 1996). The folding can be assisted by, and sometimes strictly requires, the participation of molecular chaperones. Until recently, transient assistance in the protein folding has been ascribed exclusively to specific protein molecules. However, indications exist that a lipid molecule like, PE, might function as a chaperone, similarly as described above for Lac Y permease in the bacterial system. The studies carried out with a mutant of *S. cerevisiae,* lacking all the biosynthetic pathways of PE, revealed a specific effect on the activity of proton motive force-driven uptake systems in the plasma membrane (Robl et al. 2001). It was documented that PE depletion primarily affects the delivery of, e.g. arginine permease Can1p through the secretory pathway to the plasma membrane. The effect is specific since the PE depletion causes complete arresting of Can1p in the internal compartments of the secretory pathway, whereas, it does not interfere substantially with the trafficking of other plasma membrane proteins like Pma1p and Hxt1p (Opekarová et al. 2002). This observation can be interpreted as being due to the absence of the specific chaperone, PE, where the incorrectly folded Can1p is recognized by the mechanisms of a quality control (Ellgaard et al. 1999) and is retained in, or rerouted to, the ER. It is noteworthy that PE depletion does not affect the Can1p association with rafts (Opekarová, unpublished results). A chaperone function has also been suggested for phosphatidyl serine in the trafficking of CFTR (cystic fibrosis transmembrane conductance regulator) (Eidelman et al. 2002).

8 Conclusions

During the last two decades, evidence has been accumulated documenting the specific functions of lipids that are indispensable for many regulatory functions. Thus, besides their structural function, lipids were shown to be directly involved in signal transduction (Brown and London 1998; Simons and Ikonen 1997). A definite lipid composition is required for stabilization of individual membrane proteins and for their conformational changes accompanying, e.g. substrate translocation. Specific lipids were revealed to be strictly required for proper functioning of a number of membrane proteins and, in all 3D structures available so far, several lipid species were shown to form an integral part of these protein complexes (reviewed in Opekarová and Tanner 2003).

The raft concept provided a new insight into membrane organization. The most apparent roles of rafts appeared to be in the sorting and vesicle formation stages. This review collects evidence justifying the hypothesis that the sorting and trafficking of a number of integral plasma membrane proteins is interconnected with the sorting and trafficking of sphingolipids and associated sterols. Due to the enrichment with sphingolipids and sterols, the lipid bilayer is thickened along the secretory pathway. Thermodynamic conveniences of the membrane thickness matching with the length of hydrophobic parts of a membrane protein together with putative motifs therein determining a specific association with definite lipids represent one significant determinant controlling protein sorting towards the plasma membrane.

The raft association poses a prerequisite for sorting and trafficking of a number of plasma membrane transporters, but neither is sufficient for this, nor represents an indispensable requirement for the delivery of all transporters to the cell surface. In addition to the secretory vesicle movement as the main trafficking vehicle, cytoskeletal connections of rafts and raft- associated proteins are being increasingly studied.

Acknowledgement

This study was supported by grants from Grant Agency of Czech Republic (204/02/143) and partially the Deutsche Forschungsgemeinschaft (DFG/SP1108). My grateful thanks belong to Professor Widmar Tanner (University of Regensburg) for his encouragement, fruitful discussion, and careful reading of the manuscript.

References

Akhter S, Kovbasnjuk O, Li X, Cavet M, Noel J, Arpin M, Hubbard AL, Donowitz M (2002) Na^+/H^+ exchanger 3 is in large complexes in the center of the apical surface of proximal tubule-derived OK cells. Am J Physiol Cell Physiol 283:927-940

Anderson RGW, Jacobson K (2003) A role for lipid shells in targeting proteins to caveolae, rafts, and other lipid domains. Science 296:1821-1825

Auer M, Scarborough GA, Kuhlbrandt W (1998) Three-dimensional map of the plasma membrane H^+-ATPase in the open conformation. Nature 392:840-843

Bagnat M, Chang A, Simons K (2001) Plasma membrane proton ATPase Pma1p requires raft association for surface delivery in yeast. Mol Biol Cell 12:4129-4138

Bagnat M, Keranen S, Shevchenko A, Simons K (2000) Lipid rafts function in biosynthetic delivery of proteins to the cell surface in yeast. Proc Natl Acad Sci USA 97:3254-3259

Bagnat M, Simons K (2002) Cell surface polarization during yeast mating. Proc Natl Acad Sci USA 99:14183-14188

Baumann A, Ribon V, Kanzaki M, Thurmond DC, Mora S, Shigematsu S, Bickel PE, Pessin JE, Saltiel AR (2000) CAP defines a second signalling pathway required for insulin-stimulated glucose transport. Nature 407:202-207

Bishop WR, Bell RM (1988) Functions of diacylglycerol in glycerolipid metabolism, signal transduction and cellular transformation. Oncogene Res 2:205-18

Bogdanov M, Haeacock PN, Dowhan W (2002) A polytopic membrane protein displays a reversible topology dependent on membrane lipid composition. EMBO J 21: 2107-2116

Bogdanov M, Heacock PN, Dowhan W (1998) Phospholipid-assisted protein folding: phosphatidylethanolamine is required at a late step of the conformational maturation of the polytopic membrane protein lactose permease. EMBO J 17:5255-52664

Bossuyt J, Taylor BE, James-Kracke M, Hale CC (2002) The cardiac sodium-calcium exchanger associates with caveolin-3. Ann NY Acad Sci 976:197-204

Bravo-Zehnder M, Orio P, Norambuena A, Wallner M, Meera P, Toro L, Latorre R, Gonzalez A (2000) Apical sorting of a voltage- and Ca^{2+}-activated K^+ channel alpha - subunit in Madin-Darby canine kidney cells is independent of N-glycosylation. Proc Natl Acad Sci USA 97:13114-13119

Bretscher M, Munro S (1993) Cholesterol and the Golgi apparatus. Science 261:1280-1281

Brown DA, London E (1997) Structure of detergent-resistant membrane domains: Does phase separation occur in biological membranes? Biochem Biophys Res Commun 240:1-7

Brown DA, London E (1998) Functions of lipid rafts in biological membranes. Annu Rev Cell Dev Biol 14:111-136

Brown DA, London E (2000) Structure and function of sphingolipid- and cholesterol-rich membrane rafts. J Biol Chem 275:17221-17224

Brown DA, Rose JK (1992) Sorting of GPI-anchored proteins to glycolipid-enriched membrane subdomains during transport to the apical cell surface. Cell 68:533-544

Bullis BL, Li X, Singh DN, Berthiaume LG, Fliegel L (2002) Properties of the Na^+/H^+ exchanger protein. Detergent-resistant aggregation and membrane microdistribution. Eur J Biochem 269:4887-4895

Chadwick CC, Goormaghtigh E, Scarborough GA (1987) A hexameric form of the *Neurospora crassa* plasma membrane H^+-ATPase. Arch Biochem Biophys 252:348-356

Chamberlain LH, Gold GW (2002) The vesicle- and target-SNARE proteins that mediate Glut4 vesicle fusion are localized in detergent-insoluble lipid rafts present on distinct intracellular membranes. J Biol Chem 277:49750-49754

Chiang SH, Baumann A, Kanzaki M, Thurmond DC, Watson RT, Neudauer CL, Macara IG, Pessin JE, Saltiel AR (2001) Insulin-stimulated GLUT4 translocation requires the CAP-dependent activation of TC10. Nature 410:944-948

Cornea RL, Thomas DD (1994) Effects of membrane thickness on the molecular dynamics and enzymatic activity of reconstituted Ca^{2+}-ATPase. Biochemistry 33:2912-2920

Dalbey RE, Chen M, Jiang F, Samuelson JC (2000) Understanding the insertion of transporters and other membrane proteins. Curr Opin Cell Biol 12:435-442

Daum G, Lees ND, Bard M, Dickson R (1998) Biochemistry, cell biology and molecular biology of lipids of *Saccharomyces cerevisiae*. Yeast 16:1471-1510

de Kruijff B (1997) Lipid polymorphism and biomembrane function. Curr Opin Chem Biol 1:564-569

Delling M, Wischmeyer E, Dityatev A, Sytnyk V, Veh RW, Karschin A, Schachner M (2002) The neural cell adhesion molecule regulates cell-surface delivery of G-protein-activated inwardly rectifying potassium channels via lipid rafts. J Neurosci 22:154-164

Demeule M, Jodoin J, Gingras D, Beliveau R (2000) P-glycoprotein is localized in caveolae in resistant cells and in brain capillaries. FEBS Lett 466:219-224

Dowhan W (1998) Genetic analysis of lipid-protein interactions in *Escherichia coli* membranes. Biochim Biophys Acta 1376:455-466

Dufour JP, Goffeau A (1980) Molecular and kinetic properties of the purified plasma membrane ATPase of the yeast *Schizosaccharomyces pombe*. Eur J Biochem 105:154-145

Dupre S, Haguenauer-Tsapis R (2003) Raft partitioning of the yeast uracil permease during trafficking along the endocytic pathway. Traffic 4:83-96

Eidelman O, BarNoy S, Razin M, Zhang J, McPhie P, Lee G, Huang Z, Sorscher EJ, Pollard HB (2002) Role for phospholipid interactions in the trafficking defect of Delta F508-CFTR. Biochemistry 41:11161-11170

Eisenkolb M, Zenzmaier C, Leitner E, Schneiter R (2002) A specific structural requirement fir ergosterol in long-chain fatty acid synthesis mutants important for maintaining raft domains in yest. Mol Biol Cell 13:4414-4428

Ellgaard L, Molinari M, Helenius A (1999) Setting the standards: quality control in the secretory pathway. Science 286:1882-1888

Epand RM (1998) Lipid polymorphism and protein-lipid interactions. Biochim Biophys Acta 1376:353-368

Eskandari S, Kreman M, Kavanaugh MP, Wright EM, Zampighi GA (2000) Pentameric assembly of a neuronal glutamate transporter. Proc Natl Acad Sci USA 97:8641-8646

Funato K, Riezman H (2001) Vesicular and nonvesicular transport of ceramide from ER to the Golgi in yeast. J Cell Biol 155:949-959

Futerman AH, Stieger B, Hubbard AL, Pagano RE (1990) Sphingomyelin synthesis in rat livers occurs predominantly at the *cis* and *medial* cisternae of the Golgi apparatus. J Biol Chem 265:8650-8657

Galbiati F, Razani B, Lisanti MP (2001) Emerging themes in lipid rafts and caveolae. Cell 106:403-411

Gilstring CF, Melin-Larsson M, Ljungdahl PO (1999) Shr3p mediates specific COPII coatomer-cargo interactions required for the packaging of amino acid permeases into ER-derived transport vesicles. Mol Biol Cell 10:3549-3565

Goormaghtigh E, Chadwick CC, Scarborough GA (1987) A hexameric form of the *Neurospora crassa* plasma membrane H^+-ATPase. Arch Biochem Biophys 252:348-356

Grove SN, Bracker CE, Morre DJ (1968) Cytomembrane differentiation in the endoplasmic reticulum-Golgi apparatus-vesicle complex. Science 161:171-173

Gu M (1997) Ceramide profiling of complex lipid mixtures by electrospray ionization mass spectrometry. Anal Biochem 244:347-356

Hamamoto S, Salama N, Rexach MF, Barlowe C, Orci L, Yeung T, Hosobuchi M, Ravazzola M, Amherdt M, Schekman R (1994) COPII: A membrane coat formed by sec proteins that drive vesicle budding from the endoplasmic reticulum. Cell 77:895–907

Han G, Gable K, Kohlwein SD, Beaudoin F, Napier JA, Dunn T (2002) The *Saccharomyces cerevisiae* YBR159w gene encodes the 3-ketoreductase of the microsomal fatty acid elongase. J Biol Chem 277:35440-35449

Hanwell D, Ishikawa T, Saleki R, Rotin D (2002) Trafficking and cell surface stability of the epithelial Na^+ channel expressed in epithelial Madin-Darby canine kidney cells. J Biol Chem 277:9772-9779

Hao M, Mukherjee S, Maxfield FR (2001) Cholesterol depletion induces large scale domain segregation in living cell membranes. Proc Natl Acad Sci USA 98:13072-13077

Harris DS, Slot JW, Geuze HJ, James DE (1992) Polarized distribution of glucose transporter isoforms in Caco-2 cells. Proc Natl Acad Sci USA 89:7556-7560

Hartmann MA (1998) Plant sterols and the membrane environment. Trends Plant Sci 3:170-175

Hastrup H, Karlin A, Javitch JA (2001) Symmetrical dimer of the human dopamine transporter revealed by cross-linking Cys-306 at the extracellular end of the sixth transmembrane segment. Proc Natl Acad Sci USA 98:10055-10060

Hearn JD, Lester RL, Dickson RC (2003) The uracil transporter Fur4p associates with lipid rafts. J Biol Chem 278:3679-3686

Hechtberger P, Zinser E, Saf R, Hummel K, Paltauf F, Daum G (1994) Characterization, quantification and subcellular localization of inositol-containing sphingolipids of the yeast, *Saccharomyces cerevisiae*. Eur J Biochem 225:641-649

Hill WG, An B, Johnson JP (2002) Endogenously expressed epithelial sodium channel is present in lipid rafts in A6 cells. J Biol Chem 277:33541-33544

Hirschberg K, Rodger J, Futerman AH (1993) The long chain sphingoid base of sphingolipids is acylated at the cytosolic surface of the endoplasmic reticulum in rat liver. Biochem J 290:751-757

Holthuis JCM, Pomorski T, Raggers RJ, Sprong H, van Meer G (2001) The organizing potential of sphingolipids in intracellular membrane transport. Physiol Rev 81:1689-1723

Ikonen E (2001) Roles of lipid rafts in membrane transport. Curr Opin Cell Biol 13:470-477

Jacobson K, Dietrich C (1999) Looking at lipid rafts? Trends in Cell Biol 9:87-91

Jeckel D, Karrenbauer A, Birk R, Schmidt RR, Wieland F (1990) Sphingomyelin is synthesized in the *cis* Golgi. FEBS Lett 261:155-157

Joost HG, Thorens B (2001) The extended GLUT-family of sugar/polyol transport facilitators: nomenclature, sequence characteristics, and potential function of its novel members. Mol Membr Biol 18:247-256

Keller P, Simons K (1997) Post-Golgi biosynthetic trafficking. J Cell Sci 110:3001-3009

Kent C (1995) Eukaryotic phospholipid biosynthesis. Annu Rev Biochem 64:315-343

Kilic F, Rudnick G (2000) Oligomerization of serotonin transporter and its functional consequences. Proc Natl Acad Sci USA 97:3106-3111

Kizhatil K, Albritton LM (2002) System y+ localizes to different membrane subdomains in the basolateral plasma membrane of epithelial cells. Am J Physiol Cell Physiol 283:1784-1794

Kohlwein SD, Daum G, Schneiter R, Paltauf F (1996) Phospholipids: synthesis, sorting, subcellular traffic. The yeast approach. Trends Cell Biol 6:260-266

Lange Y, Swaisgood MH, Ramos BV, Steck TL (1989) Plasma membranes contain half the phospholipid and 90% of the cholesterol and sphingomyelin in cultured human fibroblasts. J Biol Chem 264:3786-3793

Lee MC, Hamamoto S, Schekman R (2002) Ceramide biosynthesis is required for the formation of the oligomeric H+-ATPase Pma1p in the yeast endoplasmic reticulum. J Biol Chem 277:22395-22401

Levine TP, Wiggins CA, Munro S (2000) Inositol phosphorylceramide synthase is located in the Golgi apparatus of *Saccharomyces cerevisiae*. Mol Biol Cell 11:2267-2281

Li X, Galli T, Leu S, Wade JB, Weinman EJ, Leung G, Cheong A, Louvard D, Donowitz M (2001) Na+-H+ exchanger 3 (NHE3) is present in lipid rafts in the rabbit ileal brush border: a role for rafts in trafficking and rapid stimulation of NHE3. J Physiol 537:537-552

Ljungdahl PO, Gimeno CJ, Styles CA, Fink GR (1992) SHR3: a novel component of the secretory pathway specifically required for localization of amino acid permeases in yeast. Cell 71:463-478

Malínská K, Malínský J, Opekarová M, Tanner W (2003) Visualization of protein compartmentation within the plasma membrane of living yeast cells. Mol Biol Cell 14:4427-4436

Malkus P, Jiang F, Scheckman R (2002) Concentrative sorting of secretory cargo proteins into COPII vesicles. J Cell Biol 159:915-921

Marchant JS, Subramanian VS, Parker I, Said HM (2002) Intracellular trafficking and membrane targeting mechanisms of the human reduced folate carrier in mammalian epithelial cells. J Biol Chem 277:33325-33333

Martens JR, Navarro-Polanco R, Coppock EA, Nishiyama A, Parshley L, Grobaski TD, Tamkun MM (2000) Differential targeting of Shaker-like potassium channels to lipid rafts. J Biol Chem 275:7443-7446

Mayor S, Maxfield FR (1995) Insolubility and redistribution of GPI-anchored proteins at the cell surface after detergent treatment. Mol Biol Cell 6:929-944

Muniz M, Morsomme P, Riezman H (2001) Protein sorting upon exit from the endoplasmic reticulum. Cell 104:313-320

Muniz M, Riezman H (2000) Intracellular transport of GPI-anchored proteins. EMBO J 19:10-15

Munro S (1998) Localization of proteins to the Golgi apparatus. Trends Cell Biol 8:11-15

Oh CS, Toke DA, Mandala S, Martin CE (1997) *ELO2* and *ELO3*, homologues of the *Saccharomyces cerevisiae ELO1* gene, function in fatty acid elongation and are required for sphingolipid formation. J Biol Chem 272:17376-17384

Opekarová M, Robl I, Tanner W (2002) Phosphatidyl ethanolamine is essential for targeting the arginine transporter Can1p to the plasma membrane of yeast. Biochim Biophys Acta 1564:9-13

Opekarová M, Tanner W (2003) Specific lipid requirements of membrane proteins-a putative bottleneck in heterologous expression. Biochim Biophys Acta 1610:11-22

Pascoe WS, Inukai K, Oka Y, Slot JW, James DE (1996) Differential targeting of facilitative glucose transporters in polarized epithelial cells. Am J Physiol 271:547-554

Patton JL, Lester RL (1991) The phosphoinositol sphingolipids of *Saccharomyces cerevisiae* are highly localized in the plasma membrane. J Bacteriol 173:3101-3108

Rayner J, Pelham HR (1997) Transmembrane domain dependent sortingof proteins to the ER and plasma membrane in yeast. EMBO J 16:1832-1841

Rietveld A, Simons K (1998) The differential miscibility of lipids as the basis for the formation of functional membrane rafts. Biochim Biophys Acta 1376:467-479

Robl I, Grassl R, Tanner W, Opekarová M (2001) Construction of phosphatidyl ethanolamine-less strain of *Saccharomyces cerevisiae*. Effect on amino acid transport. Yeast 18:251-260

Rodgers W, Zavzavadjian J (2001) Glycolipid-enriched membrane domains are assembled into membrane patches by associating with the actin cytoskeleton. Exp Cell Res 276:173-183

Roper K, Corbeil D, Huttner WB (2000) Retention of prominin in microvilli reveals distinct cholesterol-based lipid micro-domains in the apical plasma membrane. Nat Cell Biol 2:582-592

Ros-Baro A, Lopez-Iglesias C, Peiro S, Bellido D, Palacin M, Zorzano A, Camps M (2001) Lipid rafts are required for GLUT4 internalization in adipose cells. Proc Natl Acad Sci USA 98:12050-12055

Rothman JE, Wieland FT (1996) Protein sorting by transport vesicles. Science 272:227–234

Runembert I, Queffeulou G, Federici P, Vrtovsnik F, Colucci-Guyon E, Babinet C, Briand P, Trugnan G, Friedlander G, Terzi F (2002) Vimentin affects localization and activity of sodium-glucose cotransporter SGLT1 in membrane rafts. J Cell Sci 115:713-724

Sakyo T, Kitagawa T (2002) Differential localization of glucose transporter isoforms in non-polarized mammalian cells: distribution of GLUT1 but not GLUT3 to detergent-resistant membrane domains. Biochim Biophys Acta 1567:165-175

Sankaram MB, Thompson TE (1990) Interaction of cholesterol with various glycerophospholipids and sphingomyelin. Biochemistry 29:10670-10675

Shlyonsky VG, Mies F, Sariban-Sohraby S (2003) Epithelial sodium channel activity in detergent-resistant membrane microdomains. Am J Physiol Renal Physiol 284:182-188

Schekman R, Orci L (1996) Coat proteins and vesicle budding. Science 271:1526–1533

Schmid JA, Scholze P, Kudlacek O, Freissmuth M, Singer EA, Sitte HH (2001) Oligomerization of the human serotonin transporter and of the rat GABA transporter 1 visualized by fluorescence resonance energy transfer microscopy in living cells. J Biol Chem 276:3805-3810

Schroeder RJ, Ahmed SN, Zhu Y, London E, Brown DA (1998) Cholesterol and sphingolipid enhance the Triton X-100 insolubility of glycosylphosphatidylinositol-anchored proteins by promoting the formation of detergent-insoluble ordered membrane domains. J Biol Chem 273:1150-1157

Schuck S, Honsho M, Ekroos K, Shevchenko A, Simons K (2003) Resistance of cell membranes to different detergents. Proc Natl Acad Sci USA 100:5795-5800

Simons K, Ikonen E (1997) Functional rafts in cell membranes. Nature 387:569-572

Singer SJ, Nicolson GL (1972) The fluid mosaic model of the structure of cell membranes. Science 175:720-731

Sprong H, van den Sluijs P, van Meer G (2001) How proteins move lipids and lipids move proteins. Nature Rev Mol Cell Biol 2:504-513

Starling AP, East JM, Lee AG (1993) Effects of phosphatidylcholine fatty acid chain length on calcium binding and other functions of the $(Ca^{2+}-Mg^{2+})$-ATPase. Biochemistry 32:1593-1600

Subramanian VS, Marchant JS, Parker I, Said HM (2003) Cell biology of the human thiamine transporter-1 (hTHTR1). Intracellular trafficking and membrane targeting mechanisms. J Biol Chem 278:3976-3984

Torres GE, Carneiro A, Seamans K, Fiorentini C, Sweeney A, Yao W-D, Caron MG (2003) Oligomerization and trafficking of the human dopamine transporter. J Biol Chem 278:2731-2739

Umebayashi K, Nakano A (2003) Ergosterol is required for targeting of tryptophan permease to the yeast plasma membrane. J Cell Biol 161:1117-1131

van Meer G (1989) Lipid traffic in animal cells. Annu Rev Cell Biol 5:247-275

van Meer G (1998) Lipids of the Golgi membrane. Trends Cell Biol 8:29-33

van Meer G, Holthuis JCM (2000) Sphingolipid transport in eukaryotic cells. Biochim Biophys Acta 1486:145-170

van Voorst F, de Kruijff B (2000) Role of lipids in the translocation of proteins across membranes. Biochem J 347:601-612

Veenhoff LM, Heuberger EH, Poolaman B (2002) Quartery structure and function of transport proteins. Trends Biochem Sci 27:242-249

Veenhoff LM, Heuberger EH, Poolman B (2001) The lactose transport protein is cooperative dimer with two sugar translocatioon pathways. EMBO J 20:3056-3062

Zanolari B, Friant S, Kunato K, Sutterlin C, Stevenson BJ, Riezman H (2000) Sphingolipid base synthesis requirement for endocystosis in *Saccharomyces cerevisiae*. EMBO J 19:2824-2833

Zottola RJ, Cloherty EK, Coderre PE, Hansen A, Hebert DN, Carruthers A (1995) Glucose transporter function is controlled by transporter oligomeric structure. A single intramolecular disulfide promotes Glut1 tetramerization. Biochemistry 34:9734-9741

Opekarová, Miroslava

Institute of Microbiology, Czech Academy of Sciences, 142 20 Prague, Czech Republic

opekaro@biomed.cas.cz

Trafficking of vesicular transporters to secretory vesicles

Vania F. Prado, Marc G. Caron, and Marco A.M. Prado

Abstract

Three families of proteins that mediate the uptake of neurotransmitters into secretory vesicles have been identified. One family includes two vesicular monoamine transporters (VMAT1 and VAMT2) and the vesicular acetylcholine transporter (VAChT). A second includes the GABA transporter (VGAT), whereas, the third includes three isoforms of the vesicular glutamate transporter (VGLUT1, VGLUT2, and VGLUT3). VAChT, VGAT, and the three VGLUTs are present in axon terminals, predominantly in synaptic vesicles (SVs), conversely VMAT2 can be found in SVs and in large dense core vesicles. The cell biological mechanisms by which VAChT and VMAT2 traffic to different populations of neurosecretory vesicles have started to be uncovered. In this review, we discuss the cellular routes taken by proteins targeted to different secretory organelles and summarize the sequence motifs that have been identified in the cytoplasmic domains of VMAT2 and VAChT that are involved both in endocytosis and in targeting of these transporters to secretory vesicles.

1 Classes of vesicular transporters

Loading of classical neurotransmitters into specialized secretory vesicles is an essential step for proper function of neural, endocrine, and even some immune/inflammatory cells. Vesicular transport allows storage and subsequent exocytosis of these messengers in response to specific stimulation (Burgoyne and Morgan 2003). The loading process depends on the activity of vesicular transporter proteins, which recognize neurotransmitters and move them from the cytosol into the vesicle.

Molecular cloning techniques have identified three families of proteins that mediate the uptake of neurotransmitters into secretory vesicles. One family of proteins includes two vesicular monoamine transporters (VMAT1 and VMAT2) (Erickson et al. 1992; Liu et al. 1992) and the vesicular acetylcholine transporter (VAChT) (Alfonso et al. 1993; Bejanin et al. 1994; Erickson et al. 1994; Roghani et al. 1994). A second family includes the γ-aminobutyric acid (GABA) transporter (VGAT) (McIntire et al. 1997; Chaudhry et al. 1998), whereas, the third family includes three isoforms of the vesicular glutamate transporter (VGLUT1, VGLUT2 and VGLUT3) (Takamori et al. 2000, 2001, 2002; Herzog et al. 2001;

Topics in Current Genetics, Vol. 9
E. Boles, R. Krämer (Eds.): Molecular Mechanisms Controlling Transmembrane Transport
DOI 10.1007/b95777 / Published online: 9 March 2004
© Springer-Verlag Berlin Heidelberg 2004

Fremeau Jr et al. 2001, 2002; Varoqui et al. 2002; Schafer et al. 2002). Although the transport activity of these three classes of proteins depends on a proton electrochemical gradient generated by a vacuolar H^+-ATPase, differences in bioenergetics are observed. Vesicular monoamine and ACh transporters exchange luminal protons for cytoplasmic transmitter, and their transport activity depends primarily on the chemical component (Δ pH) of the electrochemical gradient. In contrast, vesicular glutamate transporter activity depends primarily on the electrical component ($\Delta \Psi$) of the electrochemical gradient, and vesicular GABA transporter activity seems to depend more equally on both Δ pH and $\Delta \Psi$ (Schuldiner et al. 1995; Liu and Edwards 1997b). Differences in bioenergetics probably result from their different structure, since these three families of proteins show no sequence similarity.

2 Classes of secretory vesicles

Neurons contain different types of secretory vesicles that store neurotransmitters and undergo regulated exocytosis. Synaptic vesicles (SVs) are small organelles (40-50 nm in diameter) that cluster at the nerve terminal and store neurotransmitters, such as acetylcholine, dopamine, GABA and glutamate. These organelles mediate the fast and localized transmitter release necessary for information processing. In order to maintain sustained levels of neurotransmitter release, SVs have to go through a rapid process of recycling and refilling that occur locally at the nerve terminal (Hannah et al. 1999; Sudhof 2000).

Large dense core vesicles (LDCVs; 100-200 nm in diameter) appear in the cell body, dendrites and also at the nerve terminal and mediate the regulated secretion of neuropeptides and in some cases also of small messengers such as monoamines (Bruns and Jahn 1998). LDCVs exocytosis has been demonstrated to be slower than that of SVs and controlled by different stimuli (Burgoyne and Morgan 2003). Due to its protein cargo, LDCVs biogenesis and refilling depends on the trans-Golgi network (TGN). Therefore, recycling of LDCV membrane components is much slower than recycling of SVs, and it is not known whether multiple rounds of cycling can take place for these secretory vesicles (Tooze 1998; Tooze et al. 2001; Burgoyne and Morgan 2003).

A third class of secretory organelle has been demonstrated to store monoamines in peripheral catecholaminergic neurons. They are bigger than SVs, ranging in size from 50-60 nm, show an electron dense core after certain chemical fixations (Thureson-Klein 1983), and therefore they are named small dense core vesicles (SDCVs). Very little is currently known about the biogenesis of SDCV. It is likely that they are formed in endosomal compartments at the nerve terminal, using membrane components of both LDCVs and SVs that have been retrieved after exocytosis (Bauerfeind et al. 1995; Winkler 1997; Partoens et al. 1998).

3 Cellular routes taken by vesicular proteins

3.1 SV proteins

The specialized role of SVs on neuronal transmission relies on a particular set of membrane protein components and most of these proteins have been characterized. In addition to vesicular transporters, SV membrane contains proteins, such as a H^+-ATPase, which is responsible for intravesicular acidification, synaptobrevin II/vesicle associated membrane protein-2 (VAMP-2) that mediate membrane fusion, synaptotagmins that are involved in transmitting the Ca^{2+} signal to the exocytotic machinery, and small GTPases of the Rab family (Rab-3 and possibly Rab-5) that are probably involved in vesicle attachment to the plasma membrane. Other proteins, such as synaptophysin, synaptogyrin, synapsin, and secretory carrier membrane proteins, also have important functions in synaptic vesicle cycle (Sudhof 1995, 2000; Fernandez-Chacon and Sudhof 1999). Several of these proteins are specific to SVs; however, their trafficking pathways and the mechanisms responsible for their targeting to these organelles remain unknown. Inspection of amino acid sequence and predicted secondary structure of these proteins has not revealed any common sorting motif that could be responsible for their specific targeting to SVs, suggesting that these motifs are not easily recognized, or that SV proteins rely on distinct signals to reach their final destination.

It is generally accepted that SVs integral membrane proteins are synthesized in the rough endoplasmic reticulum, pass through the Golgi complex, and are packaged into carrier vesicles, which deliver their content to the plasma membrane by exocytic fusion (Fig. 1). There is evidence demonstrating that these carrier vesicles are different from mature SVs both in morphology and protein composition (Tsukita and Ishikawa 1980; Okada et al. 1995; Nakata et al. 1998), giving support to the hypothesis that SVs are synthesized at the nerve terminal. At this sorting step, one important question is not completely understood. Are these carrier vesicles selectively sorted to axons and reach the plasma membrane only at the nerve terminal or are they also delivered to somatodendritic plasma membrane? Although earlier studies suggested that SV precursors that bud from the TGN are directly sorted to axons (Okada et al. 1995; Nakata et al. 1998), a more recent study indicates that VAMP-2 is delivered to the surface of axons, soma, and dendrites, and that rapid endocytosis from somatodendritic regions is important for the efficient sorting of VAMP-2 to SVs (Sampo et al. 2003). Furthermore, in neuronal cell models, synaptophysin and VAChT have been shown to be delivered to the plasma membrane at the cell body prior to targeting to SVs (Regnier-Vigouroux et al. 1991; Santos et al. 2001; Barbosa et al. 2002). These observations suggest that sorting mechanisms acting on somatodendritic endosomes may be important to target newly synthesized SV proteins to nerve terminals. Further studies are going to be necessary to determine whether other SV proteins follow the same trafficking pathway of VAMP-2, VAChT and synaptophysin. Interestingly, Ahmari et al. (2000) have suggested that vesicles containing plasma membrane proteins, vesicles containing SV proteins, and cytoplasmic components, travel together in axons

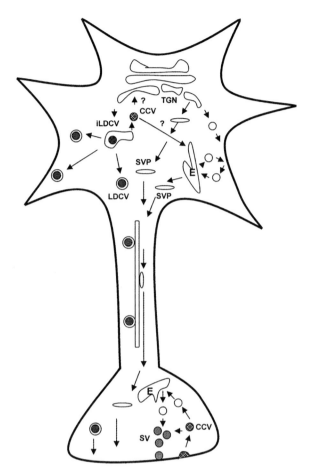

Fig. 1. Cellular routes taken by large dense core vesicle (LDCV) and synaptic vesicle (SV) membrane proteins. Membrane proteins that are going to be targeted to SVs or LDCVs traffic through different routes inside the cell. The initial sorting step occurs probably at the trans-Golgi network (TGN) where LDCVs proteins are sorted to the regulated secretory pathway and synaptic vesicles go into synaptic vesicle precursors (SVP). SVP seem to be delivered to the surface of axons, soma and dendrites, suggesting that rapid endocytosis from somatodendritic regions is importing for efficient sorting of SV proteins. Transport of SVP along axonal processes depends on different motor proteins of the kinesin superfamily. At least two pathways are responsible for SVP maturation on the nerve terminal. An extended pathway where vesicles are formed via an endosomal intermediate (E) after endocytosis, and a shorter pathway, where vesicles can be generated without a trafficking intermediate. LDCVs bud from the TGN as immature vesicles (iLDCV). Removal of certain soluble proteins, peptides and membrane proteins from iLDCV through sorting into CCV is necessary for the formation of a mature LDCV. Whether iLDCVs derived from CCVs are targeted to endosomes or recycle back to the TGN is currently unknown. LDCVs are sorted to both axonal and somatodendritic pathways (for additional details see text).

from cultured neurons as a "packet" that represent a "preformed" pre-synaptic terminal. How synaptic vesicle proteins are sorted to these conglomerates is currently unknown, but most likely depend on sorting mechanisms at the neuronal cell body.

The transport of SV precursors along axonal processes to the nerve terminal depends on motor proteins and different motor proteins of the kinesin superfamily (KIFs) carry distinct synaptic vesicle proteins to nerve-endings (Okada et al. 1995; Hirokawa et al. 1998). How the different vesicles that take the axonal route associate with their motor proteins is still poorly understood. Members of the c-Jun N-terminal kinase (JNK)-interacting protein (JIP) group, including JIP1, JIP2, and JIP3 have been suggested to be the link between certain vesicles and motor proteins and also regulate the activity of the motor by bringing JNK signaling components in close proximity to the motor (Stockinger et al. 2000; Verhey et al. 2001; Taru et al. 2002; Matsuda et al. 2003; Inomata et al. 2003). JIP3, in particular, seems to be involved with trafficking of vesicles containing SV proteins, since mutations in the *C. elegans* JIP3 homolog, UNC-16, results in mislocalization of synaptic vesicle proteins (Byrd et al. 2001).

How SV precursors mature after their delivery to the nerve terminal is still a matter of debate. Strong evidence suggests the existence of at least two pathways for SV biogenesis. An extended pathway, where the vesicles are formed via an endosomal intermediate after endocytosis, and a shorter pathway, where vesicles are generated directly without a trafficking intermediate. (Schmidt et al. 1997; Shi et al. 1998; Faundez et al. 1998; Hannah et al. 1999; Sudhof 2000). Adaptor protein (AP) complexes (AP-2 from the plasma membrane and AP-3 from endosomes) have been implicated in SV biogenesis and recycling (Gonzalez-Gaitan and Jackle 1997; Shi et al. 1998; Faundez et al. 1998; Faundez and Kelly 2000).

3.2 LDCV proteins

Even though SVs and LDCVs share a number of membrane components and are both involved in regulated exocytosis, they show different biogenesis. Consequently, membrane proteins that are going to be targeted to SVs or LDCV traffic through different routes inside the cells. The "crossroad" point is the TGN where LDCV proteins are sorted to vesicles that are going to follow the secretory pathway while SV proteins go into vesicles that possibly take the constitutive pathway.

LDCVs that bud from the TGN are classified initially as immature (iLDCVs) and they need to go through a maturation process to become LDCV (Fig. 1). No cytoplasmic coat has yet been demonstrated to be associated with budding iLDCV from the TGN. It has been proposed that lateral self-aggregation of integral membrane proteins at lipid microdomains rich in cholesterol and sphingolipids (rafts) may contribute to the formation of the nascent iLDCV (Thiele and Huttner 1998; Wang et al. 2000). The maturation process seems to involve homotypic fusion of iLDCVs and removal of certain soluble proteins, peptides and membrane proteins (Tooze 1998; Tooze et al. 2001). Sorting events at the iLDCV seem to be mediated by the AP-1 adaptor complex and clathrin-coated vesicles (CCVs). Membrane proteins that are going to be excluded from iLDCV, such as furin, mannose-6-phosphate receptors, and carboxypeptidase D recruit AP-1 in a casein kinase II phosphorylation-dependent manner (Dittie et al. 1996, 1997, 1999; Varlamov et al. 1999; Kalinina et al. 2002). Recent data suggest that iLDCV derived CCVs are targeted to endosomes (Turner and Arvan 2000), although it remains possible that the iLDCV-derived CCVs can also be targeted to the TGN. Mature LDCVs are sorted to both axonal and somatodendritic pathways where they reach the plasma membrane only after the neuron receives external stimuli.

4 Localization of vesicular transporters in neurons

The role of neurotransmitters such as ACh, GABA and glutamate on information processing depends on their release from SVs at the nerve terminal. In agreement with that, immunoelectron microscopy and biochemical fractionation have demonstrated that VAChT, VGAT, VGLUT1, VGLUT2 and VGLUT3 are present in axon terminals, predominantly in synaptic vesicles (Gilmor et al. 1996; Weihe et al. 1996; McIntire et al. 1997; Roghani et al. 1998; Schafer et al. 1998, 2002; Chaudhry et al. 1998; Takamori et al. 2000, 2001, 2002; Hayashi et al. 2001; Herzog et al. 2001; Fremeau Jr et al. 2001, 2002; Varoqui et al. 2002; Gras et al. 2002).

On the other hand, monoamines are released from both SVs and LDCVs (Bruns and Jahn 1998). Consistent with these observations, the neuronal isoform of the vesicular monoamine transporter (VMAT2), which is expressed in serotoninergic, noradrenergic, dopaminergic, histaminergic, and adrenergic neurons can be found in both secretory vesicles. In monoaminergic neurons of the rat solitary tract, VMAT2 was observed predominantly on LDCV albeit it was also present in SVs (Nirenberg et al. 1995). In dopaminergic neurons of the rat dorsolateral striatum, VMAT2 was observed almost exclusively in SVs in axon terminals (Nirenberg et al. 1997), whereas in peripheral monoaminergic neurons, VMAT2 was observed in SDCVs (Bauerfeind et al. 1995). These observations suggest that, depending on the cellular background, VMAT2 can be preferentially sorted to LDCVs, SDCVs or SVs, which may indicate that cell-specific events govern targeting of VMAT2. The VMAT1 isoform is expressed in various endocrine and neuroendocrine cells and is present in LDCV (Liu et al. 1994).

In addition to their presence in SVs (and LDCVs in the case of VMAT2), immunocytochemical analysis of brain sections show that vesicular transporters can also be found in membranous organelles at the cell body and dendrites. Although these organelles may contain newly synthesized vesicular transporters in transit to the nerve terminal, it calls attention the fact that the pattern of expression of different transporters in these organelles vary considerably. For instance, VAChT and VMAT2 immunoreactivity occur at high levels in organelles at cell body and dendrites whereas VGAT is poorly detected in these sites (Peter et al. 1995; Gilmor et al. 1996; Nirenberg et al. 1996; Chaudhry et al. 1998). Robust cell body immunostaining for VGLUT3 could be detected in the hypothalamus but in all other brain regions, VGLUT3 was only detected in varicose fibers (Schafer et al. 2002). These different patterns of expression may indicate that different sorting signals/trafficking routes may be used by the vesicular transporters to get to SVs, and suggest that cell-specific components may be important to govern the trafficking of all vesicular transporters.

5 Sorting of vesicular transporters in PC12 cells

Most of the studies trying to identify sorting motifs necessary for correct targeting to SVs and LDCVs have used a neuroendocrine cell line, the rat pheochromocytoma-derived cell line PC12 (reviewed by Liu et al. 1999; Prado et al. 2002; Prado and Prado 2003). PC12 cells are particularly useful due to the fact that they can store and release ACh from synaptic-like microvesicles (SLMV) and norepinephrine from LDCV (Greene and Rein 1977a, 1977b; Melega and Howard 1981, 1984). Subcellular fractionation and immunofluorescence of PC12 cells show that the endocrine-specific VMAT1 isoform is found in LDCVs, being mostly excluded from synaptic-like microvesicles (SLMVs; the counterpart of SVs in neuroendocrine cells) (Liu et al. 1994), and that VAChT is found predominantly in SLMV (Liu and Edwards 1997a; Varoqui and Erickson 1998). Interestingly, when VMAT2 was expressed in PC12 cells, the protein was sorted predominantly to LDCVs (Weihe et al. 1996; Varoqui and Erickson 1998) suggesting that in these cells VMAT2 and the endogenous VMAT1 are sorted using similar mechanisms, and that VAChT and VMATs may be sorted through a different pathway. Moreover, the fact that in transfected PC12 cells VMAT2 is not preferentially target to SLMVs, as it is in some neurons, may suggest that these cells lack components that are required for VMAT2 targeting to SVs.

PC12 cells have also been used to investigate sorting of the three VGLUT isoforms (Fremeau Jr et al. 2001, 2002). Subcellular fractionation indicates that they are predominantly expressed on SLMVs and immunofluorescence analysis confirms that VGLUT1, VGLUT2, and VGLUT3 colocalize with synaptophysin in the processes of nerve growth factor differentiated PC12 cells. However, in the cell bodies, all three transporters show very little colocalization with synaptophysin. In addition, while VGLUT2 and VGLUT3 are present in membranous compartments that are spread throughout the cytoplasm, VGLUT1 show a more peripheral distribution, and is concentrated in organelles close to the plasma membrane suggesting that different sorting signals may influence their targeting to SLMV in PC12 cells (Fremeau Jr et al. 2001, 2002). In fact, trafficking patterns of the VGLUT transporters may also be different in neurons, since subcellular fractionation of brain extracts demonstrates that although these three transporters localize to SVs, their pattern of expression in membranes other than SVs is different (Fremeau Jr et al. 2001, 2002). Interestingly, VGLUT1 and VGLUT2 are distributed in a complementary fashion to distinct populations of excitatory neurons, that is, VGLUT1 is mostly present in telencephalic regions and VGLUT2 expressed predominantly in the diencephalic regions of the brain and in the spinal cord

(Takamori et al. 2000, 2001; Bellocchio et al. 2000; Herzog et al. 2001; Fremeau Jr et al. 2001; Varoqui et al. 2002). This differential expression of VGLUT1 and VGLUT2 appear to correlate with an important property of the synapses in these regions, the probability of transmitter release. VGLUT1 appears to be expressed in synapses with low probability of release and VGLUT2 in synapses with high probability of release (Fremeau Jr et al. 2001). It has been suggested that differences in trafficking may be responsible for the differences in the probability of transmitter release. Although VGLUT1 and VGLUT2 show 82% amino acid identity, they clearly differ at the C-terminal tail, a region generally involved with traffic and sorting of membrane proteins, indicating that they could interact with different proteins. VGLUT3 pattern of expression is much less abundant and clearly separated from the other VGLUTs. VGLUT3 is expressed in subpopulations of inhibitory neurons, in cholinergic interneurons, monoamine neurons and glia (Schafer et al. 2002; Takamori et al. 2002; Fremeau Jr et al. 2002). Expression of VGLUT3 in subsets of GABAergic, cholinergic and monoaminergic neurons in adult brain defies Dale's principle, and indicates that vesicular storage of glutamate occur at these synapses, suggesting the possibility of co-release of neurotransmitters. How glutamatergic co-release affects neuronal communication awaits further investigation. In the case of ACh, dopamine and serotonin neurons, it could have profound implications for the role of these neurons in motor control, reward pathways, and neuropsychiatric diseases (Fremeau Jr et al. 2002).

6 Sorting motifs identified in vesicular transporters

As has been discussed above, membrane proteins that are going to be targeted to SVs or LDCV would have to traffic through different routes inside the cells. Which trafficking route a membrane protein is going to follow is usually controlled by short specific sequences (sorting motifs) located within its cytoplasmic domains. These motifs may associate directly or indirectly with adaptor protein (AP) complexes such as AP-1, AP-2 and AP-3, which assemble with clathrin during vesicular budding allowing the sorting of the membrane protein to different vesicles (Kirchhausen et al. 1997, 1999, 2002; Bonifacino and Traub 2003).

Recent studies have begun to uncover the mechanisms by which VAChT and VMAT2 traffic to different populations of neurosecretory vesicles (Liu and Edwards 1997a; Tan et al. 1998; Varoqui and Erickson 1998; Cho et al. 2000; Krantz et al. 2000; Santos et al. 2001; Waites et al. 2001; Barbosa et al. 2002). The molecular determinants involved in the VAChT trafficking to SLMVs in PC12 cells reside mainly at its C-terminal tail. This conclusion was based on the observation that a chimeric VMAT2 with the C–terminal tail exchanged with that of VAChT was sorted predominantly to SLMV (Varoqui and Erickson 1998). Likewise, it has also been suggested that the C-terminal tail of VMAT2 is sufficient to target VAChT to LDCVs (Liu et al. 1999). In fact, VAChT and VMATs show a high degree of homology among their predicted 12 transmembrane

Table 1. Sorting motifs

Protein	Motif sequence	Function
VMAT2-human (NP_003045)	KEEKMA**IL**M	Internalization; Sorting at the TGN?
VAChT-human (AAB92675)	R<u>S</u>ERDV**LL**D	Internalization; Sorting at the TGN?
Synaptotagmin I -human (BMHU1Y)	EEEVDA**ML**A	Sorting from endosomes to synaptic-like microvesicles in an AP-3 dependant manner
Tyrosinase-human (NP_000363)	PEEKQP**LL**M	Sorting from endosomes to synaptic-like microvesicles in an AP-3 dependant manner
VMAT2-human (NP_003045)	**DDEE<u>S</u>E<u>S</u>D**	Prevent VMAT2 removal from iLDCV
Furin-human (KXHUF)	**<u>S</u>D<u>S</u>EEDE**	Retrieval of the protein from iLDCV upon binding of PACS-1.

Key residues are indicated in bold type. Underlined serines are sites regulated by phosphorylation. Parentheses contain GenBank accession numbers.

segments but their C-terminal tails, which reside in the cytoplasm, are largely unrelated, supporting the idea that they could carry different sorting signals.

A dileucine like motif present at the C-terminal tail of VMAT2 (Ile-483/Leu-484; Table I) is responsible for its internalization in cultured cells (Tan et al. 1998). Studies of VMAT2 internalization in COS1 and CHO cells demonstrated that replacement of both Ile-483 and Leu-484 by alanine interferes with endocytosis and results in accumulation of the transporter at the cell surface. Amino acid residues surrounding the Ile-Leu motif (Table I) were also investigated by alanine scanning and did not show to influence internalization (Tan et al. 1998). Interestingly, single point mutants I483A and L484A showed normal internalization.

Dileucine motifs have been demonstrated to mediate several targeting events in the cell, such as internalization, sorting from TGN to endosomes, targeting to lysosomes or specialized compartments such as stimulus-responsive storage vesicles, premelanosomes and melanosomes and even targeting to synaptic vesicles from an endosomal compartment (Bonifacino and Traub 2003). Both leucines and a cluster of acidic amino acid residues preceding the leucines were found to be important for intracellular sorting of many proteins. However, the acidic residues do not seem to be always important for internalization (Pond et al. 1995; Sandoval et al. 2000). Consistent with these broad range of functions, dileucine motifs have been found to bind different AP complexes (AP-1, AP-2, AP-3) and adaptor related proteins (GGA-1, GGA-2, GGA-3) in various *in vitro* assays (Dietrich et al. 1997; Honing et al. 1998; Hofmann et al. 1999; Puertollano et al. 2001; Zhu et al. 2001; Peden et al. 2001; Nielsen et al. 2001; Shiba et al. 2002). Surrounding amino acid residues seem to influence the preference for different AP complexes and adaptor proteins (Rodionov et al. 2002; Doray et al. 2002). The fact that only one of the di-hydrophobic residues was enough to maintain endocytosis of

VMAT2 in fibroblast cells makes the VMAT2 signal somewhat different from the dileucine motifs previously described. Since no direct or indirect interaction between the dileucine like motif in VMAT2 and AP-2 has yet been described, it is not clear how this endocytic motif influences the internalization of the transporter.

The VAChT C-terminal also contains a dileucine motif (Leu485/Leu486; Table I) required for internalization (Tan et al. 1998; Santos et al. 2001; Barbosa et al. 2002). Trafficking studies of an endocytic mutant of VAChT tagged with the green fluorescent protein (GFP-VAChT L485A/L486A) in a neuronal differentiated cholinergic cell line have shown that this endocytic mutant is localized at the plasma membrane and exhibits reduced neuronal processes targeting (Santos et al. 2001). Both leucines on VAChT signal seem to be important for efficient internalization, since GFP-VAChT mutants with L485 or L486 mutated to alanine also localized to the plasma membrane (Ferreira LT, Santos MS, Prado VF, and Prado MAM, unpublished results). GST pull-down and yeast-two-hybrid experiments have shown that VAChT C-terminal domain interacts with the AP-2 adaptor complex. Moreover, the interaction was not observed when the dileucine motif was mutated to alanine (L485A/L486A-mutant, Barbosa et al. 2002). Together these results point to a crucial role for a clathrin mediated endocytic process for VAChT trafficking.

Although the presence of negative charges upstream of the dileucine like motif of VAChT and VAMT2 do not seem to influence internalization of these transporters, it has been demonstrated that negative charges at positions -4 and -5 relative to the dileucine-like motif can influence sorting to LDCVs in PC12 cells (Krantz et al. 2000). VAChT dileucine signal has a glutamic acid at position -4 relative to the dileucine and a serine at position -5 (SERDVLL), which undergoes regulated phosphorylation by protein kinase C (Cho et al. 2000; Krantz et al. 2000). Replacement of S480 by glutamate, to mimic the phosphorylation event, increases the localization of VAChT to LDCV in PC12 cells (Krantz et al. 2000). On the other hand, VMAT2 dileucine-like signal has two glutamates at these positions (Table I). Replacement of these glutamates by alanine partially redistributes VMAT2 to lighter membranes and away from LDCV (Krantz et al. 2000). Although the sorting event(s) affected by this motif has not been determined, one possibility is that a negative charge at position -5 of the dileucine motif would be important for sorting at the TGN. In this case, instead of taking the route to reach SV precursors, phosphorylated VAChT might sort to immature secretory granules and end up in LDCV (Krantz et al. 2000). However, the possibility that this motif may influence other sorting events should not be ruled out. In fact, a dileucine motif was shown to mediate sorting of synaptotagmin I and tyrosinase from endosomes to synaptic-like microvesicles in an AP-3 dependant manner (Table I) (Blagoveshchenskaya et al. 1999). VAChT is phosphorylated in nerve endings by protein kinase C (Barbosa Jr et al. 1997), suggesting that trafficking events at the nerve terminal could also be modulated. It is unlikely that in nerve endings phosphorylation of VAChT could interfere with its distribution to LDCV, since, as it was discussed previously, incorporation of proteins in LDCVs occurs at the TGN. The role of phosphorylation on VAChT activity and trafficking in nerve terminals remains unknown.

An acidic cluster present at the end of the C-terminal domain of VMAT2 that includes two serines phosphorylated by casein kinase 2 (DDEESESD) is important to target the transporter to LDCVs (Waites et al. 2001). This motif seems to influence targeting of VMAT2 after the transporter has left the TGN and is located at iLDCV (Fig. 1, Table I). Mutants that lack the acidic motif are removed from the iLDCV and, therefore, have reduced expression on LDCVs, suggesting that this motif acts as a retention signal, preventing VMAT2 removal from iLDCV. Although this motif has no counterpart in VAChT, this type of acidic cluster is found in many transmembrane proteins that are localized to the TGN at steady state (Bonifacino and Traub 2003). The TGN endoprotease furin, for instance, contains a similar acidic cluster (Table I) with two serines phosphorylated by casein kinase 2. Phosphorylation of this motif in furin promotes binding of PACS-1 and retrieval of the protein from iLDCV (Dittie et al. 1997; Wan et al. 1998). PACS-1 appears to function as a connector between phosphorylated acidic clusters and AP-1 (and possibly AP-3) (Crump et al. 2001). Likewise, in VMAT2 , replacement of the serines in the acidic motif by aspartate to mimic phosphorylation promotes the removal of VMAT2 from iLDCVs and therefore, reduces its expression on LDCVs (Waites et al. 2001). Since VAMT2 is constitutively phosphorylated by casein kinase 2 in PC12 cells (Krantz et al. 1997), its sorting to LDCVs in these cells appears to depend on a dephosphorylation event. Similarly to furin, PACS-1 was shown to bind strongly to the phosphorylated acidic cluster in VMAT2 (Waites et al. 2001). Interestingly VMAT1, which is preferentially sorted to LDCV in neuroendocrine cells, have a somewhat different acidic cluster (ENSDDPSSGE) and is not phosphorylated by casein kinase 2 (Krantz et al. 1997). Although in PC12 cells the iLDCV-derived vesicles seem to be targeted to TGN, there is no information about the trafficking of these vesicles in neurons. If in neuronal cells iLDCV-derived vesicles are destined to endosomes, as it has been described for pancreatic β-cells (Turner and Arvan 2000), phosphorylated VMAT2 could be diverted to the constitutive secretory pathway and hence to SVs.

7 Concluding remarks

Packaging neurotransmitter into SVs for rapid release at nerve terminal is essential for neuronal communication. Consistent with that, most vesicular neurotransmitter transporters are sorted to axon terminals, predominantly to SVs. Analysis of localization of the different vesicular transporters expressed in brain, however, suggests that different vesicular transporter may use different trafficking signals/ routes to get to SVs and that these differences may have an important role in the normal function of these proteins. Furthermore, for some neurons, such as the midbrain dopaminergic neurons of the substantia nigra and the ventral tegmental area, sorting of vesicular transporter to LDCVs may allow regulated release of neurotransmitters from somatodendritic sites, which might be important for the function of these neurons. Analysis of the C-terminal region of VAChT and VMAT2 has reveled important signals responsible for trafficking of these transporters to SVs and

LDCVs respectively. However, several sorting steps are only poorly understood and the cellular mechanisms responsible for recognizing the signals identified are still not determined. Trafficking studies on the VGAT and the three VGLUT could probably help to uncover the different sorting steps responsible for SVs targeting. Learning how to interfere with vesicular transporters traffic (for example by phosphorylation) in order to alter the concentration of transmitter in SVs membranes would help to understand the role of the transporters in signaling and behavior.

Acknowledgements

Marco A.M. Prado and Vania F. Prado acknowledge the support of CAPES, FAPEMIG, CNPq, PRONEX, PADCT, and IBRO.

References

Ahmari SE, Buchanan J, Smith SJ (2000) Assembly of presynaptic active zones from cytoplasmic transport packets. Nat Neurosci 3:445-451

Alfonso A, Grundahl K, Duerr JS, Han HP, Rand JB (1993) The *Caenorhabditis elegans* unc-17 gene: a putative vesicular acetylcholine transporter. Science 261:617-619

Barbosa J Jr, Clarizia AD, Gomez MV, Romano-Silva MA, Prado VF, Prado MA (1997) Effect of protein kinase C activation on the release of [3H]acetylcholine in the presence of vesamicol. J Neurochem 69:2608-2611

Barbosa J, Ferreira LT, Martins-Silva C, Santos MS, Torres GE, Caron MG, Gomez MV, Ferguson SSG, Prado MAM, Prado VF (2002) Trafficking of the vesicular acetylcholine transporter in SN56 cells: a dynamin-sensitive step and interaction with the AP-2 adaptor complex. J Neurochem 82:1221-1228

Bauerfeind R, Jelinek R, Hellwig A, Huttner WB (1995) Neurosecretory vesicles can be hybrids of synaptic vesicles and secretory granules. Proc Natl Acad Sci USA 92:7342-7346

Bejanin S, Cervini R, Mallet J, Berrard S (1994) A unique gene organization for two cholinergic markers, choline acetyltransferase and a putative vesicular transporter of acetylcholine. J Biol Chem 269:21944-21947

Bellocchio EE, Reimer RJ, Fremeau RT Jr, Edwards RH (2000) Uptake of glutamate into synaptic vesicles by an inorganic phosphate transporter. Science 289:957-960

Blagoveshchenskaya AD, Hewitt EW, Cutler DF (1999) Di-leucine signals mediate targeting of tyrosinase and synaptotagmin to synaptic-like microvesicles within PC12 cells. Mol Biol Cell 10:3979-3990

Bonifacino JS, Traub LM (2003) Signals for sorting of transmembrane proteins to endosomes and lysosomes. Annu Rev Biochem 72:395-447

Bruns D, Jahn R (1998) Monoamine transmitter release from small synaptic and large dense-core vesicles. Adv Pharmacol 42:87-90

Burgoyne RD, Morgan A (2003) Secretory granule exocytosis. Physiol Rev 83:581-632

Byrd DT, Kawasaki M, Walcoff M, Hisamoto N, Matsumoto K, Jin Y (2001) UNC-16, a JNK-signaling scaffold protein, regulates vesicle transport in *C. elegans*. Neuron 32:787-800

Chaudhry FA, Reimer RJ, Bellocchio EE, Danbolt NC, Osen KK, Edwards RH, Storm-Mathisen J (1998) The vesicular GABA transporter, VGAT, localizes to synaptic vesicles in sets of glycinergic as well as GABAergic neurons. J Neurosci 18:9733-9750

Cho GW, Kim MH, Chai YG, Gilmor ML, Levey AI, Hersh LB (2000) Phosphorylation of the rat vesicular acetylcholine transporter. J Biol Chem 275:19942-19948

Crump CM, Xiang Y, Thomas L, Gu F, Austin C, Tooze SA, Thomas G (2001) PACS-1 binding to adaptors is required for acidic cluster motif-mediated protein traffic. EMBO J 20:2191-2201

Dietrich J, Kastrup J, Nielsen BL, Odum N, Geisler C (1997) Regulation and function of the CD3gamma DxxxLL motif: a binding site for adaptor protein-1 and adaptor protein-2 *in vitro*. J Cell Biol 138:271-281

Dittie AS, Hajibagheri N, Tooze SA (1996) The AP-1 adaptor complex binds to immature secretory granules from PC12 cells, and is regulated by ADP-ribosylation factor. J Cell Biol 132:523-536

Dittie AS, Klumperman J, Tooze SA (1999) Differential distribution of mannose-6-phosphate receptors and furin in immature secretory granules. J Cell Sci 112 Pt 22:3955-3966

Dittie AS, Thomas L, Thomas G, Tooze SA (1997) Interaction of furin in immature secretory granules from neuroendocrine cells with the AP-1 adaptor complex is modulated by casein kinase II phosphorylation. EMBO J 16:4859-4870

Doray B, Bruns K, Ghosh P, Kornfeld S (2002) Interaction of the cation-dependent mannose 6-phosphate receptor with GGA proteins. J Biol Chem 277:18477-18482

Erickson JD, Eiden LE, Hoffman BJ (1992) Expression cloning of a reserpine-sensitive vesicular monoamine transporter. Proc Natl Acad Sci USA 89:10993-10997

Erickson JD, Varoqui H, Schafer MK, Modi W, Diebler MF, Weihe E, Rand J, Eiden LE, Bonner TI, Usdin TB (1994) Functional identification of a vesicular acetylcholine transporter and its expression from a "cholinergic" gene locus. J Biol Chem 269:21929-21932

Faundez V, Horng JT, Kelly RB (1998) A function for the AP3 coat complex in synaptic vesicle formation from endosomes. Cell 93:423-432

Faundez VV, Kelly RB (2000) The AP-3 complex required for endosomal synaptic vesicle biogenesis is associated with a casein kinase Ialpha-like isoform. Mol Biol Cell 11:2591-2604

Fernandez-Chacon R, Sudhof TC (1999) Genetics of synaptic vesicle function: toward the complete functional anatomy of an organelle. Annu Rev Physiol 61:753-776

Fremeau RT, Jr., Burman J, Qureshi T, Tran CH, Proctor J, Johnson J, Zhang H, Sulzer D, Copenhagen DR, Storm-Mathisen J, Reimer RJ, Chaudhry FA, Edwards RH (2002) The identification of vesicular glutamate transporter 3 suggests novel modes of signaling by glutamate. Proc Natl Acad Sci USA 99:14488-14493

Fremeau RT, Jr., Troyer MD, Pahner I, Nygaard GO, Tran CH, Reimer RJ, Bellocchio EE, Fortin D, Storm-Mathisen J, Edwards RH (2001) The expression of vesicular glutamate transporters defines two classes of excitatory synapse. Neuron 31:247-260

Gilmor ML, Nash NR, Roghani A, Edwards RH, Yi H, Hersch SM, Levey AI (1996) Expression of the putative vesicular acetylcholine transporter in rat brain and localization in cholinergic synaptic vesicles. J Neurosci 16:2179-2190

Gonzalez-Gaitan M, Jackle H (1997) Role of *Drosophila* alpha-adaptin in presynaptic vesicle recycling. Cell 88:767-776

Gras C, Herzog E, Bellenchi GC, Bernard V, Ravassard P, Pohl M, Gasnier B, Giros B, El Mestikawy S (2002) A third vesicular glutamate transporter expressed by cholinergic and serotoninergic neurons. J Neurosci 22:5442-5451

Greene LA, Rein G (1977a) Release of (3H)norepinephrine from a clonal line of pheochromocytoma cells (PC12) by nicotinic cholinergic stimulation. Brain Res 138:521-528

Greene LA, Rein G (1977b) Synthesis, storage and release of acetylcholine by a noradrenergic pheochromocytoma cell line. Nature 268:349-351

Hannah MJ, Schmidt AA, Huttner WB (1999) Synaptic vesicle biogenesis. Annu Rev Cell Dev Biol 15:733-798

Hayashi M, Otsuka M, Morimoto R, Hirota S, Yatsushiro S, Takeda J, Yamamoto A, Moriyama Y (2001) Differentiation-associated Na+-dependent inorganic phosphate cotransporter (DNPI) is a vesicular glutamate transporter in endocrine glutamatergic systems. J Biol Chem 276:43400-43406

Herzog E, Bellenchi GC, Gras C, Bernard V, Ravassard P, Bedet C, Gasnier B, Giros B, El Mestikawy S (2001) The existence of a second vesicular glutamate transporter specifies subpopulations of glutamatergic neurons. J Neurosci 21:RC181

Hirokawa N, Noda Y, Okada Y (1998) Kinesin and dynein superfamily proteins in organelle transport and cell division. Curr Opin Cell Biol 10:60-73

Hofmann MW, Honing S, Rodionov D, Dobberstein B, von Figura K, Bakke O (1999) The leucine-based sorting motifs in the cytoplasmic domain of the invariant chain are recognized by the clathrin adaptors AP1 and AP2 and their medium chains. J Biol Chem 274:36153-36158

Honing S, Sandoval IV, von Figura K (1998) A di-leucine-based motif in the cytoplasmic tail of LIMP-II and tyrosinase mediates selective binding of AP-3. EMBO J 17:1304-1314

Inomata H, Nakamura Y, Hayakawa A, Takata H, Suzuki T, Miyazawa K, Kitamura N (2003) A scaffold protein JIP-1b enhances amyloid precursor protein phosphorylation by JNK and its association with kinesin light chain 1. J Biol Chem 278:22946-22955

Kalinina E, Varlamov O, Fricker LD (2002) Analysis of the carboxypeptidase D cytoplasmic domain: implications in intracellular trafficking. J Cell Biochem 85:101-111

Kirchhausen T (1999) Adaptors for clathrin-mediated traffic. Annu Rev Cell Dev Biol 15:705-732

Kirchhausen T (2002) Clathrin adaptors really adapt. Cell 109:413-416

Kirchhausen T, Bonifacino JS, Riezman H (1997) Linking cargo to vesicle formation: receptor tail interactions with coat proteins. Curr Opin Cell Biol 9:488-495

Krantz DE, Peter D, Liu Y, Edwards RH (1997) Phosphorylation of a vesicular monoamine transporter by casein kinase II. J Biol Chem 272:6752-6759

Krantz DE, Waites C, Oorschot V, Liu Y, Wilson RI, Tan PK, Klumperman J, Edwards RH (2000) A phosphorylation site regulates sorting of the vesicular acetylcholine transporter to dense core vesicles. J Cell Biol 149:379-396

Liu Y, Edwards RH (1997a) Differential localization of vesicular acetylcholine and monoamine transporters in PC12 cells but not CHO cells. J Cell Biol 139:907-916

Liu Y, Edwards RH (1997b) The role of vesicular transport proteins in synaptic transmission and neural degeneration. Annu Rev Neurosci 20:125-156

Liu Y, Krantz DE, Waites C, Edwards RH (1999) Membrane trafficking of neurotransmitter transporters in the regulation of synaptic transmission. Trends Cell Biol 9:356-363

Liu Y, Peter D, Roghani A, Schuldiner S, Prive GG, Eisenberg D, Brecha N, Edwards RH (1992) A cDNA that suppresses MPP+ toxicity encodes a vesicular amine transporter. Cell 70:539-551

Liu Y, Schweitzer ES, Nirenberg MJ, Pickel VM, Evans CJ, Edwards RH (1994) Preferential localization of a vesicular monoamine transporter to dense core vesicles in PC12 cells. J Cell Biol 127:1419-1433

Matsuda S, Matsuda Y, D'Adamio L (2003) Amyloid b protein precursor (AbPP), but not AbPP like protein-2, is bridged to the kinesin light chain by the scaffold JNK-interacting protein 1. J Biol Chem

McIntire SL, Reimer RJ, Schuske K, Edwards RH, Jorgensen EM (1997) Identification and characterization of the vesicular GABA transporter. Nature 389:870-876

Melega WP, Howard BD (1981) Choline and acetylcholine metabolism in PC12 secretory cells. Biochemistry 20:4477-4483

Melega WP, Howard BD (1984) Biochemical evidence that vesicles are the source of the acetylcholine released from stimulated PC12 cells. Proc Natl Acad Sci USA 81:6535-6538

Nakata T, Terada S, Hirokawa N (1998) Visualization of the dynamics of synaptic vesicle and plasma membrane proteins in living axons. J Cell Biol 140:659-674

Nielsen MS, Madsen P, Christensen EI, Nykjaer A, Gliemann J, Kasper D, Pohlmann R, Petersen CM (2001) The sortilin cytoplasmic tail conveys Golgi-endosome transport and binds the VHS domain of the GGA2 sorting protein. EMBO J 20:2180-2190

Nirenberg MJ, Chan J, Liu Y, Edwards RH, Pickel VM (1996) Ultrastructural localization of the vesicular monoamine transporter-2 in midbrain dopaminergic neurons: potential sites for somatodendritic storage and release of dopamine. J Neurosci 16:4135-4145

Nirenberg MJ, Chan J, Liu Y, Edwards RH, Pickel VM (1997) Vesicular monoamine transporter-2: immunogold localization in striatal axons and terminals. Synapse 26:194-198

Nirenberg MJ, Liu Y, Peter D, Edwards RH, Pickel VM (1995) The vesicular monoamine transporter 2 is present in small synaptic vesicles and preferentially localizes to large dense core vesicles in rat solitary tract nuclei. Proc Natl Acad Sci USA 92:8773-8777

Okada Y, Yamazaki H, Sekine-Aizawa Y, Hirokawa N (1995) The neuron-specific kinesin superfamily protein KIF1A is a unique monomeric motor for anterograde axonal transport of synaptic vesicle precursors. Cell 81:769-780

Partoens P, Slembrouck D, Quatacker J, Baudhuin P, Courtoy PJ, De Potter WP (1998) Retrieved constituents of large dense-cored vesicles and synaptic vesicles intermix in stimulation-induced early endosomes of noradrenergic neurons. J Cell Sci 111 Pt 6:681-689

Peden AA, Park GY, Scheller RH (2001) The Di-leucine motif of vesicle-associated membrane protein 4 is required for its localization and AP-1 binding. J Biol Chem 276:49183-49187

Peter D, Liu Y, Sternini C, de Giorgio R, Brecha N, Edwards RH (1995) Differential expression of two vesicular monoamine transporters. J Neurosci 15:6179-6188

Pond L, Kuhn LA, Teyton L, Schutze MP, Tainer JA, Jackson MR, Peterson PA (1995) A role for acidic residues in di-leucine motif-based targeting to the endocytic pathway. J Biol Chem 270:19989-19997

Prado M, Reis R, Prado V, de Mello M, Gomez M, de Mello F (2002) Regulation of acetylcholine synthesis and storage. Neurochem Int 41:291

Prado VF, Prado MAM (2003) Signals involved in targeting membrane proteins to synaptic vesicles. Cell Mol Neurobiol: in press

Puertollano R, Aguilar RC, Gorshkova I, Crouch RJ, Bonifacino JS (2001) Sorting of mannose 6-phosphate receptors mediated by the GGAs. Science 292:1712-1716

Regnier-Vigouroux A, Tooze SA, Huttner WB (1991) Newly synthesized synaptophysin is transported to synaptic-like microvesicles via constitutive secretory vesicles and the plasma membrane. EMBO J 10:3589-3601

Rodionov DG, Honing S, Silye A, Kongsvik TL, von Figura K, Bakke O (2002) Structural requirements for interactions between leucine-sorting signals and clathrin-associated adaptor protein complex AP3. J Biol Chem 277:47436-47443

Roghani A, Feldman J, Kohan SA, Shirzadi A, Gundersen CB, Brecha N, Edwards RH (1994) Molecular cloning of a putative vesicular transporter for acetylcholine. Proc Natl Acad Sci USA 91:10620-10624

Roghani A, Shirzadi A, Butcher LL, Edwards RH (1998) Distribution of the vesicular transporter for acetylcholine in the rat central nervous system. Neurosci 82:1195-1212

Sampo B, Kaech S, Kunz S, Banker G (2003) Two distinct mechanisms target membrane proteins to the axonal surface. Neuron 37:611-624

Sandoval IV, Martinez-Arca S, Valdueza J, Palacios S, Holman GD (2000) Distinct reading of different structural determinants modulates the dileucine-mediated transport steps of the lysosomal membrane protein LIMPII and the insulin-sensitive glucose transporter GLUT4. J Biol Chem 275:39874-39885

Santos MS, Barbosa J, Jr., Veloso GS, Ribeiro F, Kushmerick C, Gomez MV, Ferguson SS, Prado VF, Prado MA (2001) Trafficking of green fluorescent protein tagged-vesicular acetylcholine transporter to varicosities in a cholinergic cell line. J Neurochem 78:1104-1113

Schafer MK, Eiden LE, Weihe E (1998) Cholinergic neurons and terminal fields revealed by immunohistochemistry for the vesicular acetylcholine transporter. II. The peripheral nervous system. Neurosci 84:361-376

Schafer MK, Varoqui H, Defamie N, Weihe E, Erickson JD (2002) Molecular cloning and functional identification of mouse vesicular glutamate transporter 3 and its expression in subsets of novel excitatory neurons. J Biol Chem 277:50734-50748

Schmidt A, Hannah MJ, Huttner WB (1997) Synaptic-like microvesicles of neuroendocrine cells originate from a novel compartment that is continuous with the plasma membrane and devoid of transferrin receptor. J Cell Biol 137:445-458

Schuldiner S, Shirvan A, Linial M (1995) Vesicular neurotransmitter transporters: from bacteria to humans. Physiol Rev 75:369-392

Shi G, Faundez V, Roos J, Dell'Angelica EC, Kelly RB (1998) Neuroendocrine synaptic vesicles are formed in vitro by both clathrin-dependent and clathrin-independent pathways. J Cell Biol 143:947-955

Shiba T, Takatsu H, Nogi T, Matsugaki N, Kawasaki M, Igarashi N, Suzuki M, Kato R, Earnest T, Nakayama K, Wakatsuki S (2002) Structural basis for recognition of acidic-cluster dileucine sequence by GGA1. Nature 415:937-941

Stockinger W, Brandes C, Fasching D, Hermann M, Gotthardt M, Herz J, Schneider WJ, Nimpf J (2000) The reelin receptor ApoER2 recruits JNK-interacting proteins-1 and -2. J Biol Chem 275:25625-25632

Sudhof TC (1995) The synaptic vesicle cycle: a cascade of protein-protein interactions. Nature 375:645-653

Sudhof TC (2000) The synaptic vesicle cycle revisited. Neuron 28:317-320

Takamori S, Malherbe P, Broger C, Jahn R (2002) Molecular cloning and functional characterization of human vesicular glutamate transporter 3. EMBO Rep 3:798-803

Takamori S, Rhee JS, Rosenmund C, Jahn R (2000) Identification of a vesicular glutamate transporter that defines a glutamatergic phenotype in neurons. Nature 407:189-194

Takamori S, Rhee JS, Rosenmund C, Jahn R (2001) Identification of differentiation-associated brain-specific phosphate transporter as a second vesicular glutamate transporter (VGLUT2). J Neurosci 21:RC182

Tan PK, Waites C, Liu Y, Krantz DE, Edwards RH (1998) A leucine-based motif mediates the endocytosis of vesicular monoamine and acetylcholine transporters. J Biol Chem 273:17351-17360

Taru H, Iijima K, Hase M, Kirino Y, Yagi Y, Suzuki T (2002) Interaction of Alzheimer's beta -amyloid precursor family proteins with scaffold proteins of the JNK signaling cascade. J Biol Chem 277:20070-20078

Thiele C, Huttner WB (1998) Protein and lipid sorting from the trans-Golgi network to secretory granules-recent developments. Semin Cell Dev Biol 9:511-516

Thureson-Klein A (1983) Exocytosis from large and small dense cored vesicles in noradrenergic nerve terminals. Neurosci 10:245-259

Tooze SA (1998) Biogenesis of secretory granules in the trans-Golgi network of neuroendocrine and endocrine cells. Biochim Biophys Acta 1404:231-244

Tooze SA, Martens GJ, Huttner WB (2001) Secretory granule biogenesis: rafting to the SNARE. Trends Cell Biol 11:116-122

Tsukita S, Ishikawa H (1980) The movement of membranous organelles in axons. Electron microscopic identification of anterogradely and retrogradely transported organelles. J Cell Biol 84:513-530

Turner MD, Arvan P (2000) Protein traffic from the secretory pathway to the endosomal system in pancreatic beta-cells. J Biol Chem 275:14025-14030

Varlamov O, Eng FJ, Novikova EG, Fricker LD (1999) Localization of metallocarboxypeptidase D in AtT-20 cells. Potential role in prohormone processing. J Biol Chem 274:14759-14767

Varoqui H, Erickson JD (1998) The cytoplasmic tail of the vesicular acetylcholine transporter contains a synaptic vesicle targeting signal. J Biol Chem 273:9094-9098

Varoqui H, Schafer MK, Zhu H, Weihe E, Erickson JD (2002) Identification of the differentiation-associated Na+/PI transporter as a novel vesicular glutamate transporter expressed in a distinct set of glutamatergic synapses. J Neurosci 22:142-155

Verhey KJ, Meyer D, Deehan R, Blenis J, Schnapp BJ, Rapoport TA, Margolis B (2001) Cargo of kinesin identified as JIP scaffolding proteins and associated signaling molecules. J Cell Biol 152:959-970

Waites CL, Mehta A, Tan PK, Thomas G, Edwards RH, Krantz DE (2001) An acidic motif retains vesicular monoamine transporter 2 on large dense core vesicles. J Cell Biol 152:1159-1168

Wan L, Molloy SS, Thomas L, Liu G, Xiang Y, Rybak SL, Thomas G (1998) PACS-1 defines a novel gene family of cytosolic sorting proteins required for trans-Golgi network localization. Cell 94:205-216

Wang Y, Thiele C, Huttner WB (2000) Cholesterol is required for the formation of regulated and constitutive secretory vesicles from the trans-Golgi network. Traffic 1:952-962

Weihe E, Tao-Cheng JH, Schafer MK, Erickson JD, Eiden LE (1996) Visualization of the vesicular acetylcholine transporter in cholinergic nerve terminals and its targeting to a specific population of small synaptic vesicles. Proc Natl Acad Sci USA 93:3547-3552

Winkler H (1997) Membrane composition of adrenergic large and small dense cored vesicles and of synaptic vesicles: consequences for their biogenesis. Neurochem Res 22:921-932

Zhu Y, Doray B, Poussu A, Lehto VP, Kornfeld S (2001) Binding of GGA2 to the lysosomal enzyme sorting motif of the mannose 6-phosphate receptor. Science 292:1716-1718

Caron, Marc G.
Department of Cell Biology and Howard Hughes Medical Institute, Duke University Medical Center, Durham NC

Prado, Marco A.M.
Departamento de Farmacologia, ICB, Universidade Federal de Minas Gerais, Av. Antonio Carlos 6627, Belo Horizonte, 31270-901 Brazil
mprado@icb.ufmg.br

Prado, Vania F.
Departamento de Bioquímica-Imunologia, ICB, Universidade Federal de Minas Gerais, Av. Antonio Carlos 6627, Belo Horizonte, 31270-901 Brazil

Membrane trafficking of yeast transporters: mechanisms and physiological control of downregulation

Rosine Haguenauer-Tsapis and Bruno André

Abstract

Of the 125 plasma membrane transporters thus far identified in the yeast *S. cerevisiae*, a growing number is reported to be subject to tight control at membrane trafficking level, in addition to control at transcriptional level. Typical physiological conditions inducing these controls include changes of substrate concentration and availability of alternative nutrients. These changes of conditions often provoke the downregulation of specific transporters eventually accompanied by upregulation of others, which are more appropriate for the new conditions. Downregulation generally involves the onset or acceleration of endocytosis of the transporter and subsequent targeting to the vacuole where it is degraded. In many cases, the same physiological signals also induce diversion of the neosynthesized transporter from the Golgi apparatus to the endosomal/vacuolar degradation pathway without passing through the plasma membrane. In the present review, we first describe the main factors – including the ubiquitin ligase Rsp5p and its partners – involved in downregulation of yeast transporters, at endocytosis and endosomal/vacuolar targeting steps. We then summarize the wide variety of physiological situations reported to induce the downregulation of specific yeast transporters. Finally, we discuss possible mechanisms responsible for a specific transporter to be ubiquitylated and downregulated only under particular conditions. Evidence is accumulating that these mechanisms involve association of the protein with specific lipid environments, possible phosphorylation by protein kinases, and specific binding of the transporter to its own substrates.

1 Introduction

About 282 transmembrane solute transporters have been inventoried in the yeast *Saccharomyces cerevisiae*, a minimum of 125 of which are known to be located at the plasma membrane (Van Belle and André 2001; Huh et al. 2003). Some of these transporters, such as the proton ATPase Pma1p, are essential in all growth conditions. Such proteins are continuously synthesized and are present in an active and stable form at the cell surface. There is growing evidence that these "constitutive" transporters make up only a small fraction of the yeast *transportome*. Indeed,

Topics in Current Genetics, Vol. 9
E. Boles, R. Krämer (Eds.): Molecular Mechanisms Controlling Transmembrane Transport
DOI 10.1007/b97215 / Published online: 9 March 2004
© Springer-Verlag Berlin Heidelberg 2004

the number of yeast transporters reported to be subject to tight regulation according to environmental conditions is increasing all the time. The mechanisms by which cells control transporters act essentially at three levels: gene transcription, intracellular membrane trafficking of the synthesized proteins, and modulation of intrinsic activity. As the same principles seem to govern the regulation of transporters in the cells of plants and animals, yeast constitutes a powerful model for dissecting the underlying mechanisms in detail.

In the last ten years, a large number of studies have focused on the regulation of yeast transporters at the level of membrane trafficking. At least three types of control have been described to date. First, a widespread and efficient way of eliminating a transporter is to trigger its specific sorting into endocytic vesicles and to target the internalized protein to the endosome/vacuolar degradation pathway. Second, the fate of newly synthesized transporters within the secretory pathway is also frequently regulated, the protein being either targeted to the cell surface or diverted to the vacuole without passing through the plasma membrane. Third, some transporters may be stored in intracellular membranes and recycled to the plasma membrane in response to specific signals. Typical physiological situations inducing these control mechanisms include changes of substrate concentration, availability of alternative nutrients, or the transfer of cells to adverse conditions.

We review here current knowledge concerning the mechanisms controlling the membrane trafficking of yeast plasma membrane transporters, making reference to similarities with the membrane trafficking of mammalian transporters. As we still know little about the regulation of internal membrane transporters (about 90 proteins in yeast), this topic will not be covered here.

2 Trafficking of plasma membrane transporters along the secretory pathway

Studies of thermosensitive secretory (*sec*) mutants provided an early demonstration that plasma membrane transporters follow the secretory pathway to the plasma membrane, with no inducible permease activity detected after induction in *sec* mutants placed at restrictive temperature (first reports by Ferro-Novick et al. 1984; Tschopp et al. 1986; Jund et al. 1988) (Fig. 1). These initial reports were followed by other studies providing further insight into the trafficking of yeast transporters to the plasma membrane. However, this was not the object of intense investigations as many of the genes/proteins involved in this process are also involved in the trafficking of soluble proteins along the secretory pathway. We will briefly summarize some of the general and specific features of transporter trafficking along the secretory pathway.

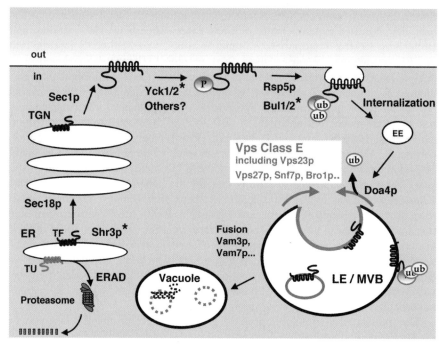

Fig. 1. Intracellular trafficking of a transporter along the secretory and endocytic pathways. After ER insertion, misfolded transporters (TU) undergo ERAD and proteasomal degradation. Correctly folded transporters (TF) are packaged in COPII vesicles, with the aid of chaperones, such as Shr3p in the case of amino acid permeases. ER to Golgi and Golgi to plasma membrane trafficking is dependent on a number of *SEC* genes. Whether trafficking from Golgi to plasma membrane occurs directly via secretory vesicles, or via late endosomes and then secretory vesicles is still unknown. Many transporters undergo phosphorylation (P), in some instances a plasma membrane event dependent of Yck1/2 kinases. Specific physiological signals trigger Rsp5p-dependent ubiquitylation (ub), also dependent of the Bul proteins, as described for some transporters. This modification acts as a signal triggering transporters internalization followed by targeting to early endosomes (EE). At the late endosome/MVB, transporters undergo sorting to internal vesicles, an event dependent on Vps class E proteins, followed by Doa4p-dependent deubiquitylation. Vam-dependent fusion of MVB with the vacuole leads to delivery of internal MVB vesicles to the vacuolar lumen accompanied by transporters degradation by vacuolar proteases.

2.1 ER-associated events

Transporters are first inserted into the endoplasmic reticulum (ER) membrane. They are then targeted to the Golgi apparatus and packaged into secretory vesicles that ultimately fuse with the plasma membrane at the incipient bud. Most yeast transporters have no cleavable signal sequence. They are targeted to the ER by the first hydrophobic segment, which functions as an internal signal sequence, as has been demonstrated experimentally in a few cases (Silve et al. 1991; Kolling and Hollenberg 1994b). Transporters are thought to be co-translationally inserted into the ER membrane, acquiring their final topology. Computer-aided sequence analysis based on hydropathy profile analysis has provided useful topological predictions for transporters. However, whereas extensive studies have been carried out for bacterial and mammalian transporters, the topology of yeast transporters has received much less attention. A few studies based on protease protection (Silve et al. 1991; Kolling and Hollenberg 1994b), or glycosylation scanning experiments (Ahmad and Bussey 1988; Garnier et al. 1996; Gilstring and Ljungdahl 2000) have been reported. The only complete *in vivo* analysis of the topology of a yeast transporter published to date is that for the general amino acid permease (Gap1p), which has twelve transmembrane spans and N- and C-termini oriented towards the cytoplasm (Gilstring and Ljungdahl 2000). Most plasma membrane transporters appear to have cytoplasmic N-termini, often longer than those of their bacterial orthologs, and shorter than those of their mammalian orthologs. Exceptions include the ammonium Mep2p transporter, which has luminal/external N-terminus (Marini and André 2000). Structure/function studies have identified signals in the cytoplasmic regions of transporters — in the N- and C-terminal regions in particular — that are important for internalization.

The translocation of soluble proteins across the ER and the ER insertion of membrane-bound proteins are accompanied by the folding of these proteins. Misfolded proteins undergo ER-associated degradation (ERAD), which involves a proteolytic pathway including retrotranslocation (soluble proteins) or dislocation (membrane-bound proteins) of the cargo proteins and their concomitant ubiquitylation by cytosolic enzymes, followed by their degradation by the proteasome (Hampton 2002; Kostova and Wolf 2003). A certain point mutation (ΔF508) in the mammalian cystic fibrosis transmembrane conductance regulator Cl channel (CFTR) leads to ERAD of the mutant protein. This leads to a major decrease in steady-state plasma membrane levels of CFTR, the basis of the common inherited disease cystic fibrosis. The ERAD of CFTR is regarded as the prototype of ERAD of membrane-bound proteins and the events involved in this process have been extensively studied. The ERAD of several yeast plasma membrane proteins has also been reported, including the case of the uracil permease Fur4p (Galan et al. 1997), the ABC transporter Pdr5p (Plemper et al. 1998), the a-factor transporter Ste6p (Loayza et al. 1998), and the plasma membrane (H^+) ATPase Pma1p (Wang et al. 1999). Early studies indicated that yeast was potentially a good model system for studying the ERAD of CFTR (Paddon et al. 1996). One critical step in ERAD is the recognition, by specific ubiquitin protein ligases, of substrates that have to be ubiquitylated. A RING-finger ubiquitin ligase, Hrd1/Der3, was initially reported to be responsible for the ubiquitylation of a subset of soluble and membrane-bound ERAD substrates, but this enzyme played no part in the degradation of misfolded uracil permease, or mammalian CFTR produced in yeast (Hampton 2002). Another ER membrane-embedded ubiquitin ligase involved in the degradation of a few ERAD substrates, Doa10p, was recently identified (Swanson et al. 2001), but little is known about its function.

The transporters leave the ER once they are correctly folded. Efficient transport of membrane and secretory proteins from the ER requires concentrative and signal-mediated sorting into coat protein complex II (COPII)-coated vesicles. In the case of Gap1p, this sorting involves a C-terminal di-acidic signal (Malkus et al. 2002) recognized by the COPII protein Sec24p, that has multiple independent sites of cargo recognition (Miller et al. 2003). Correct ER exit of transporters in COPII-coated vesicles requires ER-associated proteins, which help packaging of specific class of transporters (Ljungdahl, this issue) (Fig. 1).

2.2 Posttranslational modifications

Many soluble proteins display posttranslational modifications in their trafficking along the secretory pathway, in both the ER and Golgi apparatus (glycosylation, formation of disulfide bonds, proteolytic cleavages) (Table 1). Only a few yeast transporters have been reported to undergo glycosylation. Indeed, several yeast plasma membrane transporters, including the uracil permease Fur4p, the a-factor transporter Ste6p, or the Hxt2p high affinity glucose transporter, and a putative anion exchanger (Ynl275p), have clearly been shown not to be glycosylated (Volland et al. 1992; Kuchler et al. 1993; Wendell and Bisson 1993; Zhao and Reithmeier 2001). Unglycosylated transporters include a number of proteins with potential glycosylation sites (Asn/X/Ser or Thr). These sites may be cytoplasmic, within transmembrane segments, or luminal, but positioned too close to the plane of the membrane for oligosaccharyl transferases to have access (Garnier et al. 1996). The few reported examples of glycosylated transporters include the Mep2p ammonium transporter, which is N-glycosylated in its N-terminal domain (Marini and André 2000), the endosomal transporter Nhx1p (Wells and Rao 2001), or the copper transporter Ctr1p, which displays extensive O-glycosylation (Dancis et al. 1994). An apparently common posttranslational modification of many plasma membrane transporters is the phosphorylation of serine residues (Table 1 and list in Rotin et al. 2000). The plasma membrane [H^+] ATPase, Pma1p (Chang and Slayman 1991), uracil permease (Volland et al. 1992), and the a-factor transporter Ste6p (Kuchler et al. 1993) were among the first serine-phosphorylated transporters to be identified. Experiments in *sec* mutants demonstrated that several residues of Pma1p are phosphorylated at different steps of the secretory pathway (Chang and Slayman 1991), whereas uracil permease is mainly phosphorylated after its arrival at the plasma membrane (Volland et al. 1992). C-terminal phosphorylation of Pma1p has been shown to be involved in the activity of this protein (Goossens et al. 2000), whereas the role played by phosphorylation remains unclear for most transporters. However, phosphorylation has been shown to control the rate of endocytosis of transporters in some cases (Marchal et al. 1998; Medintz et al. 2000; Kolling 2002).

Table 1. A list of yeast transporters with identified post-translational modifications on specific residues

Protein	Description	Predicted topology and post-translationally modified residues[1]	References[2]
Alr1	Transporter of magnesium and other divalent cations	$[N^{1-771}, S^{190}]_1-TM_{1-2}-[C^{795-859}]_i$	Peng et al. (2003)
Bap2	Broad-specificity amino-acid permease – important role in branched-chain amino-acid uptake – member of the AAP/YAT family of amino-acid permeases	$[N^{1-98}, K^{12}K^{13}K^{29}K^{42}K^{44}]_i-TM_{1-12}-[C^{557-609}]_i$	Omura et al. (2001b)
Cot1	Vacuolar zinc (and possibly other metals) transporter	$[N^{1-12}]_c-TM_{1-4}-[L^{135-242}]_c-TM_{5-6}-[C^{298-439}, K^{301}]_c$	Hitchcock et al. (2003)
Ena5	Plasma membrane P-type ATPase involved in Na$^+$ and Li$^+$ efflux	$[N^{1-64}]_1-TM_{1-2}-[L^{111-285}, K^{196}]_1-TM_{3-4}-[L^{336-814}]_1-TM_{5-10}-[C^{1045-1091}]_1$	Hitchcock et al. (2003)
Fur4	Uracil permease	$[N^{1-109}, K^{38}K^{41}g^{42}g^{43}S^{45}S^{55}S^{56}]_i-TM_{1-12}-[C^{580-633}]_i$	Marchal et al (1998, 2000); Soetens et al. (2001); Peng et al. (2003); Hitchcock et al. (2003)
Gap1	General amino acid permease – member of the AAP/YAT family of amino-acid permeases	$[N^{1-96}, K^9K^{16}K^{76}]_1-TM_{1-12}-[C^{547-602}]_i$	
Gnp1	Broad-specificity amino-acid permease – important role in threonine uptake – member of the AAP/YAT family of amino-acid permeases	$[N^{1-183}, K^{34}K^{41}K^{132}g^{114}g^{116}]_i-TM_{1-12}-[C^{609-663}]_i$	Peng et al. (2003)
Hxt1	Low-affinity hexose facilitator	$[N^{1-59}]_1-TM_{1-6}-[L^{257-348}]_i-TM_{7-12}-[C^{513-570}, Y^{531}]_i$	Peng et al. (2003)
Hxt4	Hexose facilitator of moderately low affinity	$[N^{1-64}, K^{45}]_1-TM_{1-6}-[L^{263-364}]_i-TM_{7-12}-[C^{519-560}]_i$	Hitchcock et al. (2003)
Hxt5	Hexose facilitator of moderately low affinity (Buziol et al. 2002)	$[N^{1-82}, K^{61}]_1-TM_{1-6}-[L^{278-350}]_1-TM_{7-12}-[C^{534-592}]_i$	Peng et al (2003)
Hxt6	High-affinity hexose facilitator	$[N^{1-58}]_1-TM_{1-6}-[L^{257-358}]_i-TM_{7-12}-[C^{513-570}, K^{560}]_i$	Peng et al (2003)
Hxt7	High-affinity hexose facilitator	$[N^{1-58}, K^{40}]_1-TM_{1-6}-[L^{257-358}, K^{318}]_i-TM_{7-12}-[C^{513-570}, K^{560}]_i$	Peng et al (2003)
Itr1	Inositol permease	$[N^{1-84}]_1-TM_{1-6}-[L^{267-334}]_1-TM_{7-12}-[C^{530-587}, K^{573}]_i$	Peng et al. (2003); Hitchcock et al. (2003)
Jen1	Lactate and pyruvate permease	$[N^{1-141}, K^9S^{83}]_1-TM_{1-6}-[L^{317-363}, K^{338}]_1-TM_{7-12}-[C^{556-616}]_i$	Peng et al. (2003)
Lyp1	Lysine permease – member of the AAP/YAT family of amino-acid permeases	$[N^{1-113}, T^{92}]_1-TM_{1-12}-[C^{571-611}]_i$	Peng et al. (2003)
Mal61	Maltose permease – member of the sugar transporter superfamily	$[N^{1-83}]_1-TM_{1-6}-[L^{301-364}]_1-TM_{7-12}-[C^{545-614}, S^{581}]_i$	
Mep2	High affinity ammonium permease - proposed role also as ammonium sensor involved in pseudohyphal growth – member of the Amt/Mep family of ammonium transporters	$[N^{1-33}, N^4]_o-TM_{1-11}-[C^{414-499}]_o$	Marini and André (2001)

Table 1. continued

Protein	Description	Predicted topology and post-translationally modified residues[1]	References[2]
Nhx1	Na+/H+ exchanger of the prevacuolar compartment - involved in salt tolerance	$[N^{1-58}]_i-TM_{1-12}-[C^{481-633}, \mathbf{N^{515}N^{550}}]_i$	Wells and Rao (2001)
Pdr5	(NBD-TM)₂-type ABC transporter involved in multidrug resistance	$[N^{1-519}]_i-TM_{1-6}-[L^{794-1205}, \mathbf{K^{825}S^{851}S^{860}S^{867}}]_i-TM_{7-12}-[C^{1500-1511}]_i$	Peng et al. (2003); Hitchcock et al. (2003)
Pdr12	(NBD-TM)₂-type ABC transporter involved in excretion of weak organic acids	$[N^{1-594}, \mathbf{K^{412}}]_i-TM_{1-6}-[L^{787-1176}]_i-TM_{7-12}-[C^{1476-1511}]_i$	Peng et al. (2003)
Pho84	Inorganic phosphate transporter – member of the phosphate:proton symporter (PHS) family	$[N^{1-133}, \mathbf{K^6}]_i-TM_{1-6}-[L^{271-349}, \mathbf{K^{298}T^{304}S^{306}T^{319}S^{324}}]_i-TM_{7-12}-[C^{542-587}, \mathbf{S^{581}}]_i$	Peng et al (2003)
Pho87	Low affinity phosphate transporter – proposed role also as phosphate sensor	$[N^{1-457}, \mathbf{K^{102}S^{230}}]_i-TM_{1-12}-[C^{915-923}]_i$	Peng et al (2003)
Pho91	Low affinity phosphate transporter – Pho87 homolog	$[N^{1-424}, \mathbf{S^{313}S^{215}}]_i-TM_{1-12}-[C^{890-894}]_i$	Peng et al (2003)
Pma1	Plasma membrane H+-transporting P-type ATPase	$[N^{1-111}]_i-TM_{1-2}-[L^{161-288}]_i-TM_{3-4}-[L^{354-690}, \mathbf{K^{555}S^{566}K^{644}}]_i-TM_{5-10}-[C^{875-918}]_i$	Peng et al. (2003); Hitchcock et al. (2003)
Ste6	(TM-NBD)₂ type ABC transporter responsible for the export of the "a" factor mating pheromone	$[N^{1-25}]_i-TM_{1-6}-[L^{279-715}, \mathbf{T^{613}S^{623}}]_i-TM_{7-12}-[C^{1003-1290}, \mathbf{K^{1022}}]_i$	Kolling (2002) ; Peng et al. (2003)
Tat1	Broad-specificity amino-acid permease – important role in tyrosine and leucine uptake – member of the AAP/YAT family of amino-acid permeases	$[N^{1-98}, \mathbf{K^{39}}]_i-TM_{1-12}-[C^{549-619}]_i$	Hitchcock et al. (2003)
Tat2	Broad-specificity amino-acid permease – important role in tryptophan uptake – member of the AAP/YAT family of amino-acid permeases	$[N^{1-85}, \mathbf{K^{10}K^{17}K^{20}K^{29}K^{31}}]_i-TM_{1-12}-[C^{538-592}]_i$	Beck et al (1999); Umebayashi & Nakano (2003)
Thi10	Thiamine permease	$[N^{1-42}]_i-TM_{1-12}-[C^{507-598}, \mathbf{S^{562}}]_i$	Peng et al (2003)
Tna1	Nicotinic acid permease	$[N^{1-89}]_i-TM_{1-6}-[L^{272-321}, \mathbf{K^{283}}]_i-TM_{7-12}-[C^{497-534}]_i$	Peng et al (2003)
Ycf1	(TM-NBD)₂-type vacuolar transporter mediating uptake into the vacuole of glutathione-S-conjugates	$[N^{1-27}]_i-TM_{1-5}-[L^{186-345}]_c-TM_{6-11}-[L^{590-949}, \mathbf{S^{910}T^{914}}]_c-TM_{12-17}-[L^{1198-1515}]_c$	Peng et al (2003)
Zrt1	High-affinity zinc transporter – member of the ZIP family of metal transporters	$[N^{1-47}]_o-TM_{1-3}-[L^{147-218}, \mathbf{K^{195}}]_i-TM_{4-8}-[C^{373-376}]_o$	Gitan et al. (2003)

[1]: N, C, and L (between brackets) designate, respectively N-terminal tails, C-terminal tails, and loops. These regions may be intracellular/cytosolic (i,c), extracellular (o), or lumenal (l). TM designates successive transmembrane domains. The position numbers of residues are in upper case. The modified residues are indicated in bold : K (ubiquitylation), S, T and Y (phosphorylation), N (N-glycosylation). Hence, $[N^{1-96}, \mathbf{K^9x^1K^{16}K^{76}}]_i-TM_{1-12}-[C^{547-602}]_i$ designates a protein successively made of an internal N-terminal tail of 96 residues containing three lysine (at positions 9, 16 and 96) shown to be ubiquitylated, a hydrophobic core made of 12 successive transmembrane domains, and an internal C-terminal tail.

[2]: Only publications reporting post-translational modifications of specific residues are indicated

2.3 ER-to-Golgi trafficking, and beyond

Various studies have emphasized the critical role played by lipids in the correct trafficking of plasma membrane proteins in the late steps of the secretory pathway. Transporters are no exception (Opekarova, this issue). Therefore, we will focus exclusively on certain general features of the requirement for lipids, paying particular attention to the role played by rafts in transporter trafficking. Rafts were recently described as dynamic assemblies of cholesterol and sphingolipids that form in the exoplasmic leaflet of cellular bilayer membranes (Ikonen 2001). Yeast has rafts composed of sphingolipids and of the structural plasma membrane sterol, ergosterol. They have been identified experimentally as fractions resistant to Triton X-100 at 4°C. Fur4p and Pma1p have been reported to be associated with rafts in the ER (Dupre and Haguenauer-Tsapis 2003) and the Golgi apparatus (Bagnat et al. 2001), respectively. Impairment of association with rafts due to the inhibition of sphingolipid synthesis delays the delivery of Fur4p to the plasma membrane (Dupre and Haguenauer-Tsapis 2003). The inhibition of sphingolipids or of ergosterol synthesis has been reported to lead to the routing of Pma1p (Bagnat et al. 2001) or the Tat2p tryptophan permease (Umebayashi and Nakano 2003) directly from the Golgi apparatus to the vacuole, rather than the plasma membrane. Association with rafts seems to depend upon the conformational features of nascent proteins, and mutant forms of Pma1p that fail to associate with rafts are also directly targeted from the Golgi apparatus for vacuolar degradation (Arvan et al. 2002). Thus, during the synthesis of membrane-bound proteins, quality control mechanisms may operate in both the ER and Golgi compartments.

Some of the signals important in protein trafficking in yeast, such as di-lysine, which governs the retrieval of membrane proteins from the Golgi back to the ER, or acidic cluster signals, which are involved in recycling from endosomes back to the Golgi apparatus, have been identified (Bonifacino and Traub 2003). However, we still know little about the signals involved in yeast in anterograde trafficking from the Golgi apparatus to the plasma membrane as opposed to Golgi to vacuolar pathway trafficking. For example, we still do not know why transporters of the same family sometimes display different final intracellular locations. This is true for the Smf, Zrt, and Ftr families of metal transporters, which include proteins resident in the plasma or vacuolar membranes, calcium P-type ATPases located in the Golgi (Pmr1p) or vacuolar (Pmc1p) membrane, and some closely related ABC transporters present in the plasma (Yor1p) or vacuolar (Ycf1p) membranes (Van Belle and André 2001). Studies of the trafficking of yeast Golgi, endosomal, or vacuolar transporters are just beginning and will, therefore, not be covered here. However, we will cite the recent example of Ycf1p (yeast cadmium factor 1), a prototype of the ABC transporter superfamily. This vacuolar transporter undergoes posttranslational processing by vacuolar proteases. It has an N-terminal hydrophobic extension (with five transmembrane spans) that appears to be required for vacuolar targeting (Mason and Michaelis 2002).

Although much is known about the routes and actors involved in ER insertion, ER-to-Golgi, and Golgi-to-vacuole trafficking, we know little about Golgi-to-cell surface sorting. Exocytic cargos have been reported to be transported to the yeast plasma membrane by at least two routes, involving two vesicle populations with different densities, carrying different types of cargo (Harsay and Bretscher 1995; David et al. 1998). It was recently shown that at least one branch of the secretory pathway passes through endosomes before reaching the cell surface. The plasma membrane ATPase seems to be packaged directly in low-density secretory vesicles (LDSV), which bud from the Golgi apparatus, whereas the soluble glycoproteins invertase and acid phosphatase seem to be transported from the Golgi to an endosomal compartment, and then packaged in high-density secretory vesicles (HDSV) (Gurunathan et al. 2002; Harsay and Schekman 2002). Clearly, defects in the targeting of a cargo *via* one of these pathways may lead to its sorting *via* the other pathway. This situation has for instance been reported for a mutant Pma1p protein missorted from the Golgi apparatus directly to the vacuole in wild type cells but retargeted to the plasma membrane in mutants defective in Golgi-to-endosome targeting (Arvan et al. 2002). It is unclear whether most plasma membrane transporters are packaged directly in LDSV, like Pma1p, or whether they pass through endosomes on their way to the plasma membrane. As yet, information is currently available only for the trafficking of the tryptophan permease Tat2p, which may pass through early endosomes on its way to the plasma membrane (Umebayashi and Nakano 2003). However, it is unknown whether this protein is finally recovered in LDSV or HDSV.

3 Downregulation of plasma membrane transporters and channels

3.1 Discovery of ubiquitin-dependent internalization of yeast plasma membrane transporters

Studies on endocytosis in mammalian cells have focused primarily on receptors (Mukherjee et al. 1997), with far fewer dealing with transporters. In yeast, major insight into the mechanisms of endocytosis have been provided by studies of two G-protein-coupled receptors (GPCRs), the α- and a-factor receptors, Ste2p and Ste3p, respectively (Riezman 1998; D'Hondt et al. 2000), but a number of reports concerning the endocytosis of transporters have also made a major contribution to advances in our understanding of endocytosis. In the last few years, it has become clear that, like ligand-induced receptor endocytosis, the internalization of transporters is highly regulated by numerous factors, including the substrates and nutrient availability. We will first describe the actors and mechanisms generally involved in the internalization of transporters and will then deal with subsequent vacuolar targeting *via* endosomes and the mechanisms regulating internalization, focusing mainly on yeast. The discovery of the role played by ubiquitin, and the identification of the ubiquitin ligase Rsp5p as a key player in these processes in particular, is clearly one of the most important advances in our understanding of the downregulation of transporters (Rotin et al. 2000; Hicke and Dunn 2003).

Protein ubiquitylation is a posttranslational modification in which a 76-amino acid polypeptide, ubiquitin, is attached to the ε-amino group of lysine residues in target proteins. This process involves sequential E1 (ubiquitin-activating), E2 (ubiquitin-conjugating or UBC), and E3 (ubiquitin-protein ligase) enzyme activities with the ligase carrying out the important task of target recognition. E3 are classified into several families. RING-finger E3s act as a scaffold, bringing together E2 and the substrate. HECT (homologous to E6AP C-terminus) domain E3s are the only class of E3 directly involved in catalysis: the activated ubiquitin is transferred from E2 to an internal Cys residue on E3 and is then conjugated to a lysine residue within the target. Proteins may be modified by monoubiquitylation: the conjugation of a single ubiquitin to one lysine or to each of several lysines in the target. Monoubiquitylation is involved in histone modification, and in several trafficking events. As ubiquitin itself carries several conserved acceptor lysines — notably Lys48 and Lys63 — multiubiquitin chains are frequently formed (Fig. 2). K48-linked multiubiquitin chains, at least four ubiquitin units long, are potent targeting signals leading to the recognition and subsequent degradation of target proteins by the 26S proteasome, a large multisubunit protease complex. K63-linked ubiquitin chains are involved in other functions including DNA repair, the activation of translation and of specific kinases, and endocytosis. Ubiquitylation is a dynamic process. Deubiquitylating enzymes (DUBs) cleave ubiquitin from proteins and disassemble ubiquitin chains. Several such enzymes are involved in endocytosis (Weissman 2001; Glickman and Ciechanover 2002).

The pheromone transporter Ste6p was the first yeast plasma membrane transporter shown to be ubiquitylated. As ubiquitylated forms of this transporter accumulate in mutants defective in endocytosis, it was suggested that ubiquitin may serve as a signal triggering the endocytosis of Ste6p (Kolling and Hollenberg 1994a). Subsequent work on two other yeast transporters fully supported this model. Early studies by Grenson and coworkers showed that ammonium triggers the rapid downregulation ("inactivation") of several yeast amino acid permeases, including the general amino acid permease Gap1p. Several *npi* (nitrogen permease inactivator) mutants have been isolated in which Gap1p is not inactivated by ammonium (Grenson 1992). Cloning and sequencing of the *NPI1* gene showed it to encode the ubiquitin protein ligase Rsp5p (Hein et al. 1995), a member of the Hect family of E3s (Scheffner et al. 1995). *NPI2* was found to encode Doa4p (Springael et al. 1999b), a ubiquitin isopeptidase (Papa and Hochstrasser 1993). It was then shown that both Gap1p and another transporter, the Fur4p uracil permease, undergo Rsp5p-dependent ubiquitylation, and that this modification is required for their internalization (Galan et al. 1996; Springael and André 1998) (Fig. 3) and subsequent proteasome-independent vacuolar degradation (Galan et al. 1996). Another study carried out at about the same time showed that a ubiquitin ligase of the same family, the mammalian Nedd4, plays a critical role in the ubiquitylation and downregulation of a plasma membrane channel, ENaC (Schild et al. 1996 ; Staub et al. 1997).

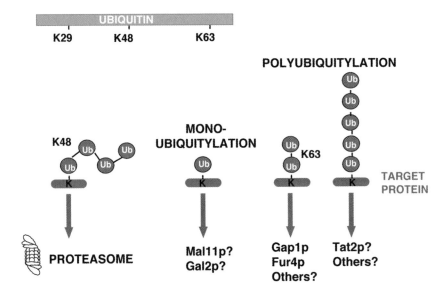

Fig. 2. Type of ubiquitylation of yeast transporters. Ubiquitin carries 11 Lys, some of which notably Lys29, 48 and 63.Lys, act as target Lys for linkage to other ubiquitin residues. Lys residues in target proteins can receive a single ubiquitin (monoubiquitylation), or chains of ubiquitin moieties, linked either via ubiquitin Lys48, or Lys63. Lys48-linked ubiquitin chains, at least four residues long target proteins for proteasomal degradation. Some yeast plasma membrane transporters were described to display primarily monoubiquitylation. Fur4p and Gap1p receive on their two target Lys two-three ubiquitin residues linked via ubiquitin Lys63. Other transporters appear to display polyubiquitylation, with undetermined type of ubiquitin linkage, notably during direct Golgi to vacuolar targeting.

Rsp5p and Nedd4 belong to a subfamily of Hect domain ligases that harbour a C2 domain, followed by two to four WW domains in their N-terminal regions (Rotin et al. 2000). The C2 domain is a 120-amino acid sequence that has been shown to bind phospholipids and proteins in a Ca^{++}-dependent manner in several proteins, including Nedd4 (Plant et al. 1997, 2000). WW domains are 40-amino acid protein-protein interaction modules that bind Pro-rich ligands of various types. WW domains can be classified into four different subtypes on the basis of binding specificity. Two-hybrid and systematic phage display experiments indicated that the Rsp5p and Nedd4 WW domains are group I WW domains that bind PPxY motifs (Chang et al. 2000; Rotin et al. 2000). Rsp5p, the only member of this protein family in *S. cerevisiae,* also has functions unrelated to endocytosis. Its essential function is the ubiquitylation of two transcription factors, Spt23p and Mga2p, leading to their processing by the proteasome and subsequent nuclear import (Hoppe et al. 2000; Shcherbik et al. 2003). As Rsp5p is required for viability, some of its functions have been studied using thermosensitive mutants. However, for many studies on the endocytosis of transporters, the original *npi1* mutant, which grows well despite having a modified promoter that decreases Rsp5p protein levels by a factor of > 10 (Hein et al. 1995; Springael and André 1998) proved to be a very useful tool.

It soon became clear that ubiquitylation is a prerequisite for the internalization of many transporters, and for the ligand-induced and constitutive internalization of the α- and a–factor receptors, respectively, and that Rsp5p is the only ubiquitin protein ligase involved in this posttranslational modification (Rotin et al. 2000; Hicke and Dunn 2003; Horak 2003). Evidence that ubiquitin acts as an internalization signal is provided by the following observations (Rotin et al. 2000; Hicke and Dunn 2003; Horak 2003): 1) ubiquitin conjugates of plasma membrane proteins accumulate in mutants deficient for the internalization step of endocytosis, as reported in precursor studies on Ste6p, the a-factor transporter (Kolling and Hollenberg 1994a). 2) Internalization is impaired in *rsp5 / npi1* mutant cells, in cells deficient in E2s of the Ubc1/4/5 family that may be involved in transferring ubiquitin to Rsp5p, and after the mutation of target lysines in various transporters. 3) The N-terminal in-frame fusion of ubiquitin restores some internalization of mutant transporters (Blondel et al. 2004) and receptors (Terrell et al. 1998; Roth and Davis 2000) lacking their own usual targets Lys residues, and of a wild type transporter in *npi1* cells (Blondel et al. 2004) (representative example Fig. 3). Similarly, the in-frame fusion of ubiquitin increases the rate of turnover of the stable Pma1p (Shih et al. 2000). 4) The signal mediating the ubiquitylation and downregulation of one particular plasma membrane protein (Ste6p) has also been shown to be functional for another plasma membrane protein (Kolling and Losko 1997).

3.2 Ubiquitin-dependent internalization of mammalian channels

At about the same time that it was discovered that Rsp5p was crucial for the downregulation of yeast plasma membrane proteins, it was demonstrated that Nedd4 plays a key role in downregulation of the amiloride-sensitive sodium channel, ENaC. This demonstration was the result of studies investigating the origin of a specific disease, Liddle syndrome, an inheritable severe form of hypertension caused by an increase in the cell surface activity of ENaC. ENaC is essential for sodium management in the kidney. This channel consists of three subunits (α, β, γ), all of which have two transmembrane spans. Patients with Liddle syndrome were found to have deletion or point mutations affecting the C-terminal tails of the α or β ENaC subunits, in a region containing a short Pro-rich conserved motif, PPxY (PY) (Schild et al. 1996). A two-hybrid approach, with the C-terminus of the β-subunit as bait, was used to search for partners binding this motif. This approach led to identification of the WW domains of the ubiquitin protein ligase Nedd4. Consistent with this observation, ENaC has been shown to be regulated by ubiquitylation, primarily on a cluster of lysine residues at the N-terminus of the γ subunit. Mutations affecting these lysines impaired channel ubiquitylation and increased channel density at the cell surface (Staub et al. 1997). Moreover, the overproduction of a catalytically active form of Nedd4 in a *Xenopus* oocyte expression system results in the inhibition of ENaC activity, whereas the overproduction of an inactive form does not; this inhibition was found to depend on ENaC PPxY motifs (Staub et al. 1997). Thus, Nedd4 WW domains interact with ENaC via its PY motifs, leading to the ubiquitylation and subsequent downregulation and lysosomal degradation of this protein (reviewed in Rotin et al. 2000). Nedd4 exists as two isoforms: Nedd4-1 and Nedd4-2. Subsequent studies showed that Nedd4-2, which can be coimmunoprecipitated with ENaC, is the isoform responsible for ENaC regulation (Kamynina and Staub 2002). It is tempting to speculate that ENaC downregulation results from Nedd4-2-dependent cell-surface ubiquitylation. However, this remains to be demonstrated and ubiquitylation at other cellular locations is possible given the role demonstrated for ubiquitylation at other steps of intracellular trafficking (see below).

Two other channels also appear to display ubiquitin-dependent downregulation controlled by E3s of the Nedd4 family. The cardiac voltage-gated Na+ channel (rH1), which contains PY motifs, is also negatively regulated by Nedd4 when produced in *Xenopus* oocytes (Abriel et al. 2000). The chloride channel ClC5 has a PY motif that is critical for its downregulation in response to interaction with an E3 of the Nedd4 family (Schwake et al. 2001). The involvement of ubiquitin ligases of the Nedd4 family in the downregulation of plasma membrane proteins is not limited to channels: some receptors are also substrates.

Although no mammalian plasma membrane transporters have yet been described to undergo ubiquitylation, or ubiquitin-dependent downregulation, posttranslational modifications of this type have been reported for two mammalian glucose transporters, GLUT1 and GLUT4. Two-hybrid systems demonstrated interaction between a conserved 11-amino acid sequence in these transporters and a Ubc-like enzyme, Ubc9, responsible for conjugation with the ubiquitin-like protein, sentrin. Indeed, both transporters were found to be modified by sentrin, and were regulated in opposite ways by Ubc9. The overproduction of Ubc9 resulted in an increase in the amount of GLUT1 and a decrease in the amount of GLUT4. The authors suggested that these effects might be due to effects on intracellular pools of these transporters, rather than on plasma membrane-located transporters (Giorgino et al. 2000). Additional information is now required to determine the precise role in trafficking of this currently unique example of sentrin modification of a transporter.

3.3 Mechanisms involved in the ubiquitylation of yeast plasma membrane transporters

We still know little about the structural motifs mediating the recognition of various substrates - including transporters - by their specific ubiquitin ligases. The interaction may be direct or mediated by accessory proteins. The recognition motif on the substrate may be part of or separate from the region carrying the target Lys (Glickman and Ciechanover 2002). Thus, the interaction between ENaC and

Nedd4 appears to be a textbook example, in which the C2 and WW domains are involved in enzyme localization and PPxY-dependent substrate recognition, respectively, with the target Lys distinct from the recognition site, and located elsewhere in the substrate. However, this is not a general rule concerning the functioning of Nedd4. The interaction of Nedd4 with another of its substrates, the insulin-like growth factor receptor, is mediated by an adapter, Grb10, which interacts with the Nedd4 C2 domain (Vecchione et al. 2003). Nedd4 has also been reported to interact in a WW-dependent way both *in vivo* and *in vitro,* with a substrate lacking PPxY, it has also been shown to interact with PPxY-containing proteins that are unlikely to be substrates (Murillas et al. 2002).

Although much is already known about the putative ubiquitylation signals in yeast transporters and, to a lesser extent, the role played by Rsp5p domains in this process, we still have no clear overall picture of the way in which Rsp5p interacts with transporters. We summarize below what is currently known and what remains to be determined.

3.3.1 Factors required for ubiquitylation: Rsp5p domains and partners

The C2 domain of Rsp5p has been shown to be required for location of the protein within the cell, in both plasma membrane invaginations and endosomes (Wang et al. 2001). Deletion of the C2 domain has been reported to impair fluid-phase endocytosis (Dunn and Hicke 2001) and to delay the internalization of both Gap1p and Fur4p, with no major effect on the ubiquitylation of these transporters (Springael et al. 1999a; Wang et al. 2001). This suggests that Rsp5p may, in addition to its enzymatic function, play a role in endocytosis. WW domains in Rsp5p interact with certain soluble substrates in a PPxY-dependent way (Wang et al. 1999). Point mutations affecting conserved residues in WW domains have shown that the WW3 domain and, to a lesser extent, the WW2 domain of Rsp5p are critical for Fur4p ubiquitylation and internalization, whereas the WW1 domain is not required (Gajewska et al. 2001; Marchal and Urban-Grimal, unpublished data). These findings suggest that the WW2 and WW3 domains of Rsp5p are involved, directly or indirectly, in Fur4p recognition. However, neither Fur4p nor any other yeast plasma membrane Rsp5p substrate has an obvious PPxY motifs that could accommodate the direct binding of Rsp5p WW domains. Thus, there may be other, currently unknown motifs involved in the interaction of Rsp5p via its WW domains, or Rsp5p may interact with its plasma membrane substrates via its other domains, or with the assistance of adaptors.

Candidate adaptors for the yeast Rsp5p protein include the Bul proteins. Bul1p interacts physically with Rsp5p via its PPxY motif, and Bul2p is a homolog (Yashiroda et al. 1996, 1998). Deletion of *BUL1* and *BUL2* impairs cell-surface ubiquitylation and downregulation of several transporters, Gap1p (Soetens et al. 2001) (Fig. 1), and Fui1p (C. Volland, personal communication). Thus, Bul proteins may be involved in the recognition of some Rsp5p substrates. However, it remains unclear whether these proteins are involved in the ubiquitylation of all plasma membrane Rsp5p substrates. The two Bul proteins clearly differ in their properties. The low- and high-affinity tryptophan permeases, Tat2p and Tat1p, both display Rsp5p-dependent downregulation if yeast cells are subjected to high pressure. The deletion of *BUL2* protected Tat2p against this downregulation, whereas the deletion of *BUL1* did not and Tat1p downregulation was not prevented by the deletion of either gene (Abe and Iida 2003).

3.3.2 Ubiquitylation signals and target lysines

Whether Rsp5p interacts with transporters via the Bul proteins or via other partners, this interaction must also involve sequence motifs in the transporters. Investigations aiming to define the motifs in yeast transporters required for their ubiquitylation have revealed that an acidic stretch in the linker region connecting the two halves of the ABC-transporter Ste6p (Kolling and Losko 1997), and an N-terminal acidic PEST-like sequence in Fur4p (Marchal et al. 1998) and in the maltose permease Mal61p (Medintz et al. 2000) are essential. Many yeast transporters are phosphorylated (Table 1), and phosphorylation, notably within PEST sequences, is frequently linked to ubiquitylation (Glickman and Ciechanover 2002). This has led to investigation of the effect of transporter phosphorylation on downregulation. A point mutation affecting a potential PKC phosphorylation site in Mal61p (S295A) outside a PEST sequence has been shown to protect against internalization and degradation (Brondijk et al. 1998). The PEST sequence of Fur4p displays serine phosphorylation, a modification required for permease ubiquitylation at nearby lysines (Marchal et al. 1998, 2000). Phosphorylation in this sequence is partly dependent on the redundant Yck1p/Yck2p casein kinase I homologs (Fig. 1) (Marchal et al. 2000). A critical threonine in the linker region of Ste6p is also important for phosphorylation, efficient ubiquitylation, and internalization (Kolling 2002). Therefore, the acidity of ubiquitylation signals may be affected by phosphorylation events. There is no general simple relationship between transporter phosphorylation and downregulation, and the role of the Yck kinases does not seem to be simple. The ABC transporter, Pdr5p, which has been reported to undergo *in vitro* phosphorylation mediated by Yck1/2, was found to be unstable in Yck1/2-deficient cells (Decottignies et al. 1999), as was the Fur4p variant lacking phosphoacceptor sites in the PEST sequence (Marchal et al. 2000). This suggests that Yck1/2 may be involved in the negative regulation of certain components of the endocytic machinery.

In addition to these sequences, potentially important for recognition by the ubiquitylation machinery -Rsp5p or its partners- the critical Lys residues involved in Rsp5p-dependent ubiquitylation have been identified in a number of cases (Table 1). Fur4p and Gap1p are ubiquitylated on two Lys in their N-terminal regions (Marchal et al. 2000; Soetens et al. 2001) (Fig. 1). Zrt1 is ubiquitylated on a single Lys in an intracellular loop (Gitan and Eide 2000), and in the Tat2p and Bap2p amino acid permeases, the replacement of five and six Lys, respectively, in the N-terminal region protects against ubiquitylation (Beck et al. 1999; Omura et al. 2001b). Strikingly, in Fur4p, the Ser (Ser42, 43, 45, 55, 56) that must be phosphorylated for subsequent Fur4p ubiquitylation (Marchal et al. 1998) lie very close to the target Lys (K38, 41) (Marchal et al. 2000). A situation similar to that described for Ste2p, in which an acidic, Ser-rich sequence, SINNDAKSS, is the target of both phosphorylation and ubiquitylation (Hicke et al. 1998). Similarly, the unique target Lys in the Zrt1p zinc transporter is 14 amino acids away from a short acidic sequence containing several serines required for Zrt1p ubiquitylation (Gitan et al. 2003).

The data described above were obtained by extensive mutagenesis (introduction of deletion and point mutations) of individual genes, for analysis of the downregulation of the corresponding proteins in defined physiological conditions known to trigger their ubiquitylation. Advances in this field were recently accelerated by the use of proteomics (Peng et al. 2003). Attempts have been made to identify all the ubiquitylated proteins in cells with 6His-tagged ubiquitin as their sole ubiquitin. Ubiquitylated conjugates were purified by retention on nickel, and analyzed by multidimensional liquid chromatography coupled with tandem mass spectrometry. In a total of 1075 potentially ubiquitylated proteins, 110 precise ubiquitylation sites were identified in 72 ubiquitin-protein conjugates. These proteins included 12 plasma membrane transporters (Table 1), with this specific group of proteins comprising a strikingly high proportion of the total. These transporters include three proteins that have already been shown to be ubiquitylated (Ste6p, Pdr5p and Gap1p) (Kolling and Hollenberg 1994a; Egner and Kuchler 1996; Springael and André 1998) and three others shown to undergo Rsp5p- and/or ubiquitin-dependent endocytosis (Hxt6p, Hxt7p, and Jen1p) (Krampe et al. 1998; Paiva et al. 2002). This bias probably reflects the higher abundance of the corresponding ubiquitin conjugates. This is consistent with the observation that the vacuolar degradation of proteins in which ubiquitylation acts as an endocytic signal is much slower than the proteasome-dependent degradation of known short-lived ubiquitylated proteins, such as cyclins, which were not recovered in this study. Strikingly, the target Lys identified in these transporters lie in very acidic motifs, often rich in Ser residues. Sixteen phosphorylation sites (mostly Ser, several Thr, and 2 Tyr) were also identified within nine transporters. Some of these sites are clustered (for instance Ser306, Thr304, Ser324 and Thr319 in Pho84p). For two of these transporters, Pho84p and Pdr5p, the ubiquitylated target Lys lie very close to the identified phosphorylated amino acids, as reported for Fur4p. Precise analysis and comparison of the sequences surrounding the target Lys may help us to understand how these motifs fit into the catalytic site of the corresponding E3, probably Rsp5p, and would undoubtedly stimulate new investigations. Data for cells cultured in standard conditions, should be complemented with data obtained in defined physiological conditions. It will be necessary to define, for instance, whether the identified target Lys are ubiquitylated at the plasma membrane, as ubiquitylation is also involved in other trafficking steps. The target Lys in Gap1p identified in this study (Lys76) differs from the two target Lys (Lys9, Lys16) shown to be ubiquitylated following the addition of ammonium to cells cultured in nitrogen-poor medium (Soetens et al. 2001), and its function remains to be determined. It may well correspond to an ubiquitylation occurring at an intracellular location. In another proteomics project based on the comparative analysis of wild type cells and mutant cells deficient in an ERAD component, a large number of ubiquitylated proteins were identified, a number of which correspond to endogenous

ERAD substrates (Hitchcock et al. 2003). About 40 ubiquitylation sites were defined including 10 in plasma membrane or vacuolar transporters (Table 1).

Although the proteomic approach requires complementary analyses, it constitutes an important step in determining whether the ubiquitylation sites in plasma membrane transporters display some signature. The data currently available highlight the link between the phosphorylation and ubiquitylation of plasma membrane transporters lacking PPxY motifs. It is tempting to think that Rsp5p may be able to recognize any of its target lysines provided that they are uncovered following a change in the conformation of the substrate transporter due to phosphorylation at a nearby site. The roles of the various domains of Rsp5p in this process, including the catalytic domain that interacts with the ubiquitylation sites, remain to be defined. Analysis of the potential interactions of plasma membrane transporters with the entire Rsp5p protein, its domains, or partners, is required.

3.3.3 Mono- and/or polyubiquitylation of yeast plasma membrane transporters

Yeast plasma membrane transporters are classical endocytic cargoes that are targeted for vacuolar degradation after internalization (Kolling and Hollenberg 1994a; Lai and McGraw 1994; Volland et al. 1994; Egner et al. 1995). The first studies on ubiquitin-dependent endocytosis indicated that despite their ubiquitylation, transporters were not recognized and degraded by the proteasome (Galan et al. 1996; Horak and Wolf 1997; Kolling and Losko 1997). This finding was accounted for in some cases by the addition to target Lys of one ubiquitin moiety, or of short chains of ubiquitin residues only two or three residues long (Fur4p, Gap1p, Mal61p, Zrt1p) as deduced from ubiquitylation patterns on western blots probed with transporter-specific antibodies (Galan and Haguenauer-Tsapis 1997; Medintz et al. 1998; Springael et al. 1999b; Gitan and Eide 2000). It was then shown that for Fur4p and Gap1p, chain extension occurred by means of a ubiquitin Lys63 bond (Galan and Haguenauer-Tsapis 1997; Springael et al. 1999b) (Fig. 2). This was demonstrated by comparing the ubiquitylation patterns of transporters in cells with wild type ubiquitin, or ubiquitin unable to make Lys63-linked ubiquitin chains as the sole source of ubiquitin (Galan and Haguenauer-Tsapis 1997). In an alternative approach, ubiquitylation patterns were determined in npi2/doa4 cells that have low intracellular ubiquitin level (Galan and Haguenauer-Tsapis 1997; Springael et al. 1999b; Swaminathan et al. 1999), and in which ubiquitin-dependent processes were rescued by expression of plasmid-encoded ubiquitin variants incompetent for the formation of Lys29-, 48- or 63-linked ubiquitin chains (Galan and Haguenauer-Tsapis 1997; Springael et al. 1999b). For Gap1p and Fur4p, it is thus clear that there are too few ubiquitin conjugates for proteasome recognition, and the ubiquitin chains present are of an inappropriate type. However, multiubiquitylated transporter species have been observed in many cases (Kolling and Hollenberg 1994a; Egner and Kuchler 1996; Horak and Wolf 1997; Beck et al. 1999), but it is unclear whether these forms result from the monoubiquitylation of several Lys or the addition of chains of ubiquitin. In studies in which ubiquitin conjugates were detected by immunoprecipitation followed by western blotting, including a recent proteomic study, high molecular weight smears were observed (Kolling and Hollenberg 1994a; Egner and Kuchler 1996; Horak and Wolf 1997; Beck et al. 1999; Peng et al. 2003). However, it is unclear whether these observations reflect extensive multiubiquitylation or the aggregation of highly hydrophobic ubiquitin conjugates. We need to determine whether Fur4p and Gap1p, two of the very few (<10) known ubiquitylated substrates carrying Lys63-linked ubiquitin residues, are representative of a larger class of Rsp5p plasma membrane substrates. Recent attempts to identify yeast ubiquitylated substrates by means of proteomics have shown that Lys63 chains are more abundant than previously thought (Peng et al. 2003).

The ubiquitylation of yeast plasma membrane transporters has also been characterized to determine the type of ubiquitylation required for internalization (Fig. 2). For Fur4p and Gap1p, the generation of Lys63-linked ubiquitin chains is required for efficient endocytosis (Galan and Haguenauer-Tsapis 1997; Springael et al. 1999b). Consistent with these data, a single ubiquitin fused in-frame at the N-terminus of Fur4p triggers the internalization of a variant Fur4p lacking its own target Lys (Fig. 3), but this mutant protein is internalized at a fifth the rate for the wild type, fully ubiquitylated permease (Blondel et al. 2004). In contrast, based on the efficient rescue of Gal2p (Horak and Wolf 2001) and Mal11p (Lucero et al. 2000) internalization in *doa4Δ* cells by the production of ubiquitin incompetent for conjugation, it was concluded that monoubiquitylation is the relevant signal for these two transporters. The reasons for this difference remain unclear. It is also unclear whether Rsp5p, which may mediate the monoubiquitylation of a membrane-bound ER substrate, Spt23p (Hoppe et al. 2000), or Lys63-linked ubiquitin modification in at least two transporters, and which is required for polyubiquitylation leading to proteasomal degradation of the soluble Rbp1p (Beaudenon et al. 1999), makes use of different adapters for these different ubiquitin modifications.

3.4 Plasma membrane-to-vacuole targeting of yeast transporters

3.4.1 Internalization

Major progress in deciphering the role of ubiquitin as a sorting signal at various steps in intracellular trafficking was achieved with the discovery of ubiquitin-binding proteins that also contained variable effector domains (Katzmann et al. 2002; Bonifacino and Traub 2003). These proteins include Ede1p (the closest homolog of the mammalian Eps15), and the homologs Ent1p and Ent2p (yeast epsins). Ede1p was initially shown to be involved in the internalization of Fur4p (Gagny et al. 2000), whereas Ent1/2 were identified as responsible for the internalization of Ste6p (Wendland et al. 1999). Ede1p has one UBA (ubiquitin-associated) motif and Ent1/2 carry two UIM (ubiquitin-interacting motif) (Shih et al. 2002). Both these domains are small (20-40 amino acids) and are known or predicted to be α-helical. The Ent1/2 UIM domains have been reported to bind ubiquitin *in vitro* (Shih et al. 2002), and both Ent1p and Ede1p have been reported to bind membranes in a ubiquitin-dependent fashion (Aguilar et al. 2003). Several

UBA and UIM domains have been reported to bind polyubiquitin chains with an affinity higher than that with which they bind to monoubiquitin. The nature of the ubiquitin chains recognized has generally not been investigated, but the UBA domain of Rad23p has been shown to bind Lys48-linked ubiquitin chains preferentially (Raasi and Pickart 2003). Thus, the possible influence of the type of ubiquitylation of membrane-bound endocytic cargoes on their potential interaction with ubiquitin-binding proteins is unknown. The Ent1/2 proteins also carry a clathrin-binding motif and an N-terminal ENTH (epsin N-terminal homology) domain. This type of domain has been shown to bind and to penetrate into membranes, causing them to curve. Therefore, it has been suggested that epsins act as adaptors in the selection of ubiquitylated cargos, clathrin binding, and membrane bending (Wendland 2002). This proposed role is similar to that of the main AP2 adaptors for non-ubiquitylated cargos in mammals. Homologs of AP2 subunits have been identified in yeast, but play no role in internalization (Huang et al. 1999).

Despite these recent advances in the characterization of several proteins involved in the internalization step of endocytosis, and in other investigations in this field, this process remains poorly understood, for both transporters and other cargoes (recent review in D'Hondt et al. 2000). Internalization is thought to require correct organization of the actin cytoskeleton, and many transporters accumulate as ubiquitylated conjugates in various mutants with impaired cytoskeleton organization (D'Hondt et al. 2000). There is still some debate about the function of clathrin in the internalization of yeast receptors because point mutations or deletions in the clathrin heavy chain gene (*CHC1*) decrease receptor internalization by 30-50% (reviewed in Baggett and Wendland 2001). Similarly, internalization is partly inhibited in at least one transporter, maltose permease (Penalver et al. 1999). The role played by clathrin in the internalization of other transporters is unknown. In contrast, a number of reports have described the clathrin-dependent internalization of plasma membrane transporters in mammals. However, little is known about the type of adaptor involved in the recognition of internalization signals in latter proteins.

3.4.2 Trafficking to endosomes and entry into multivesicular bodies (MVB)

Following their internalization, all endocytic ubiquitylated yeast plasma membrane cargoes are transported through several endocytic intermediates known as early (or post-Golgi endosomes, PGE), and late endosomes (or prevacuolar endosomes, PVE), where they meet membrane proteins from the Golgi apparatus, and are then delivered to the vacuole lumen for degradation. One of the first post-internalization events appears to be segregation into specific lipid domains distinct from plasma membrane lipid domains. For example, Fur4p, which is primarily present in rafts when at the plasma membrane, whether ubiquitylated or not, is mostly recovered in non-raft domains once it has entered endosomes (Dupre and Haguenauer-Tsapis 2003).

Many of the factors involved in late steps of endosome–to-vacuole trafficking have been identified by genetic screens based on the missorting in the medium of carboxypeptidase Y (CPY), a vacuolar protease. In some cases, screens based on several transporters were used. Some thermosensitive mutations affecting the essential plasma membrane ATPase Pma1p, or the arginine permease Can1p, result in the direct delivery of these proteins, at 37°C, from the Golgi complex to endosomes/vacuoles for degradation. Genetic selection for suppressors restoring normal growth at 37°C, or the reacquisition of normal sensitivity at 37°C to canavanin, a toxic analog of arginine, led to the identification of known or new *vps* (vacuolar protein sorting) mutants (Luo and Chang 1997; Li et al. 1999). For example, *SPT22/VPS23*, encoding a protein conserved from yeast to man carrying a Ubc-like domain, which plays a crucial role in the generation of and sorting to multivesicular bodies (MVBs) was first identified by screening for the reacquisition of normal sensitivity at 37°C to canavanin (Li et al. 1999).

Understanding the molecular mechanisms involved in the formation of and sorting to MVBs has been a major goal in the last three years (Piper and Luzio 2001; Katzmann et al. 2002). MVBs form in late endosomes, when the limiting membrane invaginates and buds into the lumen of the organelle to form internal vesicles (Fig. 1). During this process, a subset of the membrane proteins within the limiting membrane of the endosome is sorted to these invaginating vesicles. Subsequent fusion of the mature MVB with the lysosome/vacuole results in the delivery of the internal vesicles and their associated cargoes to the lumen of the lysosome/vacuole, where they are degraded by vacuolar proteases and lipases. Proteins that remain in the limiting membrane of the MVBs are delivered to the limiting membrane of the lysosome/vacuole. This process, thereby, separates proteins destined for the lumen of the lysosome/vacuole from proteins destined for the limiting membrane of this organelle. MVB formation is impaired in a subset of *vps* mutants, the *vps* class E mutants, which accumulate an exaggerated late endosome known as the class E compartment, in which endocytic and membrane-bound Golgi proteins remain trapped. Class E Vps proteins, conserved from yeast to man, have been shown to be involved in both the formation of MVBs and sorting to MVBs.

Endocytic cargoes, including plasma membrane transporters, are sorted to the internal vesicles of MVBs (Fig. 1). Therefore, they are directed to the vacuolar lumen (representative example Fig. 3), together with a subset of biosynthetic cargoes, such as carboxypeptidase S (Cps1p), a transmembrane protein that is processed to generate its active enzymatic form in the vacuolar lumen. Various studies have demonstrated that the ubiquitylation of MVB cargoes plays a critical role in sorting to the internal vesicles of MVBs. Early data concerning the fate of plasma membrane or vacuolar transporters significantly contributed to this demonstration. Ste6p was shown to accumulate at the vacuolar membrane rather than being released into the vacuolar lumen in *doa4* cells, which have low levels of ubiquitin, and to be correctly sorted if ubiquitin was overproduced (Losko et al. 2001). Fth1p, a vacuolar iron transporter with multiple transmembrane spans is normally delivered to the vacuolar membrane but was shown to be sorted to the vacuolar lumen following the in-frame fusion of ubiquitin (Urbanowski and Piper 2001).

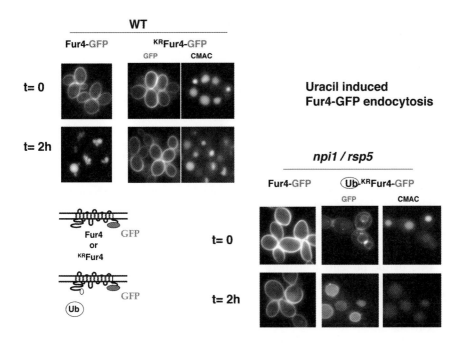

Fig. 3. Substrate-induced endocytosis of the uracil permease Fur4p: ubiquitin-dependent, Rsp5p-dependent plasma membrane internalization and MVB sorting. GFP-tagged Fur4p or a variant Fur4p mutated in its two target Lys (Lys38, 41) fused or not to a variant ubiquitin lacking Lys acceptor sites (Ub) were expressed in wild type (WT) or *npi* (*npi1/rsp5*) mutant cells. The fate of presynthesized permease was followed before and two hours (2H) after addition of uracil in the medium by visualizing GFP fluorescence. Vacuolar lumen was visualized using a dye, CMAC. Fur4-GFP undergoes uracil-induced endocytosis in wild type cells: fluorescence disappears from the plasma membrane, and is recovered in the vacuole lumen (low rate of degradation of the GFP tag). Internalization is impaired by mutation of Fur4 target Lys, or by the *npi1* mutation in *RSP5* gene. In frame fusion of ubiquitin lacking its own target Lys restores some internalization, and triggers partial direct vacuolar sorting, but MVB sorting is not entirely efficient, and both neosynthesized permease and internalized permease are partially retained at the vacuolar membrane. Selected data from experiments reported in Blondel et al. (2004); with permission by the American Society for Cell Biology.

Cps1p, which has a single transmembrane span, primarily undergoes monoubiquitylation and this modification is sufficient for correct sorting to MVBs (Katzmann et al. 2001). Little is known about the type of ubiquitylation required for the sorting to MVBs of transporters with multiple transmembrane spans originating from the plasma membrane. However, the ubiquitylation status of the protein acquired at the plasma membrane clearly leads to efficient MVB sorting. In-frame ubiquitin has been shown to restore the plasma membrane internalization of Fur4p lacking its normal target Lys to some extent. Nonetheless, some of this chimeric protein was retained at the vacuolar membrane, suggesting that monoubiquitylation is not sufficient for the efficient MVB sorting of this class of proteins (Fig. 3) (Blondel et al. 2004).

Ubiquitylated cargoes arriving from the plasma membrane and the Golgi apparatus are thought to interact successively with several class E proteins. The first are two Vps class E proteins carrying UIM domains, Hse1p and Vps27p, which form a complex (Bilodeau et al. 2002). Cargoes are then recognized successively by other class E proteins that form three high–molecular weight complexes, referred to as ESCRT I, II and III (endosomal sorting complexes required for transport I, II and III) (Katzmann et al. 2002). Vps23p, which has been shown to interact with Vps27p (Bilodeau et al. 2003), is the only component of the ESCRT complexes with a ubiquitin-binding region, the Ubc-like domain (Katzmann et al. 2002). Most of the proteins comprising these complexes were identified by the cloning of identified class E *vps* gene, systematic purification in the native state of the various corresponding proteins, and analysis of coimmunoprecipitated partners.

One of the proteins identified in specific screens for Vps class E proteins, Vps31p/Bro1p, which is associated with the ESCRT III complex (Odorizzi et al. 2003), was also isolated in two genetic screens based on amino acid uptake. Mutations in *BRO1* have been identified as suppressors of phenotypes associated with a lack of the amino acid sensor Ssy1p (Forsberg et al. 2001). Curiously, mutations in *BRO1* were also identified long ago as an *npi* mutant, *npi3*, impaired in the NH_4^+-induced downregulation of cell-surface Gap1p. NH_4^+-induced ubiquitylation and downregulation of Gap1p was strongly inhibited in *npi3* cells (Springael et al. 2002). Although these observations initially suggested a possible role for Bro1p at both the plasma and late endosome membranes, further analysis of the *npi3/bro1* mutant revealed that Gap1p was ubiquitylated and internalized by endocytosis but that its sorting to MVB was defective. Thus, a fraction of the internalized permease is missorted to the limiting membrane of the vacuole. However, most of the internalized Gap1p is recycled back to the plasma membrane, explaining why Gap1p remains largely active in this mutant after the addition of NH_4^+. Interestingly, this recycling seems to be accompanied by de-ubiquitylation of the permease (Nikko et al. 2003).

Although most Vps class E proteins appear to be involved in the sorting of all MVB cargoes, the possibility of some specificity cannot be ruled out, as shown recently by the identification of new class E *vps* genes by means of a strategy based on the fate of the Ste6p transporter. A colony assay was developed to identify genes involved in late steps of Ste6p endocytosis. This assay was used to search for genes that, when overexpressed, caused impaired degradation of a Ste6-lacZ fusion protein. A number of known class E *vps* genes were recovered, and *MOS10* (more of Ste6), a gene of previously unknown function was also identified (Kranz et al. 2001). Mutations in *MOS10* were also recovered as suppressors of *ssy1Δ* phenotypes (Forsberg et al. 2001). The overexpression and deletion of *MOS10* both gave a class E phenotype. *MOS10* encodes a coiled-coil protein, homologous to two proteins of the ESCRT III complex, Vps20p and Snf7p. These three proteins, which form a small family, display subtle differences in function.

The phenotypes of the deletion strains are not identical. For instance, *snf7Δ* cells display impaired glucose signalling, whereas *vps20Δ* and *snf7Δ* cells display heat sensitivity for growth. These phenotypes are not displayed by *mos10Δ* cells, but *mos10Δ* / *mos10Δ* cells have specifically impaired filament maturation during pseudohyphal growth (Kohler 2003). Although the precise reasons for these differences in phenotype are unknown, these observations suggest that these three proteins may well preferentially affect the sorting of specific MVB cargoes.

The fate of the attached ubiquitin during the final step of MVB sorting has also been analyzed. Genetic data suggested that the Doa4p ubiquitin isopeptidase may be involved in the trafficking of endocytic cargoes. Intracellular ubiquitin concentrations are low in *doa4Δ* cells (Springael et al. 1999b; Swaminathan et al. 1999), and this phenotype is suppressed in mutants defective for the internalization step of endocytosis (Swaminathan et al. 1999). The potential role of Doa4p in the deubiquitylation of plasma membrane proteins was tested by following the fate of Fur4p-ubiquitin conjugates. Fur4p accumulated entirely in the form of plasma membrane ubiquitin conjugates over several hours in cells in which the internalization step of endocytosis was impaired, but was recovered in deubiquitylated form in vacuoles of *pep4Δ* cells, deficient in vacuolar proteases activities. In contrast, *pep4Δ doa4Δ* cells accumulated vacuolar Fur4p exclusively in the ubiquitin-conjugated form. Thus, Ub-Fur4p undergoes Doa4p-dependent deubiquitylation during plasma membrane-to-vacuole trafficking (Fig. 1). A lack of deubiquitylation did not prevent the MVB sorting of Fur4p (Dupré and Haguenauer-Tsapis 2001). Doa4p, which has been shown to accumulate in endosomes in class E *vps* mutants (Amerik et al. 2000), has also been shown to be involved in deubiquitylation of MVB cargoes from the Golgi apparatus (Katzmann et al. 2001). This protein is probably the central ubiquitin isopeptidase acting after the ESCRT III complex, just before the sorting of cargoes to MVBs. The delivery to the vacuole of ubiquitin-conjugated endocytic and biosynthetic cargoes in *doa4Δ* cells leads to the massive proteolytic degradation of ubiquitin, resulting in defective ubiquitin homeostasis in these cells.

In addition to the role played by ubiquitin as a sorting signal in early and late steps in the endocytosis of plasma membrane transporters, at least one additional signal has been shown to be involved in the endocytic pathway of a yeast transporter. The normal NH_4^+-induced downregulation of Gap1p is also dependent on a di-leucine motif present in a predicted coiled domain of the cytosolic C-terminal tail of the permease (Hein and André 1997). Interestingly, the sequence context of this di-leucine fits the structural requirements for recognition by the VHS domain of Gga proteins, recently identified as adaptors (Bonifacino and Traub 2003). Substitution mutations affecting the di-leucine or neighbouring acidic residues do not prevent NH_4^+-induced Gap1p ubiquitylation but do impair the subsequent endocytosis and degradation of Gap1p (Springael and André 1998). It remains to be determined whether the di-leucine is required for early and/or late steps of internalization of the Gap1p permease. A role for the di-leucine late in endocytosis would suggest that Gap1p is recycled back to the cell surface if the di-leucine motif is not accessible to some VHS-containing proteins.

4 Regulation of transporters and channels at membrane trafficking levels: signals and mechanisms

We summarize below the diverse physiological situations reported to induce adaptive changes in the membrane sorting of specific transporters. We will begin by considering the numerous cases of accelerated endocytosis of a transporter in response to environmental stimuli. We will then describe situations involving controlled sorting of transporters present in the Golgi and/or endosomal membranes. We will then comment briefly on similar control systems for mammalian transporters. Finally, we will discuss models of the integration of upstream signals for control of the global trafficking of transporters.

4.1 Physiological control of the rate of internalization of transporters

Changes in environmental conditions often lead to drastic increases in the rates of internalization of specific transporters by endocytosis, and to the targeting of these transporters to the endosome/vacuolar pathway for degradation. Furthermore, the downregulation of transporters is often accompanied by the induction of other transport systems more appropriate for the new conditions. This dynamic control of transporters has been studied in detail in both yeast and mammalian cells.

4.1.1 Regulated endocytosis of amino acid permeases

Yeast cells possess about twenty different amino acid permeases, the substrate specificities of which overlap in many cases. It is now established that these proteins are not simultaneously active in the cells. Instead, most are differentially regulated according to the nitrogen and/or amino acid content of the growth medium. For instance, the general amino acid permease (Gap1p) (Grenson et al. 1970; Jauniaux and Grenson 1990) is the most active amino acid permease in conditions of limiting nitrogen supply, such as NH_4^+ at a low level, or urea being the sole available sources of nitrogen. Gap1p is also highly active if cells are grown on proline, another poor nitrogen source. Under these conditions, the *GAP1* gene is strongly transcribed (Jauniaux and Grenson 1990) and the newly synthesized Gap1p is targeted to the plasma membrane, where it is active and highly stable (Grenson 1983a; De Craene et al. 2001). If a preferential source of nitrogen, such as NH_4^+, is added at a sufficiently high concentration, the *GAP1* gene is repressed and pre-synthesized Gap1p is progressively removed from the cell surface by endocytosis and sorting to the MVB/vacuolar degradation pathway (Grenson 1983a; Hein et al. 1995; Springael and André 1998; De Craene et al. 2001; Nikko et al. 2003). This downregulation may also be triggered by the addition of high concentrations of amino acids (Stanbrough and Magasanik 1995). These changes of nitrogen supply conditions are thought to inactivate Npr1p, a protein kinase playing a pivotal role in the control of trafficking of several permeases including Gap1p and which is specifically active when cells grow under poor nitrogen supply conditions (Grenson 1983b; Vandebol et al. 1987, 1990; Schmidt et al. 1998). Inactivation of Nprlp by shifting an $nprl^{ts}$ mutant to the non-permissive temperature also leads to downregulation of Gap1p (Grenson 1983b; De Craene et al. 2001). As discussed below, inactivation of Nprlp in response to good nitrogen supply conditions is dependent on the TOR signalling pathway (Schmidt et al. 1998). The Rsp5p- and Bul-dependent ubiquitylation of Gap1p on its two most N-terminal lysine residues (at positions 9 and 16) is essential for the downregulation of this protein (Hein et al. 1995; Springael and André 1998; Soetens et al. 2001) (Fig. 1). Under conditions of poor nitrogen supply, only a small fraction of Gap1p is monoubiquitylated. The addition of NH_4^+ concomitantly increases the monoubiquitylation and induces the polyubiquitylation of Gap1p via Lys63-linked chains (Springael and André 1998; Springael et al. 1999b).

During the downregulation of Gap1p following the addition of high concentrations of amino acids, Gap1p degradation seems to be preceded by dephosphorylation (Stanbrough and Magasanik 1995). In contrast to Gap1p, other broad-specificity amino acid permeases such as Agp1p, Tat2p, Tat1p, and Bap2p are induced, with increases in transcription of the corresponding genes in response to the addition of amino acids (Iraqui et al. 1999). These permeases, therefore, take on the functions of Gap1p. This induction is dependent on the Ssy1p permease-like sensor of external amino acids (Boles and André, this issue). Unlike Gap1p, these permeases are synthesized and active in cells grown on YPD medium. However, Bap2p and Tat2p are not exclusively located at the cell surface under these conditions. Instead, a significant proportion of these permeases are present in internal membranes, probably those of endosomes (Beck et al. 1999; Omura et al. 2001a; Umebayashi and Nakano 2003). The presence of large amounts of these permeases within cells may be due to their direct sorting for degradation because their amino acid substrates are present at high concentrations in the medium.

Shifting cells from YPD medium to nitrogen starvation conditions results in the derepression of Gap1p and the rapid downregulation of Tat2p and Bap2p. This downregulation process may also be induced by adding rapamycin or possibly other compounds such as anesthetics (Palmer et al. 2002) or the FTY720 immuno-suppressive agent (Welsch et al. 2003) to the medium. These compounds interfere with the TOR signalling pathway, which is active in favourable nutrient supply conditions (Crespo and Hall 2002). The requirements for ubiquitin, Rsp5p, and lysines in the downregulation of Tat2p and Bap2p have been extensively studied (Beck et al. 1999; Omura et al. 2001b; Umebayashi and Nakano 2003). Lysines in the N-terminus of Bap2p, and the C-terminal tail of this protein have been shown to play a key role in determining its fate in the endocytic pathway (Omura et al. 2001a).

The Gap1p, Bap2p and Tat2p permeases all belong to the AAP/YAT family of closely related amino acid permeases. So, what are the structural features in these permeases responsible for determining their pattern of regulation according to nitrogen and amino acid availability? Although the C-terminal tail of Gap1p plays an essential role in NH_4^+-induced downregulation, it is not sufficient to confer a similar pattern of regulation on the NH_4^+-insensitive Can1p permease (Hein and André 1997). However, replacement of the N-terminal tail of Gap1p with that of

Bap2p confers partial sensitivity to rapamycin-induced degradation. This suggests that the N-terminal tail, which is subject to ubiquitylation, contains sufficient information to confer nitrogen-dependent control of endocytosis (Omura et al. 2001b).

4.1.2 Regulated endocytosis of sugar and monocarboxylate transporters

Although glucose at high concentration is the preferred carbon source, yeast cells can also use alternative sugars such as galactose or maltose, or monocarboxylates such as lactate or pyruvate. The permeases involved in the uptake of these compounds (Gal2p, Mal61p, Jen1p) are typically induced by their own substrates and are subject to glucose-mediated inhibition (Horak 2003). If high concentrations of glucose are added to cells grown under inducing conditions, the genes encoding the induced permeases are repressed and the permeases already present at the cell surface undergo ubiquitin-dependent downregulation (Horak and Wolf 1997; Lucero and Lagunas 1997; Medintz et al. 2000; Paiva et al. 2002). The same fate has been observed for two high-affinity glucose transporters (Hxt6p, -7) and the high- to intermediate-affinity transporter Hxt2p, which are specifically induced in response to low extracellular glucose concentrations (Kruckeberg et al. 1999; Krampe and Boles 2002). For unknown reasons, the glucose-triggered downregulation of sugar permeases is more pronounced if cells are simultaneously starved of nitrogen (Lucero et al. 2002). It was recently suggested that this is due to a contribution, in this induced degradation, of the autophagy pathway (Krampe and Boles 2002).

4.1.3 Regulated endocytosis of nucleobase transporters

The high-affinity uracil permease (Fur4p) undergoes ubiquitylation and subsequent internalization and vacuolar degradation at basal rates in exponentially growing cells (Volland et al. 1994; Galan et al. 1996). As a result, the permease can be visualized both at the plasma membrane, and in endosomes, *en route* for vacuolar degradation (Dupré and Haguenauer-Tsapis 2001). Exposing cells to various adverse conditions — nitrogen, carbon, or phosphate starvation, inhibition of protein synthesis by cycloheximide — and the lead-up to the stationary phase of growth result in increases in the rates of ubiquitylation, endocytosis and degradation (Volland et al. 1994; Galan et al. 1996). The addition of high concentrations of uracil to the medium also decreases uracil permease levels, partly by destabilizing *FUR4* mRNA (Séron et al. 1999). This type of regulation, therefore, differs from the more usual transcriptional regulation displayed in the case of other permeases. Uracil also triggers rapid degradation of the existing permease by accelerating ubiquitylation and endocytosis (Séron et al. 1999) (Fig. 3). As the adverse conditions that accelerate permease turnover also lead to ribosome degradation (Warner 1999), they are likely to trigger an increase in the internal pool of uracil of catabolic origin, which may be the main signal for Fur4p downregulation.

Interestingly, the uracil-induced control of Fur4p trafficking requires direct binding of uracil to the permease, as a mutant permease with an abnormally low affinity for uracil displays impaired transport activity (Urban-Grimal et al. 1995) and is not internalized following the addition of uracil to the medium (Séron et al. 1999). These observations suggest that the recognition by transporters of their own substrates and the susceptibility of these proteins to be ubiquitylated may be intimately linked.

4.1.4 Regulated endocytosis of metal transporters

Metal transporters constitute another class of tightly regulated transport systems. This tight regulation is accounted for by the need for metals, at particular concentrations, and their potential toxicity at high concentration. The maintenance of correct metal homeostasis is, therefore, critical.

Mg^{2+}, essential for the activation of hundreds of enzymes, is the most abundant divalent cation in cells (concentrations in the millimolar range). Total intracellular Mg^{2+} concentrations are tightly controlled, varying by a factor of no more than four, whereas external concentrations may change by four orders of magnitude. The yeast permease Alr1p was the first candidate Mg^{2+} transporter to be identified in eukaryotic cells. Alr1p is essential for the growth of yeast cells in all media except those with high Mg^{2+} concentrations. The exposure of cells to even standard Mg^{2+} concentrations leads to a dramatic decrease in the stability of this protein, which undergoes ubiquitin-dependent rapid internalization (Graschopf et al. 2001).

A similar situation has been described for *ZRT1*, which encodes the high-affinity zinc transporter Zrt1p. *ZRT1* is repressed in cells replete with zinc, and spectacular Rsp5p-dependent ubiquitylation of the transporter is triggered within minutes of exposure of cells to high levels of zinc (Gitan et al. 1998; Gitan and Eide 2000).

The highly conserved Smf1p and Smf2p proteins display broad specificity for both essential and nonessential metals (manganese, cadmium, copper, iron etc.). Smf1p is rapidly internalized in the presence of manganese (Liu and Culotta 1999b).

Finally, copper is yet another essential metal required for the activity of various enzymes involved in critical areas of metabolism. Copper is also potentially toxic if it accumulates beyond cellular needs, and intracellular levels of this nutrient must, therefore, be precisely regulated. Copper is transported in yeast and mammals by homologous Ctr1 proteins with three predicted membrane-spanning regions that form oligomeric complexes. The yeast Ctr1p is highly responsive to copper availability, being upregulated by the Mac1p transcription factor if extracellular copper levels are low. In addition to this transcriptional regulation, the levels of Ctr1p protein at the plasma membrane are posttranslationally regulated by copper availability. Two different mechanisms have been described. At high copper concentrations (in the range of 10 µM), Ctr1p is degraded *in situ* at the plasma membrane, via a mechanism independent of endocytosis — an unusual situation for a plasma membrane protein. At low micromolar levels of copper,

Ctr1p undergoes accelerated endocytosis (Ooi et al. 1996), a process dependent on the same transcription factor as *CTR1* induction (Yonkovich et al. 2002). For both Smf1p and Ctr1p, the potential role of ubiquitylation events in these processes has yet to be determined.

4.1.5 Regulated endocytosis of other yeast transporters

One of the first illustrated cases of regulation of a transporter by its own substrates was provided in the case of Itr1p, an inositol permease. Upon addition of excess inositol, the *ITR1* gene is repressed and pre-synthesized Itr1p undergoes accelerated endocytosis and degradation (Lai et al. 1995). Substrate-induced endocytosis has also been reported for the spermidine transporter (Kaouass et al. 1998). The high-affinity phosphate transporter, Pho84p, is also regulated by its own substrate. In the presence of low external phosphate concentrations, the *PHO84* gene and other *PHO* genes involved in phosphate metabolism are derepressed, and Pho84p is targeted to the plasma membrane. If phosphate is added at high concentrations, the *PHO84* gene is repressed and accelerated Pho84p endocytosis is observed (Martinez et al. 1998; Petersson et al. 1999).

4.1.6 Regulated endocytosis of mammalian transporters

A number of mammalian plasma membrane transporters and channels undergo regulated trafficking, involving regulated internalization in particular as discussed in reviews (Lee (2000) and Royle and Murrell-Lagnado (2003)). Therefore, we provide only significant examples of different types of regulation, some of which are similar to that described for yeast plasma membrane transporters.

Substrate-regulated internalization is a common feature of mammalian metal transporters. The human copper transporter, hCtr1, which undergoes copper-stimulated endocytosis and degradation like its yeast counterpart (Petris et al. 2003) has been studied in detail. Interest in yeast and human Ctr1 recently increased with the demonstration that these transporters mediate uptake of the anti-cancer drug, cysplatin, which also accelerates endocytosis (Ishida et al. 2002).

The endocytosis of neurotransmitter transporters is highly regulated and enables the neuron to fine-tune transport capacity to match demands. The cocaine- and amphetamine-sensitive dopamine (DA) transporter (DAT), a presynaptic plasma membrane protein responsible for the regulation of extracellular DA concentrations, undergoes endocytosis in response to protein kinase C (PKC) activation. Activated PKC then catalyzes DAT phosphorylation. Several transporters belonging to the monoamine branch of the same family of Na^+/Cl^--dependent plasma membrane transporters — the serotonin (SERT), norepinephrine (NET), and epinephrine transporters — display similar PKC-dependent downregulation (Buckley et al. 2000). For a number of these transporters, PKC-dependent phosphorylation is triggered by external substrates. DAT substrates, including amphetamine, induce DAT internalization (Saunders et al. 2000). This effect of substrates is the opposite of that described for SERT (Ramamoorthy and Blakely 1999), suggesting markedly different mechanisms of regulation for two closely related transporters.

Another related transporter, the γ-aminobutyric acid transporter (GAT1), undergoes substrate-induced tyrosine phosphorylation, reducing the rate of GAT1 internalization (Whitworth and Quick 2001). It is unclear how phosphorylation by different kinds of kinases can regulate the rates of internalization of target transporters in opposite ways. Similarly, although substrate-induced regulation of the rate of internalization appears to be a common property of yeast and mammalian plasma membrane transporters, it is unknown whether the underlying mechanisms are similar and whether ubiquitin is important for the downregulation of mammalian transporters.

Some transporters are downregulated by hormones. Extensive studies of the insulin-regulated trafficking of the Glut glucose-transporters have been carried out (Hasani, this issue). In addition, ENaC remains a rare example of a plasma membrane protein involved in transport for which hormonal downregulation is mediated by a ubiquitylation event. ENaC activity is known to be regulated by a number of hormones, including aldosterone. This regulation seems to involve Sgk1 kinase (serum and glucocorticoid regulated kinase), a member of the Akt family of Ser/Thr kinases, which is induced by aldosterone and stimulates ENaC. It was recently shown that Sgk1 acts directly on the Nedd4-2 /ENaC interaction. Based on the observation that Sgk1 has a PPxY motif and that Nedd4-2 includes two consensus sites for phosphorylation by Sgk1, Staub and coworkers showed that Sgk1 phosphorylates Nedd4-2 in a PPxY-dependent manner in *Xenopus* oocytes, and that this phosphorylation reduces the interaction between Nedd4-2 and ENaC, leading to high levels of ENaC at the cell surface, possibly as a result of impaired internalization (Debonneville et al. 2001). It is unknown whether similar mechanisms control the interaction of yeast Rsp5p with any of its plasma membrane substrates. However, such regulation appears unlikely, given the increasing number of known substrates of this single member of the Nedd4 protein family at various subcellular locations in yeast.

4.2 Physiological control of the sorting and recycling of newly synthesized transporters

The fate of newly synthesized transporters present in the secretory pathway may also depend on environmental conditions: the protein may be targeted to the cell surface, or directly diverted to the MVB/vacuole for degradation without passing through the plasma membrane. The membrane compartment in which this control is achieved is probably the late Golgi compartment and/or a post-Golgi endosome, although additional controls in the late endosome compartment cannot be ruled out. Ubiquitin and lipid rafts appear to play crucial roles in regulating transporter sorting. On the other hand, a few yeast transporters present in internal membranes have been reported to be recycled back to the plasma membrane in response to environmental stimuli. However, in these situations, it is important to distinguish between the targeting to the cell surface of newly synthesized transporters and the recycling of internalized transporters previously present at the plasma membrane.

4.2.1 Regulated sorting of amino acid permeases

The first evidence for control at the sorting level of a newly synthesized yeast transporter was reported for the Gap1p permease (Roberg et al. 1997a, 1997b). In cells grown on glutamate as the sole nitrogen source, Gap1p is synthesized but inactive and is located principally in the internal membranes. Most of the Gap1p synthesized in these cells is diverted from the secretory pathway to the vacuole without passing through the plasma membrane (Fig. 4). If the cells are transferred back onto urea, Gap1p activity increases slightly, even if cycloheximide is added to the medium (Roberg et al. 1997b). This increase in Gap1p activity may be due to Gap1p proteins present in the secretory pathway that are targeted to the cell surface instead of being sorted to the vacuolar degradation pathway. Alternatively, as suggested by the authors, this increase may be due to Gap1p proteins stored in internal membranes and recycled back to the plasma membrane once the cells are transferred to conditions of poor nitrogen supply. In a subsequent study, Helliwell *et al.* (2001) identified Bul1p and Bul2p as redundant proteins required for the direct sorting of Gap1p to the vacuole. These proteins were also found to be specifically required for the polyubiquitylation of Gap1p, which suggests that polyubiquitylation is essential for direct sorting of the permease to the vacuole, monoubiquitylation being insufficient. In the *bul1 bul2* and *rsp5* mutants, newly synthesized Gap1p normally directed to the vacuole is retargeted to the cell surface (Fig. 4).

Gap1p permease produced artificially in cells growing in the presence of high concentrations of NH_4^+ is also directly sorted to the MVB/vacuole without passing through the cell surface. Ubiquitylation of Gap1p on lysines 9 and 16 is required for this direct sorting to the vacuole, as are the Rsp5p ubiquitin ligase and Bul proteins. Mutations altering any of these components target the Gap1p permease to the cell surface, regardless of the quality of the nitrogen source (Soetens et al. 2001). A similar phenotype has been observed in *doa4* cells, which have a much smaller than normal internal ubiquitin pool, and this phenotype was suppressed by ubiquitin overproduction. Furthermore, a similar suppression was observed following the overproduction of a ubiquitin variant unable to form polyubiquitin chains (Ub^{RRR}). Thus, at least under certain conditions, the monoubiquitylation of Gap1p seems to be sufficient for direct sorting of the permease to the vacuole (Soetens et al. 2001). Further investigations are clearly required to elucidate the actual roles of the Bul proteins and of mono vs. polyubiquitylation in the regulated sorting of Gap1p.

It also remains to be determined whether ubiquitylation of the newly synthesized permease indeed takes place in the secretory pathway and serves as a signal for the nitrogen-induced diversion of the permease to the MVB/vacuolar degradation pathway (Helliwell et al. 2001). Alternatively, in cells growing on glutamate or high concentrations of NH_4^+, newly synthesized Gap1p may be missorted to the late endosome without ubiquitin involvement. The permease would then undergo ubiquitylation, leading to its sorting to the MVB/vacuolar degradation pathway. If not ubiquitylated, the protein would instead be retargeted to the cell surface.

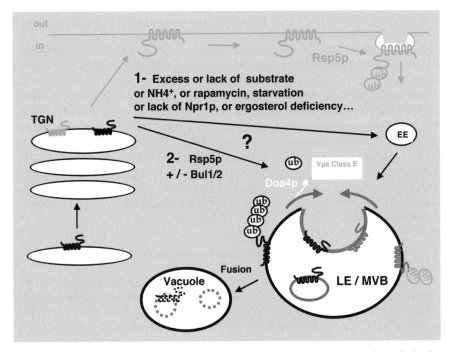

Fig. 4. Direct vacuolar targeting of plasma membrane transporters. A number of physiological (nutrient changes, excess substrate) or pathological situations (mutation in transporters, or in Nprlp kinase, deficiencies in lipid metabolism) lead to direct Golgi to vacuolar sorting of transporters bypassing plasma membrane delivery. These transporters undergo Rsp5p-dependent ubiquitylation, in some cases with the help of the Bul proteins, possibly required for polyubiquitylation. The route followed is either first Golgi to early endosome (EE), or Golgi to late endosome/MVB, where ubiquitylation results in sorting into MVB vesicles. MVB fusion with the vacuole is followed by vacuolar degradation.

The sorting of Tat2p is also regulated by cellular nitrogen status (Beck et al. 1999). If cells growing on rich medium are shifted to nitrogen starvation conditions or if rapamycin is added to the medium, Tat2p proteins present in the secretory pathway are preferentially sorted to the vacuole for degradation. Mutations impairing Tat2p ubiquitylation (*npi1/rsp5*, *npi2/doa4*, replacement of five lysine residues in the N-terminal tail of the permease) or preventing access of the permease to the late endosome (*pep12*) or its sorting in the MVB pathway (*vps27*) protect it against degradation. It was recently shown that tryptophan availability in the medium plays an important role in controlling the fate of newly synthesized Tat2p (Umebayashi and Nakano 2003). In the presence of low concentrations of tryptophan, Tat2p is targeted to the cell surface, and this process is thought to involve transit of the permease through a post-Golgi or early endosome compartment. On media containing high concentrations of tryptophan, Tat2p is instead sorted from this compartment to the vacuole. Lipid rafts and ubiquitin play key roles in this regulated permease sorting. A defect in the *ERG6* gene encoding S-adenosylmethionine-delta-24-sterol c-methyltransferase, an enzyme involved in a late step of ergosterol biosynthesis, leads to vacuolar sorting of the permease even in the presence of low concentrations of tryptophan. However, if access to the late endosome is further restricted by means of a *pep12* mutation, Tat2p is retargeted to the cell surface. Thus, the association of Tat2p with rafts is required for the normal targeting of this protein to the cell surface. Tat2p may be unable to associate with rafts if tryptophan is abundant in the cell, leading to its missorting to the MVB/vacuolar degradation pathway. Vacuolar sorting of Tat2p in the presence of high concentrations of tryptophan or in *erg6* mutants grown in the presence of low concentrations of tryptophan is defective in the *bul1* mutant, in which the ubiquitylation (probably polyubiquitylation) of Tat2p is defective. Ubiquitylation seems to occur after Tat2p has left the Golgi apparatus, but it is unknown whether it occurs in the post-Golgi endosome or only once Tat2p has reached the late endosome membrane.

4.2.2 Regulated sorting of nucleobase transporters

Similar nutrient conditions trigger both the onset or acceleration of amino acid permease endocytosis and the direct vacuolar routing of newly synthesized proteins. A similar dual regulation system has been reported for nucleobase permeases. Uracil and uridine permeases, both of which undergo substrate-induced plasma membrane internalization, display direct vacuolar routing if synthesized *de novo* in the presence of their respective substrates (Séron et al. 1999; Blondel et al. 2004). The molecular mechanisms underlying this process have been studied in more detail for uracil permease. Most of the experiments were carried out with GFP-tagged wild type or variant versions of Fur4p.

A mutant version of Fur4p with a very low affinity for uracil could not be diverted to the vacuolar pathway when synthesized *de novo* in the presence of uracil. Instead, the permease was targeted to the cell surface. These experiments were conducted on cells producing a GFP-tagged version of the mutant permease, together with a wild type untagged permease to permit uracil uptake. In contrast to the fate of this mutant permease, in cases of defective ubiquitylation, and in *rsp5/npi1* cells in particular, Fur4p synthesized *de novo* in the presence of uracil was able to leave the Golgi apparatus and to reach the vacuole, but was missorted to the vacuolar membrane (Fig. 3). Correct luminal delivery was restored by the in-frame N-terminal fusion of ubiquitin, which also resulted in the partial diversion of Fur4p to the vacuolar pathway (Blondel et al. 2004). These data suggest that the binding of intracellular uracil to the permease promotes a conformational change preventing transport of the permease to the plasma membrane. Missorted permease then undergoes ubiquitylation, leading to its delivery to the vacuolar lumen (Blondel et al. 2004). It should be noted that defective ubiquitylation (*npi1* cells) does not seem to prevent Fur4p from leaving the Golgi apparatus and reaching the late endosome. It is the subsequent MVB sorting of the permease that is specifically impaired. Note that all the available data are consistent with a similar model for the control by amino acids of the Gap1p permease, except that this protein is mainly re-targeted to the cell surface if not correctly ubiquitylated.

The requirements for Fur4p ubiquitylation at the plasma membrane and during Golgi–to-MVB sorting have been compared. Both events are Rsp5p-dependent. The ubiquitylation required for MVB sorting did not require phosphorylation of the PEST sequence of the permease (Blondel et al. 2004), whereas plasma membrane ubiquitylation did (Marchal et al. 1998). Fur4p Lys38 and Lys41, the only target Lys for plasma membrane ubiquitylation (Marchal et al. 2000) were also sites of ubiquitylation for MVB sorting, together with other Lys (Blondel et al. 2004). The type of ubiquitylation occurring during Fur4p Golgi-to-vacuole targeting has not yet been determined. Experiments conducted with an in-frame ubiquitin showed that a single ubiquitin, at least in the form of a non removable, fused ubiquitin, could restore some MVB sorting, but was not as efficient as normal permease ubiquitylation, which involves the addition of several ubiquitin moieties, for MVB sorting (Blondel et al. 2004). The deletion of *BUL1/2* does not seem to prevent the uracil-induced sorting of Fur4 to the MVB pathway, suggesting that Bul1p/Bul2p may be adaptors involved in the efficient ubiquitylation and sorting of only a subset of Rsp5p transporter substrates.

4.2.3 Regulated sorting of metal transporters

Several metal transporters also display regulated sorting during biosynthesis. *S. cerevisiae* takes up iron bound to siderophores by two separate systems, one of which requires the ARN family of siderophore–iron transporters. Arn1p is located in endosome-like intracellular vesicles when produced in the absence of its specific substrate, ferrichrome. If cells are exposed to low concentrations of ferrichrome, Arn1p is stably redistributed to the plasma membrane (Kim et al. 2002). Arn1p has two surface binding sites for ferrichrome: a site with low-nanomolar affinity and a site with low-micromolar affinity. Ferrichrome binding to the high-affinity site is required for Arn1p transfer from endosomes to the plasma membrane, as mutations affecting this site also affect the intracellular distribution of Arn1p (Moore et al. 2003). A model has been put forward in which ferrichrome enters cells by fluid-phase endocytosis and binds to high-affinity binding sites on endosomal Arn1p, inducing a conformational change and relocalisation of the protein to plasma membrane. Arn1p also displays a second type of trafficking regulation: high ferrichrome concentrations induce rapid internalization of the protein at the plasma membrane (Kim et al. 2002).

Most newly synthesized Smf1p and Smf2p is targeted to the vacuole if manganese is plentiful, and this trafficking involves Bsd2p (Liu et al. 1997), a protein also required for the vacuolar targeting of a misfolded Pma1p variant (Luo and Chang 1997). Manganese starvation triggers the targeting of Smf1p to the plasma membrane, whereas Smf2p is redistributed to intracellular vesicles (Portnoy et al. 2000). Extensive mutational analysis of Smf1p led to the identification of conserved residues in Smf1p that appear to be required for both Smf1p activity (as defined by complementation of an *smf1* mutation), and Smf1p trafficking to the plasma membrane in cases of manganese depletion (Liu and Culotta 1999a). Thus, for both Arn1p and Smf1p, much is already known about the residues involved in substrate interaction, and substrate-induced regulated trafficking. The potential involvement of ubiquitylation events in these processes was not yet reported.

4.2.4 Recycling of yeast transporters?

Constant recycling between the plasma membrane and internal compartments is known to be a characteristic feature of a few yeast plasma membrane proteins: the yeast receptor Ste3p (Chen and Davis 2000), and the v-SNARE protein Snc1p (Lewis et al. 2000). These proteins are recycled to the plasma membrane for ligand binding and plasma membrane fusion, respectively, in a process that requires the fusion of endosome-derived vesicles with the trans-Golgi network (TGN), dependent on the small rab *Yt6p* or the VFT complex (Vps51p, Vps52p...) (Siniossoglou and Pelham 2001). It is unknown whether yeast transporters recycle between internal compartments and the plasma membrane. Permanent recycling was suggested years ago for the ABC transporter Ste6p, which is present in the TGN/endosomes in steady state but functions at the plasma membrane (Kuchler et al. 1993). Recent investigations with *end* and *sec* mutants indeed showed that Ste6p can be recycled to the plasma membrane after internalization (Losko et al. 2001).

Several studies have reported the potential recycling of receptors and transporters in class E *vps* mutants, which would account for the accumulation of these proteins in large amounts at the plasma membrane of these cells (Davis et al. 1993; Li et al. 1999; Forsberg et al. 2001). However, an alternative interpretation of some of these observations is that the observed phenotypes may result from impaired direct vacuolar sorting of transporters synthesized *de novo* (Forsberg et al. 2001).

Strong evidence has been obtained for effective recycling of Gap1p after its internalization in class E *vps31/bro1* mutant cells (Nikko et al. 2003). This recycling is impaired in cells mutated in genes (*YPT6, TLG1, VPS52*) involved in the fusion of endosome-derived vesicles with the TGN. Interestingly, such recycling is also defective in *vam3* and *vam7* mutants impaired in fusion between MVB and the vacuole, raising the possibility that Gap1p is recycled to the cell surface once it has reached the limiting membrane of the vacuole (Nikko et al. 2003). Similar observations have been reported for Fur4p. This permease seems to be stable at the plasma membrane in class E *vps* mutants under conditions leading to rapid endocytosis in wild type cells (nitrogen starvation), suggesting internalization followed by recycling. A formal proof of Fur4p recycling from the class E compartment was recently obtained in experiments in which severe stress led to the internalization of GFP-tagged Fur4p and its trapping in the class E compartment. Subsequent release of the stress concerned (in the absence of new permease synthesis) led to recycling of the fluorescent transporter back to the plasma membrane. Strikingly, in this case, recycling did not occur via the Golgi apparatus (no impairment in *vps51Δ* cells) (Bugnicourt and Galan, personal communication). Therefore, permeases seem to be able to use several routes for recycling to the plasma membrane in class E *vps* mutants. It is unknown whether the routes followed by Gap1p and Fur4p in these recycling pathways are similar to or different from the normal sorting of these permeases along the secretory pathway. It is also unclear whether, in wild type cells, rapid transfer to environmental conditions favouring the plasma membrane localization of a given transporter can lead to the recycling of an internalized transporter.

4.2.5 Regulated sorting and recycling of transporters in higher eukaryotes

Plasma membrane sorting is also regulated for a number of mammalian transporters, and some diseases are associated with deficiencies in such regulation. Two such diseases involve P-type ATPases — the Menkes protein (MNK), and the Wilson's disease protein (WND) — both of which are copper transporters, and targets involved in copper disorders. In basal medium, both MNK and WND reside in the TGN, where they transport copper to the lumen of the organelle, providing various copper-dependent enzymes with copper (Petris et al. 2002). Heterologous expression of the MNK gene in yeast can complement the function of the yeast homolog, Ccc2p, which is also located in the TGN (Petris et al. 2002). In conditions of high copper concentration, MNK is rapidly transported from the Golgi apparatus to the plasma membrane, where it is involved in copper efflux. In conditions of high copper concentration, WND is relocalised to cytoplasmic vesicles (Hung et al. 1997). In both cases, copper-induced relocalisation occurs independently of new protein synthesis (Petris et al. 1996; Hung et al. 1997). The copper-induced trafficking of MNK and WND presumably switches the function of these ATPases from a nutritional role to a protective role — the export of excess copper from post-Golgi compartments, or across the plasma membrane. Several disease-causing mutations in the MNK and WND genes prevent copper-induced trafficking from the TGN. A conserved feature of all P-type ATPases is the formation of an acyl-phosphate intermediate (Lutsenko and Kaplan 1995), which occurs as part of the catalytic cycle during cation transport. Extensive mutagenesis, including mutations in the residues affected in patients, has shown that mutations blocking the formation of phosphorylated catalytic intermediates also prevent the copper-induced relocalisation of MNK from the TGN. Furthermore, mutations in the phosphatase domain resulting in the hyperphosphorylation of MNK result in constitutive trafficking from the TGN to the plasma membrane. A phosphatase mutation has a similar effect on trafficking in patients with Wilson's disease (Petris et al. 2002). These data suggest a catalysis-dependent trafficking model, although they do not predict TGN-to-plasma membrane sorting at each catalytic cycle.

In addition to these examples of substrate-induced sorting at the plasma membrane or internal compartments, many transporters in higher eukaryotes display steady-state localization in both endosomal compartments and at the plasma membrane. Rapid changes in surface expression are then achieved by regulation of the relative rates of endocytosis and recycling from endosome compartments to the plasma membrane. In many cases, the proportion of cell-surface versus internal transporters is controlled by hormonal regulation. This is the case for the Glut glucose transporters, the intracellular distribution of which is regulated by insulin. Regulation of this type has been studied in detail (Al-Hasani, this issue). Similarly, the trafficking of the water channel AQP2 is regulated by vasopressin (reviewed in Royle and Murrell-Lagnado 2003 and Nielsen, this issue). Some ABC transporters in the rat liver are stored in intracellular pools, and presynthesized transporters are sorted from endosomes to the canalicular plasma membrane following stimulation, with cAMP for example (Kipp et al. 2001). In other cases, there is constitutive recycling between endosomes and the plasma membrane, resulting in endosomal and plasma membrane pools of equivalent sizes, as observed for the mammalian Nramp2 metal transporter (Touret et al. 2003), and the Na^+/H^+ neuronal exchanger NHE5 (Szaszi et al. 2002). Similar situations have also been described for plant transporters. The active, polar transport of auxin through the plant is thought to be due to the coordinated polar distribution of carriers in the plasma membrane: the auxin uptake carrier AUX1 and the auxin efflux carrier PIN1 are distributed apically and basolaterally, respectively. PIN1 undergoes continuous cycling between endosomes and the basolateral membrane (reviewed in Jurgens and Geldner 2002).

4.3 Mechanisms and signalling pathways governing the regulated trafficking of transporters

Although the same machineries seem to be involved in the membrane trafficking and degradation of most, if not all, yeast transporters (e.g. Rsp5p ubiquitin ligase, ESCRT complexes, actin cytoskeleton, etc.), the physiological conditions inducing the downregulation of these transporters or favouring their stabilization at the plasma membrane differ markedly according to transporter considered. Even transporters belonging to the same highly conserved family may display radically different regulation profiles. So, how is the specific control of transporters achieved at the level of membrane trafficking? Recent studies have suggested that substrate recognition, association with lipid rafts, and phosphorylation mediated by protein kinases such as Npr1p are central components determining the fate of transporters.

4.3.1 Role of Npr1p kinase and the TOR signalling pathway

Studies of amino acid permeases have suggested that the phosphorylation events catalyzed by the Npr1p kinase also play a central role in the regulation of transporter sorting. This kinase (Grenson 1983b; Vandenbol et al. 1987, 1990) prevents Gap1p at the plasma membrane from being sorted to the endocytic and endosome/vacuolar degradation pathways in conditions of nitrogen limitation. It also prevents newly synthesized Gap1p present in the late secretory pathway from being missorted to the vacuole (Fig. 4). If Npr1p is inactivated, Gap1p at the plasma membrane undergoes ubiquitylation, endocytosis and degradation, whereas Gap1p present in the secretory pathway is diverted to the vacuole without passing through the plasma membrane (De Craene et al. 2001). Furthermore, Npr1p seems to regulate the Tat2p and Bap2p permeases in an inverse manner: if cells are transferred from YPD medium to conditions of limiting nitrogen supply, Npr1p simultaneously favours the accumulation of Gap1p at the plasma membrane and promotes the downregulation of Tat2p and Bap2p. In support of a negative effect of Npr1p on Tat2p permease, the overexpression of *NPR1* on rich medium promotes Tat2p inactivation. Furthermore, *npr1* mutants grown on YPD medium are partially resistant to the negative effect of rapamycin on Tat2p activity (Schmidt et al. 1998). However, the mechanisms underlying the inverse control of the fate of Gap1p, Tat2p and Bap2p by Npr1p remain unknown. Gap1p may be phosphorylated, even in an *npr1* mutant (De Craene et al. 2001), but this does not rule out the possibility that Npr1p is involved in the phosphorylation of the permease in normal conditions. Glutathione-S-transferase fused to the N- and C-terminal cytosolic tails of Bap2p was recently shown to undergo Npr1p-dependent phosphorylation in response to rapamycin (conditions promoting Npr1 activity), suggesting that Npr1p directly phosphorylates the permeases via their tails (Omura and Kodama 2003).

The apparently key role of Npr1p in the nitrogen-regulated sorting of amino acid permeases raises a further question: how is Npr1p regulated by nitrogen? Npr1p is phosphorylated in conditions in which nitrogen is abundant (in which Npr1p is apparently inactive) and is dephosphorylated if cells are transferred to poor nitrogen conditions or if rapamycin is added to the medium (i.e. when Npr1p is active). This dephosphorylation requires Tap42p, a phosphatase under control of the TOR signalling pathway (Beck and Hall 1999). Tor1p and Tor2p (targets of rapamycin) are phosphatidylinositol kinase-related protein kinases controlling multiple cellular functions in response to nutrients (Crespo and Hall 2002). Recent studies have shown that the TOR proteins are subunits of two large protein complexes (Loewith et al. 2002). Both these complexes include Lst8p, the partial deficiency of which leads to the loss of Gap1p activity (Roberg et al. 1997a, 1997b). It was proposed that Npr1p is controlled by phosphorylation via the TOR pathway, which is itself regulated by nutrients (Beck and Hall 1999). However, it remains unclear how nutrients control TOR, both in yeast and higher organisms (Crespo and Hall 2002; Manning and Cantley 2003). Starting from the observation that the internal pool of amino acids has a strong influence on Gap1p sorting (Chen and Kaiser 2002), C. Kaiser and coworkers proposed an alternative model. According to this model, the influence of TOR and rapamycin on the regulation of Gap1p is indirect, with TOR affecting only the internal pool of amino acids through its effects on transcription of the genes involved in amino acid biosynthesis. This internal pool of amino acids then controls the fate of Gap1p: Gap1p is downregulated if the internal pool of amino acids is large and sorted to the plasma membrane if this pool is small (Chen and Kaiser 2003). However, the means by which the amino acid pool controls Gap1 trafficking remain unknown. Internal amino acids may inhibit Npr1p activity. Alternatively, they may bind to Gap1p, thereby, inducing a conformational change promoting sorting of the permease to the MVB/vacuolar degradation pathway. However, this second model must also take into account the observation that a lack of Npr1p kinase is sufficient to induce Gap1p degradation (De Craene et al. 2001).

4.3.2 Role of substrate binding and association with lipid rafts

There is growing evidence that the binding of substrates to transporters plays a central role in their controlled membrane trafficking. For instance, we have previously reported that a mutant form of Fur4p, unable to recognize its substrate uracil even though it reaches the cell surface normally (Urban-Grimal et al. 1995), is insensitive to uracil-induced endocytosis (Séron et al. 1999) and sorting to the vacuole (Blondel et al. 2004). This observation suggests a model in which Fur4p exists in two conformations according to the relative abundance of uracil: one amenable (when uracil is abundant) and the other insensitive (when uracil is present in low concentrations) to ubiquitylation. As the phosphorylation of Fur4p is a prerequisite for its ubiquitylation at the plasma membrane (Marchal et al. 1998), internal uracil may act by altering the phosphorylation status of the permease. This would modulate the sensitivity of the permease to ubiquitylation, unmasking Fur4p domains interacting directly or indirectly with Rsp5p domains. Intracellular uracil is also the critical signal triggering the direct sorting of newly synthesized Fur4p to the endosomal pathway, ultimately leading to its Rsp5p-dependent ubiquitylation and MVB sorting, an event that does not require prior Fur4p phosphorylation, at least as far as the PEST sequence is concerned (Blondel et al. 2004). Other possibilities could be suggested in this case, such as a substrate-induced conformational change leading to segregation into specific lipid domains, required for Rsp5p-dependent ubiquitylation. Here too, the relevant substrate/enzyme mode of interaction remains to be defined.

A link between lipid-raft association and substrate availability is also suggested by recent work on control of the sorting of the Tat2p permease by one of its main substrates, tryptophan (Umebayashi and Nakano 2003) (see section 3.2). These data are consistent with Tat2p binding to tryptophan (when tryptophan is abundant), leading to its dissociation from lipid rafts and sorting to the MVB/vacuole in a ubiquitylation-dependent manner. For instance, tryptophan may cause Tat2p to undergo a conformational change incompatible with raft association. The overproduction of Npr1p may inhibit Tat2p activity (Schmidt et al. 1998) by direct phosphorylation of the permease, thereby, inducing a conformational change mimicking that normally adopted on high-tryptophan medium.

The best illustration of the role of substrate binding in controlling the membrane trafficking of transporters is probably that of the Arn1p and Arn3p ferrichrome (FC) transporters (Kim et al. 2002; Moore et al. 2003) (see section 4.2). Only the substrates of each Arn protein promote their sorting from endosomes to the plasma membrane. Furthermore, in the model proposed by C. Philpott and coworkers, FC enters the cell by fluid-phase endocytosis and interacts with Arn domains exposed to the lumen of the early endosome. This binding leads the Arn proteins to change conformation and induces their sorting to the plasma membrane. The possible involvement of an association between Arn proteins and lipid rafts in the sorting of these proteins to the cell surface has not been demonstrated. The proposed model also raises the general question of the role of fluid-phase endocytosis in the substrate-mediated controlled sorting of transporters present in internal compartments. The binding of FC to a second, lower-affinity binding site exposed by Arn1p at the plasma membrane may induce a second conformational change, promoting the endocytosis of Arn1p (Moore et al. 2003).

5 Conclusion

Most membrane transporters are tightly regulated at the membrane trafficking level and studies carried out over the last ten years have generated a considerable body of novel data, revealing some of the general features of the underlying mechanisms. It has been demonstrated that the covalent attachment of ubiquitin to the cytosolic regions of transporters is an essential prerequisite for the sorting of transporters located at the plasma membrane to endocytic vesicles and for the sorting of transporters present in endosomal compartments to the MVB/vacuolar degradation pathway (Fig. 1 and 4). The ubiquitin ligase Rsp5p, acting together or not with its Bul partners, mediates the ubiquitylation of proteins at the plasma membrane (Rotin et al. 2000; Hicke and Dunn 2003) and also in the Golgi to MVB pathway (Helliwell et al. 2001; Soetens et al. 2001; Umebayashi and Nakano 2003; Blondel et al. 2004). The main components of the cellular machineries involved in the recognition and sorting of ubiquitylated cargos have now been identified. Identification of the upstream molecular events triggering the ubiquitylation of specific transporters in response to appropriate environmental stimuli is a key aim of current research. Evidence is accumulating that at least three events are linked to the likelihood of a transporter undergoing ubiquitylation: association of the protein with specific lipid environments, possible phosphorylation by protein kinases, and specific binding of the transporter to its own substrates. It is tempting to suggest that transporters may adopt different conformational states, dictating their association with specific lipid environments and their probability of being ubiquitylated. The conformation adopted by the transporters may, in turn, be modulated by the binding of their own substrates, and possibly also by phosphorylation. Transporters may also have distinct substrate-binding domains, oriented either towards the extracellular medium (i.e. the lumen of compartments if the protein is internal) or the cytosol. These simple molecular mechanisms would readily generate a large diversity of controls adapted to the physiological roles of each transporter. Future work will be needed to improve our understanding of the complex relationships between the processes involved and to determine whether the conclusions drawn for yeast proteins also apply to the transporters of mammalian cells, which display as well substrate-induced trafficking regulation.

Acknowledgements

We thank the members of André's and Haguenauer-Tsapis's labs for fruitful discussions and critical reading of this manuscript. We are grateful to Alex Edelman & Associates for editorial assistance. The work performed in our laboratories is supported by an EU program (EFFEXPORT, contract QLRT-2001-00533). Work of the laboratory of R. Haguenauer-Tsapis is supported by the *Centre National de la Recherche Scientifique*, the Universities Paris 6 and Paris 7, and a grant from the *Association pour la recherche contre le Cancer* (ARC, grant no. 5681), and work of the laboratory of B. André is supported by a grant FRSM 3.4597.00 for Medical Scientific Research, Belgium.

References

Abe F, Iida H (2003) Pressure-induced differential regulation of the two tryptophan permeases Tat1 and Tat2 by ubiquitin ligase Rsp5 and its binding proteins Bul1 and Bul2. Mol Cell Biol 23:7566-7584

Abriel H, Kamynina E, Horisberger J-D, Staub O (2000) Regulation of the cardiac voltage-gated Na+ channel (rH1) by the ubiquitin-protein ligase Nedd4. FEBS Lett 466:377-380

Aguilar RC, Watson HA, Wendland B (2003) The yeast Epsin Ent1 is recruited to membranes through multiple independent interactions. J Biol Chem 278:10737-10743

Ahmad M, Bussey H (1988) Topology of membrane insertion *in vitro* and plasma membrane assembly *in vivo* of the yeast arginine permease. Mol Microbiol 2:627-635

Amerik AY, Nowak J, Swaminathan S, Hochstrasser M (2000) The Doa4 deubiquitinating enzyme is functionally linked to the vacuolar protein-sorting and endocytic pathways. Mol Biol Cell 11:3365-3380

Arvan P, Zhao X, Ramos-Castaneda J, Chang A (2002) Secretory pathway quality control operating in Golgi, plasmalemmal, and endosomal systems. Traffic 3:771-780

Baggett JJ, Wendland B (2001) Clathrin function in yeast endocytosis. Traffic 2:297-302

Bagnat M, Chang A, Simons K (2001) Plasma membrane proton ATPase pma1p requires raft association for surface delivery in yeast. Mol Biol Cell 12:4129-4138

Beaudenon SL, Huacani MR, Wang G, McDonnell DP, Huibregtse JM (1999) Rsp5 ubiquitin-protein ligase mediates DNA damage-induced degradation of the large subunit of RNA polymerase II in *Saccharomyces cerevisiae*. Mol Cell Biol 19:6972-6979

Beck T, Hall MN (1999) The TOR signalling pathway controls nuclear localization of nutrient-regulated transcription factors. Nature 402:689-692

Beck T, Schmidt A, Hall MN (1999) Starvation induces vacuolar targeting and degradation of the tryptophan permease in yeast. J Cell Biol 146:1227-1238

Bilodeau PS, Urbanowski JL, Winistorfer SC, Piper RC (2002) The Vps27p Hse1p complex binds ubiquitin and mediates endosomal protein sorting. Nat Cell Biol 4:534-539

Bilodeau PS, Winistorfer SC, Kearney WR, Robertson AD, Piper RC (2003) Vps27-Hse1 and ESCRT-I complexes cooperate to increase efficiency of sorting ubiquitinated proteins at the endosome. J Cell Biol 163:237-243

Blondel MO, Morvan J, Dupré S, Urban-Grimal D, Haguenauer-Tsapis R, Volland C (2004) Direct sorting of the yeast uracil permease to the endosomal system is controlled by uracil binding and Rsp5p-dependent ubiquitylation. Mol Biol Cell 15:884-895

Bonifacino JS, Traub LM (2003) Signals for sorting of transmembrane proteins to endosomes and lysosomes. Annu Rev Biochem 72:395-447

Brondijk TH, van der Rest ME, Pluim D, de Vries Y, Stingl K, Poolman B, Konings WN (1998) Catabolite inactivation of wild-type and mutant maltose transport proteins in *Saccharomyces cerevisiae.* J Biol Chem 273:15352-15357

Buckley KM, Melikian HE, Provoda CJ, Waring MT (2000) Regulation of neuronal function by protein trafficking: a role for the endosomal pathway. J Physiol 525:11-19

Buziol S, Becker J, Baumeister A, Jung S, Mauch K, Reuss M, Boles E (2002) Determination of *in vivo* kinetics of the starvation-induced Hxt5 glucose transporter of *Saccharomyces cerevisiae.* FEMS Yeast Res 2:283-291

Chang A, Cheang S, Espanel X, Sudol M (2000) Rsp5 WW domains interact directly with the carboxyl-terminal domain of RNA polymerase II. J Biol Chem 275:20562-20571

Chang A, Slayman CW (1991) Maturation of the yeast plasma membrane [H+]ATPase involves phosphorylation during intracellular transport. J Cell Biol 115:289-295

Chen EJ, Kaiser CA (2003) *LST8* negatively regulates amino acid biosynthesis as a component of the TOR pathway. J Cell Biol 161:333-347

Chen L, Davis NG (2000) Recycling of the yeast a-factor receptor. J Cell Biol 151:731-738

Crespo JL, Hall MN (2002) Elucidating TOR signaling and rapamycin action: lessons from *Saccharomyces cerevisiae.* Microbiol Mol Biol Rev 66:579-591

D'Hondt K, Heese-Peck A, Riezman H (2000) Protein and lipid requirements for endocytosis. Annu Rev Genet 34:255-295

Dancis A, Haile D, Yuan DS, Klausner RD (1994) The *Saccharomyces cerevisiae* copper transporter protein (Ctr1p). J Biol Chem 269:25660-25667

David D, Sundarababu S, Gerst JE (1998) Involvement of long chain fatty acid elongation in the trafficking of secretory vesicles in yeast. J Cell Biol 143:1167-1182

Davis NG, Horecka JL, Sprague JGF (1993) *Cis-* and *trans-* acting functions required for endocytosis of the yeast pheromone receptors. J Cell Biol 122:53-65

De Craene JO, Soetens O, André B (2001) The Npr1 kinase controls biosynthetic and endocytic sorting of the yeast Gap1 permease. J Biol Chem 276:43939-44348

Debonneville C, Flores SY, Kamynina E, Plant PJ, Tauxe C, Thomas MA, Munster C, Chraibi A, Pratt JH, Horisberger JD, Pearce D, Loffing J, Staub O (2001) Phosphorylation of Nedd4-2 by Sgk1 regulates epithelial Na(+) channel cell surface expression. EMBO J 20:7052-7059

Decottignies A, Owsianik G, Ghislain M (1999) Casein kinase-I-dependent phosphorylation and stability of the yeast multidrug transporter Pdr5p. J Biol Chem 274:37139-37146

Dunn R, Hicke L (2001) Domains of the Rsp5 ubiquitin-protein ligase required for receptor-mediated and fluid-phase endocytosis. Mol Biol Cell 12:421-435

Dupré S, Haguenauer-Tsapis R (2003) Raft partitioning of the yeast uracil permease during trafficking along the endocytic pathway. Traffic 4:83-96

Dupré S, Haguenauer-Tsapis R (2001) Deubiquitination step in the endocytic pathway of yeast plasma membrane proteins: crucial role of Doa4p ubiquitin isopeptidase. Mol Cell Biol 21:4482-4494

Egner R, Kuchler K (1996) The yeast multidrug transporter Pdr5 of the plasma membrane is ubiquitinated prior to endocytosis and degradation in the vacuole. FEBS Lett 378:177-181

Egner R, Mahé Y, Pandjaitan R, Kuchler K (1995) Endocytosis and vacuolar degradation of the plasma membrane-localized Pdr5 ATP-binding cassette multidrug transporter in *Saccharomyces cerevisiae.* Mol Cell Biol 15:5879-5887

Ferro-Novick S, Novick P, Field C, Schekman R (1984) Yeast secretory mutants that block the formation of active cell surface enzymes. J Cell Biol 98:35-43

Forsberg H, Hammar M, Andreasson C, Moliner A, Ljungdahl PO (2001) Suppressors of ssy1 and ptr3 null mutations define novel amino acid sensor-independent genes in *Saccharomyces cerevisiae*. Genetics 158:973-988

Gagny B, Wiederkehr A, Dumoulin P, Winsor B, Riezman H, Haguenauer-Tsapis R (2000) A novel EH domain protein of *Saccharomyces cerevisiae* involved in endocytosis. J Cell Sci 113:3309-3319

Gajewska B, Kaminska J, Jesionowska A, Martin NC, Hopper AK, Zoladek T (2001) WW domains of Rsp5p define different functions: determination of roles in fluid phase and uracil permease endocytosis in *Saccharomyces cerevisiae*. Genetics 157:91-101

Galan J-M, Haguenauer-Tsapis R (1997) Ubiquitin Lys63 is involved in ubiquitination of a yeast plasma membrane protein. EMBO J 16:5847-5854

Galan JM, Cantegrit B, Garnier C, Namy O, Haguenauer-Tsapis R (1997) "ER degradation" of a mutant yeast plasma membrane protein by the ubiquitin-proteasome pathway. FASEB J 12:315-323

Galan JM, Moreau V, André B, Volland C, Haguenauer-Tsapis R (1996) Ubiquitination mediated by the Npi1p/Rsp5p ubiquitin-protein ligase is required for endocytosis of the yeast uracil permease. J Biol Chem 271:10946-10952

Garnier C, Blondel MO, Hagenauer-Tsapis R (1996) Membrane topology of the yeast uracil permease. Mol Microbiol 21:1061-1073

Gilstring CF, Ljungdahl PO (2000) A method for determining the *in vivo* topology of yeast polytopic membrane proteins demonstrates that Gap1p fully integrates into the membrane independently of Shr3p. J Biol Chem 275:31488-31495

Giorgino F, de Robertis O, Laviola L, Montrone C, Perrini S, McCowen KC, Smith RJ (2000) The sentrin-conjugating enzyme mUbc9 interacts with GLUT4 and GLUT1 glucose transporters and regulates transporter levels in skeletal muscle cells. Proc Natl Acad Sci USA 97:1125-1130

Gitan RS, Eide DJ (2000) Zinc-regulated ubiquitin conjugation signals endocytosis of the yeast *ZRT1* zinc transporter. Biochem J 346:329-336

Gitan RS, Luo H, Rodgers J, Broderius M, Eide D (1998) Zinc-induced inactivation of the yeast *ZRT1* zinc transporter occurs through endocytosis and vacuolar degradation. J Biol Chem 273:28617-28624

Gitan RS, Shababi M, Kramer M, Eide DJ (2003) A cytosolic domain of the yeast Zrt1 zinc transporter is required for its post-translational inactivation in response to zinc and cadmium. J Biol Chem 278:39558-39564

Glickman MH, Ciechanover A (2002) The ubiquitin-proteasome proteolytic pathway: destruction for the sake of construction. Physiol Rev 82:373-428

Goossens A, de La Fuente N, Forment J, Serrano R, Portillo F (2000) Regulation of yeast H(+)-ATPase by protein kinases belonging to a family dedicated to activation of plasma membrane transporters. Mol Cell Biol 20:7654-7661

Graschopf A, Stadler JA, Hoellerer MK, Eder S, Sieghardt M, Kohlwein SD, Schweyen RJ (2001) The yeast plasma membrane protein Alr1 controls Mg2+ homeostasis and is subject to Mg2+-dependent control of its synthesis and degradation. J Biol Chem 276:16216-16222

Grenson M (1983a) Inactivation-reactivation process and repression of permease formation regulate several ammonia-sensitive permeases in the yeast *Saccharomyces cerevisiae*. Eur J Biochem 133:135-139

Grenson M (1983b) Study of the positive control of the general amino-acid permease and other ammonia-sensitive uptake systems by the product of the *NPR1* gene in the yeast *Saccharomyces cerevisiae.* Eur J Biochem 133:141-144

Grenson M (1992) Amino acid transporters in yeast: structure, function and regulation. In Pont, D (ed.), Molecular Aspects of Transport Proteins. Elsevier Science Publishers B. V, Amsterdam, pp. 219-245.

Grenson M, Hou C, Crabeel M (1970) Multiplicity of the amino acid permeases in *Saccharomyces cerevisiae.* IV. Evidence for a general amino acid permease. J Bacteriol 103:770-777

Gurunathan S, David D, Gerst J (2002) Dynamin and clathrin are required for the biogenesis of a distinct class of secretory vesicles in yeast. EMBO J 21:602-614

Hampton RY (2002) ER-associated degradation in protein quality control and cellular regulation. Curr Opin Cell Biol 14:476-482

Harsay E, Bretscher A (1995) Parallel secretory pathways to the cell surface in yeast. J Cell Biol 131:297-310

Harsay E, Schekman R (2002) A subset of yeast vacuolar protein sorting mutants is blocked in one branch of the exocytic pathway. J Cell Biol 156:271-285

Hein C, André B (1997) A C-terminal di-leucine motif and nearby sequences are required for NH4(+)-induced inactivation and degradation of the general amino acid permease, Gap1p, of *Saccharomyces cerevisiae.* Mol Microbiol 24:607-616

Hein C, Springael JY, Volland C, Haguenauer-Tsapis R, André B (1995) NPI1, an essential yeast gene involved in induced degradation of Gap1 and Fur4 permeases, encodes the Rsp5 ubiquitin-protein ligase. Mol Microbiol 18:77-87

Helliwell SB, Losko S, Kaiser CA (2001) Components of a ubiquitin ligase complex specify polyubiquitination and intracellular trafficking of the general amino acid permease. J Cell Biol 153:649-662

Hicke L, Dunn R (2003) Regulation of membrane protein transport by ubiquitin and ubiquitin-binding proteins. Annu Rev Cell Dev Biol 19:141-172

Hicke L, Zanolari B, Riezman H (1998) Cytoplasmic tail phosphorylation of the alpha-factor receptor is required for its ubiquitination and internalization. J Cell Biol 141:349-358

Hitchcock AL, Auld K, Gygi SP, Silver PA (2003) A subset of membrane-associated proteins is ubiquitinated in response to mutations in the endoplasmic reticulum degradation machinery. Proc Natl Acad Sci USA 100:12735-12740

Hoppe T, Matuschewski K, Rape M, Schlenker S, Ulrich HD, Jentsch S (2000) Activation of a membrane-bound transcription factor by regulated ubiquitin/proteasome-dependent processing. Cell 102:577-586

Horak J (2003) The role of ubiquitin in down-regulation and intracellular sorting of membrane proteins: insights from yeast. Biochim Biophys Acta 1614:139-155

Horak J, Wolf DH (1997) Catabolite inactivation of the galactose transporter in the yeast *Saccharomyces cerevisiae:* ubiquitination, endocytosis, and degradation in the vacuole. J Bacteriol 179:1541-1549

Horak J, Wolf DH (2001) Glucose-induced monoubiquitination of the *Saccharomyces cerevisiae* galactose transporter is sufficient to signal its internalization. J Bacteriol 183:3083-3088

Huang KM, D'Hondt K, Riezman H, Lemmon SK (1999) Clathrin functions in the absence of heterotetrameric adaptors and AP180-related proteins in yeast. EMBO J 18:3897-3908

Huh WK, Falvo JV, Gerke LC, Carroll AS, Howson RW, Weissman JS, O'Shea EK (2003) Global analysis of protein localization in budding yeast. Nature 425:686-691

Hung IH, Suzuki M, Yamaguchi Y, Yuan DS, Klausner RD, Gitlin JD (1997) Biochemical characterization of the Wilson disease protein and functional expression in the yeast *Saccharomyces cerevisiae*. J Biol Chem 272:21461-21466

Ikonen E (2001) Roles of lipid rafts in membrane transport. Curr Opin Cell Biol 13:470-477

Iraqui I, Vissers S, Bernard F, de Craene JO, Boles E, Urrestarazu A, André B (1999) Amino acid signaling in *Saccharomyces cerevisiae*: a permease-like sensor of external amino acids and F-Box protein Grr1p are required for transcriptional induction of the AGP1 gene, which encodes a broad-specificity amino acid permease. Mol Cell Biol 19:989-1001

Ishida S, Lee J, Thiele DJ, Herskowitz I (2002) Uptake of the anticancer drug cisplatin mediated by the copper transporter Ctr1 in yeast and mammals. Proc Natl Acad Sci USA 99:14298-14302

Jauniaux JC, Grenson M (1990) *GAP1*, the general amino acid permease gene of *Saccharomyces cerevisiae*. Nucleotide sequence, protein similarity with the other bakers yeast amino acid permeases, and nitrogen catabolite repression. Eur J Biochem 190:39-44

Jund R, Weber E, Chevallier MR (1988) Primary structure of the uracil transport protein of *Saccharomyces cerevisiae*. Eur J Biochem 171:417-424

Jurgens G, Geldner N (2002) Protein secretion in plants: from the trans-Golgi network to the outer space. Traffic 3:605-613

Kamynina E, Staub O (2002) Concerted action of ENaC, Nedd4-2, and Sgk1 in transepithelial Na(+) transport. Am J Physiol Renal Physiol 283:F377-387

Kaouass M, Gamache I, Ramotar D, Audette M, Poulin R (1998) The spermidine transport system is regulated by ligand inactivation, endocytosis, and by the Npr1p Ser/Thr protein kinase in *Saccharomyces cerevisiae*. J Biol Chem 273:2109-2117

Katzmann D, Odorizzi G, Emr S (2002) Receptor downregulation and multivesicular-body sorting. Nat Rev Mol Cell Biol 3:893-905

Katzmann DJ, Babst M, Emr SD (2001) Ubiquitin-dependent sorting into the multivesicular body pathway requires the function of a conserved endosomal protein sorting complex, ESCRT-I. Cell 106:145-155

Kim Y, Yun CW, Philpott CC (2002) Ferrichrome induces endosome to plasma membrane cycling of the ferrichrome transporter, Arn1p, in *Saccharomyces cerevisiae*. EMBO J 21:3632-3642

Kipp H, Pichetshote N, Arias IM (2001) Transporters on demand: intrahepatic pools of canalicular ATP binding cassette transporters in rat liver. J Biol Chem 276:7218-7224

Kohler JR (2003) Mos10 (Vps60) is required for normal filament maturation in *Saccharomyces cerevisiae*. Mol Microbiol 49:1267-1285

Kolling R (2002) Mutations affecting phosphorylation, ubiquitination and turnover of the ABC-transporter Ste6. FEBS Lett 531:548-552

Kolling R, Hollenberg CP (1994a) The ABC-transporter Ste6 accumulates in the plasma membrane in a ubiquitinated form in endocytosis mutants. EMBO J 13:3261-3271

Kolling R, Hollenberg CP (1994b) The first hydrophobic segment of the ABC-transporter, Ste6, functions as a signal sequence. FEBS Lett 351:155-158

Kolling R, Losko S (1997) The linker region of the ABC-transporter Ste6 mediates ubiquitination and fast turnover of the protein. EMBO J 16:2251-2261

Kostova Z, Wolf DH (2003) For whom the bell tolls: protein quality control of the endoplasmic reticulum and the ubiquitin-proteasome connection. EMBO J 22:2309-2317

Krampe S, Boles E (2002) Starvation-induced degradation of yeast hexose transporter Hxt7p is dependent on endocytosis, autophagy and the terminal sequences of the permease. FEBS Lett 513:193-196

Krampe S, Stamm O, Hollenberg CP, Boles E (1998) Catabolite inactivation of the high affinity hexose transporters Hxt6 and Hxt7 of *Saccharomyces cerevisiae* occurs in the vacuole after internalization by endocytosis. FEBS Lett 441:343-347

Kranz A, Kinner A, Kolling R (2001) A family of small coiled-coil-forming proteins functioning at the late endosome in yeast. Mol Biol Cell 12:711-723

Kruckeberg AL, Ye L, Berden JA, van Dam K (1999) Functional expression, quantification and cellular localization of the Hxt2 hexose transporter of *Saccharomyces cerevisiae* tagged with the green fluorescent protein. Biochem J 339 Pt 2:299-307

Kuchler K, Dohlman HG, Thorner J (1993) The a-factor transporter (*STE6* gene product) and cell polarity in the yeast *Saccharomyces cerevisiae*. J Cell Biol 120:1203-1215

Lai K, Bolognese CP, Swift S, McGraw P (1995) Regulation of inositol transport in *Saccharomyces cerevisiae* involves inositol-induced changes in permease stability and endocytic degradation in the vacuole. J Biol Chem 270:2525-2534

Lai K, McGraw P (1994) Dual control of inositol transport in *Saccharomyces cerevisiae* by irreversible inactivation of permease and regulation of permease synthesis by INO2, INO4, and OPI1. J Biol Chem 269:2245-2251

Lee VH (2000) Membrane transporters. Eur J Pharm Sci 11 Suppl 2:S41-S50

Lewis MJ, Nichols BJ, Prescianotto-Baschong C, Riezman H, Pelham HR (2000) Specific retrieval of the exocytic SNARE Snc1p from early yeast endosomes. Mol Biol Cell 11:23-38

Li Y, Kane T, Tipper C, Spatrick P, Jenness DD (1999) Yeast mutants affecting possible quality control of plasma membrane proteins. Mol Cell Biol 19:3588-3599

Liu XF, Culotta VC (1999a) Mutational analysis of *Saccharomyces cerevisiae* Smf1p, a member of the Nramp family of metal transporters. J Mol Biol 289:885-891

Liu XF, Culotta VC (1999b) Post-translation control of Nramp metal transport in yeast. Role of metal ions and the BSD2 gene. J Biol Chem 274:4863-4868

Liu XF, Supek F, Nelson N, Culotta VC (1997) Negative control of heavy metal uptake by the *Saccharomyces cerevisiae* BSD2 gene. J Biol Chem 272:11763-11669

Loayza D, Tam A, Schmidt WK, Michaelis S (1998) Ste6p mutants defective in exit from the endoplasmic reticulum (ER) reveal aspects of an ER quality control pathway in *Saccharomyces cerevisiae*. Mol Biol Cell 9:2767-2784

Loewith R, Jacinto E, Wullschleger S, Lorberg A, Crespo JL, Bonenfant D, Oppliger W, Jenoe P, Hall MN (2002) Two TOR complexes, only one of which is rapamycin sensitive, have distinct roles in cell growth control. Mol Cell Biol 10:457-468

Losko S, Kopp F, Kranz A, Kolling R (2001) Uptake of the ATP-binding cassette (ABC) transporter Ste6 into the yeast vacuole is blocked in the doa4 Mutant. Mol Biol Cell 12:1047-1059

Lucero P, Lagunas R (1997) Catabolite inactivation of the yeast maltose transporter requires ubiquitin-ligase npi1/rsp5 and ubiquitin-hydrolase npi2/doa4. FEMS Microbiol Lett 147:273-277

Lucero P, Moreno E, Lagunas R (2002) Catabolite inactivation of the sugar transporters in *Saccharomyces cerevisiae* is inhibited by the presence of a nitrogen source. FEMS Yeast Res 1:307-314

Lucero P, Penalver E, Vela L, Lagunas R (2000) Monoubiquitination is sufficient to signal internalization of the maltose transporter *in Saccharomyces cerevisiae*. J Bacteriol 182:241-243

Luo WJ, Chang A (1997) Novel genes involved in endosomal traffic in yeast revealed by supression of a targeting-defective plasma membrane ATPase mutant. J Cell Biol 138:731-746

Lutsenko S, Kaplan JH (1995) Organization of P-type ATPases: signficance of structural diversity. Biochem 14:15607-15613

Malkus P, Jiang F, Schekman R (2002) Concentrative sorting of secretory cargo proteins into COPII-coated vesicles. J Cell Biol 159:915-921

Manning BD, Cantley LC (2003) Rheb fills a GAP between TSC and TOR. Trends Biochem Sci 28:573-576

Marchal C, Haguenauer-Tsapis R, Urban-Grimal D (1998) A PEST-like sequence mediates phosphorylation and efficient ubiquitination of the yeast uracil permease. Mol Cell Biol 18:314-321

Marchal C, Haguenauer-Tsapis R, Urban-Grimal D (2000) Casein kinase I-dependent phosphorylation within a PEST sequence and ubiquitination at nearby lysines, signal endocytosis of yeast uracil permease. J Biol Chem 275:23608-23614

Marini AM, André B (2000) *In vivo* N-glycosylation of the mep2 high-affinity ammonium transporter of *Saccharomyces cerevisiae* reveals an extracytosolic N-terminus. Mol Microbiol 38:552-564

Martinez P, Zvyagilskaya R, Allard P, Persson BL (1998) Physiological regulation of the derepressible phosphate transporter in *Saccharomyces cerevisiae*. J Bacteriol 180:2253-2256

Mason DL, Michaelis S (2002) Requirement of the N-terminal extension for vacuolar trafficking and transport activity of yeast Ycf1p, an ATP-binding cassette transporter. Mol Biol Cell 13:4443-4455

Medintz I, Jiang H, Michels C (1998) The role of ubiquitin conjugation in glucose-induced proteolysis of *S. cerevisiae* maltose permease. J Biol Chem 273:34454-34462

Medintz I, Wang X, Hradek T, Michels CA (2000) A PEST-like sequence in the N-terminal cytoplasmic domain of *Saccharomyces* maltose permease is required for glucose-induced proteolysis and rapid inactivation of transport activity. Biochemistry 182:4518-4526

Miller EA, Beilharz TH, Malkus PN, Lee MC, Hamamoto S, Orci L, Schekman R (2003) Multiple cargo binding sites on the COPII subunit Sec24p ensure capture of diverse membrane proteins into transport vesicles. Cell 114:497-509

Moore RE, Kim Y, Philpott CC (2003) The mechanism of ferrichrome transport through Arn1p and its metabolism in *Saccharomyces cerevisiae*. Proc Natl Acad Sci USA 100:5664-5669

Mukherjee S, Ghosh RN, Maxfield FR (1997) Endocytosis. Physiol Rev 77:759-803

Murillas R, Simms KS, Hatakeyama S, Weissman AM, Kuehn MR (2002) Identification of developmentally expressed proteins that functionally interact with Nedd4 ubiquitin ligase. J Biol Chem 277:2897-2907

Nikko E, Marini AM, André B (2003) Permease recycling and ubiquitination status reveal a particular role for Bro1 in the multivesicular body pathway. J Biol Chem 278(50):50732-50743

Odorizzi G, Katzmann DJ, Babst M, Audhya A, Emr SD (2003) Bro1 is an endosome-associated protein that functions in the MVB pathway in *Saccharomyces cerevisiae*. J Cell Sci 116:1893-1903

Omura F, Kodama Y, Ashikari T (2001a) The basal turnover of yeast branched-chain amino acid permease Bap2p requires its C-terminal tail. FEMS Microbiol Lett 194:207-214

Omura F, Kodama Y, Ashikari T (2001b) The N-terminal domain of the yeast permease Bap2p plays a role in its degradation. Biochem Biophys Res Commun 287:1045-1050

Omura F, Kodama Y (2003) The Npr1p kinase phosphorylates the N-terminal domain of amino acid permease Bap2p in response to rapamycin. Yeast 20:S142

Ooi CE, Rabinovich E, Dancis A, Bonifacino JS, Klausner RD (1996) Copper-dependent degradation of the *Saccharomyces cerevisiae* plasma membrane copper transporter Ctr1p in the apparent absence of endocytosis. EMBO J 15:3515-3523

Paddon C, Loayza D, Vangelista L, Solari R, Michaelis S (1996) Analysis of the localization of STE6/CFTR chimeras in a *Saccharomyces cerevisiae* model for the cystic fibrosis defect CFTR delta F508. Mol Microbiol 19:1007-1017

Paiva S, Kruckeberg A, Casal M (2002) Utilization of green fluorescent protein as a marker for studying the expression and turnover of the monocarboxylate permease Jen1p of *Saccharomyces cerevisiae*. Biochem J 363:737-744

Palmer LK, Wolfe D, Keeley JL, Keil RL (2002) Volatile anesthetics affect nutrient availability in yeast. Genetics 161:563-574

Penalver E, Lucero P, Moreno E, Lagunas R (1999) Clathrin and two components of the COPII complex, Sec23p and Sec24p, could be involved in endocytosis of the *Saccharomyces cerevisiae* maltose transporter. J Bacteriol 181:2555-2563

Peng J, Schwartz D, Elias JE, Thoreen CC, Cheng D, Marsischky G, Roelofs J, Finley D, Gygi SP (2003) A proteomics approach to understanding protein ubiquitination. Nat Biotechnol 21:921-926

Petersson J, Pattison J, Kruckeberg AL, Berden JA, Persson BL (1999) Intracellular localization of an active green fluorescent protein-tagged Pho84 phosphate permease in *Saccharomyces cerevisiae*. FEBS Lett 462:37-42

Petris MJ, Mercer JF, Culvenor JG, Lockhart P, Gleeson PA, Camakaris J (1996) Ligand-regulated transport of the Menkes copper P-type ATPase efflux pump from the Golgi apparatus to the plasma membrane: a novel mechanism of regulated trafficking. EMBO J 15:6084-6095

Petris MJ, Smith K, Lee J, Thiele DJ (2003) Copper-stimulated endocytosis and degradation of the human copper transporter, hCtr1. J Biol Chem 278:9639-96346

Petris MJ, Voskoboinik I, Cater M, Smith K, Kim BE, Llanos RM, Strausak D, Camakaris J, Mercer JF (2002) Copper-regulated trafficking of the Menkes disease copper ATPase is associated with formation of a phosphorylated catalytic intermediate. J Biol Chem 277:46736-46742

Piper RC, Luzio JP (2001) Late endosomes: sorting and partitioning in multivesicular bodies. Traffic 2:612-621

Plant PJ, Lafont F, Lecat S, Verkade P, Simons K, Rotin D (2000) Apical membrane targeting of Nedd4 is mediated by an association of its C2 domain with annexin XIIIb. J Cell Biol 149:1473-1484

Plant PJ, Yeger H, Staub O, Howard P, Rotin D (1997) The C2 domain of the ubiquitin protein ligase Nedd4 mediates Ca2+- dependent plasma membrane localization. J Biol Chem 272:32329-32336

Plemper RK, Egner R, Kuchler K, Wolf DH (1998) Endoplasmic reticulum degradation of a mutated ATP-binding cassette transporter Pdr5 proceeds in a concerted action of Sec61 and the proteasome. J Biol Chem 273:32848-32856

Portnoy ME, Liu XF, Culotta VC (2000) Saccharomyces cerevisiae expresses three functionally distinct homologues of the nramp family of metal transporters. Mol Cell Biol 20:7893-7902

Raasi S, Pickart CM (2003) Rad23 ubiquitin-associated domains (UBA) inhibit 26 S proteasome-catalyzed proteolysis by sequestering lysine 48-linked polyubiquitin chains. J Biol Chem 278:8951-8959

Ramamoorthy S, Blakely RD (1999) Phosphorylation and sequestration of serotonin transporters differentially modulated by psychostimulants. Science 285:763-766

Riezman H (1998) Downregulation of yeast G protein-coupled receptors. Semin Cell Dev Biol 9:129-134

Roberg KJ, Bickel S, Rowley N, Kaiser CA (1997a) Control of amino acid permease sorting in the late secretory pathway of Saccharomyces cerevisiae by SEC13, LST4, LST7 and LST8. Genetics 147:1569-1584

Roberg KJ, Rowley N, Kaiser CA (1997b) Physiological regulation of membrane sorting late in the secretory pathway of Saccharomyces cerevisiae. J Cell Biol 137:1469-1482

Roth AF, Davis NG (2000) Ubiquitination of the PEST-like endocytosis signal of the yeast a-factor receptor. J Biol Chem 275:8143-8153

Rotin D, Staub O, Haguenauer-Tsapis R (2000) Ubiquitination and endocytosis of plasma membrane proteins: role of Nedd4/Rsp5p family of ubiquitin-protein ligases. J Membr Biol 176:1-17

Royle SJ, Murrell-Lagnado RD (2003) Constitutive cycling: a general mechanism to regulate cell surface proteins. Bioessays 25:39-46

Saunders C, Ferrer JV, Shi L, Chen J, Merrill G, Lamb ME, Leeb-Lundberg LM, Carvelli L, Javitch JA, Galli A (2000) Amphetamine-induced loss of human dopamine transporter activity: an internalization-dependent and cocaine-sensitive mechanism. Proc Natl Acad Sci USA 97:6850-6855

Schild L, Lu Y, Gautschi I, Schneeberger E, Lifton RP, Rossier BC (1996) Identification of a PY motif in the epithelial Na channel subunits as a target sequence for mutations causing channel activation found in Liddle syndrome. EMBO J 15:2381-2387

Schmidt A, Beck T, Koller A, Kunz J, Hall M (1998) The TOR nutrient signalling pathway phosphorylates NPR1 and inhibits turnover of the tryptophan permease. EMBO J 17:6924-6931

Schwake M, Friedrich T, Jentsch TJ (2001) An internalization signal in ClC-5, an endosomal Cl-channel mutated in dent's disease. J Biol Chem 276:12049-12054

Séron K, Blondel M-O, Haguenauer-Tsapis R, Volland C (1999) Uracil-induced down regulation of the yeast uracil permease. J Bacteriol 181:1793-1800

Shcherbik N, Zoladek T, Nickels JT, Haines DS (2003) Rsp5p is required for ER bound Mga2p120 polyubiquitination and release of the processed/tethered transactivator Mga2p90. Curr Biol 13:1227-1233

Shih SC, Katzmann DJ, Schnell JD, Sutanto M, Emr SD, Hicke L (2002) Epsins and Vps27p/Hrs contain ubiquitin-binding domains that function in receptor endocytosis. Nat Cell Biol 4:389-393

Shih SC, Sloper-Mould KE, Hicke L (2000) Monoubiquitin carries a novel internalization signal that is appended to activated receptors. EMBO J 19:187-198

Silve S, Volland C, Garnier C, Jund R, Chevallier MR, Haguenauer-Tsapis R (1991) Membrane insertion of uracil permease, a polytopic yeast plasma membrane protein. Mol Cell Biol 11:1114-1124

Siniossoglou S, Pelham HR (2001) An effector of Ypt6p binds the SNARE Tlg1p and mediates selective fusion of vesicles with late Golgi membranes. EMBO J 20:5991-5998

Soetens O, De Craene JO, André B (2001) Ubiquitin is required for sorting to the vacuole of the yeast general amino acid permease, Gap1. J Biol Chem 276:43949-43957

Springael J-Y, De Craene J-O, André B (1999a) The yeast Npi1/Rsp5 ubiquitin-ligase lacking its N-terminal C2 domain is competent for ubiquitination but not for subsequent endocytosis of the Gap1 permease. Biochem Biophys Res Comm 257:561-566

Springael J-Y, Galan J-M, Haguenauer-Tsapis R, André B (1999b) NH4+-induced downregulation of the *Saccharomyces cerevisiae* Gap1p permease involves its ubiquitination with lysine-63-linked chains. J Cell Sci 112:1375-1383

Springael JY, André B (1998) Nitrogen-regulated ubiquitination of the Gap1 permease of *Saccharomyces cerevisiae*. Mol Biol Cell 9:1253-1263

Springael JY, Nikko E, André B, Marini AM (2002) Yeast Npi3/Bro1 is involved in ubiquitin-dependent control of permease trafficking. FEBS Lett 517:103-109

Stanbrough M, Magasanik B (1995) Transcriptional and posttranslational regulation of the general amino acid permease of *Saccharomyces cerevisiae*. J Bacteriol 177:94-102

Staub O, Gautschi I, Ishikawa T, Breitschop K, Ciechanover A, Schild L, Rotin D (1997) Regulation of stability and function of the epithelial Na+ channel (ENaC) by ubiquitination. EMBO J 16:6325-6336

Swaminathan S, Amerik AY, Hochstrasser M (1999) The Doa4 deubiquitinating enzyme is required for ubiquitin homeostasis in yeast. Mol Biol Cell 10:2583-2594

Swanson R, Locher M, Hochstrasser M (2001) A conserved ubiquitin ligase of the nuclear envelope/endoplasmic reticulum that functions in both ER-associated and Matalpha2 repressor degradation. Genes Dev 15:2660-2674

Szaszi K, Paulsen A, Szabo EZ, Numata M, Grinstein S, Orlowski J (2002) Clathrin-mediated endocytosis and recycling of the neuron-specific Na+/H+ exchanger NHE5 isoform. Regulation by phosphatidylinositol 3'-kinase and the actin cytoskeleton. J Biol Chem 277:42623-42632

Terrell J, Shih S, Dunn R, Hicke L (1998) A function for monoubiquitination in the internalization of a G protein-coupled receptor. Mol Cell 1:193-202

Touret N, Furuya W, Forbes J, Gros P, Grinstein S (2003) Dynamic traffic through the recycling compartment couples the metal transporter Nramp2 (DMT1) with the transferrin receptor. J Biol Chem 278:25548-25557

Tschopp JF, Emr SD, Field C, Schekman R (1986) *GAL2* codes for a membrane-bound subunit of the galactose permease in *Saccharomyces cerevisiæ*. J Bacteriol 166:313-318

Umebayashi K, Nakano A (2003) Ergosterol is required for targeting of tryptophan permease to the yeast plasma membrane. J Cell Biol 161:1117-11131

Urban-Grimal D, Pinson B, Chevallier J, Haguenauer-Tsapis R (1995) Replacement of Lys by Glu in a transmembrane segment strongly impairs the function of the uracil permease from *Saccharomyces cerevisiae*. Biochem J 308:847-851

Urbanowski J, Piper RC (2001) Ubiquitin sorts proteins into the lumenal degradative compartment of the late endosome/vacuole. Traffic 2:622-630

Van Belle D, André B (2001) A genomic view of yeast membrane transporters. Curr Opin Cell Biol 13:389-398

Vandenbol M, Jauniaux JC, Grenson M (1990) The *Saccharomyces cerevisiae* NPR1 gene required for the activity of ammonia-sensitive amino acid permeases encodes a protein kinase homologue. Mol Gen Genet 222:393-399

Vandenbol M, Jauniaux JC, Vissers S, Grenson M (1987) Isolation of the NPR1 gene responsible for the reactivation of ammonia-sensitive amino-acid permeases in *Saccharomyces cerevisiae*. RNA analysis and gene dosage effects. Eur J Biochem 164:607-612

Vecchione A, Marchese A, Henry P, Rotin D, Morrione A (2003) The Grb10/Nedd4 complex regulates ligand-induced ubiquitination and stability of the insulin-like growth factor I receptor. Mol Cell Biol 23:3363-3372

Volland C, Garnier C, Haguenauer-Tsapis R (1992) *In vivo* phosphorylation of the yeast uracil permease. J Biol Chem 267:23767-23771

Volland C, Urban-Grimal D, Géraud G, Haguenauer-Tsapis R (1994) Endocytosis and degradation of the yeast uracil permease under adverse conditions. J Biol Chem 269:9833-9841

Wang G, McCaffery JM, Wendland B, Dupre S, Haguenauer-Tsapis R, Huibregtse JM (2001) Localization of the Rsp5p ubiquitin-protein ligase at multiple sites within the endocytic pathway. Mol Cell Biol 21:3564-3575

Wang G, Yang J, Huibregste J (1999) Fonctional domains of the RSP5 ubiquitin-protein ligase. Mol Cell Biol 19:342-352

Warner JR (1999) The economics of ribosome biosynthesis in yeast. Trends Biochem Sci 24:437-440

Weissman AM (2001) Themes and variations on ubiquitylation. Nat Rev Mol Cell Biol 2:169-178

Wells KM, Rao R (2001) The yeast Na+/H+ exchanger Nhx1 is an N-linked glycoprotein. Topological implications. J Biol Chem 276:3401-3407

Welsch CA, Hagiwara S, Goetschy JF, Movva NR (2003) Ubiquitin pathway proteins influence the mechanism of action of the novel immunosuppressive drug FTY720 in *Saccharomyces cerevisiae*. J Biol Chem 278:26976-26982

Wendell DL, Bisson LF (1993) Physiological characterization of putative high-affinity glucose transport protein Hxt2 of *Saccharomyces cerevisiae* by use of anti-synthetic peptide antibodies. J Bacteriol 175:7689-7696

Wendland B (2002) Epsins: adaptors in endocytosis? Nat Rev Mol Cell Biol 3:971-977

Wendland B, Steece KE, Emr SD (1999) Yeast epsins contain an essential N-terminal ENTH domain, bind clathrin and are required for endocytosis. EMBO J 18:4383-43893

Whitworth TL, Quick MW (2001) Substrate-induced regulation of gamma-aminobutyric acid transporter trafficking requires tyrosine phosphorylation. J Biol Chem 276:42932-42937

Yashiroda H, Kaida D, Toh-e A, Kikuchi Y (1998) The PY-motif of Bul1 protein is essential for growth of *Saccharomyces cerevisiae* under various stress conditions. Gene 225:39-46

Yashiroda H, Oguchi T, Yasuda Y, Toh-e A, Kikuchi Y (1996) Bul1, a new protein that bind to the Rsp5 ubiquitin ligase. Mol Cell Biol 16:3255-3263

Yonkovich J, McKenndry R, Shi X, Zhu Z (2002) Copper ion-sensing transcription factor Mac1p post-translationally controls the degradation of its target gene product Ctr1p. J Biol Chem 277:23981-23984

Zhao R, Reithmeier RA (2001) Expression and characterization of the anion transporter homologue YNL275w in *Saccharomyces cerevisiae*. Am J Physiol Cell Physiol 281:C33-45

André, Bruno
Laboratoire de Physiologie Cellulaire, Institut de Biologie et de Médecine Moléculaires, Université Libre de Bruxelles, P.O. Box 300, Rue des Professeurs Jeener et Brachet 12, 6041 Gosselies, Belgium
bran@ulb.ac.be

Haguenauer-Tsapis, Rosine
Institut Jacques Monod-CNRS, Universités Paris VI and VII, 2 place Jussieu 75251 Paris Cedex 05, France
haguenauer@ijm.jussieu.fr

Regulated transport of the glucose transporter GLUT4

Hadi Al-Hasani

Abstract

In adipose and muscle cells, insulin induces the translocation of the glucose transporter isoform GLUT4 from intracellular storage compartments to the plasma membrane, where it catalyzes the facilitated diffusion of glucose into the cells. Despite recent progress that has been made in the dissection of the signaling pathways involved in GLUT4 translocation, and the mechanisms of certain aspects of exocytosis and endocytosis of GLUT4, many critical questions still remain unanswered: What is the nature of the GLUT4 storage compartment? What is the molecular basis for the unique sorting of GLUT4 into this compartment? How are the signals emerging from the insulin receptor translated into the directional movement of GLUT4 vesicles?

1 Mammalian glucose transport proteins

For most mammalian cells, glucose is the primary source of energy. Since hydrophilic sugar molecules cannot penetrate the plasma membrane, specific transport proteins are required to catalyze the uptake of glucose into the cell. In mammals, cellular glucose uptake is mediated by two types of non-related glucose transporters. The sodium-coupled glucose transporters (SGLTs) are expressed only in epithelial cells of the gut and kidney, and mediate active transport of glucose (and other hexoses) against a concentration gradient (Wood and Trayhurn 2003). In contrast, the second type of glucose transporters, termed GLUT (glucose transport protein) is expressed in all cell types and catalyzes the facilitated diffusion of glucose and/or other hexoses (Mueckler 1994). To date, fourteen mammalian GLUT genes have been cloned, nine of them within the last five years (Joost and Thorens 2001). Even though only eight of the GLUTs have been studied on the protein level so far, all members of the GLUT family are predicted to share a common topology of twelve membrane spanning domains where both N- and C-termini are located in the cytosol. In addition to their complex tissue distribution, the individual GLUT proteins differ in their substrate specificity, kinetic properties, subcellular localization, and levels of regulation. While the expression of some GLUT isoforms is tightly restricted to one certain cell type, many types of cells contain more than one GLUT isoform.

Topics in Current Genetics, Vol. 9
E. Boles, R. Krämer (Eds.): Molecular Mechanisms Controlling Transmembrane Transport
DOI 10.1007/b95778 / Published online: 9 March 2004
© Springer-Verlag Berlin Heidelberg 2004

2 Insulin regulated glucose uptake

A fundamental action of insulin is the stimulation of glucose uptake and metabolism in fat and muscle tissue that is essential for whole-body glucose homeostasis (Kahn 1996). Biochemical analyses and studies of transgenic mice have shown that glucose uptake in muscle and fat tissue is the rate-limiting step in peripheral glucose disposal (Wallberg-Henriksson and Zierath 2001; Minokoshi et al. 2003). Failure of muscle and fat tissue to take up appropriate amounts of glucose from the blood in response to insulin is a fundamental characteristic of type-2 diabetes mellitus.

In isolated muscle fibers and adipose cells, insulin stimulation results in a ~10-20 fold increase in cellular glucose uptake. Early studies have shown that insulin stimulation of isolated adipose and muscle cells results in a dramatic increase in the maximum rate (V_{max}) of glucose transport without affecting the affinity of the transporters for the substrate (Vinten et al. 1976). Subsequent analyses using subcellular fractionation combined with a ligand binding assay revealed that the insulin-induced increase in the V_{max} for glucose transport results from a rapid, reversible and insulin concentration-dependent translocation of glucose transporters from a low-density intracellular membrane pool to the plasma membrane (Cushman and Wardzala 1980; Suzuki and Kono 1980). Since then, an overwhelming amount of experimental evidence supporting this mechanism has accumulated. Nine years after the initial proposition of the translocation mechanism, the GLUT4 isoform was identified as the glucose transporter that is responsible for most of the acute increase in insulin-stimulated glucose transport in muscle and fat tissue (Birnbaum 1989; Fukumoto et al. 1989; James et al. 1989).

3 The GLUT4 translocation cycle

The GLUT4 isoform is expressed exclusively in white and brown adipose cells, heart, and skeletal muscle. The mechanism underlying the hormone-regulated translocation is best understood in adipose cells (Fig. 1). In non-stimulated adipose cells, the majority (>90%) of GLUT4 resides in intracellular vesicles and membrane tubules that appear to constitute a unique insulin-sensitive storage compartment, morphologically related to the trans-Golgi network. In the absence of insulin, GLUT4 slowly recycles between this storage compartment and the plasma membrane. However, because the rate of endocytosis exceeds the rate of exocytosis, GLUT4 is effectively sequestered within the intracellular compartment. Through not completely defined signal transduction pathways, insulin induces a large increase (~10-fold) in the rate of exocytosis of GLUT4-containing vesicles, and a smaller (~2-fold) decrease in the rate of GLUT4 endocytosis (Satoh et al. 1993; Lee et al. 2000). These changes in the rate constants result in a rapid shift in the steady-state distribution of GLUT4 to the plasma membrane, such that about 50% of the total cellular GLUT4 can contribute to the glucose

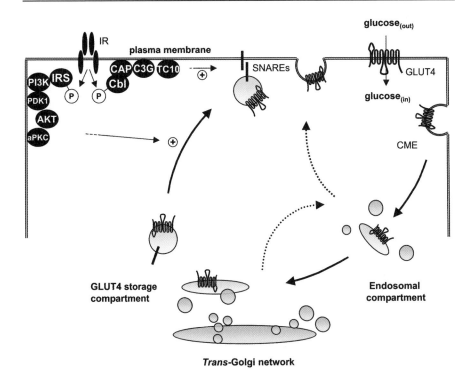

Fig. 1. Proposed model for the insulin-stimulated GLUT4 translocation in adipose cells. Insulin receptor (IR) catalyzed phosphorylation of IRS and Cbl leads to the activation of protein kinases AKT and atypical PKC isoforms (aPKC), and to the activation of the cdc42-like GTPase TC10. Through an unknown mechanism, this leads to the translocation of GLUT4-containing vesicles from an intracellular storage compartment to the plasma membrane, where the vesicles fuse via a SNARE-dependent process. GLUT4 in the plasma membrane catalyzes glucose uptake, and is eventually internalized by clathrin-mediated endocytosis (CME) that requires the activity of dynamin GTPase. Finally, GLUT4 is sorted back into the storage compartment by an unknown mechanism. Dotted arrows: proposed alternative pathways of GLUT4, that involve regulated recycling of GLUT4 between the trans-Golgi network/GLUT4 storage compartment, endosomes, and the plasma membrane (Bryant et al. 2002).

transport across the cell surface in the insulin-stimulated state. Eventually, GLUT4 is internalized and recycles back into the intracellular storage compartment by an unknown mechanism (Fig. 1).

Numerous studies have found that the acute insulin-stimulated glucose transport in muscle and fat tissue is directly proportional to the amount of extracellularly exposed GLUT4 in the plasma membrane, indicating that the intrinsic activity of the glucose transporter is not regulated by the hormone. However, even though this concept has been challenged recently (Furtado et al. 2002), it is generally accepted that the glucose transport activity in insulin-target cells is essentially regulated by the dynamics of GLUT4 recycling. The cellular machinery responsible for the recycling of GLUT4 and its hormone regulated movement between multiple membrane compartments is exceedingly complex and involves a vast number of enzymes, docking proteins, and regulatory components. Accompanying GLUT4 on its itinerary that begins in the GLUT4 storage compartment, the insulin-stimulated GLUT4 recycling can be divided up into four consecutive steps: i) signal transduction, ii) exocytosis of GLUT4 vesicles and fusion with the plasma membrane, iii) endocytosis of GLUT4, and iv) sorting of GLUT4 back into the storage compartment. This review will summarize pertinent findings for each of these different steps.

3.1 The insulin-sensitive GLUT4 storage compartment

Early morphological studies by electron microscopy of cryosections from quiescent adipose and muscle cells revealed that GLUT4 resides in multiple distinct cellular compartments, including the trans-Golgi network (TGN), early endosomes, clathrin- and non-clathrin coated vesicles, and cytoplasmic tubulovesicular structures (Fig. 2; Slot et al. 1991a, 1991b). The localization of GLUT4 in cardiac myocytes was found to be essentially the same as in brown adipose cells, skeletal muscle, and white adipose cells, indicating that GLUT4 compartments are similar in all insulin-sensitive cells (Malide et al. 2000). Very little GLUT4 was found in the plasma membrane in non-stimulated adipose and muscle cells (Fig. 2). Insulin stimulation resulted in a rapid redistribution of GLUT4 from nearly all internal GLUT4-positive structures to the plasma membrane, where the largest depletion of GLUT4 was observed in the TGN and tubulo-vesicular structures (Slot et al. 1991b; Malide et al. 2000). In addition, insulin stimulated cells also showed increased amounts of GLUT4 in coated vesicles and endosomes indicating an ongoing recycling of the transporter in the presence of the hormone (Slot et al. 1991a, 1991b). Since then, the concept of GLUT4 being recycled through multiple pools within the cell has been confirmed by a wealth of studies using morphological and biochemical methods that include immunocytochemistry, chemical ablation and inhibition studies, and subcellular fractionation in combination with photolabeling of GLUT4 (Simpson et al. 2001).

Even though the evidence is compelling that GLUT4 in muscle and adipose cells is targeted to a unique, highly specialized storage compartment, it is unclear how these GLUT4 containing storage vesicles are related to the TGN and the endosomal system. Insulin causes a modest increase in the rate of recycling of endosomal proteins such as transferrin receptors (TfR) that are targeted to the constitutive recycling pathway (Tanner and Lienhard 1987). However, inhibition of endosomal recycling by chemical ablation of endosomes with TfR-HRP had only a small effect on the GLUT4 translocation (Martin et al. 1998). Moreover, in non-stimulated adipose cells, the rate of exocytosis of GLUT4 is an order of magnitude lower that of the TfR (Tanner and Lienhard 1987; Holman et al. 1994), clearly indicating that in basal cells GLUT4 is excluded from recycling endosomes. Finally, Brefeldin A mediated destruction of the TGN had no effect on the intracellular sequestration of GLUT4, and its insulin-dependent movement to the cell surface

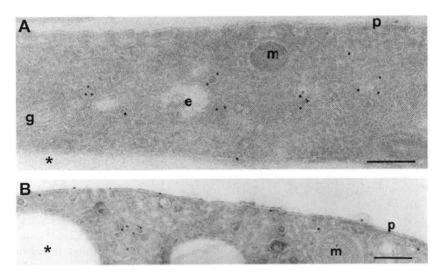

Fig. 2. Localization of GLUT4 by immunogold-EM on ultrathin cryosections of isolated rat white adipose cells. In basal cells (A), GLUT4 labeling is visible over small tubular and vesicular structures (50-70 nm) clustered near Golgi stacks and endosomal intermediates, or scattered throughout the cytoplasm. The plasma membrane is virtually devoid of gold particles. In insulin-stimulated cells (B), an increased number of gold particles are associated with the plasma membrane. Intracellular tubulovesicles and endosomes are also labeled. p, plasma membrane; m, mitochondria; g, Golgi; e, endosome; asterisks, lipid droplets. Bars, 200 nm. Images courtesy of Dr. Daniela Malide and Dr. Samuel W. Cushman, National Institutes of Health, Bethesda, USA.

(Kono-Sugita et al. 1996; Martin et al. 2000). These findings demonstrate that in basal cells GLUT4 partitions between Brefeldin A-resistant subdomains of the TGN and some sort of slowly recycling endosomes, but the individual contribution of these compartments to the insulin-induced movement of GLUT4 vesicles to the plasma membrane remains unknown.

3.2 Retention of GLUT4 in the insulin-sensitive compartment

Because introduction of a peptide derived from the C-terminus of GLUT4 into rat adipose cells was reported to result in an insulin-independent translocation of GLUT4 to the plasma membrane (Lee and Jung 1997), it was speculated that the retention mechanism of GLUT4 is dependent on the protein itself, e.g. via an intracellular GLUT4 binding protein. Thus, GLUT4 overexpression has been proposed to saturate the intracellular retention mechanism. Consequently, the overexpressed GLUT4 was supposed to become missorted to the plasma membrane even in the absence of insulin. Apparently consistent with this concept, transgenic overexpression of GLUT4 in adipose cells 20-fold above normal resulted in a dramatic reduction in the insulin fold response of the GLUT4-translocation (Shepherd et al.

1993). However, whereas basal glucose transport was increased 20-fold, insulin-stimulated glucose transport was only about threefold greater in cells from transgenic mice compared to the controls. Hence, the insulin-stimulated glucose transport activity was not even remotely related to the increased amount of GLUT4 in the adipose cells from the transgenics. In order to investigate the mechanism of retention, we performed quantitative measurements of cell-surface epitope-tagged GLUTs in rat adipose cells. On the cDNA level, a GLUT reporter construct was engineered to contain an epitope tag (such as the hemagglutinin (HA) peptide or the FLAG-tag) in the first exofacial loop. The construct was then introduced into isolated primary rat adipose cells by electroporation, and the newly expressed protein was detected on the cell surface of intact cells by an antibody binding assay (Al-Hasani et al. 1999).

In the study, we have overexpressed the ubiquitously expressed isoform GLUT1 approximately 200-fold over endogenous levels without observing any changes in the subcellular distribution of the protein (Al-Hasani et al. 1999). In contrast, a 20-fold overexpression of GLUT4 in adipose cells resulted in reduced insulin fold responses of the GLUT4-translocation, comparable to that observed in the transgenic animals overexpressing GLUT4. However, we recognized that the insulin-stimulated cell-surface HA-GLUT4 levels did not increase proportionally to the expression levels. In fact, the amount of GLUT4 translocated in response to insulin appeared to reach a constant value. Apparently, the transfected cells failed to target overexpressed GLUT4 to the plasma membrane upon stimulation with insulin rather than failing to retain the excess GLUT4 inside the cell in the absence of hormone. This conclusion was confirmed by tracer experiments where two differently epitope-tagged GLUT4s (FLAG, HA) were cotransfected to monitor the cell-surface distribution of the reporter HA-GLUT4 (i.e. tracer) under conditions of both low (minus FLAG-GLUT4) and high (plus FLAG-GLUT4) expression levels of total GLUT4. Overexpression of FLAG-GLUT4 led to a drastic decrease of cell-surface tracer HA-GLUT4 in the insulin-stimulated state, without changing the cell-surface levels of the basal HA-GLUT4. These data clearly demonstrate that the cellular translocation machinery has a limited capacity to recruit excess GLUT4 to the plasma membrane upon insulin stimulation. In addition, we showed that 200-fold overexpression of GLUT1 did not affect the targeting of coexpressed GLUT4, suggesting that the two transporters traffic through separate pathways (Al-Hasani et al. 1999).

Thus, a possible explanation for this observed effect might be that such accessory proteins required for the formation of translocation-competent GLUT4 vesicles become limiting when GLUT4 is overexpressed. Likewise, these proteins may be directly involved in the insulin-stimulated translocation of the GLUT4-containing vesicles. This latter concept is supported by the finding that the protein composition of the vesicles containing GLUT4 from adipose cells overexpressing GLUT4 was unchanged except for a marked increase in GLUT4 itself (Tozzo et al. 1996).

3.3 Recycling of GLUT4 in heterologous cells

The question whether the insulin-sensitive GLUT4 compartment is unique to adipose and muscle cells has not been answered conclusively. Ectopic expression of GLUT4 leads to intracellular sequestration of the protein.But in contrast to adipose and muscle cells, insulin stimulation of e.g. fibroblasts does not lead to an efficient translocation of GLUT4 to the cell surface (Haney et al. 1991). However, a distinct subcellular targeting of GLUT4 has also been observed in cells that do not express these proteins endogenously. Recent analyses show that in Chinese hamster ovary (CHO) cells, GLUT4 is sequestered in recycling endosomes and that insulin treatment causes a partial redistribution (of rather low amounts) of GLUT4 into a novel class of vesicles that translocate to the plasma membrane in response to the hormone (Lampson et al. 2001). Moreover, we have observed that *Xenopus* oocytes expressing GLUT4 after being preincubated with high doses of insulin display a rapid, fully reversible and insulin concentration-dependent translocation of glucose transporters from intracellular vesicles to the plasma membrane (Rumsey et al. 2000). Without the pre-incubation step that likely activates the *Xenopus* insulin-like growth factor (xIGF-1) receptors, virtually no insulin-stimulated translocation of GLUT4 was observed. Hence, although the molecular mechanism of this insulin-induced sensitivity to the hormone remains to be further characterized, these data clearly demonstrate that an insulin-dependent GLUT4 translocation can take place in cells that do not contain endogenous GLUT4.

A model recently gathering momentum is that in non-insulin sensitive cells, GLUT4 is targeted to a specialized compartment that is derived from the endosomal system (Simpson et al. 2001). Even though the evidence at present is limited, these observations might indicate that insulin action is in the formation of translocation-competent vesicles, where a budding process recruits GLUT4 from multiple storage pools into some sort of transport vesicle that is eventually relocated to the plasma membrane. Thus, it is tempting to speculate that the GLUT4 storage compartments in insulin-sensitive and non-sensitive cells are related, but differ mainly in the extent to which the specialized GLUT4 compartment is preformed prior to insulin-stimulated translocation.

3.4 Components of GLUT4 vesicles

In an attempt to characterize the GLUT4 storage compartment by analyzing the protein composition of GLUT4 vesicles, several groups have applied biochemical methods such as density fractionation and/or immunoadsorption of membrane fractions with immobilized GLUT4 antibodies. These isolated GLUT4 containing vesicles appear to be moderately homogenous in size and density, and according to the majority of the studies, exclude markers for the TGN and the endosomal system (Simpson et al. 2001). However, it is unclear whether these vesicles do reflect the GLUT4 storage compartment as it is found in intact cells, or if they are formed during the process of the homogenization/purification procedure, where different membrane components may fuse and mix with each other. Unfortunately, the lack of an *in vitro* assay for the translocation of GLUT4 does not allow a functional characterization of the vesicles based upon their ability to fuse with plasma membrane.

Despite these limitations, several proteins have been identified as components of GLUT4 vesicles. IRAP, a membrane spanning aminopeptidase and receptor for angiotensin IV constitutes a major constituent of GLUT4-containing vesicles, and a large body of biochemical and morphological evidence now exists that both GLUT4 and IRAP share the very same trafficking routes in adipose cells (Keller et al. 1995; Malide et al. 1997; Albiston et al. 2001). However, the biological significance for the colocalization of IRAP and GLUT4 remains unknown.

v-SNARE proteins are vesicular membrane proteins that are involved in the docking and fusion of vesicles by binding to their cognate t-SNAREs on the respective target membrane (Sollner 2003). Two v-SNAREs, VAMP2 and cellubrevin have been found in GLUT4 vesicles (Cain et al. 1992; Volchuk et al. 1995), and it has been demonstrated that they interact with the t-SNAREs Syntaxin 4 and SNAP-23 at the plasma membrane prior to the fusion of GLUT4 vesicles. VAMP2 but not cellubrevin function is required for insulin-stimulated GLUT4 translocation in adipose and muscle cells (Randhawa et al. 2000).

Several other ubiquitously expressed membrane proteins were found in isolated GLUT4 vesicles that in contrast to GLUT4, IRAP, and v-SNAREs are not translocated to the plasma membrane upon stimulation with insulin, and may have a generic role in vesicle trafficking. SCAMPs (secretory carrier-associated membrane proteins), pantophysin, and cellugyrin belong to the family of tetraspan vesicle membrane proteins (TVPs) that are abundant components of perhaps all known types of vesicles and shuttle between various membranous compartments of the biosynthetic and endocytotic pathways (Laurie et al. 1993; Brooks et al. 2000; Kupriyanova and Kandror 2000). Sortilin, a mammalian homolog for the yeast protein Vps10p has also been identified as a major protein component of GLUT4 vesicles (Lin et al. 1997). In yeast, Vps10p is the sorting receptor for secreted carboxypeptidase Y (CPY) and shuttles between the TGN and a prevacuolar endosome-like compartment. Thus, sortilin, that binds a variety of unrelated ligands, has been implicated as a sorting receptor for proteins in the synthetic pathway and TGN-endosome transport (Nielsen et al. 2001), but similar to the TVPs, its function in GLUT4 recycling remains to be established. In summary, several proteins have been identified in GLUT4 vesicles that play roles in vesicle trafficking, but so far, none of these proteins has been proven to be unique for the recycling of GLUT4 in insulin-target cells.

4 Signal transduction for the insulin-stimulated GLUT4-translocation

Upon insulin binding, the activated insulin receptor (IR) tyrosine phosphorylates a variety of cellular substrates including members of the insulin receptor substrate (IRS) family, and the proto-oncoproteins Cbl and Shc (White 1998). Previously, it was thought that the insulin-induced translocation of GLUT4 is triggered exclusively by tyrosine phosphorylation of IRS proteins and the subsequent activation of the lipid kinase phosphatidylinositol 3-kinase (PI3K). However, recent studies have led to the discovery of a second signaling pathway for the insulin-stimulated movement of GLUT4 to the plasma membrane, which is independent of PI3K (Fig. 1). This novel pathway is initiated with the phosphorylation of the adaptor protein Cbl by the IR and results in the activation of a Rho-like GTPase TC10 that is found in lipid rafts of the plasma membrane (Khan and Pessin 2002). Lipid rafts are stable microdomains in the plasma membrane that are highly enriched in sphingolipids, cholesterol, GPI-anchored and palmitoylated proteins, and caveolin, the latter being the main constituent of caveoli, flask-like invaginations of the plasma membrane that are highly abundant in adipose cells (Smart et al. 1999).

Thus, the insulin-stimulated translocation of GLUT4 is the result of a concerted activation of at least two independent and perhaps spatially separated signaling pathways. However, at present it is still unknown how activation of all these enzymes eventually translates into the directional movement of GLUT4-containing vesicles from the intracellular storage pool to the cell surface.

4.1 The IRS\PI3K\AKT\aPKC pathway

Tyrosine phosphorylated IRS proteins bind and activate src-homology 2 (SH2) domain containing enzymes such as PI3K, SHP2, and Grb2 (White 1998). Genetic ablation studies in mice have demonstrated that IRS1 and IRS2, the predominant IRS isoforms in fat and muscle appear to be most important for the regulation of insulin-mediated glucose uptake (Previs et al. 2000). Binding of the SH2 domain containing regulatory subunit of PI3K (p85) to tyrosine phosphorylated docking sites in IRS1 and IRS2 results in activation of the catalytic PI3K subunit (p110), leading to the production of phosphatidylinositol 3,4,5, trisphosphate (PIP3) in the cytosolic leaflet of the membrane bilayer. A large body of evidence now supports a crucial role for PI3K in mediating GLUT4 translocation (Khan and Pessin 2002).

The insulin-stimulated production of PIP3 directly leads to the subsequent activation of the protein kinase AKT (also known as protein kinase B) that has been implicated as a major mediator of insulin-stimulated GLUT4 translocation. AKT exists as three isoforms, where the AKT2 isoform (PKBβ) is highly enriched in insulin-responsive tissues (Hill et al. 1999). Expression of constitutively active mutants of AKT in rat adipose cells (Tanti et al. 1997; Chen et al. 2003), 3T3-L1 adipocytes (Kohn et al. 1996), and L6 muscle cells (Hajduch et al. 1998) stimulates GLUT4 translocation, whereas, overexpression of kinase-inactive mutants, and microinjection of an AKT substrate peptide or an antibody to AKT results in a substantial inhibition of the insulin stimulated GLUT4 translocation (Hill et al. 1999; Wang et al. 1999).

4.1.1 Insulin-stimulated activation of AKT

Upon stimulation with insulin, AKT is recruited to cellular membranes by binding of its amino terminal pleckstrin homology (PH) domain to membrane bound PIP3 (Vanhaesebroeck and Alessi 2000). The membrane bound form of AKT then becomes phosphorylated on two regulatory residues, Thr308 within the activation loop and Ser473 in the C-terminus of the enzyme, and both phosphorylations are considered to be required for AKT to reach maximum kinase activity. The kinase responsible for the phosphorylation of Thr308 has been identified as phosphoinositide-dependent kinase 1 (PDK1), an enzyme that is predominantly localized at the plasma membrane of quiescent cells in its active form (Vanhaesebroeck and Alessi 2000). Thus, AKT phosphorylation on Thr308 is regulated via PI3K mediated PIP3 formation, i.e. recruitment of AKT to the plasma membrane. In fact, artificial membrane targeting of AKT also leads to phosphorylation of both Thr308 and Ser473 in AKT, independent of the levels of PIP3 (Newton 2003). However, despite intensive experimental efforts the kinase responsible for the phosphorylation of Ser473 in AKT, tentatively dubbed PDK2, remains elusive. Currently, as many as four different, albeit controversially discussed, models are proposed to explain phosphorylation of AKT's Ser473. Recent studies have linked PDK2 activity to PDK1, Integrin-linked kinase (ILK), and an unidentified kinase distinct from PDK1 and ILK (Balendran et al. 1999; Hill et al. 2002; Hresko et al. 2003; Troussard et al. 2003). Intriguingly, it has been shown that phosphorylation of AKT on Ser473 *in vitro* occurs as the result of an autophosphorylation event after PDK1 mediated phosphorylation of Thr308 (Toker and Newton 2000). Consistent with these findings, we have recently demonstrated that phosphorylation of Ser473 of a membrane targeted AKT fusion protein requires catalytic activity of the kinase, thus, arguing for an autophosphorylation (Chen et al. 2003). Moreover, a specific targeting of AKT to the plasma membrane as well as prephosphorylation of Thr308 were not prerequisites for Ser473 phosphorylation of AKT in rat adipose cells.

4.1.2 Insulin-stimulated activation of atypical PKC isoforms

In adipose and muscle cells, insulin also stimulates membrane recruitment and activation of the closely related atypical PKC isoforms PKCλ/ζ/ι (PKC-lambda/zeta/iota; aPKC) via a PI3K dependent pathway. Activation of aPKC requires PDK1 mediated phosphorylation of a conserved threonine in the activation loop of the kinase and/or direct interaction with PIP3 that is followed by an autophosphorylation of a threonine residue in the C-terminus (Farese 2001). Similar to the effects observed for AKT, expression of active aPKC provokes strong insulin-like effects on GLUT4 translocation and/or glucose transport in rat adipose cells, 3T3-L1 adipocytes, L6 mytotubes, and rat skeletal muscle (Bandyopadhyay et al. 1997a, 1997b; Standaert et al. 1997; Etgen et al. 1999). Likewise, expression of kinase-inactive mutants of aPKC (Kotani et al. 1998; Bandyopadhyay et al. 1999) leads to an inhibition of GLUT4 translocation, indicating an involvement of the kinase in the regulation of insulin-stimulated glucose uptake. Thus, the atypical

PKC isoforms appear to be interchangeable in their ability to stimulate GLUT4 translocation, and the biological effects of the various aPKC mutants are quite similar to that of the corresponding AKT constructs when overexpressed in adipose and muscle cells. Surprisingly, PKCζ acts as a negative regulator of AKT, but not vice versa (Doornbos et al. 1999). It was found that overexpressed PKCζ binds to and inhibits PIP3 mediated activation of AKT. In contrast, expression of kinase-inactive PKCζ augmented the kinase activity of AKT. Mutation of both regulatory phosphorylation sites in AKT, Thr308 and Ser473, to acidic residues (AKT^{DD}) mimics phosphorylation of these residues and renders the mutant kinase constitutively active, thereby, overriding the requirement for membrane targeting, PIP3 formation, and PDK1 action. Because expression of PKCζ also showed an inhibitory effect on AKT^{DD}, it is evident that the inhibitory activity of PKCζ results from a mechanism that is independent of PI3K and PDK1. Perhaps PKCζ-mediated inhibition of AKT results from binding-induced conformational changes of AKT in the PKCζ/AKT complex. Interestingly, both aPKC and AKT have been reported to associate with GLUT4-containing vesicles after stimulation with insulin (Kupriyanova and Kandror 1999; Braiman et al. 2001), and it is tempting to speculate that AKT/aPKC action takes place in the GLUT4 storage compartment where phosphorylation of some ill-defined targets might induce the movement of GLUT4 vesicles to the plasma membrane. However, insulin-stimulated PI3K activation and PIP3 formation appears to occur primarily in the plasma membrane (Venkateswarlu et al. 1998), and so far, little evidence is available in living cells that the AKT/aPKC complex is sequentially targeted from the cytosol to GLUT4 vesicles via the plasma membrane. Obviously, the crosstalk between aPKC and AKT, and the apparent redundancy within this signaling pathway requires further investigation.

4.2 The Cbl-CAP\C3G\TC10 pathway

Several observations led to the conclusion that in insulin target cells, activation of the IRS/PI3K/AKT/PKCλ/ζ pathway may not be the sole signal transduction mechanism that leads to the translocation of GLUT4 from an intracellular storage compartment to the plasma membrane. Perhaps the most obvious example is that contraction in skeletal muscle induces GLUT4 translocation and glucose transport without activating PI3K (Lund et al. 1995). Because the effects of both contraction and insulin are additive with respect to the GLUT4 translocation, it is assumed that in skeletal muscle GLUT4 is recruited from two different compartments, one that is insulin-sensitive and another that is sensitive for contraction.

Even though overexpression of constitutively active PI3K in 3T3-L1 adipocytes stimulates glucose uptake (Egawa et al. 1999), a membrane permeable PIP3 analogue fails to stimulate glucose transport in the same cell line (Jiang et al. 1998). In addition, it has been reported that growth factor receptors such as platelet derived growth factor (PDGF) receptors, activate PI3K to a similar extent compared to insulin, but fail to elicit GLUT4 translocation (Isakoff et al. 1995). Nevertheless, PDGF induced GLUT4 translocation was observed in adipose cells overexpressing PDGF receptors (Quon et al. 1996). These discrepancies have been attributed to differential activation of various PI3K isoforms, differential targeting of signaling molecules such as PI3K, and the requirement for an additional PI3K-independent signaling pathway that it exclusive activated by insulin. And indeed, such a pathway has been proposed recently.

In addition to IRS proteins, the insulin receptor tyrosine kinase phosphorylates the adaptor protein Cbl (Ribon and Saltiel 1997) that is constitutively bound to the src-homology 3 (SH3) domain-containing adaptor protein CAP (Cbl associated protein). The phosphorylated Cbl-CAP complex binds the guanyl nucleotide exchange factor C3G and translocates to lipid rafts in the plasma membrane, where C3G is thought to activate the Rho-like GTPase TC10 by catalyzing GDP exchange for GTP (Chiang et al. 2001). Overexpression of both wild type and a dominant negative mutant of TC10 or inhibition of the membrane recruitment of the Cbl-CAP complex has an inhibitory effect on the insulin-stimulated GLUT4 translocation (Baumann et al. 2000; Chiang et al. 2001; Kimura et al. 2001).

The finding that insulin stimulates activation of TC10 under conditions where PI3K is inactive supports the idea that activation of this branch of the signaling cascade occurs in parallel to the classical IRS-mediated signaling of the IR. Interestingly, Cbl displays E3 ubiquitin ligase activity and has been described as a negative regulator for a variety of growth hormone receptors (including the IR) that are inactivated by ubiquitin dependent desensitation. Thus, Cbl appears to play a dual role, acting both as multivalent adaptor and as inhibitor of the insulin receptor (Tsygankov et al. 2001).

4.2.1 Possible downstream effectors of TC10 and AKT/aPKC

TC10 is highly homologous to the Golgi-localized GTPase cdc42 that has been implicated to play roles in the regulation of cell polarity, motility, cell cycle progression, and in the regulation of the actin cytoskeleton (Ridley 2001). Similar to cdc42, constitutively active TC10 was found to stimulate actin polymerization in 3T3-L1 adipocytes. This finding is of particular interest because of the ample evidence that GLUT4 translocation requires the actin cytoskeleton (Tsakiridis et al. 1994; Omata et al. 2000; Kanzaki and Pessin 2001; Jiang et al. 2002). Moreover, a recent study provided evidence that insulin stimulates the association of GLUT4-containing vesicles with the actin-associated Myo1C motor protein that appears to be involved in the movement of these vesicles to the plasma membrane (Bose et al. 2002). However, a direct link between components of the insulin signaling cascade and the actin cytoskeleton remains to be established.

Recently, a novel AKT substrate of 160 kDa (AS160) has been identified in 3T3-L1 adipocytes (Kane et al. 2002). AS160 contains two phosphotyrosine binding (PTB) domains and a putative GTPase activating (GAP) domain for Rab-like GTPases. Most strikingly, expression of an AS160 mutant lacking the AKT phosphorylation sites markedly inhibited insulin-stimulated GLUT4 translocation in 3T3-L1 adipocytes (Sano et al. 2003). Moreover, an intact GAP domain seems to be required for AS160 function, indicating an important role for Rab-like GTPases in insulin-stimulated translocation of GLUT4. The Rab GTPases form the largest group within the Ras superfamily, and around 60 mammalian Rab proteins are known. Rabs are involved in membrane trafficking events such as endocytosis, intracellular sorting, transport to lysosomes, and recycling to the plasma membrane. Two particular Rabs, Rab4 and Rab11 have been implicated to play roles in GLUT4 vesicle trafficking (Cormont and Le Marchand-Brustel 2001), and AS160 may very well be the missing link that connects PI3K to the movement of GLUT4 vesicles. Nevertheless, the identification of the Rab isoform that interacts with AS160 in an insulin-regulated manner is still pending.

5 Insulin-stimulated exocytosis of GLUT4

A major step forward in understanding GLUT4 trafficking has been the identification of protein components of the machinery that regulates the docking and fusion of GLUT4 vesicles with the plasma membrane. A general hypothesis called the SNARE hypothesis (soluble NSF attachment protein receptors, where NSF stands for N-ethylmaleimide-sensitive fusion protein) postulates that the specificity of secretory vesicle targeting is generated by complexes that form between membrane proteins on the transport vesicle (v-SNAREs) and membrane proteins located on the target membrane (t-SNAREs) (Sollner 2003). The binding of cognate v- and t-SNAREs between two opposing lipid bilayers has been shown to drive spontaneous membrane fusion *in vitro* (Weber et al. 1998), but in intact cells this process is regulated by an ever growing number of accessory proteins that interact with the SNAREs (Sollner 2003).

In analogy to synaptic vesicles, membrane associated SNARE proteins, VAMP2, Syntaxin 4, SNAP-23, and NSF have been shown to be involved in the docking and fusion of GLUT4 vesicles with the plasma membrane (St-Denis and Cushman 1998). Upon insulin stimulation, VAMP2, the v-SNARE that is present in GLUT4 vesicles has been shown to interact with its cognate t-SNAREs in the plasma membrane, Syntaxin-4 and SNAP-23, to form a ternary complex that is required for membrane fusion. Inhibition of this complex formation abolishes fusion of GLUT4 vesicles with the plasma membrane (St-Denis and Cushman 1998). Two particular regulators of SNARE-complex formation have recently been identified in insulin target cells. Munc18c and Synip are both expressed in the plasma membrane and bind tightly to Syntaxin-4 thereby preventing the interaction of VAMP2 with its t-SNAREs (Tamori et al. 1998; Min et al. 1999). Interestingly, insulin stimulation causes the disruption of the complexes, and these findings provide evidence that insulin might directly regulate the fusion events at the plasma membrane. Further support for insulin's involvement in vesicle docking/fusion comes from a recent report that describes insulin stimulated recruitment of the Exo70 protein to the plasma membrane that is mediated by the GTPase TC10 (Inoue et al. 2003). Exo70 is part of an octameric protein complex known as the exocyst or Sec6/8 complex (Lipschutz and Mostov 2002). The exocyst was initially discovered as a required component for the exocytosis of secretory vesicles in yeast. A number of studies suggest that in mammalian cells the exocyst is also involved in the exocytosis of various (but not all) types of vesicles by mediating tethering of the vesicles to their target membranes before the actual fusion event occurs. In 3T3-L1 adipocytes, overexpression of an Exo70 mutant blocked by 50% insulin-stimulated glucose uptake and extracellular exposure of GLUT4, but not GLUT4 translocation to the plasma membrane (Inoue et al. 2003). Therefore, it was concluded that insulin engages the exocyst by recruiting Exo70 to the plasma membrane to catalyze the tethering/docking step required for the SNARE-mediated fusion of GLUT4 vesicles with the plasma membrane.

In renal collecting duct principal cells, the water channel aquaporin 2 (AQP2) resides in intracellular vesicles, and is translocated to the apical plasma membrane upon stimulation of the cells with vasopressin (Knepper and Inoue 1997). The striking similarity between AQP2 and GLUT4 trafficking is reflected by similar pathways used for exocytosis and endocytosis of both proteins. For instance, VAMP2 also resides in AQP2 vesicles and is critical for the exocytosis (Gouraud et al. 2002). SNAP-23, Syntaxin-4, and NSF have also been found in AQP2-expressing cells, indicating that these proteins might constitute a generic framework for the exocytotic route of both AQP2 and GLUT4. Thus, comparative proteomics of GLUT4 and AQP2 expressing cells might give some clues about the specificity of the proteins involved in docking and fusion of GLUT4/AQP2 vesicles.

6 Mechanism of GLUT4 endocytosis

While insulin predominantly acts on the exocytosis of GLUT4 vesicles, endocytosis is equally important for the recycling of glucose transporters. In eukaryotic cells, the formation of transport vesicles involves the interaction of 'coat' protein complexes with proteins in the source membrane. These coats are required for cargo selection and also to catalyze the budding process of the forming vesicle (Schmid 1997). To date, four unrelated types of coat protein assemblies are known (clathrin, COP-1, COP-2, and caveolae), each mediating different transport steps. In particular, clathrin-coated vesicles, and caveolae have been implicated in the formation of transport vesicles originating from the plasma membrane. However, ultrastructural analyses of adipose cells have failed to localize GLUT4 exclusively to any particular type of coated vesicle and thus the mechanism of cytoplasmic GLUT4-vesicle formation remains unknown.

Important insights into the mechanism of vesicle formation came from earlier studies of a temperature-sensitive mutation in *Drosophila melanogaster*, termed shibire[ts]. At the restrictive temperature, the flies become paralyzed due to an impaired endocytic retrieval of recycling synaptic vesicles (Henley et al. 1999). Subsequent cloning revealed an inactivating defect in a GTPase termed 'dynamin' that causes an arrest in a late step during clathrin-coated vesicle budding from the plasma membrane. Experimental evidence suggests that dynamin is required to pinch off nascent clathrin-coated vesicles from the source membrane, but the exact role of dynamin remains controversial as to whether it behaves as a molecular switch or as a mechanochemical enzyme (Thompson and McNiven 2001).

6.1 Role of dynamin GTPase in the endocytosis of GLUT4

We coexpressed a GTPase-inactive dynamin mutant (K44A) together with an HA-tagged GLUT4 reporter construct in primary rat adipose cells (Al-Hasani et al. 1998). Dynamin-K44A expression resulted in accumulation of almost the entire pool of GLUT4 on the plasma membrane in non-stimulated cells, and insulin had no further effect. Accordingly, kinetic analyses revealed that even low expression levels of the dynamin mutant caused a striking block of GLUT4 internalization. These data, and similar work performed by another group (Kao et al. 1998), have demonstrated that the pathway for GLUT4 endocytosis resembles that of the dynamin-dependent synaptic vesicle recycling in nerve cells. High levels of dynamin overexpression appear to accelerate the endocytosis of GLUT4, suggesting that the dynamin activity may be rate-limiting in adipose cells. The GTPase activity of dynamin is regulated by a variety of stimuli, such as phosphorylation, and SH3- or PH-domain binding (Scaife and Margolis 1997). Thus, it might be speculated that the observed (rather small) insulin-induced changes in the rate of endocytosis are mediated through changes in the GTPase activity of dynamin. However, the sensitivity of GLUT4 endocytosis to a dominant-negative mutant dynamin did not reveal whether GLUT4 is internalized via a clathrin-dependent or independent, e.g. caveolae-mediated pathway, since, during our studies, the GTPase was shown to be involved in both processes (Henley et al. 1998). The results from morphological and biochemical studies did not explicitly favor either of the two pathways, and because caveolae are highly abundant in adipose cells, an involvement of this coat protein in GLUT4 trafficking seemed possible.

6.2 Clathrin but not caveolin mediate GLUT4 internalization

To distinguish between clathrin-dependent and -independent pathways for GLUT4 endocytosis, we have constructed interfering mutants of proteins that were previously identified to be involved in the regulation of clathrin-coated vesicle formation (Zhao et al. 2001). The AP180 assembly protein binds to clathrin and drives coat assembly (Schmid 1997). Knockout studies of LAP, the AP180 homolog in *Drosophila* and unc-11, the AP180 gene homolog in *C. elegans*, revealed that AP180 regulates the size of clathrin-derived vesicles. Auxilin, another clathrin-binding protein has been shown to participate in the removal of clathrin coats in conjunction with the cytoplasmic ATPase Hsc70 (Schmid 1997). Overexpression of the clathrin binding domains (CBD) of AP180 and auxilin (Zhao et al. 2001) in cultured cell lines led to a redistribution of cellular clathrin, and to aggregation of clathrin in the cytosol. In parallel, the clathrin mediated endocytosis of transferrin receptors was markedly reduced. Likewise, coexpression of epitope-tagged GLUT4 and CBDs in rat adipose cells led to an increase in the basal cell-surface levels of GLUT4 that was similar to that observed in the insulin-stimulated state or after overexpression of mutant dynamin. In contrast, expression of wild type AP180 and auxilin, or clathrin-binding deficient mutants had little if any effect on GLUT4 recycling in adipose cells. Hence, these results clearly demonstrate that GLUT4 internalization is mediated by a clathrin-dependent pathway. Furthermore, since overexpression of AP180/auxilin or their clathrin binding domains had no inhibitory effect on the insulin-stimulated translocation GLUT4 to the plasma membrane, the exocytosis of GLUT4 vesicles is unlikely to involve a clathrin-dependent mechanism.

7 Targeting signals in GLUT4

It is evident that the GLUT4 isoform contains several specific recognition sites within the amino acid sequence that are responsible for its subcellular targeting. During the last decade, numerous studies have been performed to determine the domains of GLUT4 responsible for the unique subcellular sorting of this isoform (Table 1). The common experimental strategy of all these studies was to introduce cDNAs for GLUT4 reporter molecules into cells, followed by the analyses of the reporter's subcellular distribution by either morphological, biochemical, or immunological methods, or by a combination of these methods. The reporters utilized in these studies were either chimeric molecules (i.e. hybrid proteins containing fragments of GLUT4 fused to GLUT1 or the transferrin receptor (TfR)) or various mutants of GLUT4. Lastly, the reporter constructs were introduced into a large variety of different cell types (Table 1), including non-insulin responsive cells lacking endogenous GLUT4, and cultured cell lines resembling insulin responsive cells after artificially induced differentiation (3T3-L1 adipocytes, L6 myoblasts). Perhaps, not surprisingly, all these studies came to contradictory conclusions (Table 1). Several reports showed evidence that the N-terminus of GLUT4 confers intracellular sequestration whereas others suggested that the C-terminus (via a dileucine motif) is responsible for the subcellular distribution of GLUT4. One group concluded that neither the N-terminus nor the C-terminus of GLUT4 confers intracellular sequestration, whereas others found that both the N-terminus and C-terminus of GLUT4 contain important trafficking signals.

The conflicting data concerning the role of the N- and C-terminus could reflect the use of different reporter constructs, different transfection methods, different cell types, and/or different levels of expression. The specific targeting of a protein is believed to result from protein-protein interactions of the respective targeting motif with its cognate receptor (i.e. sorting adaptor). Thus, overexpression of a low affinity binding site for an adaptor protein may increase the cellular concentration of ligand-receptor (i.e. targeting motif-adaptor protein) complexes, leading to an overestimation of the role of the respective targeting motif.

Perhaps most importantly, however, different cell types appear to display preferences for different targeting motifs, as has been shown for the targeting potential of the dileucine motif in 3T3-L1 fibroblasts (intracellular sequestration) versus

Table 1. Targeting signals found in GLUT4

Cell system[a]	Reporter Construct[c]	Targeting domain[d]	Amino acid motif[e]	Function	Reference
CHO	TfR/GLUT4-Chimera	NT	F^5	Internalization	(Garippa et al. 1994)
CHO	TfR/GLUT4-Chimera	CT	$LL^{489/490}$	Internalization	(Garippa et al. 1996)
CHO	GLUT4	NT	F^5	Internalization	(Araki et al. 1996)
CHO	GLUT4	CT	$LL^{489/490}$	Retention	(Piper et al. 1993)
CHO	GLUT4/1-Chimera	NT	F^5	Retention	(Asano et al. 1992)
CHO	GLUT4/1-Chimera	TM 7	n.d.	Retention	(Ishii et al. 1995)
CHO	GLUT4/1-Chimera	NT, Loop 6, CT	n.d.	Not sufficient for targeting	
NIH-3T3	GLUT4/1-Chimera	CT	$LL^{489/490}$	Retention	(Verhey and Birnbaum 1994)
COS7	GLUT4/1-Chimera	CT	n.d.	Retention	(Czech et al. 1993)
3T3L1-fibroblasts, COS7	GLUT4	CT	$LL^{489/490}$, Y^{502}, $P^{505}DEND$	TGN to the GSC[f]	(Martinez-Arca et al. 2000)
L6 myoblast[b]	GLUT4/1-Chimera	NT	-	No function	(Haney et al. 1995)
	GLUT4/1-Chimera	CT	$LL^{489/490}$	Targeting to GSC[f]	
3T3-L1 adipose[b]	GLUT4/1-Chimera	NT	n.d.	Retention	(Verhey et al. 1995)
		CT	$LL^{489/490}$	No function	
		CT	$LL^{489/490}$	Retention	
3T3 fibroblasts	GLUT4	NT	F^5QQI	Retention	(Marsh et al. 1995)
		CT	$LL^{489/490}$	Retention[g]	
3T3-L1 adipose[b]	GLUT4	NT	F^5QQI	Endosomal targeting	(Palacios et al. 2001)
3T3-L1 adipose[b]	GLUT4	CT	$T^{498}ELEYLGP$	Endosomal targeting	(Shewan et al. 2000)
3T3-L1 adipose[b]	GLUT4	CT	E^{491}, E^{493}	Endosomal targeting	(Cope et al. 2000)
Rat adipose[b]	GLUT4	NT	F^5	Endocytosis, AP-recognition	(Al-Hasani et al. 2002)
		CT	$LL^{489/490}$	No function	

3T3-L1 adipocytes (no function) (Verhey et al. 1995). Because insulin stimulation of GLUT4-transfected heterologous cells mostly leads to a small increase in cell-surface GLUT4, it remains to be established if the targeting of GLUT4 in these cells is related to the targeting of GLUT4 in insulin target cells (see Section 3.3). Consequently, more recent studies have utilized transfected insulin-responsive cell lines to elucidate targeting motifs in GLUT4 (Table 1). In summary, two particular amino acid motifs have been connected to the subcellular trafficking of GLUT4 by the majority of these studies: the phenylalanine in position 5 (F5) in the N-terminus and the dileucine motif at position 489/490 (LL489/90) in the C-terminus. However, no agreement was reached upon their specific roles in the targeting of GLUT4 to the insulin-responsive compartment in insulin target cells.

Recently, we have evaluated the roles of these motifs in GLUT4 by analyzing the trafficking of mutant HA-epitope-tagged GLUT4 in rat adipose cells (Al-Hasani et al. 2002). Freshly isolated primary rat adipose cells were transfected with the reporter constructs by electroporation and cultured for 4 hours, so that the HA-tagged GLUT4 was expressed at physiological levels. Then, the subcellular distribution of the HA-GLUT4 was analyzed by confocal microscopy, time lapse cell-surface antibody binding assays, and Western blotting. The results showed that mutation of the C-terminal dileucine motif did not affect the targeting of GLUT4 in rat adipose cells. Dileucine motifs have been shown to be involved in internalization and lysosomal targeting of a number of proteins, including the recently discovered GLUT isoforms GLUT6 and GLUT8 (Lisinski et al. 2001). On the other hand, these studies show that the presence of a LL-motif *per se* does not provide sufficient targeting information. Much to our surprise, deletion of almost the entire C-terminus of GLUT4 led to substantial decreases in cell-surface HA-GLUT4 in both the basal and insulin-stimulated state, but had no effect on insulin-stimulated translocation. However, mutation of the F5-based targeting sequence (F5→A) resulted in a redistribution of GLUT4 towards the plasma membrane due to a substantial decrease in the rate of internalization, suggesting that the F5-motif constitutes a recognition site that is important for efficient endocytosis. Most importantly, all of the mutants analyzed showed a significant insulin-stimulated translocation, indicating that neither the F5-based motif nor the C-terminus is relevant for sorting GLUT4 into the insulin-sensitive storage compartment.

The data from mutation analyses suggested that the phenylalanine in the N-terminus of GLUT4 might comprise a docking site for the endocytosis machinery, and indeed, we found interacting proteins that had been implicated in endocytosis and subcellular sorting of membrane receptors. Several medium chain (μ-) adaptin isoforms were found to bind to the N-terminus of GLUT4, and the μ-adaptin/GLUT4 interaction was strictly dependent on the presence of phenylalanine-5 in GLUT4 (Al-Hasani et al. 2002). The 50 kDa μ-adaptins are part of the ~300 kDa heterotetrameric adaptor proteins (AP's) that are involved in transport vesicle formation. To date, four adaptor complexes, known as AP-1 to AP-4, have been identified, each mediating different transport steps in the cell (Boehm and Bonifacino 2001). It had been demonstrated that the medium chains, $\mu 1$, $\mu 2$, and $\mu 3$ of the AP-1, AP-2, and AP-3 complexes, respectively, bind to tyrosine-based sorting signals (consensus sequence: YXXØ; Y: tyrosine, X: any amino acid, Ø:

bulky hydrophobic residue) in the cytoplasmic domains of membrane receptors. Because AP-1 and AP-2 also bind to clathrin, the μ-adaptin-mediated binding of APs to their YXXØ-target is believed to recruit membrane receptors to nascent clathrin-coated transport vesicles. Obviously, the N-terminal F5QQI motif in GLUT4 resembles such a YXXØ motif. The fact that the N-terminus of GLUT4 interacts with the AP medium chains likely explains the recently described *in vivo* association of GLUT4 and AP adaptor complexes (Gillingham et al. 1999). Thus, while further biochemical analyses are required, our data are consistent with the proposed binding of the F5QQI motif of GLUT4 to the μ-adaptins *in vivo*. Because this interaction occurs with several μ-adaptin isoforms, the N-terminus may be involved in the targeting of GLUT4 at multiple sites in the trafficking pathway.

8 Outlook

Substantial progress has been made in recent years in the understanding of the molecular machinery that is responsible for GLUT4 recycling. In certain aspects, GLUT4 vesicles resemble neurotransmitter-containing vesicles in neuronal cells. In analogy to synaptic vesicles, membrane associated SNARE proteins have been shown to be involved in the docking and fusion of GLUT4 vesicles with the plasma membrane, and internalization of GLUT4 has been demonstrated to require a clathrin/dynamin-dependent mechanism. These findings may provide a platform to discover novel elements involved in the docking/fusion process of GLUT4 vesicles that are subject of regulation by insulin, such as Munc18c and Synip, and perhaps are unique for insulin-responsive cells. Insulin's signaling network that drives GLUT4 translocation has turned out to be exceedingly complex, and involves an ever growing number of proteins, many of them having multiple functions in a given cell type. The ultimate question, however, that is how insulin binding to its receptor makes vesicles bud, move, and/or fuse with the plasma membrane, is far from being answered. Nevertheless, as we are homing in on the most proximal components of the signal transduction network, it becomes clear that insulin action is likely to modulate a variety of cellular processes such as cytoskeleton dynamics, motor protein-directed movement, and dynamic budding, sorting, and fusion of vesicles within multiple compartments of the cell. Novel approaches such as small-interfering (si)-RNA mediated gene silencing are likely to yield functional data for supposedly critical proteins for the GLUT4 translocation. In addition, novel candidate proteins involved in GLUT4 recycling may be found through DNA microarray-based screens for cell type specific gene expression. Evanescent-wave microscopy (a.k.a. Total Internal Reflection Fluorescence Microscopy: TIRFM) has received considerable recognition recently in cell biology, in particular in neurosciences (Sako and Uyemura 2002). TIRFM is ideally suited to gaining insight into events occurring at, or close to, the plasma membrane of live cells, and allows tracking of individual vesicles (even individual fluorophores) as they move on in their itinerary. Application of this technique to studying GLUT4 vesicle trafficking has already begun, and it is expected to have great impact on the understanding of the molecular events at the plasma membrane. After all, the future goal is to gain insight into the molecular mechanisms that might be relevant in the pathogenesis of type-2 diabetes mellitus and to develop novel treatments for this disease.

Acknowledgements

I thank Dr. Samuel W. Cushman and Dr. Julian A. Schnabel for helpful discussions and for critically reading the manuscript.

References

Albiston AL, McDowall SG, Matsacos D, Sim P, Clune E, Mustafa T, Lee J, Mendelsohn FA, Simpson RJ, Connolly LM, Chai SY (2001) Evidence that the angiotensin IV (AT(4)) receptor is the enzyme insulin-regulated aminopeptidase. J Biol Chem 276:48623-48626

Al-Hasani H, Hinck CS, Cushman SW (1998) Endocytosis of the glucose transporter GLUT4 is mediated by the GTPase dynamin. J Biol Chem 273:17504-17510

Al-Hasani H, Kunamneni RK, Dawson K, Hinck CS, Muller-Wieland D, Cushman SW (2002) Roles of the N- and C-termini of GLUT4 in endocytosis. J Cell Sci 115:131-140

Al-Hasani H, Yver DR, Cushman SW (1999) Overexpression of the glucose transporter GLUT4 in adipose cells interferes with insulin-stimulated translocation. FEBS Lett 460:338-342

Araki S, Yang J, Hashiramoto M, Tamori Y, Kasuga M, Holman GD (1996) Subcellular trafficking kinetics of GLU4 mutated at the N- and C-terminal. Biochem J 315 Pt 1:153-159

Asano T, Takata K, Katagiri H, Tsukuda K, Lin JL, Ishihara H, Inukai K, Hirano H, Yazaki Y, Oka Y (1992) Domains responsible for the differential targeting of glucose transporter isoforms. J Biol Chem 267:19636-19641

Balendran A, Casamayor A, Deak M, Paterson A, Gaffney P, Currie R, Downes CP, Alessi DR (1999) PDK1 acquires PDK2 activity in the presence of a synthetic peptide derived from the carboxyl terminus of PRK2. Curr Biol 9:393-404

Bandyopadhyay G, Standaert ML, Galloway L, Moscat J, Farese RV (1997a) Evidence for involvement of protein kinase C (PKC)-zeta and noninvolvement of diacylglycerol-sensitive PKCs in insulin-stimulated glucose transport in L6 myotubes. Endocrinology 138:4721-4731

Bandyopadhyay G, Standaert ML, Kikkawa U, Ono Y, Moscat J, Farese RV (1999) Effects of transiently expressed atypical (zeta, lambda), conventional (alpha, beta) and novel (delta, epsilon) protein kinase C isoforms on insulin-stimulated translocation of epitope-tagged GLUT4 glucose transporters in rat adipocytes: specific interchangeable effects of protein kinases C-zeta and C-lambda. Biochem J 337:461-470

Bandyopadhyay G, Standaert ML, Zhao L, Yu B, Avignon A, Galloway L, Karnam P, Moscat J, Farese RV (1997b) Activation of protein kinase C (alpha, beta, and zeta) by insulin in 3T3/L1 cells. Transfection studies suggest a role for PKC-zeta in glucose transport. J Biol Chem 272:2551-2558

Baumann CA, Ribon V, Kanzaki M, Thurmond DC, Mora S, Shigematsu S, Bickel PE, Pessin JE, Saltiel AR (2000) CAP defines a second signalling pathway required for insulin-stimulated glucose transport. Nature 407:202-207

Birnbaum MJ (1989) Identification of a novel gene encoding an insulin-responsive glucose transporter protein. Cell 57:305-315

Boehm M, Bonifacino JS (2001) Adaptins: the final recount. Mol Biol Cell 12:2907-2920

Bose A, Guilherme A, Robida SI, Nicoloro SM, Zhou QL, Jiang ZY, Pomerleau DP, Czech MP (2002) Glucose transporter recycling in response to insulin is facilitated by myosin MyoIc. Nature 420:821-824

Braiman L, Alt A, Kuroki T, Ohba M, Bak A, Tennenbaum T, Sampson SR (2001) Activation of protein kinase C zeta induces serine phosphorylation of VAMP2 in the GLUT4 compartment and increases glucose transport in skeletal muscle. Mol Cell Biol 21:7852-7861

Brooks CC, Scherer PE, Cleveland K, Whittemore JL, Lodish HF, Cheatham B (2000) Pantophysin is a phosphoprotein component of adipocyte transport vesicles and associates with GLUT4-containing vesicles. J Biol Chem 275:2029-2036

Bryant NJ, Govers R, James DE (2002) Regulated transport of the glucose transporter GLUT4. Nat Rev Mol Cell Biol 3:267-277

Cain CC, Trimble WS, Lienhard GE (1992) Members of the VAMP family of synaptic vesicle proteins are components of glucose transporter-containing vesicles from rat adipocytes. J Biol Chem 267:11681-11684

Chen X, Al-Hasani H, Olausson T, Wenthzel AM, Smith U, Cushman SW (2003) Activity, phosphorylation state and subcellular distribution of GLUT4-targeted Akt2 in rat adipose cells. J Cell Sci 116:3511-3518

Chiang SH, Baumann CA, Kanzaki M, Thurmond DC, Watson RT, Neudauer CL, Macara IG, Pessin JE, Saltiel AR (2001) Insulin-stimulated GLUT4 translocation requires the CAP-dependent activation of TC10. Nature 410:944-948

Cope DL, Lee S, Melvin DR, Gould GW (2000) Identification of further important residues within the Glut4 carboxy-terminal tail which regulate subcellular trafficking. FEBS Lett 481:261-265

Cormont M, Le Marchand-Brustel Y (2001) The role of small G-proteins in the regulation of glucose transport (review). Mol Membr Biol 18:213-220

Cushman SW, Wardzala LJ (1980) Potential mechanism of insulin action on glucose transport in the isolated rat adipose cell. Apparent translocation of intracellular transport systems to the plasma membrane. J Biol Chem 255:4758-4762

Czech MP, Chawla A, Woon CW, Buxton J, Armoni M, Tang W, Joly M, Corvera S (1993) Exofacial epitope-tagged glucose transporter chimeras reveal COOH-terminal sequences governing cellular localization. J Cell Biol 123:127-135

Doornbos RP, Theelen M, van der Hoeven PC, van Blitterswijk WJ, Verkleij AJ, van Bergen en Henegouwen PM (1999) Protein kinase Czeta is a negative regulator of protein kinase B activity. J Biol Chem 274:8589-8596

Egawa K, Sharma PM, Nakashima N, Huang Y, Huver E, Boss GR, Olefsky JM (1999) Membrane-targeted phosphatidylinositol 3-kinase mimics insulin actions and induces a state of cellular insulin resistance. J Biol Chem 274:14306-14314

Etgen GJ, Valasek KM, Broderick CL, Miller AR (1999) *In vivo* adenoviral delivery of recombinant human protein kinase C-zeta stimulates glucose transport activity in rat skeletal muscle. J Biol Chem 274:22139-22142

Farese RV (2001) Insulin-sensitive phospholipid signaling systems and glucose transport. Update II. Exp Biol Med (Maywood) 226:283-295

Fukumoto H, Kayano T, Buse JB, Edwards Y, Pilch PF, Bell GI, Seino S (1989) Cloning and characterization of the major insulin-responsive glucose transporter expressed in human skeletal muscle and other insulin-responsive tissues. J Biol Chem 264:7776-7779

Furtado LM, Somwar R, Sweeney G, Niu W, Klip A (2002) Activation of the glucose transporter GLUT4 by insulin. Biochem Cell Biol 80:569-578

Garippa RJ, Johnson A, Park J, Petrush RL, McGraw TE (1996) The carboxyl terminus of GLUT4 contains a serine-leucine-leucine sequence that functions as a potent internalization motif in Chinese hamster ovary cells. J Biol Chem 271:20660-20668

Garippa RJ, Judge TW, James DE, McGraw TE (1994) The amino terminus of GLUT4 functions as an internalization motif but not an intracellular retention signal when substituted for the transferrin receptor cytoplasmic domain. J Cell Biol 124:705-715

Gillingham AK, Koumanov F, Pryor PR, Reaves BJ, Holman GD (1999) Association of AP1 adaptor complexes with GLUT4 vesicles. J Cell Sci 112:4793-4800

Gouraud S, Laera A, Calamita G, Carmosino M, Procino G, Rossetto O, Mannucci R, Rosenthal W, Svelto M, Valenti G (2002) Functional involvement of VAMP/synaptobrevin-2 in cAMP-stimulated aquaporin 2 translocation in renal collecting duct cells. J Cell Sci 115:3667-3674

Hajduch E, Alessi DR, Hemmings BA, Hundal HS (1998) Constitutive activation of protein kinase B alpha by membrane targeting promotes glucose and system A amino acid transport, protein synthesis, and inactivation of glycogen synthase kinase 3 in L6 muscle cells. Diabetes 47:1006-1013

Haney PM, Levy MA, Strube MS, Mueckler M (1995) Insulin-sensitive targeting of the GLUT4 glucose transporter in L6 myoblasts is conferred by its COOH-terminal cytoplasmic tail. J Cell Biol 129:641-658

Haney PM, Slot JW, Piper RC, James DE, Mueckler M (1991) Intracellular targeting of the insulin-regulatable glucose transporter (GLUT4) is isoform specific and independent of cell type. J Cell Biol 114:689-699

Henley JR, Cao H, McNiven MA (1999) Participation of dynamin in the biogenesis of cytoplasmic vesicles. Faseb J 13 Suppl 2:S243-247

Henley JR, Krueger EW, Oswald BJ, McNiven MA (1998) Dynamin-mediated internalization of caveolae. J Cell Biol 141:85-99

Hill MM, Clark SF, Tucker DF, Birnbaum MJ, James DE, Macaulay SL (1999) A role for protein kinase Bbeta/Akt2 in insulin-stimulated GLUT4 translocation in adipocytes. Mol Cell Biol 19:7771-7781

Hill MM, Feng J, Hemmings BA (2002) Identification of a plasma membrane Raft-associated PKB Ser473 kinase activity that is distinct from ILK and PDK1. Curr Biol 12:1251-1255

Holman GD, Lo Leggio L, Cushman SW (1994) Insulin-stimulated GLUT4 glucose transporter recycling. A problem in membrane protein subcellular trafficking through multiple pools. J Biol Chem 269:17516-17524

Hresko RC, Murata H, Mueckler M (2003) Phosphoinositide-dependent kinase-2 is a distinct protein kinase enriched in a novel cytoskeletal fraction associated with adipocyte plasma membranes. J Biol Chem 278:21615-21622

Inoue M, Chang L, Hwang J, Chiang SH, Saltiel AR (2003) The exocyst complex is required for targeting of Glut4 to the plasma membrane by insulin. Nature 422:629-633

Isakoff SJ, Taha C, Rose E, Marcusohn J, Klip A, Skolnik EY (1995) The inability of phosphatidylinositol 3-kinase activation to stimulate GLUT4 translocation indicates additional signaling pathways are required for insulin-stimulated glucose uptake. Proc Natl Acad Sci USA 92:10247-10251

Ishii K, Hayashi H, Todaka M, Kamohara S, Kanai F, Jinnouchi H, Wang L, Ebina Y (1995) Possible domains responsible for intracellular targeting and insulin-dependent translocation of glucose transporter type 4. Biochem J 309 Pt 3:813-823

James DE, Strube M, Mueckler M (1989) Molecular cloning and characterization of an insulin-regulatable glucose transporter. Nature 338:83-87

Jiang T, Sweeney G, Rudolf MT, Klip A, Traynor-Kaplan A, Tsien RY (1998) Membrane-permeant esters of phosphatidylinositol 3,4,5-trisphosphate. J Biol Chem 273:11017-11024

Jiang ZY, Chawla A, Bose A, Way M, Czech MP (2002) A phosphatidylinositol 3-kinase-independent insulin signaling pathway to N-WASP/Arp2/3/F-actin required for GLUT4 glucose transporter recycling. J Biol Chem 277:509-515

Joost HG, Thorens B (2001) The extended GLUT-family of sugar/polyol transport facilitators: nomenclature, sequence characteristics, and potential function of its novel members (review). Mol Membr Biol 18:247-256

Kahn BB (1996) Lilly lecture 1995. Glucose transport: pivotal step in insulin action. Diabetes 45:1644-1654

Kane S, Sano H, Liu SC, Asara JM, Lane WS, Garner CC, Lienhard GE (2002) A method to identify serine kinase substrates. Akt phosphorylates a novel adipocyte protein with a Rab GTPase-activating protein (GAP) domain. J Biol Chem 277:22115-22118

Kanzaki M, Pessin JE (2001) Insulin-stimulated GLUT4 translocation in adipocytes is dependent upon cortical actin remodeling. J Biol Chem 276:42436-42444

Kao AW, Ceresa BP, Santeler SR, Pessin JE (1998) Expression of a dominant interfering dynamin mutant in 3T3L1 adipocytes inhibits GLUT4 endocytosis without affecting insulin signaling. J Biol Chem 273:25450-25457

Keller SR, Scott HM, Mastick CC, Aebersold R, Lienhard GE (1995) Cloning and characterization of a novel insulin-regulated membrane aminopeptidase from Glut4 vesicles. J Biol Chem 270:23612-23618

Khan AH, Pessin JE (2002) Insulin regulation of glucose uptake: a complex interplay of intracellular signalling pathways. Diabetologia 45:1475-1483

Kimura A, Baumann CA, Chiang SH, Saltiel AR (2001) The sorbin homology domain: a motif for the targeting of proteins to lipid rafts. Proc Natl Acad Sci USA 98:9098-9103

Knepper MA, Inoue T (1997) Regulation of aquaporin-2 water channel trafficking by vasopressin. Curr Opin Cell Biol 9:560-564

Kohn AD, Summers SA, Birnbaum MJ, Roth RA (1996) Expression of a constitutively active Akt Ser/Thr kinase in 3T3-L1 adipocytes stimulates glucose uptake and glucose transporter 4 translocation. J Biol Chem 271:31372-31378

Kono-Sugita E, Satoh S, Suzuki Y, Egawa M, Udaka N, Ito T, Sekihara H (1996) Insulin-induced GLUT4 recycling in rat adipose cells by a pathway insensitive to brefeldin A. Eur J Biochem 236:1033-1037

Kotani K, Ogawa W, Matsumoto M, Kitamura T, Sakaue H, Hino Y, Miyake K, Sano W, Akimoto K, Ohno S, Kasuga M (1998) Requirement of atypical protein kinase clambda for insulin stimulation of glucose uptake but not for Akt activation in 3T3-L1 adipocytes. Mol Cell Biol 18:6971-6982

Kupriyanova TA, Kandror KV (1999) Akt-2 binds to Glut4-containing vesicles and phosphorylates their component proteins in response to insulin. J Biol Chem 274:1458-1464

Kupriyanova TA, Kandror KV (2000) Cellugyrin is a marker for a distinct population of intracellular Glut4-containing vesicles. J Biol Chem 275:36263-36268

Lampson MA, Schmoranzer J, Zeigerer A, Simon SM, McGraw TE (2001) Insulin-regulated release from the endosomal recycling compartment is regulated by budding of specialized vesicles. Mol Biol Cell 12:3489-3501

Laurie SM, Cain CC, Lienhard GE, Castle JD (1993) The glucose transporter GluT4 and secretory carrier membrane proteins (SCAMPs) colocalize in rat adipocytes and partially segregate during insulin stimulation. J Biol Chem 268:19110-19117

Lee W, Jung CY (1997) A synthetic peptide corresponding to the GLUT4 C-terminal cytoplasmic domain causes insulin-like glucose transport stimulation and GLUT4 recruitment in rat adipocytes. J Biol Chem 272:21427-21431

Lee W, Ryu J, Spangler RA, Jung CY (2000) Modulation of GLUT4 and GLUT1 recycling by insulin in rat adipocytes: kinetic analysis based on the involvement of multiple intracellular compartments. Biochemistry 39:9358-9366

Lin BZ, Pilch PF, Kandror KV (1997) Sortilin is a major protein component of Glut4-containing vesicles. J Biol Chem 272:24145-24147

Lipschutz JH, Mostov KE (2002) Exocytosis: the many masters of the exocyst. Curr Biol 12:R212-R214

Lisinski I, Schurmann A, Joost HG, Cushman SW, Al-Hasani H (2001) Targeting of GLUT6 (formerly GLUT9) and GLUT8 in rat adipose cells. Biochem J 358:517-522

Lund S, Holman GD, Schmitz O, Pedersen O (1995) Contraction stimulates translocation of glucose transporter GLUT4 in skeletal muscle through a mechanism distinct from that of insulin. Proc Natl Acad Sci USA 92:5817-5821

Malide D, Ramm G, Cushman SW, Slot JW (2000) Immunoelectron microscopic evidence that GLUT4 translocation explains the stimulation of glucose transport in isolated rat white adipose cells. J Cell Sci 113 Pt 23:4203-4210

Malide D, St-Denis JF, Keller SR, Cushman SW (1997) Vp165 and GLUT4 share similar vesicle pools along their trafficking pathways in rat adipose cells. FEBS Lett 409:461-468

Marsh BJ, Alm RA, McIntosh SR, James DE (1995) Molecular regulation of GLUT-4 targeting in 3T3-L1 adipocytes. J Cell Biol 130:1081-1091

Martin LB, Shewan A, Millar CA, Gould GW, James DE (1998) Vesicle-associated membrane protein 2 plays a specific role in the insulin-dependent trafficking of the facilitative glucose transporter GLUT4 in 3T3-L1 adipocytes. J Biol Chem 273:1444-1452

Martin S, Ramm G, Lyttle CT, Meerloo T, Stoorvogel W, James DE (2000) Biogenesis of insulin-responsive GLUT4 vesicles is independent of brefeldin A-sensitive trafficking. Traffic 1:652-660

Martinez-Arca S, Lalioti VS, Sandoval IV (2000) Intracellular targeting and retention of the glucose transporter GLUT4 by the perinuclear storage compartment involves distinct carboxyl-tail motifs. J Cell Sci 113 Pt 10:1705-1715

Min J, Okada S, Kanzaki M, Elmendorf JS, Coker KJ, Ceresa BP, Syu LJ, Noda Y, Saltiel AR, Pessin JE (1999) Synip: a novel insulin-regulated syntaxin 4-binding protein mediating GLUT4 translocation in adipocytes. Mol Cell 3:751-760

Minokoshi Y, Kahn CR, Kahn BB (2003) Tissue-specific ablation of the GLUT4 glucose transporter and the insulin receptor challenge assumptions about insulin action and glucose homeostasis. J Biol Chem 278: 33609-33612

Mueckler M (1994) Facilitative glucose transporters. Eur J Biochem 219:713-725

Newton AC (2003) Regulation of the ABC kinases by phosphorylation: protein kinase C as a paradigm. Biochem J 370:361-371

Nielsen MS, Madsen P, Christensen EI, Nykjaer A, Gliemann J, Kasper D, Pohlmann R, Petersen CM (2001) The sortilin cytoplasmic tail conveys Golgi-endosome transport and binds the VHS domain of the GGA2 sorting protein. EMBO J 20:2180-2190

Omata W, Shibata H, Li L, Takata K, Kojima I (2000) Actin filaments play a critical role in insulin-induced exocytotic recruitment but not in endocytosis of GLUT4 in isolated rat adipocytes. Biochem J 346 Pt 2:321-328

Palacios S, Lalioti V, Martinez-Arca S, Chattopadhyay S, Sandoval IV (2001) Recycling of the insulin-sensitive glucose transporter GLUT4. Access of surface internalized GLUT4 molecules to the perinuclear storage compartment is mediated by the Phe5-Gln6-Gln7-Ile8 motif. J Biol Chem 276:3371-3383

Piper RC, Tai C, Kulesza P, Pang S, Warnock D, Baenziger J, Slot JW, Geuze HJ, Puri C, James DE (1993) GLUT-4 NH2 terminus contains a phenylalanine-based targeting motif that regulates intracellular sequestration. J Cell Biol 121:1221-1232

Previs SF, Withers DJ, Ren JM, White MF, Shulman GI (2000) Contrasting effects of IRS-1 versus IRS-2 gene disruption on carbohydrate and lipid metabolism in vivo. J Biol Chem 275:38990-38994

Quon MJ, Chen H, Lin CH, Zhou L, Ing BL, Zarnowski MJ, Klinghoffer R, Kazlauskas A, Cushman SW, Taylor SI (1996) Effects of overexpressing wild-type and mutant PDGF receptors on translocation of GLUT4 in transfected rat adipose cells. Biochem Biophys Res Commun 226:587-594

Randhawa VK, Bilan PJ, Khayat ZA, Daneman N, Liu Z, Ramlal T, Volchuk A, Peng XR, Coppola T, Regazzi R, Trimble WS, Klip A (2000) VAMP2, but not VAMP3/cellubrevin, mediates insulin-dependent incorporation of GLUT4 into the plasma membrane of L6 myoblasts. Mol Biol Cell 11:2403-2417

Ribon V, Saltiel AR (1997) Insulin stimulates tyrosine phosphorylation of the proto-oncogene product of c-Cbl in 3T3-L1 adipocytes. Biochem J 324 Pt 3:839-845

Ridley AJ (2001) Rho proteins: linking signaling with membrane trafficking. Traffic 2:303-310

Rumsey SC, Daruwala R, Al-Hasani H, Zarnowski MJ, Simpson IA, Levine M (2000) Dehydroascorbic acid transport by GLUT4 in Xenopus oocytes and isolated rat adipocytes. J Biol Chem 275:28246-28253

Sako Y, Uyemura T (2002) Total internal reflection fluorescence microscopy for single-molecule imaging in living cells. Cell Struct Funct 27:357-365

Sano H, Kane S, Sano E, Miinea CP, Asara JM, Lane WS, Garner CW, Lienhard GE (2003) Insulin-stimulated phosphorylation of a Rab GTPase-activating protein regulates GLUT4 translocation. J Biol Chem 278:14599-14602

Satoh S, Nishimura H, Clark AE, Kozka IJ, Vannucci SJ, Simpson IA, Quon MJ, Cushman SW, Holman GD (1993) Use of bismannose photolabel to elucidate insulin-regulated

GLUT4 subcellular trafficking kinetics in rat adipose cells. Evidence that exocytosis is a critical site of hormone action. J Biol Chem 268:17820-17829

Scaife RM, Margolis RL (1997) The role of the PH domain and SH3 binding domains in dynamin function. Cell Signal 9:395-401

Schmid SL (1997) Clathrin-coated vesicle formation and protein sorting: an integrated process. Annu Rev Biochem 66:511-548

Shepherd PR, Gnudi L, Tozzo E, Yang H, Leach F, Kahn BB (1993) Adipose cell hyperplasia and enhanced glucose disposal in transgenic mice overexpressing GLUT4 selectively in adipose tissue. J Biol Chem 268:22243-22246

Shewan AM, Marsh BJ, Melvin DR, Martin S, Gould GW, James DE (2000) The cytosolic C-terminus of the glucose transporter GLUT4 contains an acidic cluster endosomal targeting motif distal to the dileucine signal. Biochem J 350 Pt 1:99-107

Simpson F, Whitehead JP, James DE (2001) GLUT4--at the cross roads between membrane trafficking and signal transduction. Traffic 2:2-11

Slot JW, Geuze HJ, Gigengack S, James DE, Lienhard GE (1991a) Translocation of the glucose transporter GLUT4 in cardiac myocytes of the rat. Proc Natl Acad Sci USA 88:7815-7819

Slot JW, Geuze HJ, Gigengack S, Lienhard GE, James DE (1991b) Immuno-localization of the insulin regulatable glucose transporter in brown adipose tissue of the rat. J Cell Biol 113:123-135

Smart EJ, Graf GA, McNiven MA, Sessa WC, Engelman JA, Scherer PE, Okamoto T, Lisanti MP (1999) Caveolins, liquid-ordered domains, and signal transduction. Mol Cell Biol 19:7289-7304

Sollner TH (2003) Regulated exocytosis and SNARE function (Review). Mol Membr Biol 20:209-220

Standaert ML, Galloway L, Karnam P, Bandyopadhyay G, Moscat J, Farese RV (1997) Protein kinase C-zeta as a downstream effector of phosphatidylinositol 3-kinase during insulin stimulation in rat adipocytes. Potential role in glucose transport. J Biol Chem 272:30075-30082

St-Denis JF, Cushman SW (1998) Role of SNARE's in the GLUT4 translocation response to insulin in adipose cells and muscle. J Basic Clin Physiol Pharmacol 9:153-165

Suzuki K, Kono T (1980) Evidence that insulin causes translocation of glucose transport activity to the plasma membrane from an intracellular storage site. Proc Natl Acad Sci USA 77:2542-2545

Tamori Y, Kawanishi M, Niki T, Shinoda H, Araki S, Okazawa H, Kasuga M (1998) Inhibition of insulin-induced GLUT4 translocation by Munc18c through interaction with syntaxin4 in 3T3-L1 adipocytes. J Biol Chem 273:19740-19746

Tanner LI, Lienhard GE (1987) Insulin elicits a redistribution of transferrin receptors in 3T3-L1 adipocytes through an increase in the rate constant for receptor externalization. J Biol Chem 262:8975-8980

Tanti JF, Grillo S, Gremeaux T, Coffer PJ, Van Obberghen E, Le Marchand-Brustel Y (1997) Potential role of protein kinase B in glucose transporter 4 translocation in adipocytes. Endocrinology 138:2005-2010

Thompson HM, McNiven MA (2001) Dynamin: switch or pinchase? Curr Biol 11:R850

Toker A, Newton AC (2000) Akt/protein kinase B is regulated by autophosphorylation at the hypothetical PDK-2 site. J Biol Chem 275:8271-8274

Tozzo E, Kahn BB, Pilch PF, Kandror KV (1996) Glut4 is targeted to specific vesicles in adipocytes of transgenic mice overexpressing Glut4 selectively in adipose tissue. J Biol Chem 271:10490-10494

Troussard AA, Mawji NM, Ong C, Mui A, St-Arnaud R, Dedhar S (2003) Conditional knock-out of integrin-linked kinase demonstrates an essential role in protein kinase B/Akt activation. J Biol Chem 278:22374-22378

Tsakiridis T, Vranic M, Klip A (1994) Disassembly of the actin network inhibits insulin-dependent stimulation of glucose transport and prevents recruitment of glucose transporters to the plasma membrane. J Biol Chem 269:29934-29942

Tsygankov AY, Teckchandani AM, Feshchenko EA, Swaminathan G (2001) Beyond the RING: CBL proteins as multivalent adapters. Oncogene 20:6382-6402

Vanhaesebroeck B, Alessi DR (2000) The PI3K-PDK1 connection: more than just a road to PKB. Biochem J 346 Pt 3:561-576

Venkateswarlu K, Oatey PB, Tavare JM, Cullen PJ (1998) Insulin-dependent translocation of ARNO to the plasma membrane of adipocytes requires phosphatidylinositol 3-kinase. Curr Biol 8:463-466

Verhey KJ, Birnbaum MJ (1994) A Leu-Leu sequence is essential for COOH-terminal targeting signal of GLUT4 glucose transporter in fibroblasts. J Biol Chem 269:2353-2356

Verhey KJ, Yeh JI, Birnbaum MJ (1995) Distinct signals in the GLUT4 glucose transporter for internalization and for targeting to an insulin-responsive compartment. J Cell Biol 130:1071-1079

Vinten J, Gliemann J, Osterlind K (1976) Exchange of 3-O-methylglucose in isolated fat cells. Concentration dependence and effect of insulin. J Biol Chem 251:794-800

Volchuk A, Sargeant R, Sumitani S, Liu Z, He L, Klip A (1995) Cellubrevin is a resident protein of insulin-sensitive GLUT4 glucose transporter vesicles in 3T3-L1 adipocytes. J Biol Chem 270:8233-8240

Wallberg-Henriksson H, Zierath JR (2001) GLUT4: a key player regulating glucose homeostasis? Insights from transgenic and knockout mice (review). Mol Membr Biol 18:205-211

Wang Q, Somwar R, Bilan PJ, Liu Z, Jin J, Woodgett JR, Klip A (1999) Protein kinase B/Akt participates in GLUT4 translocation by insulin in L6 myoblasts. Mol Cell Biol 19:4008-4018

Weber T, Zemelman BV, McNew JA, Westermann B, Gmachl M, Parlati F, Sollner TH, Rothman JE (1998) SNAREpins: minimal machinery for membrane fusion. Cell 92:759-772

White MF (1998) The IRS-signalling system: a network of docking proteins that mediate insulin action. Mol Cell Biochem 182:3-11

Wood IS, Trayhurn P (2003) Glucose transporters (GLUT and SGLT): expanded families of sugar transport proteins. Br J Nutr 89:3-9

Zhao X, Greener T, Al-Hasani H, Cushman SW, Eisenberg E, Greene LE (2001) Expression of auxilin or AP180 inhibits endocytosis by mislocalizing clathrin: evidence for formation of nascent pits containing AP1 or AP2 but not clathrin. J Cell Sci 114:353-365

Abbreviations

HA: hemagglutinin
TGN: trans-Golgi network
TfR: transferrin receptor

Al-Hasani, Hadi
   German Institute for Human Nutrition, Arthur-Scheunert-Allee 114-116, D-
   14558 Potsdam-Rehbrücke, Germany
   al-hasani@mail.dife.de

Aquaporin-2 trafficking

Sebastian Frische, Tae-Hwan Kwon, Jørgen Frøkiær, and Søren Nielsen

Abstract

Regulation of the water permeability of the apical plasma membrane in collecting duct principal cells is essential for the regulation of renal water excretion, and thus, the regulation of body water balance. The water permeability is partly regulated by trafficking of aquaporin (AQP2) containing vesicles between an intracellular reservoir and the apical plasma membrane. Insertion of AQP2 molecules in the apical plasma membrane is induced by vasopressin binding to the V2-receptor at the basolateral side of principal cells. This activates G-proteins, which stimulate adenylyl cyclase resulting in increased intracellular cAMP-concentration and activation of Protein Kinase A (PKA). AQP2 contains a PKA consensus site at ser256, and phosphorylation of this serine in three out of four AQP2-molecules in an AQP2 homotetramer is involved in the regulated translocation of AQP2 to the apical plasma membrane. The elements of this cAMP mediated pathway, including the possible role of scaffolding A Kinase Anchoring Proteins (AKAPs), is still a major area of research. However, a number of other pathways, many of which are believed to relate to changes in the cortical actin network of the principal cells, are activated during vasopressin induced AQP2-trafficking. The purpose of this review is to present an overview of the currently known cellular signalling pathways and molecular mechanisms involved in controlling the abundance of AQP2 molecules in the apical plasma membrane according to the physiological needs of the body.

1 Vasopressin induced AQP2-trafficking

1.1 Historical overview

The discovery of aquaporin membrane water channels by Agre[1] and co-workers answered a longstanding biophysical question of how water crosses biological membranes specifically, and provided insight, at the molecular level, into the fundamental physiology of water balance and the pathophysiology of water balance disorders. A number of reviews of aquaporins have been published within the last five years, dealing with aquaporins in general (Agre et al. 2002), the specific roles

---

[1] Peter Agre received the 2003 Nobel Prize in Chemistry for the discovery of aquaporin water channels

Topics in Current Genetics, Vol. 9
E. Boles, R. Krämer (Eds.): Molecular Mechanisms Controlling Transmembrane Transport
DOI 10.1007/b97874 / Published online: 9 March 2004
© Springer-Verlag Berlin Heidelberg 2004

of aquaporins in the kidney (Nielsen et al. 2002), and AQP2-trafficking (Klussmann et al. 2000; Brown 2003).

Out of the thirteen aquaporins encoded by the human and mouse genomes at least six are known to be present in the kidney at distinct sites along the nephron and collecting duct. AQP2 is abundant in the collecting duct principal cells and is the chief target for the regulation of collecting duct water reabsorption by vasopressin. Acute regulation involves vasopressin-induced trafficking of AQP2 between an intracellular reservoir in vesicles and the apical plasma membrane. In addition AQP2 is involved in chronic/adaptational control of body water balance, which is achieved through regulation of AQP2 expression. Importantly, multiple studies have now underscored a critical role of AQP2 in several inherited and acquired water balance disorders.

As previously reviewed (Nielsen et al. 2002), much of the early work on vasopressin action was carried out in amphibian skin or bladder, which are functional analogues of the kidney collecting duct (Chevalier et al. 1974; Humbert et al. 1977; Kachadorian et al. 1977). The results from these and subsequent studies led Wade and colleagues to propose the "membrane shuttle hypothesis" (Wade and Kachadorian 1988), which states that water channels were stored in vesicles, and inserted exocytically into the apical plasma membrane in response to vasopressin (Fig. 1). The identification of the aquaporins and subsequently AQP2 made it possible to test this hypothesis directly. Water reabsorption in the collecting duct is determined by the apical plasma membrane of the collecting duct cells, which provides the rate limiting barrier to water movement (Flamion and Spring 1990). Regulation of the osmotic water permeability of this membrane could, in principle, occur via one or both of two distinct mechanisms: either (1) the permeability of existing channels could be altered, for example, by chemical modification (e.g. phosphorylation of the channel), or (2) the number of water channels present in the membrane could be altered by insertion or removal of channels from an intracellular store, as proposed by the membrane shuttle hypothesis. The presence of AQP2 in small vesicles, as determined by immunoelectron microscopy (Nielsen et al. 1993a) favoured the latter hypothesis and several *in vitro* and *in vivo* studies have now underscored the importance of regulated trafficking of AQP2.

In vitro studies using isolated perfused tubules allowed a direct analysis of both the on-set and off-set responses to vasopressin (Nielsen et al. 1995a). In this study, it was demonstrated that changes in AQP2 labelling density of the apical plasma membrane correlated closely with the water permeability in the same tubules, while there were reciprocal changes in the intracellular labelling for AQP2. *In vivo* studies using normal rats or vasopressin-deficient Brattleboro rats also showed a marked increase in apical plasma membrane labelling of AQP2 in response to vasopressin or dDAVP treatment (Marples et al. 1995; Sabolic et al. 1995; Yamamoto et al. 1995). The off-set response has been examined *in vivo* using acute treatment of rats with vasopressin-V2-receptor antagonist (Hayashi et al. 1994; Christensen et al. 1998) or acute water loading (to reduce endogenous vasopressin levels (Saito et al. 1997b)). These treatments (both reducing vasopressin action) resulted in a prominent internalization of AQP2 from the apical plasma membrane

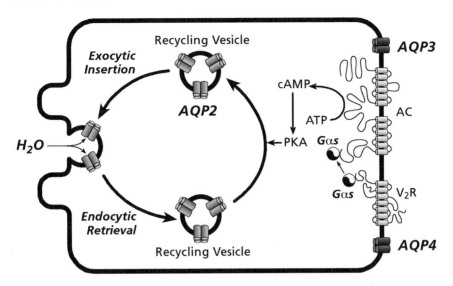

Fig. 1. Overview of vasopressin controlled AQP2 membrane shuttling in AQP2-containing collecting duct cell. Vasopressin binding to the G-protein-linked V2-receptor stimulates adenylyl cyclase leading to elevated cAMP levels and activation of protein kinase A. AQP2 is subsequently translocated to the apical plasma membrane.

to small intracellular vesicles further underscoring the role of AQP2 trafficking in the regulation of collecting duct water permeability.

PKA-induced phosphorylation of AQP2 in *Xenopus* oocytes was associated with only a very small (approximately 30%) increase in water permeability (Kuwahara et al. 1995) and consistent with this it was found that treatment with protein kinase A caused no change in the water permeability of AQP2 bearing vesicles harvested from kidney inner medulla (Lande et al. 1995). Thus, the major changes in the subcellular distribution of AQP2 in response to vasopressin or vasopressin-receptor antagonist treatment strongly support the view that collecting duct water permeability, and hence body water balance, is acutely regulated primarily by vasopressin-regulated trafficking of AQP2.

1.2 Role of AQP2-phosphorylation in vasopressin stimulated AQP2 recruitment to the plasma membrane

The signal transduction pathways leading to AQP2-recruitement to the plasma membrane have been described thoroughly in previous reviews (e.g. Knepper et al. 1994). Cyclic AMP (cAMP) levels in collecting duct principal cells are increased by binding of vasopressin to V2-receptors (Kurokawa and Massry 1973; Edwards et al. 1981). The synthesis of cAMP by adenylate cyclase is stimulated by a V2-receptor coupled heterotrimeric GTP-binding protein, Gs. Gs interconverts between an inactive GDP-bound form and an active GTP-bound form, and the vasopressin-V2-receptor complex catalyses the exchange of GTP for bound GDP on the α-subunit of Gs. This causes release of the α-subunit, Gsα-GTP, which subsequently binds to adenylate cyclase, thereby increasing cAMP production. Protein kinase A (PKA) is a multimeric protein, which is activated by cAMP, and consists in its inactive state of two catalytic subunits and two regulatory subunits. When cAMP binds to the regulatory subunits, these dissociate from the catalytic subunits, resulting in activation of the kinase activity of the catalytic subunits.

Early studies demonstrated that inhibition of PKA activity with H89 inhibits the vasopressin-induced increase in water permeability in isolated perfused rabbit collecting ducts (Snyder et al. 1992). Potentially several proteins could be substrates for phosphorylation by PKA, and early studies also indicate that several proteins indeed are phosphorylated (Dousa et al. 1972; Gapstur et al. 1988) although their nature remains to be identified. Interestingly, it has been shown that AQP2 contains a consensus site for PKA phosphorylation (RRQS) in the cytoplasmic COOH terminus at serine 256 (Fushimi et al. 1993). Recent studies using ^{32}P labelling or using an antibody specific for phosphorylated AQP2 showed a very rapid phosphorylation of AQP2 (within one minute) in response to vasopressin-treatment of slices of the kidney papilla (Nishimoto et al. 1999), consistent with the time course of vasopressin-stimulated water permeability of kidney collecting ducts (Wall et al. 1992). It was also demonstrated that vasopressin or forskolin treatment failed to induce translocation of AQP2 in AQP2-transfected LLC-PK1 cells when serine 256 was substituted by alanine (S256A) in contrast to a significant regulated trafficking of wild type AQP2 (Katsura et al. 1997). A parallel study also demonstrated the lack of cAMP-mediated exocytosis of mutated (S256A) AQP2 transfected into LLC-PK1 cells (Fushimi et al. 1997). Thus, these studies indicated a specific role of PKA-induced phosphorylation of AQP2 for regulated trafficking (Fig. 2).

In normal rats phosphorylated AQP2, identified by antibodies specifically labelling AQP2 that is phosphorylated at the PKA consensus site (S256), is present in both intracellular vesicles and in apical plasma membranes, whereas, in Brattleboro rats phosphorylated AQP2 is mainly located in intracellular vesicles as shown by immunocytochemistry and immunoblotting using membrane fractionation (Christensen et al. 2000). Surprisingly, there was no significant increase in the total amount of phosphorylated AQP2 following vasopressin treatment in normal rats (Christensen et al. 2000; Kamsteeg et al. 2000). DDAVP treatment of Brattleboro rats, however, caused a marked redistribution of phosphorylated AQP2 to the apical plasma membrane which is in agreement with an important role of PKA phosphorylation in this trafficking (Christensen et al. 2000). Conversely, treatment with V2-receptor antagonist induced a marked decrease in abundance of phosphorylated AQP2 likely to be due to reduced PKA activity and/or increased dephosphorylation of AQP2, for example, by increased phosphatase activity. It has now been demonstrated that three or four out of the four subunits of AQP2 within AQP2 homotetramers are required to be phosphorylated in order to be subjected to trafficking to the plasma membrane (Kamsteeg et al. 2000) adding yet another layer of complexity to the activating mechanisms.

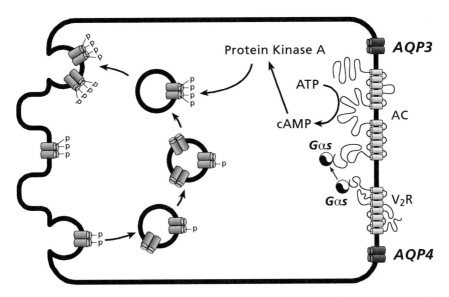

Fig. 2. Role of AQP2-phosphorylation in AQP2 recruitment to the plasma-membrane. Protein kinase A phosphorylates AQP2-monomers and phosphorylation of at least three of four AQP2 monomers in an AQP2-tetramer is associated with translocation to the plasma membrane. It is currently unknown if dephosphorylation of AQP2 is necessary for endocytosis of AQP2.

As previously reviewed (Nielsen et al. 2002; Brown 2003), blunting of the vasopressin response is seen associated with V2-receptor downregulation and desensitization, in part due to a decreased number of receptors at the cell surface. Other mechanisms also result in downregulating the vasopressin response including increased activity of phosphodiesterases (Dousa 1994), or inhibition of the vasopressin response by PGE2 (Stokes 1985), dopamine (Muto et al. 1985; Li and Schafer 1998), adenosine receptor stimulation (Edwards and Spielman 1994), adrenergic agonists (Hawk et al. 1993; Rouch and Kudo 1996), endothelin-1 (Oishi et al. 1991; Nadler et al. 1992a), and luminal Ca^{2+} (Sands et al. 1997). Direct demonstration of decreased density of apical plasma membrane labelling of AQP2 in response to washout of vasopressin from the peritubular bath of isolated perfused IMCDs provided direct evidence that endocytosis (removal) of AQP2 plays a fundamental role. Further studies have also demonstrated an increased internalization of AQP2 in response to PGE2 stimulation (in the presence of forskolin) (Zelenina et al. 2000). This indicates that the previously observed reduction in water permeability upon treatment with PGE2 in the presence of vasopressin (Stokes 1985; Nadler et al. 1992b) is due to AQP2 internalization. Also recently, this has been addressed in a study by Nejsum et al. (2003) demonstrating that the S256D mutant mimicking AQP2 phosphorylated in the Ser256 PKA phosphorylation consensus site, is also internalized in response to PGE2 or dopamine in the presence of forskolin. This demonstrates that AQP2 is endocytosed due to PGE2 or dopamine and importantly shows that this can occur in the absence of AQP2 dephosphorylation.

1.3 Constitutive AQP2-recycling and turnover rate of AQP2 in the plasma membrane

To understand the vasopressin-stimulated trafficking of AQP2, it is of importance to determine if AQP2 is recycled between the plasma membrane and intracellular compartments, or if AQP2-molecules after being retrieved from the plasma membrane are subjected to degradation. Furthermore, it is of interest to establish if vasopressin modulates constitutive AQP2 recycling or if AQP2 trafficking is entirely controlled by vasopressin. Finally, the turnover rate of AQP2 in the membrane is an important determinant of which mechanisms may control the abundance of AQP2 in plasma membrane.

In spite of inhibition of *de novo* protein synthesis using cycloheximide in LLC-PK1 cell cultures stably transfected with c-myc-tagged AQP2, repeated insertions of AQP2 into the plasma membrane occur in response to vasopressin and forskolin treatment and withdrawal of AQP2 from the membrane occurs after one hour washout of stimulating agents (Katsura et al. 1996). This provides evidence that AQP2 at least in transfected cells are able to undergo recycling between the plasma membrane and intracellular vesicles in response to forskolin stimulation and forskolin washout, respectively. In the same cell culture and in the presence of cycloheximide, cooling to 20°C or bafilomycin A1 treatment resulted in accumulation of AQP2 in a perinuclear region. Subsequent reheating lead to a translocation of AQP2 to plasma membrane domains, providing further evidence for AQP2 recycling (Gustafson et al. 2000). These effects were evident after cooling for two hours and reheating for 30 minutes. Inhibiting endocytosis, by depleting the membrane of cholesterol using methyl-β-cyclodextrin, resulted in accumulation of c-myc tagged AQP2 in the plasma membrane domain within 15 minutes, indicating a rapid turnover of AQP2 in the plasma-membrane (Lu et al. 2004). Methyl-β-cyclodextrin treatment even lead to the accumulation of AQP2 S256A mutant molecules (Lu et al. 2004), which in other cell cultures have been found not to appear in the plasma membrane (Katsura et al. 1997). This together with the previous evidence by Valenti and co-workers showing that okadaic acid treatment induced AQP2 plasma membrane insertion in the presence of the protein kinase A blocker H-89 (Valenti et al. 2000) provides evidence for a phosphorylation-independent plasma membrane accumulation of AQP2, and thus, the existence of constitutive recycling/trafficking of AQP2 as well as S256A-AQP2.

This may actually not be intuitively difficult to envision, since it is well established that there is a constitutive membrane turnover in most if not all cells. Moreover, it is well established that vasopressin stimulation as such promotes membrane trafficking including endocytosis. This has been shown by peroxidase or flourescein-dextran uptake in kidney, isolated tubules or other preparations in response to vasopressin treatment (Brown and Orci 1983; Verkman et al. 1988; Nielsen et al. 1993b). Thus, disturbance of normal membrane turnover or endocytosis is likely to interfere very dramatically with the accumulation of membrane proteins in a somewhat rather unspecific way. An example of this is the finding of increased c-myc-AQP2 abundance in the plasma-membrane domain 16 hours after transiently transfecting LLC-PK1 cells with dominant-negative dynamin (Sun et al. 2002). No changes were seen after eight hours, and the time course of the inhibition probably reflects the time taken for the mutant to replace wild type dynamin (Lu et al. 2004). Thus, these experiments demonstrate that dynamin plays a role in the endocytic pathway of AQP2, but do not provide information on AQP2 turnover rate in the plasma membrane.

1.4 Concluding remarks on regulation mechanisms in vasopressin controlled AQP2-trafficking

To reconcile all data it would be conceivable that there normally is a constitutive membrane turnover including exocytosis of AQP2 in both phosphorylated and unphosphorylated states (0-4 phosphorylated monomers in the AQP2 homotetramers). However, in the absence of significant phosphorylation (due to absence of vasopressin stimulation) AQP2 tetramers with only 0-2 phosphorylated monomers may be efficiently retrieved from the plasma membrane. This would be consistent with the absence or very low levels of AQP2 in the apical plasma membrane seen in the absence of forskolin/vasopressin/dDAVP treatment of transfected cells, isolated perfused IMCDs, and medullary collecting duct principal cells in kidneys of vasopressin-deficient Brattleboro rats. After stimulation with vasopressin, AQP2 (predominantly in a phosphorylated state) may be subjected to increased exocytosis. Alternatively, Brown and co-workers have proposed, that the constitutive recycling (i.e. non stimulated) may be sufficient to drive sufficient AQP2 to the plasma membrane and that the vasopressin induced accumulation in the plasma membrane could potentially be due to inhibition of AQP2 endocytosis after vasopressin stimulation (Lu et al. 2004). Against this hypothesis, speak the results of a recent study of step-changes in membrane capacitance in primary cultured IMCD-cells, which showed that vasopressin and cAMP generate an increase in the number of exocytotic step-changes in capacitance, which indicates increased exocytosis (Lorenz et al. 2003). Still, however, both a stimulation of AQP2 exotycosis and inhibition of AQP2 endocytosis may be involved in the vasopressin response. The relative roles of changes in these processes still remain to be established, and thus, additional studies are needed to establish whether vasopressin (*in vivo*) stimulate exocytosis or inhibit endocytosis of AQP2. Furthermore, it is at present still unknown if AQP2 is constitutively recycled in the collecting duct principal cells *in vivo*.

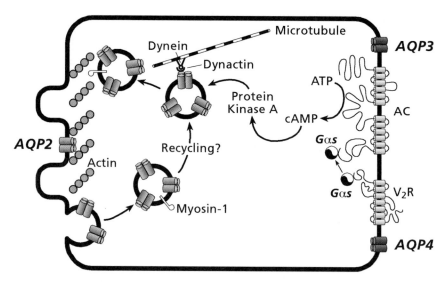

Fig. 3. Overview of cytoskeletal elements, which may be involved in AQP2-trafficking. AQP2 containing vesicles may be transported along microtubules by dynein/dynactin. The cortical actin web may act as a barrier to fusion with the plasma-membrane.

2 AQP2 trafficking and the cytoskeleton

The cytoskeleton has for a long time been known to be involved in the trafficking of AQP2 in kidney collecting duct. Within the last years, a number of studies have provided detailed knowledge of the signalling mechanisms controlling the cytoskeletal events and the mechanisms underlying the interactions between vasopressin and other hormonal stimulations (Fig. 3).

2.1 Modulation of the actin cytoskeleton

Arginine vasopressin and the cell permeant cAMP-analogue 8-Br-cAMP are well known to depolymerize the actin cytoskeleton in rat inner medullary collecting ducts (Simon et al. 1993). This and earlier observations in toad urinary bladder (Hays et al. 1987; Wade and Kachadorian 1988; Ding et al. 1991) were interpreted in accordance with the "barrier-theory", which states that the apical actin network "holds the aggregophores in place in the unstimulated cell and acts as barrier to fusion. The network would then break down in the presence of AVP, permitting aggregophores to move and fuse the apical plasma membrane." (Simon et al. 1993). These results also lead to the formulation of the "two vesicle pool" hypothesis, which states, that the AQP2 containing vesicles are divided into a pool,

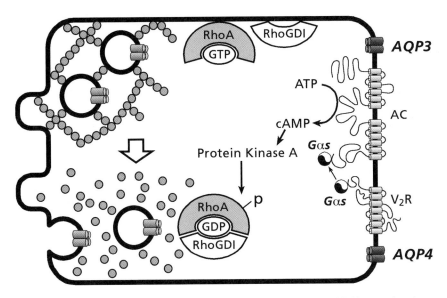

Fig. 4. Changes in the actin cytoskeleton associated with AQP2-trafficking to the plasma membrane. Inactivation of RhoA by phosphorylation and increased formation of RhoA-RhoGDI complexes seem to control the dissociation of actin fibres seen after vasopressin stimulation. See text for further details.

which can be released immediately and a pool which has longer response time, in analogy to the model of neurotransmitter release in chemical synapses (Hays et al. 1994). Within recent years, a series of studies of the connection between vasopressin induced trafficking of AQP2 and the observed changes in the actin cytoskeleton have provided valuable information concerning the role of Rho as a major regulator of these changes.

2.1.1 Rho inactivation links increased cAMP to actin depolymerisation

In primary cultures of IMCD-cells, inactivation of Rho by Clostridium difficile toxin B or Clostridium limosum C3 toxin and treatment with the Rho-kinase inhibitor Y-27632 all induced depolymerisation of actin and AQP2 translocation to the basolateral membrane in the absence of vasopressin (Klussmann et al. 2001b). Moreover, transfection of these cells with a constitutively active RhoA mutant induced actin polymerisation and inhibited AQP2 translocation to the plasma-membrane in response to Bt$_2$cAMP treatment. These results indicate that active Rho may act as a tonic inhibitor of AQP2 trafficking through its induction of actin polymerization (Fig. 4). Similar results were obtained in AQP2 transfected CD8 cells, which are RC.SV3 rabbit cortical collecting duct cells stably transfected with rat AQP2 (Valenti et al. 1998; Tamma et al. 2001). Forskolin treatment of AQP2-transfected CD8 cells resulted in a decrease in the amount of GTP bound

RhoA, increased an inactivating Ser-phosphorylation of Rho, probably catalyzed by PKA, and evoked inactivation of Rho by release of Rho-GDP dissociation inhibitor (Rho-GDI) from membranes to form increased amounts of soluble RhoA-RhoGDI complexes (Tamma et al. 2003a). This link between Rho and PKA signalling pathways may also involve the A Kinase Anchoring Protein Ht31, which co-immunoprecipitated with RhoA from IMCD cell primary cultures (Klussmann et al. 2001a).

Two conclusions emerge from these cell culture studies: 1) Rho-induced actin polymerization inhibits AQP2 accumulation in the plasma-membrane supporting the barrier-hypothesis. 2) Inactivation of Rho could be the link between increased cAMP-levels and actin depolymerisation.

2.1.2 PGE2 may exert its effect on AQP2 trafficking through Rho activation

As described above (section 1.2), PGE2 has long been known to antagonize the effect of vasopressin on collecting duct water permeability, but this effect seems to depend on prior vasopressin stimulation (see Klussman et al. 2000 for review). Inhibition of prostaglandin synthesis in toad bladders attenuated vasopressin induced actin depolymerisation (Ding et al. 1991), indicating that PGE2 is responsible for tonic stabilization of the apical actin web in this tissue. Recently, the PGE2 analogue sulprostone was shown to activate Rho in rat IMCD primary cultures, an effect probably mediated through the EP3-receptor since blocking the EP1-receptor did not affect Rho activation (Tamma et al. 2003b). Like PGE2, sulprostone had no effect on AQP2 localization in the absence of vasopressin stimulation, whereas, in the presence of vasopressin, sulprostone induced internalization of AQP2, relative to the response seen with vasopressin. The effect of sulprostone was not affected by high cAMP levels induced by forskolin or Bt_2cAMP-incubation and was not associated with changes in Ca^{2+} or IP3. Thus, although sulprostone induced a 18% reduction in adenylyl cyclase activity, the effect of sulprostone was proposed to be mediated by G proteins $G_{12/13}$ (Tamma et al. 2003b) and not through inhibition of adenylyl cyclase as normally ascribed to EP3-receptor activation (Breyer and Breyer 2001). The conclusion of these studies is that vasopressin and sulprostone exerts opposing effects on RhoA and actin polymerization paralleling their effect on AQP2 localization in primary cultures of IMCD cells.

2.1.3 PGE2 induce AQP2 internalization independent of AQP2-phosphorylation

As in the above described cell culture experiments, studies using *in vitro* PGE2 treatment of inner medulla tissue slices after vasopressin stimulation resulted in AQP2 internalization. However, no significant reduction in AQP2 Ser256 phosphorylation was observed. PGE2 had no effect on AQP2 localization (determined by cell fractionation) without pretreatment with vasopressin nor did PGE2 treatment prior to vasopressin stimulation reduce the vasopressin response (Zelenina et al. 2000). These results may be interpreted as PGE2 stabilizing the apical actin web and thus inhibiting AQP2 translocation to the plasma membrane. However, as PGE2 pretreatment did not inhibit vasopressin induced recruitment to the membrane, it was concluded, that PGE2 more likely exerts its effect by stimulating the endocytotic pathway (Zelenina et al. 2000). Recently this was addressed more directly using MDCK-C7 cells, transiently transfected with AQP2 wild type and various mutants including the S256D mutant mimicking phosphorylated AQP2 (Nejsum et al. 2003). This study confirmed the presence of phosphorylation independent PGE2-induced internalization of AQP2. However, in contrast to the studies in IMCD primary cultures, PGE2 had a much stronger effect on AQP2S256D internalization after pretreatment with forskolin, indicating that the effect of PGE2 depends on cAMP. Moreover, inhibition of PKA using H89 resulted in internalization of the AQP2 S256D mutant (Nejsum et al. 2003), further supporting the finding of phosphorylation independent internalization in kidney slices (Zelenina et al. 2000). Importantly, this finding also supports the view that PKA phosphorylation of other proteins than AQP2, for example, RhoA (Tamma et al. 2003a) is involved in PKA mediated AQP2 recruitment to the plasma membrane.

2.1.4 Concluding remarks and future perspectives on the role of the actin cytoskeleton

The inter-relationships between the effects of PGE2 and vasopressin and the signalling and cellular mechanisms involved in this need further studies. Especially the questions regarding how PGE2 in a cAMP dependent fashion is able to induce AQP2 internalization and to which extent PGE2 exerts its effect on the exocytotic or endocytic pathway still need further studies. It is likely that Rho and the actin network may play an important role in this. The role of Rho in controlling actin network remodelling accompanying vasopressin stimulation has been clearly demonstrated. It also seems clear that, at least in cell culture, actin polymerization inhibits and actin depolymerization promotes AQP2 accumulation in the plasma membrane, independently of vasopressin stimulation. However, it is unclear if the inhibitory effect on AQP2 membrane abundance due to reduced exocytosis or increased endocytosis of AQP2. In mammalian cells, two proteins have been identified, cortactin and Abp1, which may link endocytosis and actin filament formation. These proteins are able to interact with dynamin from the endocytotic machinery and the actin nucleator Arp2/3. They are, moreover, able to activate the actin nucleating activity of Arp2/3 (Schafer 2002). It would be interesting to know, if the effects of modulating actin dynamics on AQP2 localization are mediated through modulation of AQP2 endocytosis rate.

2.2 Microtubules and AQP2 vesicle transport

Disruption of microtubules inhibits the action of vasopressin (Sabolic et al. 1995). The role of microtubules in the cellular events following vasopressin stimulation is unclear, but some evidence exists, that AQP2 containing vesicles are associated with microtubule associated proteins and it has been hypothesized, that AQP2 vesicles may be transported along microtubules on their way to the apical plasma membrane (Kachadorian et al. 1979). The microtubule based motor protein dynein and the associated protein Arp1, which is part of the protein complex dynactin, was found by immunoblotting to be among the proteins associated with AQP2 immunoisolated vesicles from rat inner medulla. Immuno electron microscopy further supported the presence of both AQP2 and dynein on the same vesicles (Marples et al. 1998). It is at present unknown if AQP2 transport along microtubules is regulated as part of the response to vasopressin. However, studies performed in the urinary bladder of amphibians have provided strong evidence for a role of microtubules in vasopressin/vasotocin regulation of the osmotic water permeability (Pearl and Taylor 1985). Further studies are warranted to define the role of microtubules and associated proteins in vasopressin regulation of AQP2 trafficking.

3 Vasopressin induced intracellular Ca^{2+}-signalling

The intracellular Ca^{2+} concentration has been shown to increase upon stimulation of isolated and perfused rat inner medullary collecting ducts with 10^{-8} M vasopressin or dDAVP, i.e. ten times the concentration necessary to elicit maximum cAMP response. Treatment with 10^{-6} M 8-Br-cAMP did not elicit a calcium response, but rather lowered the basal intracellular calcium level (Star et al. 1988). These observations have been followed by a number of studies of the role of the calcium-signal in isolated IMCD's and in cell cultures.

3.1 Studies in isolated perfused IMCD's indicate a role of intracellular Ca^{2+} in AQP2 trafficking

In isolated perfused IMCD's the calcium signal was not seen to be associated with an increase in IP3 or activation of PKC (Chou et al. 1998). A few years ago it was shown, that also 10^{-10} M vasopressin and 10^{-4} M 8-4-chlorophenylthio-cAMP could induce marked increases in intracellular calcium concentration in isolated perfused IMCD. Blocking the rise in intracellular calcium with 50 μM BABTA-AM also lead to inhibition of the vasopressin induced increase in tubular water permeability (Chou et al. 2000). Blocking calmodulin with 25 μM W7 or 30 μM trifluoperazine also inhibited the effect of vasopressin on tubular water permeability and inhibited vasopressin induced AQP2 accumulation at the plasma membrane in IMCD cell primary cultures (Chou et al. 2000). Removing calcium from both bath and lumen in isolated perfused tubules did not affect the vasopressin induced calcium-signal, indicating that calcium was released from an intracellular source. This was further supported by the observations, that ryanodine inhibited the calcium signal in perfused IMCD's and inhibited AQP2 accumulation in the plasma membrane in IMCD cell primary cultures. In addition, RyR1 ryanodine receptors was localized to rat IMCD by immunohistochemistry (Chou et al. 2000).

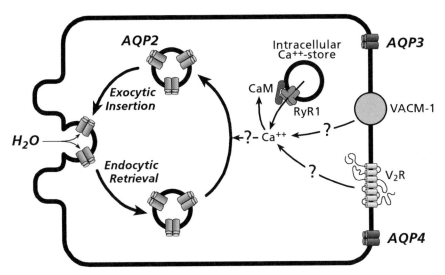

Fig. 5. Intracellular calcium signalling and AQP2-trafficking. Increases in intracellular Ca^{2+} concentration may arise from stimulation of the V2 receptor. The existence and potential role of other receptors and pathways, e.g VACM-1, in Ca^{2+} mobilization is still uncertain. The downstream targets of the calcium signal are unknown and conflicting data exist on the importance of a rise in intracellular Ca^{2+} for the hydroosmotic response to vasopressin. See text for further details.

Ryanodine receptors are generally believed to mediate a positive feedback and release Ca^{2+} from intracellular stores in response to an initial rise in intracellular calcium (Fig. 5). Thus, it is likely that another mechanism is responsible for the initialization of the rise in intracellular Ca^{2+}.

To further investigate the potential role of calcium in regulation of AQP2, Yip demonstrated that treatment with 10^{-10} M vasopressin or dDAVP elicited oscillations in intracellular calcium concentration in isolated perfused rat IMCD (Yip 2002). The calcium oscillations differed from cell to cell in frequency as well as in amplitude. In the absence of extracellular calcium no oscillations were observed, but 10^{-10} M vasopressin still induced an increase in intracellular calcium concentration (Yip 2002). In agreement with the results described above from Chou et al., 50 µM BABTA AM and 10^{-4} M ryanodine inhibited the vasopressin induced calcium increase. Using FM1-43, a membrane specific fluorescence probe, in the luminal solution, the apical plasma membrane area of isolated perfused IMCD's could be monitored at intervals of 20 s. These studies showed an increase in apical plasma membrane area following vasopressin stimulation and that preincubation with 50 µM BABTA-AM inhibited this increase (Yip 2002).

In conclusion, the results on isolated perfused IMCD's support that an intracellular increase in calcium concentration is an obligate component of the vasopressin response leading to increased AQP2 abundance in the apical plasma membrane.

3.2 Ca²⁺ is not important for AQP2 traffic in primary cultures of IMCD cells

In primary cultured IMCD cells, vasopressin stimulation induced an increase in intracellular calcium concentration to 350-400 nM (Lorenz et al. 2003). If the intracellular calcium concentration was buffered to 50 nM by applying ionomycin and EGTA/CaCl$_2$, AQP2 still accumulated at the plasma membrane in response to vasopressin. However, when the intracellular Ca^{2+}-concentration was buffered to 25 nM, vasopressin could not induce AQP2 accumulation at the plasma-membrane. Thus, very low levels of Ca^{2+} seem to be required for proper AQP2 recruitment to the plasma membrane, but the observed calcium peak seems not to be required (Lorenz et al. 2003). Elevating intracellular calcium to $\gg$ 1 mM did not induce accumulation of AQP2 in the plasma-membrane even after vasopressin stimulation (Lorenz et al. 2003), excluding an independent role of calcium in mediating AQP2-trafficking. Taken together, these results indicate, that a rise in intracellular Ca^{2+} is not necessary for the vasopressin induced accumulation of AQP2 in the plasma-membrane in this cell culture. To further support this conclusion, membrane capacitance measurements were performed to evaluate if exocytosis was affected by buffering the Ca^{2+}-level to 50 nM. These showed that with weak Ca^{2+}-buffering 300 mM cAMP alone elicited a 10% increase in membrane capacitance, which was comparable to the effect of 1 nM AVP. This effect was not affected buffering the free Ca^{2+}-concentration to 50 nM (Lorenz et al. 2003). The osmotic water permeability through primary cultured IMCD-cells was calculated from measurements of potassium concentration in the fluid adjacent to the basolateral side of the cell layer while water was driven trough the cell-layer by an osmotic gradient. 50 µM BABTA-AM did not inhibit the forskolin induced increase in osmotic water permeability, and thus, it was concluded that Ca^{2+} is not required for trafficking of AQP2 to the plasma-membrane (Lorenz et al. 2003).

3.3 Comparison of the results from IMCD's and primary cultured IMCD-cells

It is clear, that a discrepancy exists between the results from primary cultured IMCD cells described above and the findings in isolated perfused IMCD's described in the previous section with regard to the importance of calcium in vasopressin induced AQP2 trafficking. This discrepancy is clearly illustrated by the fact that 50 mM BABTA-AM inhibited the vasopressin effect on osmotic water permeability in isolated perfused IMCD's (Chou et al. 2000; Yip 2002), but had no effect in primary cultured IMCD cells (Lorenz et al. 2003). An explanation could be, that the Ca^{2+} levels in tubules is forced below resting levels by 50 µM BABTA-AM, whereas, for some unknown reason this does not happen in the cells. However, the discrepancy may also reflect other differences between tubules and the cells. For example, the osmotic water permeability differs between the primary cultured IMCD-cells and isolated rat IMCD. The osmotic water permeability was measured to only 13 µm/s in resting cells (Lorenz et al. 2003), whereas, the values reported from resting isolated rat IMCD generally are much higher, e.g 63 μm/s (Star et al. 1988) and 100 - 300 μm/s (Chou et al. 2000). The unstimulated osmotic water permeability of rat IMCD differs dramatically between the initial (21.1 μm/s) and the terminal IMCD (333.3 μm/s) (Lankford et al. 1991). Thus, with respect to basal osmotic water permeability, the primary cultured IMCD-cells resemble the initial IMCD more than the terminal IMCD. However, initial rat IMCD stimulated with 1 nM vasopressin exhibited a water permeability of 203 μm/s (Lankford et al. 1991), whereas, the water permeability of primary cultured IMCD cells only increased to 26 μm/s upon stimulation with 50 μM forskolin (Lorenz et al. 2003). In conclusion, the primary cultured IMCD cells seem to have a reduced capacity for AQP2-recruitment to the plasma membrane. Hopefully future studies can clarify the reasons for these discrepancies, which may be related to the expression levels of vasopressin receptors, AQP2 or other elements in the AQP2 trafficking system. As pointed out above, further studies are also needed on the initiation of the Ca^{2+} signal and on downstream targets of the Ca^{2+} signal.

3.4 Which vasopressin sensitive receptors are involved in calcium signalling in AQP2 containing collecting duct cells?

As reviewed in detail by Knepper et al. (1994), the studies of the role calcium in vasopressin induced AQP2-trafficking have been complicated by uncertainties about the receptors initiating the signal. In general, signals elicited by stimulation by dDAVP, have been considered to be the result of V2-receptor stimulation, since dDAVP has been considered a V2-receptor specific agonist when compared to other classical vasopressin receptors. This notion is based on the fact that dDAVP does not stimulate the V1a receptor, which is a linked to phospholipase C mediated calcium signalling. However, dDAVP has been shown also to agonize the V1b receptor (Saito et al. 1997a), and this has been proposed to explain the reports (Ecelbarger et al. 1996) of dual signalling (calcium and cAMP) pathways for the V2-receptor in IMCD (Saito et al. 2000). Furthermore, a vasopressin activated calcium mobilizing protein (VACM-1) has been cloned from rabbit kidney medulla and immunohistochemistry indicates that the protein is present in rabbit inner medullary collecting ducts (Burnatowska-Hledin et al. 1995). VACM-1 has since been shown to belong to the cullin family and is homologeous to human cul-5 (Byrd et al. 1997). RT-PCR has revealed that the protein is expressed in a wide range of organs in rat (Ceremuga et al. 2001). Interestingly, cul-5 mRNA levels in rat cerebral cortex, hypothalamus and kidney were increased following 48 hours of water deprivation, indicating a role of cul-5 in cellular response to hyperosmolality (Ceremuga et al. 2003). Vasopressin stimulation of VACM-1 transfected COS-1 cells resulted in elevated intracellular calcium concentrations (Burnatowska-Hledin et al. 1995). Although intracellular calcium has not been measured upon binding of dDAVP to this protein, competition binding studies showed that AVP, dDAVP and MeAVP competed to the same extent with binding of ^{125}I-AVP to this receptor, suggesting that dDAVP binds to the protein

(Burnatowska-Hledin et al. 1995). Additional studies in COS-cells transfected with VACM-1 have revealed that these cells exhibit reduced basal cAMP levels and no response to forskolin. This effect seems to rely on inhibition of adenylyl cyclase activity by VACM-1 (Burnatowska-Hledin et al. 2000). It is still uncertain to what extent VACM-1 is present in AQP2 containing collecting duct cells and what role it may play in the control of AQP2-trafficking. To elucidate the potential role of this protein in vasopressin induced calcium signalling in AQP2 containing collecting duct cells, studies of the effect on intracellular calcium-levels of dDAVP binding to VACM-1 are needed and the localization of this receptor in collecting ducts need to be confirmed.

3.5 Myosins as targets for calcium signalling in IMCD?

Intracellular calcium concentrations control numerous cellular processes. With regard to AQP2-trafficking, the potential role of calcium in controlling myosin motor protein activity has received some attention. Previously, immuno electron microscopy showed Myosin 1 to be present on immunoisolated AQP2 containing vesicles (Marples ASN 1997). Recently, the Myosin Light Chain Kinase (MLCK) pathway, which through calmodulin mediated calcium activation of MLCK leads to phosphorylation of Myosin Regulatory Light Chain and non-muscle myosin 2 motor activity, has attracted attention. Studies in isolated perfused rat IMCD's showed the MLCK-inhibitors ML-7 (25 µM) and ML-9 (50 µM) to reduce the vasopressin induced increased in tubule water permeability (Chou et al. 2001). These results indicate that MLCK may be a downstream target for the vasopressin induced Ca^{2+} signal. However, it may not be completely ruled out that the inhibitor concentrations might potentially also inhibit PKA (Chembank, http://chembank.med.harvard.edu), and thus, affect AQP2 trafficking through the cAMP-pathway. Recently, the calcium activated non-muscle myosin 2b and the unconventional myosin Va and Vb isoforms have been localized to AQP2-containing collecting duct cells by immunohistochemistry and immunoblotting (Frische et al. 2003). Further studies are warranted to further investigate the potential role of myosins in AQP2-trafficking.

4 Evidence for role of vesicle targeting receptors in AQP2 trafficking

Vasopressin-stimulated apical membrane accumulation of AQP2 vesicles involves several steps including: 1) translocation of vesicles from a diffuse distribution throughout the cell to the apical region of the cell, 2) translocation of AQP2 across the apical part of the cell composed of a dense filamentous actin network, 3) priming of vesicles for docking, 4) docking of vesicles, and 5) fusion of vesicles with the apical plasma membrane. Theoretically, each of these steps could be target for regulation by vasopressin. In particular, the final three steps in AQP2 vesicle trafficking – vesicle priming, docking with the plasma membrane, and fusion with the plasma membrane – are believed to involve vesicle targeting receptors, the so called "SNARE" proteins. These proteins have been identified chiefly in the context of regulated exocytosis of synaptic vesicles (Sollner et al. 1993; Bajjalieh and Scheller 1995). The SNARE proteins have been proposed to mediate specific interaction between given vesicle and its target membrane with which it is destined to fuse. One class of targeting receptors in the vesicles (v-SNARES) has been called "VAMPs" (or "synaptobrevins"). Another class of targeting receptor found predominantly in the target membrane (t-SNARES) is the syntaxins. The third family of SNARE proteins is the SNAP-25-like proteins. These were initially thought to function as t-SNAREs, but are now recognized to reside in translocating vesicles as well as the target membrane.

Members of all three of these families have been found in collecting duct cells with a subcellular localization consistent with a role in AQP2 vesicle exocytosis. First, VAMP-2 has been shown to be present in AQP2 vesicles and to be virtually absent from the apical membrane of principal cells of the inner medullary collecting ducts (Nielsen et al. 1995b). Similar evidence for an intracellular localization of VAMP-2 in collecting duct cells has been provided in other studies (Jo et al. 1995; Liebenhoff and Rosenthal 1995; Mandon et al. 1996). Second, syntaxin-4 has been localized to the apical plasma membrane of collecting duct principal cells (Mandon et al. 1996). Finally, SNAP-23 has been found to be present in both the apical plasma membrane of principal cells and in AQP2 vesicles (Inoue et al. 1998). These three proteins may potentially form a complex with the N-ethylmaleimide-sensitive factor (NSF) (Fig. 6).

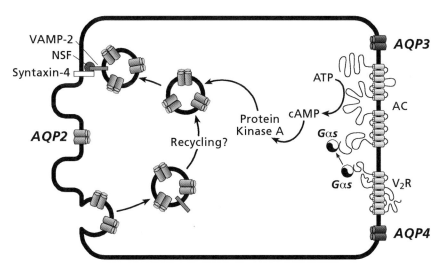

Fig. 6. Vesicle targeting receptors and AQP2-trafficking. A number of vesicle targeting receptors, for example, SNARE-proteins have been localized to the AQP2 containing collecting duct cells and cultured cells. The exact role of these remains to be established.

If the SNARE proteins are involved in the vasopressin-induced trafficking of AQP2 to the apical plasma membrane, the regulatory mechanism might involve selective phosphorylation of one of the SNARE proteins or an ancillary protein that binds to them, possibly via protein kinase A or calmodulin-dependent kinase II. Although the SNARE proteins are recognized as potential targets for phosphorylation (Shimazaki et al. 1996; Foster et al. 1998; Risinger and Bennett 1999), there is presently no evidence for the phosphorylation in the collecting duct. The establishment of a role for these SNARE proteins in regulated trafficking of AQP2 will most likely depend on the preparation of targeted knockouts for each of these genes in the collecting duct.

5 Concluding remarks

The AQP2-trafficking system in inner medullary collecting duct exemplifies how shuttling of conceptually simple "water-pores" regulates on of the most fundamental physiological parameters of the body, the water balance. The general mechanism of vasopressin controlled plasma membrane water permeability was known long before the identification of AQP2 as the water channelling protein. Now, as the basic mechanism is described, the details of the system and the exact mechanisms in the regulatory system have gained focus. These include:

a. the molecular basis of the effects of PGE2, dopamine and other signalling agents
b. the molecular events following vasopressin binding including the role and control of AQP2-phosphorylation, the role of cytoskeletal changes and the role of molecular motors
c. the relative importance of other receptors and intracellular messengers than V2 and cAMP.

Despite the large number of published studies of the AQP2-trafficking system still a number of important unanswered questions still exist. The physiological importance and pharmacological relevance of the system encourage continued research in this field.

References

Agre P, King LS, Yasui M, Guggino WB, Ottersen OP, Fujiyoshi Y, Engel A, Nielsen S (2002) Aquaporin water channels--from atomic structure to clinical medicine. J Physiol 542:3-16

Bajjalieh SM, Scheller RH (1995) The biochemistry of neurotransmitter secretion. J Biol Chem 270:1971-1974

Breyer MD, Breyer RM (2001) G protein-coupled prostanoid receptors and the kidney. Annu Rev Physiol 63:579-605

Brown D (2003) The ins and outs of aquaporin-2 trafficking. Am J Physiol Renal Physiol 284:F893-F901

Brown D, Orci L (1983) Vasopressin stimulates formation of coated pits in rat kidney collecting ducts. Nature 302:253-255

Burnatowska-Hledin M, Zhao P, Capps B, Poel A, Parmelee K, Mungall C, Sharangpani A, Listenberger L (2000) VACM-1, a cullin gene family member, regulates cellular signaling. Am J Physiol Cell Physiol 279:C266-C273

Burnatowska-Hledin MA, Spielman WS, Smith WL, Shi P, Meyer JM, Dewitt DL (1995) Expression cloning of an AVP-activated, calcium-mobilizing receptor from rabbit kidney medulla. Am J Physiol 268:F1198-F1210

Byrd PJ, Stankovic T, McConville CM, Smith AD, Cooper PR, Taylor AM (1997) Identification and analysis of expression of human VACM-1, a cullin gene family member located on chromosome 11q22-23. Genome Res 7:71-75

Ceremuga TE, Yao XL, McCabe JT (2001) Vasopressin-activated calcium-mobilizing (VACM-1) receptor mRNA is present in peripheral organs and the central nervous system of the laboratory rat. Endocr Res 27:433-445

Ceremuga TE, Yao XL, Xia Y, Mukherjee D, McCabe JT (2003) Osmotic stress increases cullin-5 (cul-5) mRNA in the rat cerebral cortex, hypothalamus and kidney. Neurosci Res 45:305-311

Chevalier J, Bourguet J, Hugon JS (1974) Membrane associated particles: distribution in frog urinary bladder epithelium at rest and after oxytocin treatment. Cell Tissue Res 152:129-140

Chou CL, de Lanerolle P, Knepper M (2001) Roles of actin, myosins and myosin light chain kinase in aquaporin-2 (AQP-2) trafficking (abstract). J Am Soc Nephrol 12:14A

Chou CL, Rapko SI, Knepper MA (1998) Phosphoinositide signaling in rat inner medullary collecting duct. Am J Physiol 274:F564-F572

Chou CL, Yip KP, Michea L, Kador K, Ferraris JD, Wade JB, Knepper MA (2000) Regulation of aquaporin-2 trafficking by vasopressin in the renal collecting duct. Roles of ryanodine-sensitive Ca2+ stores and calmodulin. J Biol Chem 275:36839-36846

Christensen BM, Marples D, Jensen UB, Frokiaer J, Sheikh-Hamad D, Knepper M, Nielsen S (1998) Acute effects of vasopressin V2-receptor antagonist on kidney AQP2 expression and subcellular distribution. Am J Physiol 275:F285-F297

Christensen BM, Zelenina M, Aperia A, Nielsen S (2000) Localization and regulation of PKA-phosphorylated AQP2 in response to V(2)-receptor agonist/antagonist treatment. Am J Physiol Renal Physiol 278:F29-F42

Ding GH, Franki N, Condeelis J, Hays RM (1991) Vasopressin depolymerizes F-actin in toad bladder epithelial cells. Am J Physiol 260:C9-C16

Dousa TP (1994) Cyclic-3',5'-nucleotide phosphodiesterases in the cyclic adenosine monophosphate (cAMP)-mediated actions of vasopressin. Semin Nephrol 14:333-340

Dousa TP, Sands H, Hechter O (1972) Cyclic AMP-dependent reversible phosphorylation of renal medullary plasma membrane protein. Endocrinology 91:757-763

Ecelbarger CA, Chou CL, Lolait SJ, Knepper MA, DiGiovanni SR (1996) Evidence for dual signaling pathways for V2 vasopressin receptor in rat inner medullary collecting duct. Am J Physiol 270:F623-F633

Edwards RM, Jackson BA, Dousa TP (1981) ADH-sensitive cAMP system in papillary collecting duct: effect of osmolality and PGE2. Am J Physiol 240:F311-F318

Edwards RM, Spielman WS (1994) Adenosine A1 receptor-mediated inhibition of vasopressin action in inner medullary collecting duct. Am J Physiol 266:F791-F796

Flamion B, Spring KR (1990) Water permeability of apical and basolateral cell membranes of rat inner medullary collecting duct. Am J Physiol 259:F986-F999

Foster LJ, Yeung B, Mohtashami M, Ross K, Trimble WS, Klip A (1998) Binary interactions of the SNARE proteins syntaxin-4, SNAP23, and VAMP-2 and their regulation by phosphorylation. Biochemistry 37:11089-11096

Frische S, Chou CL, Knepper M, Nielsen S (2003) Immunohistochemical localization of myosin IIb, Va and Vb in rat kidney collecting duct (abstract). J Am Soc Nephrol 14:309A

Fushimi K, Sasaki S, Marumo F (1997) Phosphorylation of serine 256 is required for cAMP-dependent regulatory exocytosis of the aquaporin-2 water channel. J Biol Chem 272:14800-14804

Fushimi K, Uchida S, Hara Y, Hirata Y, Marumo F, Sasaki S (1993) Cloning and expression of apical membrane water channel of rat kidney collecting tubule. Nature 361:549-552

Gapstur SM, Homma S, Dousa TP (1988) cAMP-binding proteins in medullary tubules from rat kidney: effect of ADH. Am J Physiol 255:F292-F300

Gustafson CE, Katsura T, McKee M, Bouley R, Casanova JE, Brown D (2000) Recycling of AQP2 occurs through a temperature- and bafilomycin-sensitive trans-Golgi-associated compartment. Am J Physiol Renal Physiol 278:F317-F326

Hawk CT, Kudo LH, Rouch AJ, Schafer JA (1993) Inhibition by epinephrine of AVP- and cAMP-stimulated Na+ and water transport in Dahl rat CCD. Am J Physiol 265:F449-F460

Hayashi M, Sasaki S, Tsuganezawa H, Monkawa T, Kitajima W, Konishi K, Fushimi K, Marumo F, Saruta T (1994) Expression and distribution of aquaporin of collecting duct are regulated by vasopressin V2 receptor in rat kidney. J Clin Invest 94:1778-1783

Hays RM, Ding GH, Franki N (1987) Morphological aspects of the action of ADH. Kidney Int Suppl 21:S51-S55

Hays RM, Franki N, Simon H, Gao Y (1994) Antidiuretic hormone and exocytosis: lessons from neurosecretion. Am J Physiol 267:C1507-C1524

Humbert F, Montesano R, Grosso A, de Sousa RC, Orci L (1977) Particle aggregates in plasma and intracellular membranes of toad bladder (granular cell). Experientia 33:1364-1367

Inoue T, Nielsen S, Mandon B, Terris J, Kishore BK, Knepper MA (1998) SNAP-23 in rat kidney: colocalization with aquaporin-2 in collecting duct vesicles. Am J Physiol 275:F752-F760

Jo I, Harris HW, Amendt-Raduege AM, Majewski RR, Hammond TG (1995) Rat kidney papilla contains abundant synaptobrevin protein that participates in the fusion of antidiuretic hormone-regulated water channel-containing endosomes in vitro. Proc Natl Acad Sci USA 92:1876-1880

Kachadorian WA, Ellis SJ, Muller J (1979) Possible roles for microtubules and microfilaments in ADH action on toad urinary bladder. Am J Physiol 236:F14-F20

Kachadorian WA, Levine SD, Wade JB, Di Scala VA, Hays RM (1977) Relationship of aggregated intramembranous particles to water permeability in vasopressin-treated toad urinary bladder. J Clin Invest 59:576-581

Kamsteeg EJ, Heijnen I, van Os CH, Deen PM (2000) The subcellular localization of an aquaporin-2 tetramer depends on the stoichiometry of phosphorylated and nonphosphorylated monomers. J Cell Biol 151:919-930

Katsura T, Ausiello DA, Brown D (1996) Direct demonstration of aquaporin-2 water channel recycling in stably transfected LLC-PK1 epithelial cells. Am J Physiol 270:F548-F553

Katsura T, Gustafson CE, Ausiello DA, Brown D (1997) Protein kinase A phosphorylation is involved in regulated exocytosis of aquaporin-2 in transfected LLC-PK1 cells. Am J Physiol 272:F817-F822

Klussmann E, Edemir B, Pepperle B, Tamma G, Henn V, Klauschenz E, Hundsrucker C, Maric K, Rosenthal W (2001a) Ht31: the first protein kinase A anchoring protein to integrate protein kinase A and Rho signaling. FEBS Lett 507:264-268

Klussmann E, Maric K, Rosenthal W (2000) The mechanisms of aquaporin control in the renal collecting duct. Rev Physiol Biochem Pharmacol 141:33-95

Klussmann E, Tamma G, Lorenz D, Wiesner B, Maric K, Hofmann F, Aktories K, Valenti G, Rosenthal W (2001b) An inhibitory role of Rho in the vasopressin-mediated translocation of aquaporin-2 into cell membranes of renal principal cells. J Biol Chem 276:20451-20457

Knepper MA, Nielsen S, Chou CL, DiGiovanni SR (1994) Mechanism of vasopressin action in the renal collecting duct. Semin Nephrol 14:302-321

Kurokawa K, Massry SG (1973) Interaction between catecholamines and vasopressin on renal medullary cyclic AMP of rat. Am J Physiol 225:825-829

Kuwahara M, Fushimi K, Terada Y, Bai L, Marumo F, Sasaki S (1995) cAMP-dependent phosphorylation stimulates water permeability of aquaporin-collecting duct water channel protein expressed in *Xenopus* oocytes. J Biol Chem 270:10384-10387

Lande MB, Donovan JM, Zeidel ML (1995) The relationship between membrane fluidity and permeabilities to water, solutes, ammonia, and protons. J Gen Physiol 106:67-84

Lankford SP, Chou CL, Terada Y, Wall SM, Wade JB, Knepper MA (1991) Regulation of collecting duct water permeability independent of cAMP-mediated AVP response. Am J Physiol 261:F554-F566

Li L, Schafer JA (1998) Dopamine inhibits vasopressin-dependent cAMP production in the rat cortical collecting duct. Am J Physiol 275:F62-F67

Liebenhoff U, Rosenthal W (1995) Identification of Rab3-, Rab5a- and synaptobrevin II-like proteins in a preparation of rat kidney vesicles containing the vasopressin-regulated water channel. FEBS Lett 365:209-213

Lorenz D, Krylov A, Hahm D, Hagen V, Rosenthal W, Pohl P, Maric K (2003) Cyclic AMP is sufficient for triggering the exocytic recruitment of aquaporin-2 in renal epithelial cells. EMBO Rep 4:88-93

Lu H, Sun TX, Bouley R, Blackburn K, McLaughlin M, Brown D (2004) Inhibition of endocytosis causes phosphorylation (S256)-independent plasma membrane accumulation of AQP2. Am J Physiol Renal Physiol 286:F233-F243

Mandon B, Chou CL, Nielsen S, Knepper MA (1996) Syntaxin-4 is localized to the apical plasma membrane of rat renal collecting duct cells: possible role in aquaporin-2 trafficking. J Clin Invest 98:906-913

Marples D, Knepper MA, Christensen EI, Nielsen S (1995) Redistribution of aquaporin-2 water channels induced by vasopressin in rat kidney inner medullary collecting duct. Am J Physiol 269:C655-C664

Marples D, Schroer TA, Ahrens N, Taylor A, Knepper MA, Nielsen S (1998) Dynein and dynactin colocalize with AQP2 water channels in intracellular vesicles from kidney collecting duct. Am J Physiol 274:F384-F394

Muto S, Tabei K, Asano Y, Imai M (1985) Dopaminergic inhibition of the action of vasopressin on the cortical collecting tubule. Eur J Pharmacol 114:393-397

Nadler SP, Zimpelmann JA, Hebert RL (1992a) Endothelin inhibits vasopressin-stimulated water permeability in rat terminal inner medullary collecting duct. J Clin Invest 90:1458-1466

Nadler SP, Zimpelmann JA, Hebert RL (1992b) PGE2 inhibits water permeability at a post-cAMP site in rat terminal inner medullary collecting duct. Am J Physiol 262:F229-F235

Nejsum LN, Zelenina M, Aperia A, Frokiaer J, Nielsen S (2003) Dopamine and prostaglandin stimulation and protein kinase A inhibition mediates AQP2 endocytosis in cell culture independent of AQP2 ser256 dephosphorylation (abstract). J Am Soc Nephrol 14:310A

Nielsen S, Chou CL, Marples D, Christensen EI, Kishore BK, Knepper MA (1995a) Vasopressin increases water permeability of kidney collecting duct by inducing translocation of aquaporin-CD water channels to plasma membrane. Proc Natl Acad Sci USA 92:1013-1017

Nielsen S, DiGiovanni SR, Christensen EI, Knepper MA, Harris HW (1993a) Cellular and subcellular immunolocalization of vasopressin-regulated water channel in rat kidney. Proc Natl Acad Sci USA 90:11663-11667

Nielsen S, Frokiaer J, Marples D, Kwon TH, Agre P, Knepper MA (2002) Aquaporins in the kidney: from molecules to medicine. Physiol Rev 82:205-244

Nielsen S, Marples D, Birn H, Mohtashami M, Dalby NO, Trimble M, Knepper M (1995b) Expression of VAMP-2-like protein in kidney collecting duct intracellular vesicles. Colocalization with Aquaporin-2 water channels. J Clin Invest 96:1834-1844

Nielsen S, Muller J, Knepper MA (1993b) Vasopressin- and cAMP-induced changes in ultrastructure of isolated perfused inner medullary collecting ducts. Am J Physiol 265:F225-F238

Nishimoto G, Zelenina M, Li D, Yasui M, Aperia A, Nielsen S, Nairn AC (1999) Arginine vasopressin stimulates phosphorylation of aquaporin-2 in rat renal tissue. Am J Physiol 276:F254-F259

Oishi R, Nonoguchi H, Tomita K, Marumo F (1991) Endothelin-1 inhibits AVP-stimulated osmotic water permeability in rat inner medullary collecting duct. Am J Physiol 261:F951-F956

Pearl M, Taylor A (1985) Role of the cytoskeleton in the control of transcellular water flow by vasopressin in amphibian urinary bladder. Biol Cell 55:163-172

Risinger C, Bennett MK (1999) Differential phosphorylation of syntaxin and synaptosome-associated protein of 25 kDa (SNAP-25) isoforms. J Neurochem 72:614-624

Rouch AJ, Kudo LH (1996) Alpha 2-adrenergic-mediated inhibition of water and urea permeability in the rat IMCD. Am J Physiol 271:F150-F157

Sabolic I, Katsura T, Verbavatz JM, Brown D (1995) The AQP2 water channel: effect of vasopressin treatment, microtubule disruption, and distribution in neonatal rats. J Membr Biol 143:165-175

Saito M, Tahara A, Sugimoto T (1997a) 1-desamino-8-D-arginine vasopressin (DDAVP) as an agonist on V1b vasopressin receptor. Biochem Pharmacol 53:1711-1717

Saito M, Tahara A, Sugimoto T, Abe K, Furuichi K (2000) Evidence that atypical vasopressin V(2) receptor in inner medulla of kidney is V(1B) receptor. Eur J Pharmacol 401:289-296

Saito T, Ishikawa SE, Sasaki S, Fujita N, Fushimi K, Okada K, Takeuchi K, Sakamoto A, Ookawara S, Kaneko T, Marumo F (1997b) Alteration in water channel AQP-2 by removal of AVP stimulation in collecting duct cells of dehydrated rats. Am J Physiol 272:F183-F191

Sands JM, Naruse M, Baum M, Jo I, Hebert SC, Brown EM, Harris HW (1997) Apical extracellular calcium/polyvalent cation-sensing receptor regulates vasopressin-elicited water permeability in rat kidney inner medullary collecting duct. J Clin Invest 99:1399-1405

Schafer DA (2002) Coupling actin dynamics and membrane dynamics during endocytosis. Curr Opin Cell Biol 14:76-81

Shimazaki Y, Nishiki T, Omori A, Sekiguchi M, Kamata Y, Kozaki S, Takahashi M (1996) Phosphorylation of 25-kDa synaptosome-associated protein. Possible involvement in protein kinase C-mediated regulation of neurotransmitter release. J Biol Chem 271:14548-14553

Simon H, Gao Y, Franki N, Hays RM (1993) Vasopressin depolymerizes apical F-actin in rat inner medullary collecting duct. Am J Physiol 265:C757-C762

Snyder HM, Noland TD, Breyer MD (1992) cAMP-dependent protein kinase mediates hydrosmotic effect of vasopressin in collecting duct. Am J Physiol 263:C147-C153

Sollner T, Whiteheart SW, Brunner M, Erdjument-Bromage H, Geromanos S, Tempst P, Rothman JE (1993) SNAP receptors implicated in vesicle targeting and fusion. Nature 362:318-324

Star RA, Nonoguchi H, Balaban R, Knepper MA (1988) Calcium and cyclic adenosine monophosphate as second messengers for vasopressin in the rat inner medullary collecting duct. J Clin Invest 81:1879-1888

Stokes JB 3rd (1985) Modulation of vasopressin-induced water permeability of the cortical collecting tubule by endogenous and exogenous prostaglandins. Miner Electrolyte Metab 11:240-248

Sun TX, Van Hoek A, Huang Y, Bouley R, McLaughlin M, Brown D (2002) Aquaporin-2 localization in clathrin-coated pits: inhibition of endocytosis by dominant-negative dynamin. Am J Physiol Renal Physiol 282:F998-F1011

Tamma G, Klussmann E, Maric K, Aktories K, Svelto M, Rosenthal W, Valenti G (2001) Rho inhibits cAMP-induced translocation of aquaporin-2 into the apical membrane of renal cells. Am J Physiol Renal Physiol 281:F1092-F1101

Tamma G, Klussmann E, Procino G, Svelto M, Rosenthal W, Valenti G (2003a) cAMP-induced AQP2 translocation is associated with RhoA inhibition through RhoA phosphorylation and interaction with RhoGDI. J Cell Sci 116:1519-1525

Tamma G, Wiesner B, Furkert J, Hahm D, Oksche A, Schaefer M, Valenti G, Rosenthal W, Klussmann E (2003b) The prostaglandin E2 analogue sulprostone antagonizes vasopressin-induced antidiuresis through activation of Rho. J Cell Sci 116:3285-3294

Valenti G, Procino G, Carmosino M, Frigeri A, Mannucci R, Nicoletti I, Svelto M (2000) The phosphatase inhibitor okadaic acid induces AQP2 translocation independently from AQP2 phosphorylation in renal collecting duct cells. J Cell Sci 113 (Pt 11):1985-1992

Valenti G, Procino G, Liebenhoff U, Frigeri A, Benedetti PA, Ahnert-Hilger G, Nurnberg B, Svelto M, Rosenthal W (1998) A heterotrimeric G protein of the Gi family is required for cAMP-triggered trafficking of aquaporin 2 in kidney epithelial cells. J Biol Chem 273:22627-22634

Verkman AS, Lencer WI, Brown D, Ausiello DA (1988) Endosomes from kidney collecting tubule cells contain the vasopressin-sensitive water channel. Nature 333:268-269

Wade JB, Kachadorian WA (1988) Cytochalasin B inhibition of toad bladder apical membrane responses to ADH. Am J Physiol 255:C526-C530

Wall SM, Han JS, Chou CL, Knepper MA (1992) Kinetics of urea and water permeability activation by vasopressin in rat terminal IMCD. Am J Physiol 262:F989-F998

Yamamoto T, Sasaki S, Fushimi K, Ishibashi K, Yaoita E, Kawasaki K, Marumo F, Kihara I (1995) Vasopressin increases AQP-CD water channel in apical membrane of collecting duct cells in Brattleboro rats. Am J Physiol 268:C1546-C1551

Yip KP (2002) Coupling of vasopressin-induced intracellular Ca2+ mobilization and apical exocytosis in perfused rat kidney collecting duct. J Physiol 538:891-899

Zelenina M, Christensen BM, Palmer J, Nairn AC, Nielsen S, Aperia A (2000) Prostaglandin E(2) interaction with AVP: effects on AQP2 phosphorylation and distribution. Am J Physiol Renal Physiol 278:F388-F394

Abbreviations:

AQP2: Aquaporin 2
dDAVP: 1-desamino-8-D-arginine vasopressin
cAMP: cyclic adenosine 3',5'-monophosphate
AKAP: A kinase anchoring protein
H89: N-[2-(p-Bromocinnamylamino) ethyl]-5-isoquinoline sulfonamide
LLC-PK1 cells:
Bt_2cAMP: dibutyryl cyclic adenosine 3',5'-monophosphate
8-Br-cAMP: 8-bromo cyclic adenosine 3',5'-monophosphate
IMCD: Inner medullary collecting duct
IP3: Inositol 1,4,5-trisphosphate
PGE2: Prostaglandin E2
PKA: Protein kinase A
VACM-1: Vasopressin activated calcium-mobilizing receptor
MLCK: Myosin light chain kinase
8-4-chlorophenylthio-cAMP: 8-4-chlorophenylthio   cyclic   adenosine   3',5'-monophosphate
W7: N-aminohexyl-5-chloronaphth-1-ylsulfonamide

Frische, Sebastian
    The Water and Salt Research Center, Building 233/244, Institute of Anatomy, University of Aarhus, DK-8000 Aarhus C, Denmark

Frøkiær, Jørgen
    The Water and Salt Research Center, Building 233/244, Institute of Anatomy, University of Aarhus, DK-8000 Aarhus C, Denmark

Kwon, Tae-Hwan
The Water and Salt Research Center, Building 233/244, Institute of Anatomy, University of Aarhus, DK-8000 Aarhus C, Denmark, and
Department of Biochemistry, School of Medicine, Kyungpook National University, Taegu, Korea

Nielsen, Søren
The Water and Salt Research Center, Building 233/244, Institute of Anatomy, University of Aarhus, DK-8000 Aarhus C, Denmark
sn@ana.au.dk

Molecules in motion: multiple mechanisms that regulate the GABA transporter GAT1

Michael W. Quick

Abstract

GABA is the major inhibitory neurotransmitter in brain, and its action is controlled in part by the predominant GABA transporter GAT1, which sets basal extracellular GABA concentrations, sequesters GABA away from its receptors, and prevents spillover of GABA to neighboring synapses. GAT1 is not a passive player in these processes. Rather, cells have developed multiple ways in which to control GAT1 action, and thus influence neuronal signaling. Research over the past decade has begun to address fundamental questions regarding GAT1 regulation. Is GAT1 regulated through changes in transporter trafficking, rates of GABA transport, or both? What are the molecular bases for the regulation? What are the initial triggering events and signal transduction cascades that mediate changes in GAT1 function? Over what time scale does the regulation occur? What are the functional consequences of the regulation? The present review examines the surprisingly rich results obtained from these inquiries.

1 Introduction

In the nervous system, synthesis of the major neurotransmitters occurs in the presynaptic nerve terminal, and calcium-dependent transmitter release from the nerve terminal occurs via the fusion of neurotransmitter-containing vesicles to the plasma membrane. Prior to the release event, neurotransmitter is packaged into these vesicles via a gene family of neurotransmitter transporters found on the vesicular membrane which use the electrochemical proton gradient to counter-transport transmitter against its concentration gradient into the vesicle (reviewed in Reimer et al. 1998; Gasnier 2000; Erickson, this volume). Following transmitter release into the synapse, neurotransmitter transporters from three other gene families mediate the sequestration of extracellular neurotransmitter. A family of five different transporters are responsible for removal of glutamate from excitatory synapses (reviewed in Sims and Robinson 1999; Danbolt 2001). These glutamate transporters likely have 8 transmembrane domains (Seal et al. 2000) and couple glutamate uptake to the co-transport of Na^+ and H^+, and the counter-transport of K^+ (Zerangue and Kavanaugh 1996; Levy et al. 1998). Sequestration of choline is mediated by a recently cloned Na^+- and Cl^--dependent transporter (Apparsundaram et al. 2000; Okuda et al. 2000) found in cholinergic neurons that is a member

Topics in Current Genetics, Vol. 9
E. Boles, R. Krämer (Eds.): Molecular Mechanisms Controlling Transmembrane Transport
DOI 10.1007/b95779 / Published online: 9 March 2004
© Springer-Verlag Berlin Heidelberg 2004

of the Na^+/glucose transporter family (reviewed in Jung 2002). The most abundant family of plasma membrane neurotransmitter transporters also use the co-transport of Na^+ and Cl^- to mediate neurotransmitter uptake, and these putative 12 transmembrane proteins include the transporters for the biogenic amines (norepinephrine, dopamine, serotonin), as well as the amino acids glycine, taurine, L-proline, creatine, betaine, and γ-aminobutyric acid (GABA) (reviewed in Deken et al. 2002).

The initial concept that plasma membrane transporters might exist came from work in the early 1960's that examined the fate of radiolabeled norepinephrine applied to rat tissues (Hertting and Axelrod 1961). Additional biochemical experiments using similar approaches showed that the accumulation into tissues of exogenously applied transmitter was of reasonably high affinity, saturable, and could be inhibited pharmacologically (reviewed in Iversen 1997). These results strengthened the hypothesis that neurotransmitter uptake was mediated by specific carriers, and that such mechanisms could be of physiological, pathophysiological, and therapeutic value. Subsequent to these investigations, experiments were undertaken that identified uptake processes for many of the other major neurotransmitter systems, suggesting a common process for the sequestration of released transmitter. The "molecular" era for neurotransmitter transporters was ushered in with the purification of a protein responsible for GABA uptake (Radian et al. 1986), followed by its cloning and functional expression soon thereafter (Guastella et al. 1990). Expression cloning of the norepinephrine transporter (Pacholczyk et al. 1991), comparison of its cDNA sequence to that of the GABA transporter, and similar ion dependencies of the two transporters identified them as part of the same family. Cloning by homologous sequences fleshed out this Na^+- and Cl^--dependent transporter family, which contains to date 14 transporters of known substrates and two orphans (Chen et al. 2003).

Over the past decade, molecular, genetic, biochemical, pharmacological, and physiological approaches have been brought to bear on a number of fundamental aspects of neurotransmitter transport. These include mapping transporter structure onto function, determining the kinetic steps involved in translocating substrates from one side of the membrane to the other, establishing the histological and cytological localization of each transporter, identifying the links between transporter dysfunction and disease states, and perhaps most importantly, determining the physiological roles for these transporters in the nervous system. One area of research on neurotransmitter transporters that has received significant attention over the past decade has been in understanding how transporters are regulated, especially with regard to the molecular mechanisms underlying the regulation, the triggers that initiate the regulation, and the functional consequences of such regulation. Although much work remains to be done, it is evident that clearance of neurotransmitter from the synaptic cleft is not a passive, static process. Instead, transporter function is a dynamic process under multiple forms of modulation. The goal of this review is to give the reader some flavor for the dynamic and complex mechanisms that cells use to control transporter function and, thus, transmitter levels, in brain. To illustrate these concepts, focus will be on the rat brain GABA transporter GAT1, the first neurotransmitter transporter cloned, and the predominant inhibitory neurotransmitter transporter in brain. The following section introduces general aspects of GAT1 function.

2 The rat brain GABA transporter GAT1

2.1 Molecular properties

There are several known Na^+- and Cl^--dependent GABA transporters. GAT2 encodes a transporter that shows significant expression in non-neural tissues, and may be important for osmolyte balance and transport of GABA across the blood-brain barrier (Conti et al. 1999). The GAT3 transporter is restricted to neural tissue (Clark et al. 1992), especially in glial cells near GABAergic synapses. GAT1 is the most abundant GABA transporter in brain, and high expression is evident in neocortex, hippocampus, cerebellum, basal ganglia, brainstem, spinal cord, olfactory bulb, and retina (Durkin et al. 1995). It is found in GABAergic neurons, specifically along axons and presynaptic nerve terminals (Conti et al. 1998). Thus, GAT1 is the "neuronal" transporter; however, there can be some expression of GAT1 in glia at or near GABAergic synapses (Conti et al. 1998), or in specialized glial cells such as Müeller cells of the cerebellum (Barakat and Bordey 2002). The quantity of expressed transporter is quite impressive. Using a strategy involving mice carrying a GAT1-GFP (green fluorescent protein) construct in place of wild type GAT1, and quantifying single molecule GFP fluorescence, expression of GAT1 at the cell surface of GABAergic interneurons at particular synapses in the hippocampus and cerebellum is estimated to be on the order of 1000 transporters/μm^2 (Chiu et al. 2002).

The GAT1 gene encodes a 599 amino acid protein (Guastella et al. 1990) that likely functions as an oligomer (Schmid et al. 2001). The transporter putatively contains 12 transmembrane domains, with N- and C-terminal tails oriented toward the cytoplasm. There is a large extracellular loop between TM3 and TM4 that contains sites for N-linked glycosylation. Ion substitution experiments, analysis of dose-response data, and measurements of stoichiometry suggest that two Na^+ ions are co-transported with each molecule of GABA. The role of Cl- is less clear. Cl^- is required, but may not be fully translocated; that is, it may occupy a binding site on the transporter but not unload (Loo et al. 2000). Depending on whether Cl^- translocates, each cycle of transport results in the accumulation of one or two net positive charges inside the cell, and thus GABA uptake is electrogenic and dependent on the cell's membrane potential. The apparent affinity of GAT1 for GABA is approximately 5 μM. The turnover rate of the transporter (i.e. the rate at which a single transporter can translocate GABA across the membrane), based upon electrophysiological estimates using whole-cell currents (Mager et al. 1993) or giant excised patches (Hilgemann and Lu 1999), is on the order of 10 GABA molecules per second. In terms of structure-function relationships, residues within transmembrane 1 (TM1) and TM4 are important in mediating Na^+ binding, R69 in

TM1 is important for Cl- binding, and GABA binding is mediated in part by Y140 and by residues in extracellular loops IV, V, and VI (reviewed in Kanner 2002). Residues in the N-terminal cytoplasmic tail appear to affect conformational states of the transporter, and may be important in permitting the unliganded transporter to return to its outward facing conformation after it unloads its substrates (Bennett et al. 2000).

2.2 Physiological properties

What are the roles that GAT1 plays in brain? First, GAT1 can control the ambient extracellular GABA concentration. Based upon Nernst equations and estimates of the cytoplasmic GABA concentration, the flux coupling of two Na^+ ions to one GABA molecule predicts that at steady state, the extracellular GABA concentration will be in the submicromolar range. Second, given that the vast majority of GAT1 is localized to pre-synaptic neurons, some fraction of the sequestered GABA might be re-packaged into synaptic vesicles for subsequent re-release. Third, the localization of the transporter near GABAergic synapses places it in position to limit the "spillover" of GABA from synapses where it is released. In the retina, application of GAT1 antagonists causes increased activation of extra-synaptic $GABA_C$ receptors. Thus, the transporter limits inhibitory spread during light-stimulated events (Ichinose and Lukasiewicz 2002). Limiting spillover also occurs in hippocampal preparations, although the extent of this effect depends on the amount of presynaptic stimulation and the density of GABA release sites (Overstreet and Westbrook 2003). Fourth, the transporter can transport GABA "in reverse," and such release of GABA can have functional effects. In fish and amphibian retina, efflux of GABA through the transporter is a principal means of initiating inhibitory transmission in a non-calcium-dependent manner (Schwartz 1987). In the hippocampus, high frequency stimulation can evoke depolarizations that are large enough to reverse Na^+ gradients and cause the transporter to efflux GABA onto GABA receptors (Gaspary et al 1998). Thus, efflux through GAT1 may be a means to limit hyperexcitability.

Lastly, there is evidence that GAT1 is also capable of regulating GABAergic signaling on the time scale over which synaptic transmission occurs, at least at some $GABA_A$-mediated and $GABA_B$-mediated synapses (Dingledine and Korn 1985; Hablitz and Lebeda 1985; Solis and Nicoll 1992; Isaacson et al. 1993). Many of these data come from experiments in which GABA receptor responses are recorded in the presence and absence of GAT1 antagonists. In different systems, effects of GAT1 blockade can be seen on the time course of the inhibitory post-synaptic current (IPSC), on the peak of the IPSC, or both. At synapses containing the G protein-coupled $GABA_B$ receptor, this is not surprising given that these slow synaptic responses occur on the order of 100 ms to 1 s, which is on the same time scale as the GAT1 turnover rate. However, this turnover rate is likely to be several orders of magnitude too slow to regulate synaptic transmission at synapses containing ligand-gated $GABA_A$ receptors. This observation, as well as experiments examining glutamate transporter function (Tong and Jahr 1994), has led to the idea that a significant portion of the action of GAT1 on signaling arises from the transporters acting as transmitter sinks, essentially binding transmitter away from receptor sites regardless of whether the GABA is transported or not.

Abnormal GAT1 transport may also have pathophysiological consequences. In humans with temporal lobe sclerosis, the expression of GAT1 in this brain region is reduced (During et al. 1995). Consistent with this reduction in expression, GABA efflux is decreased. This decrease in efflux could lead to a loss of inhibition during hyperexcitability (Patrylo et al. 2001). GAT1 protein levels are also abnormal in schizophrenics, although this loss may be secondary to the general loss of inhibitory terminals in neocortex (Woo et al. 1998). GAT1 is also the target of therapeutics aimed at raising extracellular GABA concentrations. For example, tiagabine is a clinically approved anti-epileptic drug that exerts its effects in part by inhibiting GABA binding to GAT1 (Soudijn and van Wijngaarden 2000).

2.3 Introduction to regulation

In general, functional regulation of GAT1 could occur either by changing the number of transporters on the plasma membrane, altering the transport process, or both. Given that some of the physiological actions of GAT1 are likely related to its ability to bind transmitter away from GABA receptors, altering the number of transporters would have a significant effect on regulating the time course of synaptic transmission and limiting spillover. Changing the rate of GABA transport, for example, would also be important in regulating synaptic transmission at "slow" synapses, and could affect the time course over which GABA is sequestered into the nerve terminal or released by efflux. Thus, a very reasonable hypothesis is that cells have developed mechanisms that control both GAT1 trafficking and transport; both of these forms of regulation are discussed in detail in the upcoming sections.

However, our initial experiments aimed at regulating the transporter were not done to test this hypothesis. After the cloning and functional expression of GAT1 in oocytes (Guastella et al. 1990), our goal was to use electrophysiology to permit high resolution analysis of GAT1 function. Initial attempts using two-electrode voltage-clamp of oocytes expressing GAT1 were unsuccessful. We reasoned that our inability to detect electrogenic currents associated with GABA transport were due to a combination of slow GAT1 turnover rates, the net movement of only one positive charge per transport cycle, and low levels of GAT1 protein expression on the plasma membrane. This experimental difficulty was eventually overcome by generating high expression levels of GAT1 in oocytes. This was accomplished by inserting the alfalfa mosaic virus 5' untranslated region upstream of GAT1's start codon, and a poly(50)-A tail downstream of GAT1's stop codon (Mager et al. 1993). In the course of attempting to produce measurable GABA-induced transporter currents, pilot experiments were performed using a variety of compounds that activate or inhibit second messenger activity in cells. The rationale for these experiments came from examination of GAT1's primary sequence, which indicated several potential consensus phosphorylation sites on serine and threonine residues, and the abundant evidence that such post-translational modifications could alter ion channel, receptor, and transporter function.

These initial experiments (Corey et al. 1994) showed that drugs, which alter protein kinase C (PKC) levels in oocytes had a dose-dependent effect on radio-labeled GABA uptake on a time scale of minutes. This result was independently shown by others, although the magnitude and direction of the regulation appeared to be cell-type dependent (Osawa et al. 1994; Sato et al. 1995). In our experiments, the PKC effect did not alter the apparent affinity of the transporter for GABA; rather, there was an alteration in the maximal transport capacity. One potential reason for a change in maximal transport capacity is a change in the number of functional transporters on the plasma membrane. Thus, we performed subcellular fractionation experiments and examined the distribution of GAT1 immunoreactivity across various subcellular compartments. PKC activation and inhibition altered the expression of GAT1 on the plasma membrane, and this change in transporter levels correlated with the changes in GABA uptake. These experiments formed the starting point for our investigations into GAT1 regulation, and its potential physiological significance.

3 GAT1 trafficking

The evidence that the subcellular distribution of GAT1 could be regulated by protein kinases led us to consider the possibility that cells actively control the functional activity of the transporter. The general hypothesis that we have been considering is that maintaining specific extracellular transmitter levels is crucial for normal neuronal activity, and that one way cells could achieve this is to regulate transporter expression. In general, this regulation could be achieved if transporter expression is linked in some way to extracellular transmitter levels, and so we have been pursuing two more specific hypotheses. First, that mechanisms exist which link the process of transmitter release to that of transmitter uptake, such that increased release is paralleled by increased surface expression of the transporter. Second, that extracellular transmitter levels directly signal the relative surface expression of the transporter. This section details what we have learned, which supports these two ideas.

3.1 Basal recycling of the transporter

Much is known about the recycling of neurotransmitter-containing synaptic vesicles (reviewed in Südhof 1995; Rothman 1996). In a simplified model, these 40-50 nm diameter, clear vesicles bud off from the endosome and are tethered to the cytoskeleton through interactions with the vesicle protein synapsin. The vesicle is freed from the cytoskeleton upon synapsin phosphorylation. The vesicle then docks with the plasma membrane at the active zone via protein-protein interactions involving the vesicle soluble NSF attachment protein receptor (SNARE) synaptobrevin (VAMP) and the plasma membrane SNAREs syntaxin 1A and SNAP-25. Following a "priming step," these vesicles then fuse with the plasma membrane upon depolarization, entry of calcium through closely-associated voltage-gated calcium channels, and binding of the calcium to the vesicle protein synaptotagmin. Membrane is then retrieved through several pathways depending on the type of fusion event, one of which involves dynamin-dependent, clathrin-mediated internalization. This vesicle is then stripped of its coat, and either returns to a recycling endosome, sorts to degradative pathways, or returns directly to the available pool of vesicles. The entire process occurs on the order of one minute. If there is a direct correlation between transmitter release and transporter expression, then one possibility is that the trafficking of GAT1 closely resembles that of transmitter release. Thus, how similar are these two pathways?

GAT1 Vesicles. Redistribution of GAT1 by PKC agonists and antagonists on a time scale of minutes implied the existence of trafficking via delivery and retrieval of transporter-containing vesicles from the plasma membrane. Immuno-electron microscopy has revealed GABA transporters (Barbaresi et al. 2001) and glycine transporters (Geerlings et al. 2001) on vesicle-like structures in presynaptic terminals. To further examine these vesicles, we isolated vesicle fractions from brain synaptosomes and found that GAT1 immunoreactivity was found in fractions of similar buoyant density to that of neurotransmitter-containing small synaptic vesicles (Deken et al. 2003). Immuno-isolation and electron microscopy of purified GAT1-containing vesicles showed them to be clear vesicles with a diameter of approximately 50 nm. The sizes of these vesicles were normally distributed, suggesting the presence of GAT1 on a homogenous population of vesicles.

The presence of GAT1 on a small clear vesicle in nerve terminals raised the intriguing possibility that GAT1 might reside on neurotransmitter containing small synaptic vesicles, and thus would be expressed on the plasma membrane precisely when GABA was being released. However, this does not appear to be the case (Deken et al. 2003). Based upon immunoblot analysis of various known vesicle proteins, GAT1-containing vesicles appear to contain little or no synaptophysin, a marker for neurotransmitter-containing vesicles. Conversely, synaptophysin-containing vesicles appear to contain little or no GAT1. In contrast to synaptophysin-containing vesicles, the GAT1-containing vesicles appear to lack SV2, synaptotagmin isoforms 1 and 2, and the vesicular GABA transporter. However, the GAT1-containing vesicles and synaptophysin-containing vesicles have in common syntaxin 1A, the small G protein rab3a, and synaptobrevin (VAMP). The lack of the vesicular GABA transporter on these GAT1-containing vesicles makes it unlikely that this population of vesicles contains GABA. We hypothesize that GAT1 resides on a vesicle separate from neurotransmitter-containing vesicles either because the transporter is recycled at regions distinct from release, or because neurons need to adjust transporter expression independent of release.

SNARE Proteins. The presence of SNARE proteins on the GAT1 vesicle provides indirect support for a role of these proteins in regulating GAT1 expression. Overexpression of syntaxin 1A or SNAP-25 in oocytes alters the subcellular distribution of the transporter (Quick et al. 1997). More direct experiments using toxins that cleave and inactivate specific SNARE proteins (Huttner 1993) show that these proteins can regulate GAT1 endogenously in neurons (Horton and Quick 2001). For example, application of botulinum toxin C1, which cleaves syntaxin 1A, causes a relative decrease in the amount of GAT1 present on the plasma membrane of neurons in hippocampal cultures, and a relative increase in cyto-plasmically-localized GAT1. Thus, as with neurotransmitter-containing synaptic vesicles, syntaxin 1A acts as a positive regulator of surface expression. Syntaxin 1A-regulated trafficking effects are also seen for glycine (Geerlings et al. 2000), norepinephrine (Sung et al. 2003), and serotonin (Haase et al. 2001; Quick 2002a) transporters.

The precise relationship between SNARE proteins and GAT1 are complex. For example, it is not clear if the trafficking effects of syntaxin 1A and SNAP-25 are due to its interaction with VAMP on the GAT1 vesicle, as might be expected if the regulation were similar to that for transmitter-containing small synaptic vesicles. Botulinum toxin B, which cleaves VAMP, has little effect on GAT1 trafficking (Deken et al. 2000; Horton and Quick 2001). It may be that syntaxin 1A and SNAP-25 interact with a different GAT1 vesicle protein. We do know that one protein with which syntaxin 1A interacts is GAT1 itself. As will be detailed below, the H3 domain of syntaxin 1A (residues 188-266) directly interacts with GAT1 and this interaction has multiple functional effects, although the contribution of this interaction to trafficking is unclear. Expression of various syntaxin 1A constructs along with GAT1 in oocytes reveals that domains in addition to the H3 domain are necessary to positively regulate the expression of GAT1 on the plasma membrane (Horton and Quick 2001). Thus, it appears that the role of syntaxin 1A on GAT1 trafficking is distinct from its direct protein-protein interactions with the transporter, although whether this role in regulating trafficking is directly compa-rable to vesicle docking has yet to be determined.

Regulated Expression and Recycling Rates. A hallmark of neurotransmitter re-lease is calcium- and depolarization-dependent exocytosis. This is also true of GAT1 (Deken et al. 2003). This conclusion is based upon experiments in which surface GAT1 protein in hippocampal neurons is labeled, permitted to internalize, and then the resurfaced, labeled GAT1 is evaluated under various experimental conditions. In the presence of extracellular Ca^{2+} and high levels of extracellular K^+ (to induce depolarization), the amount of surface GAT1 increases. This increase is prevented in the absence of extracellular Ca^{2+} or in the presence of the calcium channel blocker Cd^{2+}. There is some GAT1 expression on the cell surface even in the absence of extracellular Ca^{2+}, although whether this is due to an unregulated pathway for GAT1 expression, or an effect of calcium released from intracellular stores has yet to be determined.

This same approach, of measuring the reappearance of surface-labeled GAT1, permits estimation of the rate of GAT1 recycling. Under basal conditions of ex-tracellular Ca^{2+} and low levels of extracellular K^+, steady state levels of surface GAT1 are achieved within 5 minutes. Under depolarizing conditions, steady state levels of surface GAT1 are achieved within one minute. Thus, GAT1 recycles on a time scale similar to that seen for neurotransmitter-containing synaptic vesicles.

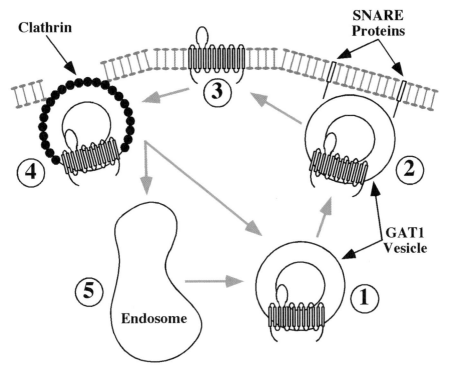

Fig. 1. Proposed model of basal GAT1 recycling. (1) GAT1 traffics on a 50 nm, clear synaptic-like vesicle. (2) The vesicle interacts at the cell surface with the SNARE proteins syntaxin 1A and SNAP-25. (3) Expression of functional GAT1 protein on the plasma membrane occurs in a calcium- and depolarization-dependent manner. (4) The transporter internalizes via a clathrin-dependent mechanism. (5) Once internalized, the transporter is able to recycle to the plasma membrane either directly or via a recycling endosome.

Clathrin Internalization and Subcellular Localization. The pathways for transporter internalization have been of significant recent interest. The dopamine transporter (DAT) is internalized via clathrin and sorted to recycling or degradative pathways (Daniels and Amara 1999; Melikian and Buckley 1999; Saunders et al. 2000). We have used multiple approaches to show that GAT1 similarly internalizes via clathrin (Deken et al. 2003). First, GAT1 and clathrin co-localize when over-expressed in various cell systems. More directly, if clathrin coat formation is blocked, the amount of GAT1 found on the plasma membrane increases. Surface GAT1 levels also increase when GAT1-expressing cells are co-transfected with a dominant-negative form of dynamin, a small G protein that participates in clathrin-mediated internalization (Damke et al. 1994). Once internalized, GAT1 is found in multiple compartments based upon subcellular fractionation and subsequent immunoblotting. GAT1 associates with the early and recycling endosomal markers rab5 and rab11, respectively, and the lysosomal marker lamp2.

Taken together, these data suggest that under basal conditions, GAT1 undergoes regulated trafficking that closely resembles, but is distinct from, the recycling of neurotransmitter-containing synaptic vesicles. Our current model of this trafficking pathway is illustrated in Figure 1.

3.2 Molecular mechanisms regulating GAT1 expression

Our present working hypothesis about GAT1 trafficking is that there is a basal recycling pathway (described in the previous section) upon which various molecules can exert modulatory effects. These modulatory mechanisms include changes in the phosphorylation state of the transporter and regulation of protein-protein interactions.

Direct Phosphorylation. Our initial investigation of direct transporter phosphorylation focused on multiple intracellular tyrosine residues within GAT1 (Law et al. 2000). In neurons that endogenously express GAT1, inhibitors of tyrosine kinases decrease GABA uptake; inhibitors of tyrosine phosphatases increase GABA uptake. The decrease in uptake seen with tyrosine kinase inhibitors correlates with a decrease in tyrosine phosphorylation of GAT1 and results in a redistribution of the transporter from the cell surface to intracellular locations. Of the five intracellular tyrosine residues present in GAT1, residues Y107 in the putative first intracellular loop and Y317 in the putative third intracellular loop can be phosphorylated *in vivo*, and these sites are responsible for the regulation of GAT1 trafficking (Whitworth and Quick 2001). Based upon experiments in which surface GAT1 is labeled and then these labeled proteins are chased into the cytoplasm, the result of tyrosine phosphorylation is a slowing of the rate of transporter internalization. Under conditions, which favor tyrosine phosphorylation, approximately 5% of GAT1 internalizes per minute, and this internalization rate increases approximately two-fold in the absence of tyrosine phosphorylation.

As discussed above, our original observation of GAT1 trafficking regulation occurred in response to changes in PKC activity. However, elimination of the consensus serine and threonine phosphorylation sites on GAT1 failed to eliminate the PKC effect (Corey et al 1994). This suggested either that PKC-mediated phosphorylation was occurring on non-consensus sites, or that the PKC was indirect. As will be discussed in the next section, there are indirect effects of PKC in regulating protein-protein interactions; however, we have recently revisited the idea of direct serine or threonine phosphorylation of GAT1. Our rationale was based on the observation that the serotonin transporter is directly phosphorylated by PKC (Ramamoorthy and Blakely 1999), and it is hypothesized that this event tags the transporter for internalization (Qian et al. 1997). In neurons, increases in PKC activity are correlated with an increase in intracellular localized GAT1 (Beckman et al. 1998; Deken et al. 2000) and, thus, the actions of PKC and the actions of tyrosine phosphorylation oppose each other. We reasoned that our failure to detect direct PKC-mediated phosphorylation might be due to its exclusion by tyrosine-mediated phosphorylation. Our most recent results (M. Quick, unpublished observations) show that GAT1 is phosphorylated on serine residues in a PKC-

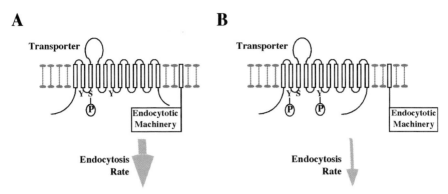

Fig. 2. A model for the regulation of GAT1 endocytosis by phosphorylation. (A) Phosphorylation of surface-localized GAT1 on serine residues places the transporter in a conformation that precludes tyrosine phosphorylation and makes it more likely to interact with internalization machinery. (B) Phosphorylation of surface-localized GAT1 on tyrosine residues places the transporter in a conformation that precludes serine phosphorylation makes it less likely to interact with internalization machinery.

dependent manner, but this state is only revealed when GAT1 tyrosine phosphorylation is eliminated or greatly reduced. The relative levels of serine phosphorylation and tyrosine phosphorylation are negatively correlated. The amount of serine phosphorylation is regulated by agents that affect tyrosine phosphorylation, and *vice versa*. The data support the idea that GAT1 exists in either of two mutually exclusive phosphorylation states, and that the relative abundance of these states determines in part the relative subcellular distribution of the transporter.

Together with the basal recycling model, we would suggest that vesicles containing GAT1 are inserted into the membrane in a calcium-dependent manner and internalized via a clathrin-dependent manner. Serine phosphorylation places the transporter in a conformation that disfavors tyrosine phosphorylation and favors interactions with internalization machinery, thus speeding endocytosis and increasing the size of the intracellular GAT1 pool. Tyrosine phosphorylation would have the opposite effects (Figure 2).

Protein-Protein Interactions. The interaction of GAT1 with SNARE proteins suggests the hypothesis that a network of protein-protein interactions could modulate GAT1 trafficking, especially since syntaxin 1A forms multiple multimeric complexes in brain. For example, in addition to interacting with SNAP-25 and VAMP, syntaxin 1A shows high affinity interactions with the cytosolic protein Munc-18 (Hata et al. 1993; Pevsner et al. 1994). In expression systems, we can reverse syntaxin 1A's effects on GAT1 trafficking by over-expression of Munc-18 (Beckman et al. 1998). Furthermore, phosphorylation of Munc-18, which inhibits its binding to syntaxin 1A, promotes syntaxin 1A's effects on GAT1. It should also be noted that GAT1 itself, through protein-protein interactions, could regulate the availability of other proteins to interact in multimeric complexes. For example, GAT1 might control the availability of syntaxin 1A to interact with calcium channels and other components of the docking and fusion apparatus. Additionally, if syntaxin 1A availability were limiting, then syntaxin 1A's participation in vesicle release might preclude it from downregulating transporter function. Thus, transporter function would be positively correlated with neurotransmitter release; such a mechanism would provide one route by which neurons could exert greater control over transmitter-mediated synaptic signaling.

3.3 Signaling cascades that regulate trafficking

In addition to identifying the end points of a modulatory signal, such as direct phosphorylation of the transporter or changes in the association of the transporter with interacting proteins, we have also been interested in identifying the starting points for these signals. In this section, several triggers for regulation of transporter trafficking are discussed.

Receptors. The evidence that direct phosphorylation of GAT1 on tyrosine or serine residues alters GAT1 trafficking suggested the likelihood that kinase-mediated changes in phosphorylation state are initiated by specific receptors known to activate these pathways. For protein kinase C activation, G protein-couple receptors are an obvious choice. G proteins transduce signals from cell surface neurotransmitter receptors to intracellular second messengers and ion channels (Simon et al. 1991). The $G\alpha_q$ family of G proteins couples receptor-mediated signals to phospholipase C activation; phospholipase C catalyzes hydrolysis of phosphotidyl 4,5-bisphosphate to create inositol trisphosphate and diacylglycerol; diacylglycerol activates PKC (Berridge 1993). In hippocampal neurons, we find that specific agonists of G protein-coupled receptors for acetylcholine, glutamate, and serotonin all downregulate GAT1 function (Beckman et al. 1999). This functional inhibition is mimicked by PKC activators and prevented by specific receptor antagonists and PKC inhibitors. And as expected, this receptor-mediated functional inhibition correlates with a redistribution of GAT1 from the plasma membrane to intracellular locations. For tyrosine kinase activation, neurotrophin receptors are an obvious possibility. In hippocampus, receptors for brain-derived neurotrophic factor (BDNF) are found on both pyramidal cells and interneurons (Kokaia et al. 1993) and, thus, are a likely choice to regulate GABA transporter trafficking. In serum-starved neuronal cultures, we found that BDNF upregulates GAT1 function, and this increase correlates with an increase surface transporter expression (Law et al. 2000)

Transporter Ligands. If cells regulate GAT1 trafficking in order to control extracellular transmitter levels, then the most direct route for this regulation would be to have the transporter sense directly extracellular transmitter levels and alter its expression. To test this hypothesis, we performed GABA uptake assays on hippocampal cultures and showed that extracellular GABA induces chronic changes in GABA transport that occur in a dose-dependent and time-dependent manner (Bernstein and Quick 1999). In addition to GABA, specific GAT1 substrates upregulate transport; GAT1 transport inhibitors that are not transporter substrates downregulate transport. These changes occur in the presence of blockers of both $GABA_A$ and $GABA_B$ receptors, occur in the presence of protein synthesis inhibitors, and are not influenced by intracellular GABA concentrations. Surface bioti-nylation experiments revealed that the increase in transport is correlated with an increase in surface transporter expression, which results from a slowing of GAT1 internalization.

This slowing of internalization was reminiscent of the slowing due to direct ty-rosine phosphorylation described above. Indeed, incubation of GAT1-expressing cells with transporter ligands alters the amount of GAT1 tyrosine phosphorylation, and substrate-induced surface expression is unchanged in a GAT1 mutant lacking tyrosine phosphorylation sites (Whitworth and Quick 2001). These data suggest a model in which substrates permit the phosphorylation of GAT1 on tyrosine resi-dues and that the phosphorylated state of the transporter is refractory for internali-zation. Overall, these data suggest that the GABA transporter fine tunes its traf-ficking in response to extracellular GABA and would act to maintain a constant level of neurotransmitter at the synaptic cleft.

4 Regulating transport

Regulating transporter trafficking represents one general method for modulating extracellular transmitter levels; another way would be to alter the transport proc-ess. Our recent work in this regard has focused upon mechanisms that regulate the GAT1 turnover rate, which is the rate of one transport cycle or the rate of GABA translocation.

Because transport through GAT1 is slow, direct measurement of the turnover rate is difficult. One indirect method is to derive the maximal transport capacity (Vmax) from saturation analysis of GABA uptake, and divide this value by the maximal binding capacity (Bmax). Alternatively, one can use electrophysiological analysis of GAT1 currents after expression of the transporter in *Xenopus* oocytes (Mager et al. 1993; Deken et al. 2000). One way is to calculate GAT1 turnover rates by measuring in the same oocyte GABA-induced peak currents (to calculate the number of charges translocated per second) and GAT1-specific charge move-ments (to calculate the number of transporters). An alternative method for calcu-lating turnover rates is to examine the relaxation of transient currents during volt-age-jump experiments. The isolated transient relaxation reflects partial functioning of the transporter and the rate of the relaxation has been interpreted as a single turnover of the transporter (Hilgemann et al. 1991; Parent et al. 1992; Mager et al. 1993). Regardless of the experimental approach, the result is that GAT1 turnover is approximately 7/sec at room temperature and 96 mM Na$^+$. Can this rate be regu-lated, and how?

4.1 Intermolecular interactions

As mentioned above, not only does the SNARE protein syntaxin 1A regulate transporter trafficking, it directly binds GAT1 and affects GAT1 permeation. The rationale for initially examining the role of syntaxin 1A on GAT1 turnover rates came from experiments showing that syntaxin 1A is able to regulate unitary properties of several different excitability proteins. For example, syntaxin 1A reduces cystic fibrosis transmembrane regulator (CFTR) Cl⁻ currents by reducing channel open probability and burst duration (Naren et al. 1999), and promotes slow inactivation gating of calcium channel currents (Bezprozvanny et al. 1995).

Immunoprecipitation experiments from rat brain using syntaxin 1A, followed by Western blotting with a GAT1 antibody reveal a physical interaction between these two proteins (Beckman et al. 1998). To map the sites of interaction on each protein, we have used multiple approaches (Deken et al. 2000): yeast two-hybrid screens, "pull-down" assays using hippocampal neuron lysates in combination with various glutathione-S-transferase (GST) fusion proteins, and *in vitro* binding assays. All assays reveal that the association of GAT1 with syntaxin 1A is mediated predominantly by amino acids 30 - 54 in the N-terminal cytoplasmic tail of GAT1 and amino acids 166 - 288 of syntaxin 1A, residues that are juxtaposed to syntaxin 1A's transmembrane domain. The binding, which has an affinity in the sub-micromolar range, is likely mediated through electrostatic interactions, since high salt concentrations inhibit binding. Site-directed mutagenesis that eliminates three aspartic acid residues (D40, D43, D45) in the N-terminal tail of GAT1 eliminates syntaxin 1A binding, further supporting the idea of electrostatic interactions.

Functionally, syntaxin 1A reduces whole-cell GABA uptake approximately two-fold (Beckman et al. 1998; Deken et al. 2000). Given that syntaxin 1A increases the whole-cell expression of the transporter two-fold (described above) means that syntaxin 1A mediates an approximate four-fold net reduction in transport per transporter. Measurements of turnover rates show that syntaxin 1A reduces GAT1 function from 7/sec to approximately 2/sec (Deken et al. 2000). The reason why syntaxin 1A is a positive modulator of GAT1 expression but a negative regulator of GAT1 transport is speculative. One possibility is that these two processes are temporally independent; another possibility is that neurons use syntaxin 1A not only to increase surface expression of GAT1 but also to keep the transporter functionally suppressed until a time when removal of GABA from the synaptic cleft becomes important.

More recently, we have attempted to identify which step in the transport process syntaxin 1A is mediating its effects (Wang et al. 2003). Functional schemes (Hilgemann and Lu 1999) suggest that GAT1 exists in a conformation in which it can bind two Na⁺, one Cl⁻, and a GABA molecule. This binding then induces conformational changes in which these molecules are translocated. After releasing its substrates, the unliganded transporter then reorients to the extracellular milieu. The inhibition of GAT1 turnover rates by syntaxin 1A could be due to effects at any of these steps. We find that syntaxin 1A causes similar reductions in forward and reverse transport that do not involve changes in apparent transport affinities for sodium, chloride, or GABA. There is also reduction in GABA exchange, a step in transport that bypasses the reorientation step (Bennett et al. 2000). These data suggest that syntaxin 1A exerts its effects at a step in the translocation process

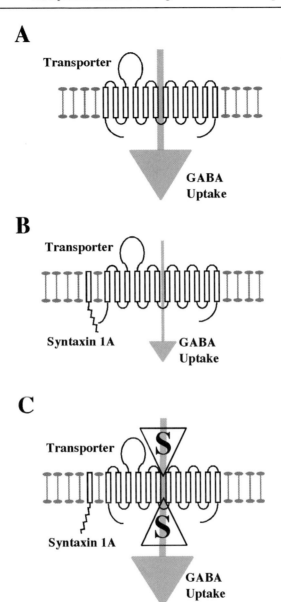

Fig. 3. A model for the regulation of GAT1 turnover rates. (A) In the basal state, GAT1 translocates GABA on the order of 10 molecules/sec. (B) The SNARE protein syntaxin 1A interacts with the N-terminal tail of GAT1, slowing turnover rates four-fold. (C) Substrates negatively influence the interaction of syntaxin 1A with GAT1, reversing the syntaxin 1A-mediated inhibition of GAT1 turnover rates.

after substrate binding but which involves both unidirectional transport and transmitter exchange.

4.2 Transporter ligands

As shown above, transporter agonists and antagonists can modulate transporter trafficking, perhaps by regulating the ability of GAT1 to be phosphorylated on tyrosine residues. Can transporter ligands also regulate GAT1 turnover rates? Indeed, in oocytes expressing GAT1 and syntaxin 1A, we have shown that application of transporter substrates increases GAT1 turnover rates, essentially reversing syntaxin 1A inhibition as described above (Quick 2002b). The substrate-induced rate change requires the presence of syntaxin 1A, and in hippocampal neurons that endogenously express both GAT1 and syntaxin 1A, substrate application results in a decrease in the fraction of syntaxin 1A that is bound to GAT1 on a time-scale comparable to the substrate-induced change in turnover rates. A model of this effect is diagrammed in Figure 3. Our data suggest that substrate translocation regulates GAT1-syntaxin 1A interactions, and provide a mechanism by which GABA transport can be increased during times of rising synaptic GABA concentrations.

5 Summary

As with any interesting area of scientific inquiry, the number of questions to be resolved in studies of GAT1 regulation is far greater than the answers we have so far obtained. For example, we know some of the mechanisms by which GAT1 function can be regulated, but we do not know where or when these forms of modulation occur *in situ*, and the extent to which they alter neuronal signaling. Additionally, little is known about GAT1 regulation on time scales involving transcription or translation; nor do we know if GAT1 trafficking and turnover rates are regulated developmentally. Also of interest will be the extent to which these forms of regulation, or inability to be regulated, contribute to disease states. And perhaps more importantly, can regulation of GAT1 by trafficking or transport rates be harnessed as an entry way into the treatment of disorders related to abnormal GABA levels in brain?

References

Apparsundaram S, Ferguson SM, George AL Jr, Blakely RD (2000) Molecular cloning of a human, hemicholinium-3-sensitive choline transporter. Biochem Biophys Res Commun 276:862-867

Barakat L, Bordey A (2002) GAT-1 and reversible GABA transport in Bergmann glia in slices. J Neurophysiol 88:1407-1419

Barbaresi P, Gazzanelli G, Malatesta M (2001) γ-aminobutyric acid transporters in the cat periaqueductal gray: a light and electron microscopic immunocytochemical study. J Comp Neurol 429:337-354

Beckman ML, Bernstein EM, Quick MW (1998) Protein kinase C regulates the interaction between a GABA transporter and syntaxin 1A. J Neurosci 18:6103-6112

Beckman ML, Bernstein EM, Quick MW (1999) Multiple G protein-coupled receptors initiate protein kinase C redistribution of GABA transporters in hippocampal neurons. J Neurosci 19:RC9

Bennett ER, Su H, Kanner BI (2000) Mutation of arginine 44 of GAT-1, a (Na(+) + Cl(-))-coupled gamma-aminobutyric acid transporter from rat brain, impairs net flux but not exchange. J Biol Chem 275:34106-34113

Bernstein EM, Quick MW (1999) Regulation of gamma-aminobutyric acid (GABA) transporters by extracellular GABA. J Biol Chem 274:889-895

Berridge MJ (1993) Inositol trisphosphate and calcium signalling. Nature 361: 315-325

Bezprozvanny I, Scheller RH, Tsien RW (1995) Functional impact of syntaxin on gating of N-type and Q-type calcium channels. Nature 378:623-626

Chen NH, Reith ME, Quick MW (2003) Synaptic uptake and beyond: the sodium- and chloride-dependent neurotransmitter transporter family SLC6. Pflugers Arch: in press

Chiu CS, Jensen K, Sokolova I, Wang D, Li M, Deshpande P, Davidson N, Mody I, Quick MW, Quake SR, Lester HA (2002) Number, density, and surface/cytoplasmic distribution of GABA transporters at presynaptic structures of knock-in mice carrying GABA transporter subtype 1-green fluorescent protein fusions. J Neurosci 22:10251-10266

Clark JA, Deutch AY, Gallipoli PZ, Amara SG (1992) Functional expression and CNS distribution of a beta-alanine-sensitive neuronal GABA transporter. Neuron 9:337-348

Conti F, Melone M, De Biasi S, Minelli A, Brecha NC, Ducati A (1998) Neuronal and glial localization of GAT-1, a high-affinity gamma-aminobutyric acid plasma membrane transporter, in human cerebral cortex: with a note on its distribution in monkey cortex. J Comp Neurol 396:51-63

Conti F, Zuccarello LV, Barbaresi P, Minelli A, Brecha NC, Melone M (1999) Neuronal, glial, and epithelial localization of gamma-aminobutyric acid transporter 2, a high-affinity gamma-aminobutyric acid plasma membrane transporter, in the cerebral cortex and neighboring structures. J Comp Neurol 409:482-494

Corey JL, Davidson N, Lester HA, Brecha N, Quick MW (1994) Protein kinase C modulates the activity of a cloned gamma-aminobutyric acid transporter expressed in *Xenopus* oocytes via regulated subcellular redistribution of the transporter. J Biol Chem 269:14759-14767

Damke H, Baba T, Warnock DE, Schmid SL (1994) Induction of mutant dynamin specifically blocks endocytic coated vesicle formation. J Cell Biol 127:915-934

Danbolt NC (2001) Glutamate uptake. Prog Neurobiol 65:1-105

Daniels GM, Amara SG (1999) Regulated trafficking of the human dopamine transporter. Clathrin- mediated internalization and lysosomal degradation in response to phorbol esters. J Biol Chem 274:35794-35801

Deken SL, Beckman ML, Boos L, Quick MW (2000) Transport rates of GABA transporters: regulation by the N-terminal domain and syntaxin 1A. Nat Neurosci 3:998-1003

Deken SL, Fremeau RT Jr, Quick MW (2002) Family of sodium-coupled transporters for GABA, glycine, proline, betaine, taurine, and creatine: pharmacology, physiology, and regulation. In: Reith M (ed) Neurotransmitter Transporters. Humana Press, Totowa, NJ, pp 193-234

Deken SL, Wang D, Quick MW (2003) Plasma membrane GABA transporters reside on distinct vesicles and undergo rapid regulated recycling. J Neurosci 23:1563-1568

Dingledine R, Korn SJ (1985) Gamma-aminobutyric acid uptake and the termination of inhibitory synaptic potentials in the rat hippocampal slice. J Physiol 366:387-409

During MJ, Ryder KM, Spencer DD (1995) Hippocampal GABA transporter function in temporal-lobe epilepsy. Nature 376:174-177

Durkin MM, Smith KE, Borden LA, Weinshank RL, Branchek TA, Gustafson EL (1995) Localization of messenger RNAs encoding three GABA transporters in rat brain: an in situ hybridization study. Brain Res Mol Brain Res 33:7-21

Gasnier B (2000) The loading of neurotransmitters into synaptic vesicles. Biochimie 82:327-337

Gaspary HL, Wang W, Richerson GB (1998) Carrier-mediated GABA release activates GABA receptors on hippocampal neurons. J Neurophysiol 80:270-281

Geerlings A, Lopez-Corcuera B, Aragon C (2000) Characterization of the interactions between the glycine transporters GLYT1 and GLYT2 and the SNARE protein syntaxin 1A. FEBS Lett 470:51-54

Geerlings A, Nunez E, Lopez-Corcuera B, Aragon C (2001) Calcium- and syntaxin 1-mediated trafficking of the neuronal glycine transporter GLYT2. J Biol Chem 276:17584-17590

Guastella J, Nelson N, Nelson H, Czyzyk L, Keynan S, Miedel MC, Davidson N, Lester HA, Kanner BI (1990) Cloning and expression of a rat brain GABA transporter. Science 249:1303-1306

Haase J, Killian AM, Magnani F, Williams C (2001) Regulation of the serotonin transporter by interacting proteins. Biochem Soc Trans 29:722-728

Hablitz JJ, Lebeda FJ (1985) Role of uptake in gamma-aminobutyric acid (GABA)-mediated responses in guinea pig hippocampal neurons. Cell Mol Neurobiol 5:353-371

Hata Y, Slaughter CA, Südhof TC (1993) Synaptic vesicle fusion complex contains unc-18 homologue bound to syntaxin. Nature 366:347-351

Hertting G, Axelrod J (1961) Fate of tritiated noradrenaline at sympathetic nerve endings. Nature 192:172–173

Hilgemann DW, Lu CC (1999) GAT1 (GABA:Na+:Cl-) cotransport function. Database reconstruction with an alternating access model. J Gen Physiol 114:459-475

Hilgemann DW, Nicoll DA, Philipson KD (1991) Charge movement during Na+ translocation by native and cloned cardiac Na+/Ca2+ exchanger. Nature 352:715-718

Horton N, Quick MW (2001) Syntaxin 1A up-regulates GABA transporter expression by subcellular redistribution. Mol Membr Biol 18:39-44

Huttner WB (1993) Cell biology: Snappy exocytoxins. Nature 365:104-105

Ichinose T, Lukasiewicz PD (2002) GABA transporters regulate inhibition in the retina by limiting GABA(C) receptor activation. J Neurosci 22:3285-3292

Isaacson JS, Solis JM, Nicoll RA (1993) Local and diffuse synaptic actions of GABA in the hippocampus. Neuron 10:165-175

Iversen LL (1997) The uptake of catechol amines at high perfusion concentrations in the rat isolated heart: a novel catechol amine uptake process. Br J Pharmacol 120(Suppl):267-282

Jung H (2002) The sodium/substrate symporter family: structural and functional features. FEBS Lett 529:73-77

Kanner BI (2002) Sodium-coupled GABA and glutamate transporters. In: Reith M (ed) Neurotransmitter Transporters. Humana Press, Totowa, NJ, pp 235-254

Kokaia Z, Bengzon J, Metsis M, Kokaia M, Persson H, Lindvall O (1993) Coexpression of neurotrophins and their receptors in neurons of the central nervous system. Proc Natl Acad Sci USA 90:6711-6715

Law RM, Stafford A, Quick MW (2000) Functional regulation of gamma-aminobutyric acid transporters by direct tyrosine phosphorylation. J Biol Chem 275:23986-23991

Levy LM, Warr O, Attwell D (1998) Stoichiometry of the glial glutamate transporter GLT-1 expressed inducibly in a Chinese hamster ovary cell line selected for low endogenous Na^+-dependent glutamate uptake. J Neurosci 18:9620-9628

Loo DD, Eskandari S, Boorer KJ, Sarkar HK, Wright EM (2000) Role of Cl- in electrogenic Na+-coupled cotransporters GAT1 and SGLT1. J Biol Chem 275:37414-37422

Mager S, Naeve J, Quick M, Labarca C, Davidson N, Lester HA (1993) Steady states, charge movements, and rates for a cloned GABA transporter expressed in *Xenopus* oocytes. Neuron 10:177-188

Melikian HE, Buckley KM (1999) Membrane trafficking regulates the activity of the human dopamine transporter. J Neurosci 19:7699-7710

Naren AP, Cormet-Boyaka E, Fu J, Villain M, Blalock JE, Quick MW, Kirk KL (1999) CFTR chloride channel regulation by an interdomain interaction. Science 286:544-548

Okuda T, Haga T, Kanai Y, Endou H, Ishihara T, Katsura I (2000) Identification and characterization of the high-affinity choline transporter. Nat Neurosci 3:120-125

Osawa I, Saito N, Koga T, Tanaka C (1994) Phorbol ester-induced inhibition of GABA uptake by synaptosomes and by *Xenopus* oocytes expressing GABA transporter (GAT1). Neurosci Res 19:287-293

Overstreet LS, Westbrook GL (2003) Synapse density regulates independence at unitary inhibitory synapses. J Neurosci 23:2618-2626

Pacholczyk T, Blakely RD, Amara SG (1991) Expression cloning of a cocaine- and antidepressant-sensitive human noradrenaline transporter. Nature 350:350-354

Parent L, Supplisson S, Loo DD, Wright EM (1992) Electrogenic properties of the cloned Na+/glucose cotransporter: I. Voltage-clamp studies. J Membr Biol 125:49-62

Patrylo PR, Spencer DD, Williamson A (2001) GABA uptake and heterotransport are impaired in the dentate gyrus of epileptic rats and humans with temporal lobe sclerosis. J Neurophysiol 85:1533-1542

Pevsner J, Hsu S-C, Scheller RH (1994) n-Sec1: a neural-specific syntaxin-binding protein. Proc Natl Acad Sci USA 91:1445-1449

Qian Y, Galli A, Ramamoorthy S, Risso S, DeFelice LJ, Blakely RD (1997) Protein kinase C activation regulates human serotonin transporters in HEK-293 cells via altered cell surface expression. J Neurosci 17:45-57

Quick MW (2002a) Role of syntaxin 1A on serotonin transporter expression in developing thalamocortical neurons. Int J Dev Neurosci 20:219-224

Quick MW (2002b) Substrates regulate gamma-aminobutyric acid transporters in a syntaxin 1A-dependent manner. Proc Natl Acad Sci USA 99:5686-5691

Quick MW, Corey JL, Davidson N, Lester HA (1997) Second messengers, trafficking-related proteins, and amino acid residues that contribute to the functional regulation of the rat brain GABA transporter GAT1. J Neurosci 17:2967-2979

Radian R, Bendahan A, Kanner BI (1986) Purification and identification of the functional sodium- and chloride-coupled gamma-aminobutyric acid transport glycoprotein from rat brain. J Biol Chem 261:15437-15441

Ramamoorthy S, Blakely RD (1999) Phosphorylation and sequestration of serotonin transporters differentially modulated by psychostimulants. Science 285:763-766

Reimer RJ, Fon EA, Edwards RH (1998) Vesicular neurotransmitter transport and the presynaptic regulation of quantal size. Curr Opin Neurobiol 8:405-412

Rothman JE (1996) The protein machinery of vesicle budding and fusion. Protein Sci 5:185-194

Sato K, Betz H, Schloss P (1995) The recombinant GABA transporter GAT1 is downregulated upon activation of protein kinase C. FEBS Lett 375:99-102

Saunders C, Ferrer JV, Shi L, Chen J, Merrill G, Lamb ME, Leeb-Lundberg LM, Carvelli L, Javitch JA, Galli A (2000) Amphetamine-induced loss of human dopamine transporter activity: an internalization-dependent and cocaine-sensitive mechanism. Proc Natl Acad Sci USA 97:6850-6855

Schmid JA, Scholze P, Kudlacek O, Freissmuth M, Singer E, Sitte HH (2001) Oligomerization of the human serotonin transporter and of the rat GABA transporter 1 visualized by fluorescence resonance energy transfer microscopy in living cells. J Biol Chem 276:3805-3810

Schwartz EA (1987) Depolarization without calcium can release g-aminobutyric acid from a retinal neuron. Science 238:350-355

Seal RP, Leighton BH, Amara SG (2000) A model for the topology of excitatory amino acid transporters determined by the extracellular accessibility of substituted cysteines. Neuron 25:695-706

Simon MI, Strathmann MP, Gautam N (1991) Diversity of G proteins in signal transduction. Science 252: 802-808

Sims KD, Robinson MB (1999) Expression patterns and regulation of glutamate transporters in the developing and adult nervous system. Crit Rev Neurobiol 13:169-197

Solis JM, Nicoll RA (1992) Pharmacological characterization of GABA(B)-mediated responses in the CA1 region of the rat hippocampal slice. J Neurosci 12:3466-3472

Soudijn W, van Wijngaarden I (2000) The GABA transporter and its inhibitors. Curr Med Chem 7:1063-1079

Südhof TC (1995) The synaptic vesicle cycle: a cascade of protein-protein interactions. Nature 375:645-653

Sung U, Apparsundaram S, Galli A, Kahlig KM, Savchenko V, Schroeter S, Quick MW, Blakely RD (2003) A regulated interaction of syntaxin 1A with the antidepressant-sensitive norepinephrine transporter establishes catecholamine clearance capacity. J Neurosci 23:1697-1709

Tong G, Jahr CE (1994) Block of glutamate transporters potentiates postsynaptic excitation.
Neuron 13:1195-1203

Wang D, Deken SL, Whitworth TL, Quick MW (2003) Syntaxin 1A inhibits both GABA flux and efflux through the rat brain GABA transporter GAT1. Molec Pharm: in press

Whitworth TL, Quick MW (2001) Substrate-induced regulation of gamma-aminobutyric acid transporter trafficking requires tyrosine phosphorylation. J Biol Chem 276:42932-42937

Woo TU, Whitehead RE, Melchitzky DS, Lewis DA (1998) A subclass of prefrontal gamma-aminobutyric acid axon terminals are selectively altered in schizophrenia. Proc Natl Acad Sci USA 95:5341-5346

Zerangue N, Kavanaugh MP (1996) Flux coupling in a neuronal glutamate transporter. Nature 383:634-637

Quick, Michael W.
Department of Biological Sciences, University of Southern California, HNB 228, 3641 Watt Way, Los Angeles CA 90089-2520, USA
mquick@usc.edu

Index

8Bromo-cAMP 13

AAP/YAT family 123, 223, 298
ABC transporter 49, 52, 158, 159
Abf1 146
ABP 10
Abp1 366
ABST 1
acetate 14
acidic cluster 267, 280
actin 364
adaptor protein 263
adenylate cyclase 197, 357
adipose cells 329
ADP-ribosylation 107
Agp 124, 146, 298
AKT 335
ALP 212
Alp1 124
Alr1 300
alternative splicing 23, 24
Amanita muscaria 132
amiloride-sensitive sodium channel 284
amino acid permease 20, 123, 146, 208,
    223, 282, 297, 303
amino acid receptor 133
amino acid sensor 127
ammonium transport 96, 277
AmMST1 132
amphiphilic 171, 174
Amt 61, 98, 100, 103, 107
analytical ultracentrifugation 160
ancillary protein 207, 221
angiotensin 30, 334
anterograde 208, 280
antitermination 58, 198
AP 263, 291
APC superfamily 123
apical 23
aquaporin 310, 340, 353
Arabidopsis 110
Arginine vasopressin 362
Arn 307, 312
ASBT 9, 34
Asi 143
Aspergillus nidulans 101
ATB^{0+} 20

autophagy 299
AVP 362
axon 259
Azospirillium brasiliense 114
Azotobacter vinelandii 105

Bacillus subtilis 97, 158, 180
bacteroid 115
bafilomycin 360
Bap 124, 146, 298
Bartter disease 27
basolateral 23
BCCT family 160
betaine 156
BetP 158, 159, 160, 172
bglPH operon 198
bicarbonate 37
bile salt transporter 1, 9
binding protein 49
Brefeldin A 329
BSC 27
Bul 286, 298, 303, 305
bulk flow 213
bumetanide-sensitive $Na^+{:}K^+{:}2Cl^-$
    cotransporter 27
BusA 158, 172, 174
butyrate 14

C2 domain 286
Caenorhabditis elegans 97
calmodulin 148, 367
cAMP 13, 28, 49, 189, 197, 356, 370
campesterol 237
Can1 124, 227, 242, 292
Candida albicans 129
CAP 338
cardiac voltage-gated Na+ channel 285
cargo receptor protein 213
casein kinase 1 135, 136, 287
casein kinase 2 267
catabolite repression 48, 49, 180, 189,
    197
caveoli 335
caveolin 341
Cbl 338
Cbl-CAP complex 338
CcpA 186, 198

Cdc53 135
cellubrevin 334
cellugyrin 334
ceramide 237
CFTR 247, 276
channel 68, 285
chaperon 236, 247
chemotaxis 180
chitin synthase III 226
cholestasis 9
cholesterol 237, 280
choline 379
Chs 226
cis-acting elements 19
cisternal progression 209, 211
c-Jun N-terminal kinase 260
class E Vps protein 292, 295
clathrin 208, 211, 261, 291, 329, 388
coated vesicles 329
coatomer 213, 214, 215, 219, 228
coiled-coil structure 173
collecting duct 354
colon 7
compatible solutes 156
cooperative interaction 75
COPI-coated vesicle 209, 237
COPII-coated vesicle 207, 216, 228,
    237, 240, 275, 277
copper transporter 301
corpus callosum syndrome 29
cortactin 366
Corynebacterium glutamicum 97, 103,
    158
Crp 189
Crypt cells 6
C-terminal tail 126
Ctr1 300
Cys1 129
cystic fibrosis 276
Cytochalasin B 84
cytoskeleton 362

DAT 301
dDAVP 367
dendrite 259
detergent 240
deubiquitylating enzymes 282
Dgt1 135
di(glucos-6-yl)PEG$_{600}$ 13
diabetes mellitus 14, 326

diacylglycerol 392
diarrhoeal conditions 14
diauxic growth 189, 193
dileucine motif 265, 296, 343
dilysine motif 280
Dip5 124, 146
disulfide bridge 75, 76
divalent metal transporter 8, 40, 43
DMT1 8, 40, 43
Doa4 282, 296
dopamine transporter 388
duodenum 7
dynactin 367
dynamin 341
dynein 367

E2 ubiquitin-conjugating enzyme 135
E3 ubiquitin ligase 135
EAAT1 148
ectoine 159
EctP 159, 172
Ede1 290
EIN2 147
electrochemical Na$^+$ potential 165
electroneutral Cl$^-$ coupled cotransporter
    25
ENaC 282, 284, 302
endocytosis 255, 259, 273, 281, 286,
    297, 313, 326, 340, 360, 391
endoproteolytic processing 143
endosomal markers 388
endosome 275, 286, 328
endosome/vacuolar degradation 274
enhancer 6
Ent 290
Enterococcus faecalis 185
enterocyte 1, 6
Enzyme II 180
epinephrine transporter 301
epsin 290
ERAD 276
ergosterol 237, 243, 280
Escherichia coli 97, 98, 105, 158, 180
escort 221
ESCRT 295
ethylene 148
exocytosis 255, 326, 339, 357, 386
exon 23

fabpi 7

F-box protein 135
Fe^{2+}/H$^+$ antiporter 40
ferroportin 1 8
fgy 230
floating 240
forskolin 357, 360, 364
FPN1 8
Fur4 242, 276, 280, 282, 286, 289, 294,
    299, 312

GABA 124, 379, 381, 394
GABA transporter 255, 381
Gal2 122, 208, 225, 290
Gap1 123, 224, 226, 242, 245, 276, 282,
    286, 289, 295, 297, 303, 308
Gas1 242
GAT1 302, 379, 381, 383, 388, 394
GFP 246, 294
Gga 265, 296
GlnB 105
GlnK 48, 61, 105, 107
global repressor 57
glp regulon 196
glucose receptor 133
glucose repression 140
glucose sensor 13, 125, 136
glucose transporter 10, 122, 299
GLUT family 302, 325
GLUT1 69, 78, 230, 242, 243, 285
GLUT2 148
GLUT4 76, 126, 230, 242, 244, 285,
    325, 326
GLUT4 recycling 328
glutamate transporter 379
glycerol kinase 196
glycine betaine 159
glycine transporter 385
glycosylation 102, 108, 277
Gnp1 124, 146
GPI anchored protein 239, 242
G-protein coupled-receptor 13, 281, 392
Grr1 135
Gsf2 225
GTP binding protein 214
GTPase activating protein 215
guanine nucleotide exchange factor 212
guide 221

H-89 360
HDSV 281
HECT 282

Henle's loop 28
hepatocyte nuclear factor 1α 10
heptamer 176
hereditary haemochromatosis 8
heteromeric complex 102
hexose transporter 122
high-density secretory vesicle 281
Hip1 123, 224, 226, 227
HNF-1 11
HPr 180
HPr kinase/phosphorylase 186
HPrK/P 185, 186
Ht31 365
Htr1 135
Htr1-23 139
Hxt 122, 136, 208, 225, 277, 288, 299
hxt-null strain 123
hydrophathy plot 98
hypercholesterolaemia 9
hyperosmolarity 30
hyperosmotic stress 140, 169
hypertonicity 28
hypoosmotic stress 155

ilbp 7
ileum 7
inducer exclusion 189, 193, 195
inner coat protein 212
inositol permease 301
inositol trisphosphate 392
insulin 326
insulin receptor 334
internalization 281, 284, 290
intestinal nutrient transporter 1
intron 23
ionomycin 370
IRAP 334
iron absorption 8
IRS proteins 335
Iss1 219, 220
Itr1 301

jejunum 7

K$^+$:Cl$^-$ cotransporter 27
KCC 27, 29
KdpDE 175
KdpFABC 158, 175
Kht 127
kidney 354
kinesin 260

Klebsiella pneumoniae 106
Kluyveromyces lactis 126

LacS 193
Lactobacillus brevis 193
Lactobacillus casei 193
Lactobacillus plantarum 166
Lactococcus lactis 158, 185
lactose permease 192, 236
LacY 189
large dense core vesicle 256, 259
large intestine 1, 14
LDCV 256, 259, 260
LDL receptor 208
LDSV 281
LevR 201
LicT 198
Liddle syndrome 284
limb 28
lipid 235, 237, 243, 247, 273, 280
lipid raft 238, 239, 302, 305, 335
liposomes 169, 173
Listeria monocytogenes 158, 166
liver receptor homologue-1 10
loop-diuretic sensitive Na$^+$:K$^+$:2Cl$^-$ cotransporter 25
low-density secretory vesicle 281
LRH-1 10
Lst1 219, 220, 240
L-type voltage-gated calcium channel 148
luminal membrane 1
Lyp1 123, 124
lysosome 292

major facilitator family 173
Mal61 127, 287
MalK 50, 52
maltodextrin 49
maltose regulon 62
maltose system 48
MCT1 1, 15, 17
MeaA 102
mechanosensitive channel 155, 176
medulla 27
membrane sequestration 59
membrane shuttle hypothesis 354
membrane strain 168
membrane surface 171
membrane-association 106

Menkes protein 309
Mep 103, 108
Mep2 276, 277
MepA 102
metal transporter 300, 301, 307
methionine sulphoxamine 114
methylammonium 103, 111
Mg^{2+} transporter 300
Mlc 48, 57
MLCK 372
Mmp1 124
monocarboxylate transporter 1, 299
monoubiquitylation 282, 283, 303
mRNA stability 5
MscL 158, 176
MscS 158, 176
Msn3 135
MSX 114
Mth1 126, 135, 136, 139
multiubiquitin 282, 289
multivesicular body 211, 291
Munc-18 391
Mup 124
MVB 292, 294, 297, 312, 313
Mycobacterium tuberculosis 158
Mycoplasma pneumoniae 188
myoblasts 342
myo-inositol 35
myosin Light Chain Kinase 372

Na$^+$/bicarbonate cotransporter 37
Na$^+$/bile acid transporter 32
Na$^+$/glucose cotransporter 1, 10
Na$^+$/*myo*-inositol cotransporter 35
Na$^+$/taurocholate cotransporter 34
NBC family 37
Nedd4 282, 284, 302
nephron 28, 354
NET 301
neuron 256, 261
neurotransmitter transporter 255, 301, 380
NifL 107
nitrogen catabolite repression 147
non-bilayer lipids 174
nonvesicular transport 237
norepinephrine transporter 380
npi1 mutant 284
Npr1 297, 305, 310
Nramp 40, 147, 310

NTCP 34
nucleobase transporter 299, 306
nutrient sensing 121

obesity 14
okadaic acid 360
oligomeric structure 68
oligomerisation 102, 246
OpuA 158, 172, 174
osmolality 155
osmolyte 35
osmoregulation 156
osmosensing 140, 156, 171, 173, 174
osmotic stress 155
outer coat protein 212
outfitter 221
ovine intestine 11

p150^{Glued} 39
packaging chaperone 207, 221, 223,
    225
PACS-1 267
pantophysin 334
Pdr5 288
pentamer 176
PEPT1 9, 20
peptide transport 9
peptidomimetics 9
PEST 287
PGE2 365
pheromone transporter 282
phloretin-sensitive urea transport 30
Pho84 225, 288, 301
Pho86 225
Pho87 147
phosphate transporter 208, 225, 301
phosphatidyl ethanolamine 236
phosphatidyl serine 247
Phosphoglycerolipid 237
phospholipase C 371, 392
phospholipid 169, 244
phosphorylation 182, 389, 391
phosphotidyl 4,5-bisphosphate 392
phosphotransferase system 48, 179
P$_{\text{II}}$ protein 105
PKC 336
Pma1 220, 240, 242, 243, 276, 292
polarization 23, 35, 41, 243
polyubiquitylation 283, 289, 303
post-transcriptional gene processing 23
PRD 198

pre-budding complexe 212, 226, 228
primary carrier 158
primed-cargo complexe 226
promoter 6
ProP 158, 159, 172
propionate 14
ProQ 174
prostaglandin 365
proteasome 135, 145, 276
protein kinase A 13, 356
protein kinase C 301, 384, 392
protein-protein interaction 389, 391
proteoliposomes 159, 173
ProU 173
pseudohyphal growth 109
Ptr 128, 143, 144
PTS 48, 57, 179
PTS regulation domain 198
P-type ATPase 175
Put4 124
PutA 59

quality control 209
quaternary structure 68

raft 240, 243, 244, 245, 280
Rag 126, 127, 136
receptor 121
reconstitution 159
recycling 309, 329, 360
renal medullary interstitum 29
retention signal 267
retrograde 208
Rgt1 135, 136
Rgt2 121, 123, 125, 132, 136
RGT2-1 125
RhAG 104
RhAK 104
Rhesus (Rh) proteins 98, 104
Rho 364
Rhodobacter capsulatus 97, 107
RING-finger 282
RNA interference 19
RNA polymerase 6
Rsp5 273, 281, 282, 286, 289, 298, 305,
    313
rumen 11
Ryanodine receptor 369

Schizosaccharomyces pombe 229

Saccharomyces cerevisiae 97, 121, 208, 239, 273
Salmonella typhimurium 158
Sam3 124
Sar1 214, 225, 228
SAT2 148
SCAMP 334
SCF^{Grr1} 135, 136, 143
sclerosis 383
SDCV 256
sec mutant 274
Sec12 214, 229
Sec13 214, 225
Sec16 216
Sec23 214, 225, 228
Sec24 214, 216, 220, 225, 227, 228, 277
Sec31 214, 225
secondary carrier 158
secretory pathway 207, 236, 237, 274
secretory vesicle 255, 275
sensing 129
sensor 121
sensor kinase 175
sentrin 285
SERT 301
Sgk1 302
SGLT 325
SGLT1 1, 10, 242, 246
SGLT3 148
SH2 domain 335
Short bowel syndrome 19
short chain fatty acids 14
Shr3 223, 227, 230, 240, 245, 275
siderophore–iron transporter 307
signal recognition particle 209
signal transduction 121, 156
signalling pathway 121
simple carrier model 74
sitosterol 237
Skp1 135
SLC10 9, 32, 34
SLC11 8, 40
SLC12 25, 27, 28, 29
SLC14 29, 32
SLC15 9
SLC16 15
SLC4 37, 38
SLC5 10, 35
SLC6 20

small dense core vesicle 256
small intestine 1
Smf 300, 307
SMIT 35
SNAP-23 373
SNAP-25 385, 386, 388
SNARE 213, 214, 216, 244, 334, 339, 373, 385, 386, 388
Snf1 kinase 138
Snf3 121, 123, 125, 132, 136
sortilin 334
sorting 219, 244, 262, 303, 342
sorting motif 216, 219, 226, 263
SP-1 11
sphingolipid 235, 237, 248, 280
spliceosome 23
splicing 41
splicing isoform 23
SPS complex 143, 144, 224
Ssn6 136, 146
Ssy1 121, 124, 127, 132, 143, 224, 295, 298
SSY1-102 128
Ssy5 143, 144
Staphylococcus xylosus 186
Staphyolococcus aureus 166
Std1 126, 135, 136, 139
Ste2 281
Ste3 281
Ste6 277, 282, 287, 295, 308
sterol 235, 237, 248
stigmasterol 237
Stp 143, 146
Streptococcus mutans 185
Streptococcus salivarius 186
Streptococcus thermophilus 193
sucrase isomaltase 7
sugar transporter 179, 208
Sulfolobus solfataricus 53
sulfonylurea herbicides 127
sulprostone 365
SUT2 147
symbiosome 115
synapse 379
synapsin 385
synaptic cleft 380
synaptic vesicle 255, 259, 382, 384
synaptic-like microvesicles 262
synaptobrevin 373, 385
synaptophysin 257, 385

synaptotagmin 257, 385
syntaxin 373, 385, 386, 388, 394
System A amino-acid transporter 148

TALH 27
targeting 342
Tat 124, 146, 242, 287, 298, 305
TATA box 11
TC10 338
tetramer 75, 84, 357
TfR 329
Thermococcus litoralis 52
thiazide-sensitive Na$^+$:Cl$^-$ cotransporter
    25
tiagabine 383
tissue distribution 23
tomato 110
topology model 100
TOR signalling 298, 311
trafficking 235, 236, 275, 280, 379,
    384, 392
transcription factor 6
transcription initiation site 6
transferrin receptor 329
trans-Golgi network 326
trehalose 156, 159
trifluoperazine 367
trimer 102, 160
Triton X-100 239, 246, 280
TSC 27
t-SNARE 213
Tup1 136, 146
turgor 156, 166, 168, 174
tyrosine kinase 389

Ubc 135
Ubc9 285
ubiquitin 135, 282, 313
ubiquitin isopeptidase 296
ubiquitin ligase 273, 276, 281, 285
ubiquitin-conjugating 282
ubiquitin-ligase RING domain 146
ubiquitylation 276, 285, 290, 299
Uga35/Dal81 143
Uga4 124
uracil permease 276, 282, 299
urea transporter 29, 32
uridylylation 105, 107
UT-A 29, 32
UT-B 29

V2-receptor 356
VAChT 255, 261, 268
VACM-1 371
vacuolar degradation 288
vacuolar targeting 305
vacuole 273, 291
VAMP 257, 373
vasopressin 28, 30, 310, 354, 360, 362,
    367
vesicular acetylcholine transporter 255
vesicular glutamate transporter 255
vesicular monoamine transporter 255
vesicular transporter 255, 261
VGAT 255, 261, 268
VGLUT 255, 261, 268
VMAT 255, 261, 268
vps mutant 292
v-SNARE 213

Wilson's disease protein 309
WW domain 286

Xenopus laevis oocytes 103
Xenopus oocyte 333
Xenopus oocytes 75, 356

Ycf1 280
Yck 135, 287
yeast 121, 208, 223, 237, 273, 285, 311

zinc transporter 300
Zrt1 287, 300

Printing: Mercedes-Druck, Berlin
Binding: Stein+Lehmann, Berlin